ADVANCES IN CONTROL

# Advances in Control

*Volume II of a selection of papers from INFO II, the Second International Conference on Information Sciences and Systems, University of Patras, Greece, July 9-14, 1979*

*Edited by*

D. G. LAINIOTIS
*Dept. of Electrical Engineering,*
*State University of New York at Buffalo, Amherst, N.Y., U.S.A.*

and

N. S. TZANNES
*University of Patras, Chair of Information Theory,*
*School of Engineering, Patras, Greece*

**D. Reidel Publishing Company**

Dordrecht : Holland / Boston : U.S.A. / London : England

**Library of Congress Cataloging in Publication Data**

International Conference on Information Sciences and Systems, 2d, University of Patras, 1979.

A selection of papers from INFO II, the Second International Conference on Information Sciences and Systems, University of Patras, Greece, July 9–14, 1979.

Includes indexes.

Contents: v. 1. Advances in communications. v. 2. Advances in control. 3. Applications of information and control systems.

1. Electronic data processing—Congresses. 2. Telecommunication—Congresses. 3. Automatic control—Congresses. 4. Pattern recognition systems—Congresses. I. Lainiotis, Demetrios G. II. Tzannes, Nicolaos, 1937– III. Title.

QA75.5.I58 1979 001.64 80–18203

ISBN 90–277–1140–2 (v. 1)

ISBN 90–277–1129–1 (v. 2)

ISBN 90–277–1143–7 (v. 3)

---

Published by D. Reidel Publishing Company
P.O. Box 17, 3300 AA Dordrecht, Holland

Sold and distributed in the U.S.A. and Canada
by Kluwer Boston Inc., Lincoln Building,
160 Old Derby Street, Hingham, MA 02043, U.S.A.

In all other countries, sold and distributed
by Kluwer Academic Publishers Group,
P.O. Box 322, 3300 AH Dordrecht, Holland

D. Reidel Publishing Company is a member of the Kluwer Group



Printed in The Netherlands

## CONFERENCE COCHAIRMEN

D.G. Lainiotis N.S. Tzannes

## SPONSORED BY

The Ministry of Culture & Sciences of Greece
The University of Patras
The IEEE Automatic Control Society
The IEEE Information Theory Group
The IEEE Systems, Man and Cybernetics Society
The World Organization of General Systems and Cybernetics
The IEEE Greece
The Greek Society for Cybernetics and Systems

## STEERING COMMITTEE

| | |
|---|---|
| D.G. Lainiotis, | U.S.A. |
| N.S. Tzannes, | Greece |
| J. Cruz, | U.S.A. |
| G.F. Ferrate-Pascual, | Spain |
| K. Kerner, | Austria |
| J.L. Lions, | France |
| Y. Sunahara, | Japan |
| C. Tapiero, | Israel |
| C.P. Tsokos, | U.S.A. |
| M. Younis, | Egypt |

## TECHNICAL PROGRAM COMMITTEE

| | |
|---|---|
| D.G. Lainiotis, | U.S.A. |
| J. Aguilar-Martin, | France |
| E. Baptista, | Brazil |
| A. Bensoussan, | France |
| E.J. Davison, | Canada |
| A. Ephremides, | U.S.A. |
| M. Gevers, | Belgium |
| H. Kwakernaak, | Holland |
| E. Mosca, | Italy |
| K. Najim, | Morocco |
| G. Saridis, | U.S.A. |
| P.D. Scott, | U.S.A. |
| C. Tapiero, | Israel |
| C.P. Tsokos, | U.S.A. |
| N.S. Tzannes, | Greece |
| M.I. Younis, | Egypt |

## LOCAL ARRANGEMENTS COMMITTEES

### GREECE

A. Sissouras, Chairman
G. Megalokonomos
P. Panagiotopoulos

### U.S.A.

C. Nikias, Chairman
E. Cosmidis
P. Groumpos
C. Holtz
S. Thomopoulos

## CONFERENCE SECRETARIES

Mrs. Niki Sarantoglou, Greece
Mrs. Helen Alexandridis, Greece
Miss Ismini Kariofili, U.S.A.

THE CONFERENCE HAS BEEN AIDED FINANCIALLY BY

The Ministry of Culture & Sciences of Greece
The University of Patras
A.B.E.E.A. Co.
A.B.E.X. Co.
ACHAIA-CLAUSS Co.
Astir Hotel
BIOSOL Co.
DINAMIKI Co.
ELLENIT Co.
KAZA Co.
Kalogerakis Co.
Koutsadonis N.
Piraiki-Patraiki Co.
Pirelli Hellas S.A.
Titan Cement Co.

# TABLE OF CONTENTS

# PREFACE

This volume contains a selection of papers presented at INFO II, the Second International Conference on Information Sciences and Systems, held at the University of Patras, Greece, from July 9th to 14th, 1979. Additional selections of papers from this Conference are published in separate volumes, also by D. Reidel and are entitled "Advances in Communications" and "Applications of Information and Control Systems", respectively.

INFO II, was hosted by the University of Patras and sponsored by, the Ministry of Culture & Sciences, IEEE Automatic Control Society, IEEE Information Theory Group, IEEE Systems -Man and Cybernetics Society, World Organization of General Systems and Cybernetics, IEEE Greece, and the Greek Society for Cybernetics and Systems. It was the second such Conference held at the University of Patras, the first one having taken place from August 19th to 24th, 1976, with its papers published by the Hemisphere Publishing Corporation in a three volume set, entitled "Applications and Research in Information Sciences and Systems".

INFO II, then, was the second high level Conference on these important topics in the Eastern Mediterranean Region. The reason that these conferences have been established is because we feel strongly that the topics covered have reached a high degree of maturity in this region, both from the point of view of theory and application. The applications areas are especially important to the Mediterranean region in view of the Common Market connections of such countries as France, Italy and Greece and the anticipated admission of Spain. In view of this, a major objective of the conferences is to focus attention on these important technical subjects and to facilitate the exchange of information and cross-fertifization of ideas between researchers of the Mediterranean region and those of the rest of the world, thereby promoting research and development by the region's engineers, researchers and academicians.

The Conference keynoted by Prof. J.L.Lions, France, Prof. L.A. Zadeh, U.S.A., and Prof. M.Dertouzos, U.S.A., was highly successful as evidenced by the attendance of over 250 scientists and engineers from approximately 30 countries, and by the presentation of over 200 technical papers by well-known scientists and engineers from around the world.

The technical program of the Conference consisted of technical sessions that covered all-important aspects of control, information processing/communications, and computers. In particular, the



*D. G. Lainiotis and N. S. Tzannes (eds.), Advances in Control, xv–xvi.*

re were numerous technical sessions covering all-important aspects of control, such as, control theory, control applications, estimation, identification, adaptive systems, linear systems, stability, cybernetics, computational methods, and simulation. The program also included many sessions on information processing/communications such as, information theory, coding, signal analysis, signal processing, communication theory, satellite communications, pattern recognition and image processing. Moreover, there were several sessions on computers, in particular on computer systems, computer communication networks, and automata. Finally, the technical program included numerous sessions on important applications of systems technology, such as, power, energy modeling and planning, earth resources, transportation, economics and management, and physiological systems.

In view of the broad scope of the Conference technical program and the extensive coverage of many important aspects of systems theory and applications by internationally known researchers, we hope that this collection of papers will be a useful supplement to the published literature and textbooks used for research and teaching.

For the success of the Conference we are indebted to a great many people and institutions, primarily to the authors of the papers, without whom the conference would not have existed, and this book would not have materialized. We are particularly indebted to the great majority of them who paid some or all of their own expenses.

We are greatly indebted to the University of Patras for hosting the Conference, and for its financial support, as well as to the various other institutions and companies, whose financial support was essential for the success of the Conference. We would also like to thank the above mentioned scientific societies for adding their name as conference co-sponsors.

Finally, in a personal vein, we would like to thank the session chairmen, and the members of various conference committees, for their enthusiasm and dedication with which they worked in making INFO II a reality.

Nicolaos S. Tzannes Demetrios G. Lainiotis

SECTION I

# A SHORT REPORT ON THE OPTIMAL CONTROL OF DISTRIBUTED SYSTEMS

The keynote paper

by

J.-L. Lions

College de France
and I.N.R.I.A.

# A SHORT REPORT ON THE OPTIMAL CONTROL OF DISTRIBUTED SYSTEMS

Jacques-Louis LIONS

Collège de France
Chaire d'Analyse Mathématique des Systèmes
et de leur Contrôle
11, Place Marcelin Berthelot, 75231 Paris Cedex 05 (F).

and I.N.R.I.A. (Institut National de Recherche en
Informatique et en Automatique)
Domaine de Voluceau, Rocquencourt
B.P. 105, 78150 Le Chesnay (France).

ABSTRACT

A short review of some of the most important problems arising in the optimal control of systems governed by Partial Differential Equations (Distributed Systems).
Connections with free boundary problems are emphasized.



*D. G. Lainiotis and N. S. Tzannes (eds.), Advances in Control, 3–18.*

## INTRODUCTION

*Distributed systems* are systems which are governed by Partial Differential Equations ; more precisely they are modelled through a set of equations which are Partial Differential Equations.

These systems depend on *control variables* (or control functions), denoted here by $v$ ; for any given $v$ , the *state of the system* $y(v)$ is given by the solution of a boundary value problem involving the Partial Differential Equations referred to above and appropriate *boundary conditions* (and, in case of evolution problems, *initial conditions*).

To every couple $\{v,y(v)\}$ we associate a real number $J(v)$ which is the *cost function.*

The problem is to minimize $J(v)$, $v$ being subject to a number of constraints (which can be "implicit", that is which can be defined through constraints on $y(v)$).

The final aim is to obtain *constructive algorithms*, which can be implemented *in real time* (in case of evolution problems).

It is quite clear that in such a general setting, these questions involve "all" the theory of Partial Differential Equations, "all" the theory of the numerical approximation of Partial Differential Equations (finite differences, finite elements), "all" the theory of calculus of variations and the theory of nonlinear programming, etc.. It is beyond the scope of this report (1).

In what follows we want first to put the above considerations in a more precise form ; this is the aim of Section 1.

We want then to give (in Section 2) a simple example where one can obtain Necessary and Sufficient conditions for optimality.

In Section 3 we present situations (which often arise in practical applications) where the control variable is of an unusual structure.

It often happens that the cost function we want to minimize is *not* differentiable ; this is in particular the case for the *optimal control* of *free surfaces* - which is also a situation which is important from the point of view of applications. Some indications on this question are given in Section 4.

*Perturbation theory* is going to play a more and more important role in these questions ; we briefly indicate some examples in Section 5.

---
(1) And of the Author of the report ...

The plan of the report is as follows :

1. General framework.
2. Optimality system in a simple example.
3. Optimum design.
4. Control of free surfaces.
5. Perturbations.
6. Final remarks.

## 1. GENERAL FRAMEWORK

Let $\mathcal{A}$ be a *partial differential operator*; $\mathcal{A}$ can be stationary (time independent) or of evolution ; it can be linear or non-linear.

We consider the state equation given by

$$\mathcal{A}(y) = \mathcal{B}(v) ; \tag{1.1}$$

in (1.1) $\mathcal{B}$ is an operator (linear or non linear) which maps the space (or the set) of *control variables* $v$ into a space where (1.1) can be solved. For the time being (1.1) is a *formal way* of writing things ; it will be made precise in the examples which will be presented below. To the equation (1.1) one should add *boundary conditions* and, in case $\mathcal{A}$ is an evolution operator, one should add *initial conditions*.

Example 1.1.

Let $\Omega$ be a bounded open set of $\mathbb{R}^n$ (1), with a smooth(2) boundary $\Gamma$. We consider the equation

$$-\Delta y + y = f \quad \text{in } \Omega \tag{1.2}$$

(where $\Delta = \frac{\partial^2}{\partial x_1^2} + \dots + \frac{\partial^2}{\partial x_n^2}$), subject to the *boundary condition*

$$\frac{\partial y}{\partial \nu} = v \quad \text{on } \Gamma \tag{1.3}$$

(where $\nu$ = unitary normal to $\Gamma$ directed toward the exterior of $\Omega$).

One can put (1.2)(1.3) in the framework of (1.1) by using Sobolev spaces and their dual. We confine ourselves here to the following : if in (1.2) $f$ is *given* in $L^2(\Omega)$ and if we choose

$$v \in L^2(\Gamma) \tag{1.4}$$

(1) In the applications, one has n=1,2 or 3.

(2) This hypothesis is not absolutely indispensable in *some* applications, and we are not going to make it precise in this presentation.

then (1.2)(1.3) *admits a unique solution in* $H^1(\Omega)$ (1).
This solution is given by the *variational formulation* : we set

$$(1.5)\quad \begin{cases} a(y,\varphi) = \sum_{i=1}^{n} \int_\Omega \frac{\partial y}{\partial x_i}\frac{\partial \varphi}{\partial x_i}\,dx + \int_\Omega y\varphi\,dx \\ (f,\varphi) = \int_\Omega f\varphi\,dx\ ; \end{cases}$$

then the variation formulation of (1.2)(1.3) is

$$(1.6)\qquad a(y,\varphi) = (f,\varphi) + \int_\Gamma v\varphi\,d\Gamma.$$

We wish to emphasize the fact that $y$ *depends on* $v$ ; *we shall denote by* $y(v) = y(x;v)$ *the solution* of (1.2)(1.3) (or equivalently of (1.6)). This is the *state* of our system. ∎

Example 1.2.

With notations similar to those of Example 1.1, we consider the equation

$$(1.7)\qquad \frac{\partial y}{\partial t} - \Delta y = f \quad \text{in } \Omega\times]0,T[,$$

subject to the *boundary condition*

$$(1.8)\qquad \frac{\partial y}{\partial \nu} = v \quad \text{on } \Sigma = \Gamma\times]0,T[,$$

and to the *initial condition*

$$(1.9)\qquad y(x,o) = 0 \quad \text{in } \Omega.$$

If we assume that $f$ is given in $L^2(\Omega\times]0,T[)$ and that

$$(1.10)\qquad v \in L^2(\Sigma)$$

then problem (1.7)(1.8)(1.9) *admits a unique solution*(2) which is denoted by $y(v) = y(x,t;v)$ ; this is again the state of our system. ∎

Let us return to the general (and formal) situation (1.1). We assume that

(1.11) *the state equation* (1.1)(3) *admits a unique solution* (which will be denoted by $y(v)$).

(1) $H^1(\Omega) = \{\varphi \mid \varphi, \frac{\partial\varphi}{\partial x_1}, \dots, \frac{\partial\varphi}{\partial x_n} \in L^2(\Omega)\}$

(2) In the space $L^2(0,T;H^1(\Omega))$, i.e. the space of functions such that $\varphi, \frac{\partial\varphi}{\partial x_1}, \dots, \frac{\partial\varphi}{\partial x_n} \in L^2(\Omega\times]0,T[)$.

(3) Subject to appropriate boundary and initial conditions which can involve the control variable.

*Remark 1.1.* The uniqueness assumption in (1.11) may not be realized. This would lead to the optimal control of *particular* solutions of the state equation, a topic that we shall not consider here.

The solution $y(v)$ is *the state of the system.*
We introduce now the *cost function* $J(v)$. It is given by

(1.12) $$J(v) = \Phi(y(v)) + \Psi(v)$$

where $\Phi$ and $\Psi$ are functionals which are real valued and which are respectively defined on the space (or set) of states and on the space (or set) of control variables.

We denote by $\mathcal{U}$ the *space of control variables* and the problem (always in a formal fashion) consists in solving

(1.13) $$\inf J(v)$$

where in (1.13) $v$ is subject to

(1.14) $$v \in \mathcal{U}_{ad} \subset \mathcal{U} \quad ;$$

$\mathcal{U}_{ad}$ is called the set of *admissible controls* or the set of *constraints*.

Example 1.1. (follows).

Let $y(v)$ be the solution of (1.2)(1.3) and let us define

(1.15) $$J(v) = \int_\Gamma (y(v)-z_d)^2 \, d\Gamma + N\int_\Gamma v^2 \, d\Gamma$$

where $z_d$ is given in $L^2(\Gamma)$ and where $N$ is given $> 0$.

In (rough) physical terms, problem (1.13) consists in this case in minimizing the "distance" from, say, the temperature $y(v)$ on $\Gamma$ to an (optimal) "desired" temperature $z_d$, taking into account the "cost of energy" expressed in (1.15) by the term $N\int_\Gamma v^2 d\Gamma$ ; the highest this cost will be, the largest we shall have to choose $N$(1).

A a *simple example of constraints*, let us mention

(1.16) $$\mathcal{U}_{ad} = \{v \mid v \in L^2(\Gamma), \quad 0 \le v \le M\} \; .$$

These constraints define a *closed convex subset of* $\mathcal{U} = L^2(\Gamma)$. ∎

---

(1) When $N$ is small (and tends to zero) one says that one deals with a *problem of cheap control*.

Example 1.2. (follows) :

Let $y(v)$ be the solution of (1.7)(1.8)(1.9) and let $J(v)$ be the cost function defined by

$$(1.17) \qquad J(v) = \int_\Sigma (y(v)-z_d)^2 d\Sigma + N\int_\Sigma v^2 d\Sigma \ ,$$

where $z_d$ is given in $L^2(\Sigma)$ and where $N$ is given $>0$. A physical interpretation of the corresponding problem can be given as for the above example. ■

*Remark 1.2*. In most of the examples to follow, $\mathcal{U}_{ad}$ *will be a closed* (non-empty) *convex subset of* $\mathcal{U}$, *where* $\mathcal{U}$ *is a Hilbert space.* ■

*The problems to consider are the following :*

(i) to see whether or not problem (1.13)(1.14) admits a solution (and if this solution is unique) ;

(ii) since we want to end up with *constructive algorithms*, we would like to obtain *necessary* (and, if possible) *necessary and sufficient conditions for* $u$ *to be optimal*, i.e. to satisfy

$$(1.18) \qquad J(u) = \inf J(v), \quad v \in \mathcal{U}_{ad} \ , \quad u \in \mathcal{U} \quad ;$$

(iii) we want to study what are the properties of the optimal control(s) $u$ (regularity, bang-bang, etc..) ;

(iv) we want to obtain *constructive algorithms* generally based on the necessary conditions obtained for $u$ to be optimal ;

(v) in order to implement the control at a time where it still makes sense to apply the "optimal control", one has to deal with the "*real-time*" *problem*. ■

*Remark 1.3*. The above program is not original ! It is, in general terms, the program followed in the classical study of Calculus of Variations and of Optimal Control for systems governed by Ordinary Differential Equations. We refer to M. HESTENES [1] and to the Bibliography therein.

The problems which are *specific* to the theory of optimal control of *distributed* systems are (essentially) the following ones :

(j) when expressing necessary (or necessary and sufficient) conditions, one meets *free boundary problems* ;

(jj) the control variable can be of a quite complicated structure (cf. Section 3 below);

(jjj) one has frequently to deal with functionals which are *not* differentiable, in particular when dealing with problems of *optimal control of free surfaces* ;

(jv) constructive algorithms (and real-time problems) rest on the *numerical solution of Partial Differential Equations*, a very broad subject in itself !

(v) the system can be "very large" - a difficulty one also meets when working with systems governed by ordinary differential equations (lumped systems), but which is particularly frequent and important for distributed systems.

*Remark 1.4.* We do not treat here the *stochastic aspects* of this kind of problem.
We do not consider here, for lack of space, the following topics :

(a) the use of techniques of *optimal control* in *numerical analysis*

(b) the use of techniques of optimal control for *system modelling*, an idea that is pursued at INRIA by G. CHAVENT and his co-workers (cf. G.CHAVENT [1] and the bibliography therein). ■

*Remark 1.5.* All the examples that we are going to mention in what follows are *motivated by applications*. ■.

## 2. OPTIMALITY SYSTEM IN A SIMPLE EXAMPLE

We consider - in order to make this exposition as little technical as possible - Example 1.1. of Section 1.

Let $\mathcal{U}_{ad}$ be a closed (non empty) convex subset of $L^2(\Gamma)$ and let $J$ be given by (1.15).
The state $y(v)$ is the unique solution of (1.6) that we write :

$$a(y(v),\varphi) = (f,\varphi) + \int_\Gamma v\varphi \, d\Gamma \qquad \forall \varphi \in H^1(\Omega). \tag{2.1}$$

It follows from standard results in the solution of boundary value problems (see for instance J.L. LIONS & E. MAGENES [1] and the bibliography therein) that

$$v \to y(v) \text{ is a continuous mapping from } L^2(\Gamma) \longrightarrow H^1(\Omega); \tag{2.2}$$

it is obvious that this mapping is "affine" (linear + constant). Therefore

$$v \to J(v) \text{ is continuous from } L^2(\Gamma) \longrightarrow \mathbb{R} \tag{2.3}$$

and $J$ is clearly *strictly convex* and $J(v) \longrightarrow +\infty$ as

$$\|v\|_{\mathcal{U}} = \left(\int_\Gamma v^2 \, d\Gamma\right)^{\frac{1}{2}} \longrightarrow \infty .$$

Consequently (cf. for instance J.L. Lions [1]) *there exists a unique element* $u \in \mathcal{U}_{ad}$ such that

(2.4) $$J(u) = \inf J(v),\ v \in \mathcal{U}_{ad}.$$

This element u - *the optimal control* - is characterized by the so-called *variational inequality*

(2.5) $$\int_\Gamma (y(u)-z_d)(y(v)-y(u))\,d\Gamma + N\int_\Gamma u(v-u)\,d\Gamma \geq 0 \quad \forall v \in \mathcal{U}_{ad}.$$

In order to transform (2.5) in a more convenient form, one introduces - this is a classical idea in the calculus of variations - the *adjoint state* p defined by

(2.6) $$\begin{cases} -\Delta p + p = 0 \text{ in } \Omega, \\ \dfrac{\partial p}{\partial \nu} = y - z_d \text{ on } \Gamma \end{cases}$$

where we have set

(2.7) $$y(u) = y \text{ (= optimal state)}.$$

If we multiply the first equation (2.6) by y(v)-y and we integrate by parts (Green's formula), we obtain

(2.8) $$0 = -\int_\Gamma \frac{\partial p}{\partial \nu}(y(v)-y)\,d\Gamma + \int_\Gamma p\left(\frac{\partial y(v)}{\partial \nu} - \frac{\partial y}{\partial \nu}\right)d\Gamma.$$

Using the second equation (2.6) and the fact that, by definition, $\frac{\partial y(v)}{\partial \nu} = v$, $\frac{\partial y}{\partial \nu} = u$, (2.8) becomes

(2.9) $$\int_\Gamma (y-z_d)(y(v)-y)\,d\Gamma = \int_\Gamma p(v-u)\,d\Gamma,$$

so that (2.5) can be rewritten :

(2.10) $$\int_\Gamma (p+Nu)(v-u)\,d\Gamma \geq 0 \quad \forall v \in \mathcal{U}_{ad}.$$

Summing up, we have obtained :

*the unique optimal control* u *is characterized by the solution* $\{y,p,u\}$ *of the optimality system, which is given by*

(2.11) $$\begin{cases} -\Delta y + y = f, \\ -\Delta p + p = 0 \quad \text{in } \Omega, \\ \dfrac{\partial y}{\partial \nu} = u\ ,\ \dfrac{\partial p}{\partial \nu} = y - z_d \text{ on } \Gamma \end{cases}$$

*and the variational inequality* (2.10).

APPLICATION (I).

Let us assume that

(2.12) $$\mathcal{U}_{ad} = \mathcal{U}\ (=L^2(\Gamma))$$

(this is the case "*without constraints*"). Then (2.10) reduces to

(2.13) $p + Nu = 0$ on $\Gamma$.

It follows that

(2.14) $u = -\frac{1}{N}p$ on $\Gamma$, *belongs to* $H^1(\Gamma)$ ,

i.e.

(2.15) $\begin{cases} \text{all the first order "tangential" derivatives of} \\ u \text{ taken on } \Gamma \text{ (1) are in } L^2(\Gamma). \end{cases}$

This is a *regularity result* on the nature of the optimal control. This type of result is useful from the viewpoint of numerical algorithms based on (2.11)(2.10). ∎

APPLICATION (II).

Let us assume that

(2.16) $\mathcal{U}_{ad} = \{ v \mid v \in L^2(\Gamma), \ v \geq 0 \ \text{ on } \Gamma \}$.

Then (2.10) is seen to be equivalent to

(2.17) $p + Nu \geq 0, \quad u \geq 0 \ , \ u(p+Nu) = 0$ on $\Gamma$

which is equivalent to

(2.18) $u = \sup \{0, -\frac{1}{N}p\}$.

This implies that

(2.19) $u \in H^1(\Gamma)$ ,

the same regularity result than above.

On the surface $\Gamma$ we have two regions : in one region $p$ is $< 0$ and $u = -\frac{1}{N}p$ ; in the other region $p$ is $\geq 0$ and $u = 0$ ; the *interface* $S$ *between these two regions is a free boundary on* $\Gamma$ ; $S$ is not given, but has to be *computed* from (2.11)(2.18) and actually, constructive algorithms can be constructed using (2.11) and (2.18). ∎

ORIENTATION : We have shown, in a very simple situation, the *structure of the optimality system*. The example is typical of what one obtains in more general situations (cf. J.L. Lions [1] [2]). In differentiable *non-linear* problems the optimality system has still a similar structure but it gives only *necessary conditions* without uniqueness of the solution. Cf. examples and applications in Brauner [1], J.P. Kernevez [1], J.P. Yvon [1].
We now proceed with problems where the structure of the control "set" is entirely different.

---

(1) This is easily made precise using local maps describing $\Gamma$.

## 3. OPTIMUM DESIGN

We confine here ourselves to defining the type of problem met in these directions.

Let $E$ be a given closed set in $\mathbb{R}^3$, of positive measure, and let $\mathcal{U}$ *be the "set" of all bounded open sets* $\Omega$ *in* $\mathbb{R}^3$ *such that*

(3.1) $$\Omega \supset E .$$

Given $\Omega \in \mathcal{U}$ ( $\Omega$ plays the role of the *"control variable"*) we define the state of the system, $y(x;\Omega) = y(\Omega)$, by

(3.2) $$\begin{cases} -\Delta y(\Omega) = f \quad \text{in } \Omega, \\ y(\Omega) = 0 \quad \text{on} \quad \partial\Omega \text{ (boundary of } \Omega) \end{cases}$$

where $f$ is a given function in $\mathbb{R}^3$, such that

(3.3) $$f \in L^2(\mathbb{R}^3) ;$$

in (3.2) $f$ really means the *restriction* of $f$ to $\Omega$.

The *cost function* is - for instance[1] -

(3.4) $$J(\Omega) = \int_E [y(x;\Omega)-z_d(x)]^2 dx + N \text{ meas}(\Omega)$$

and the problem is to find $\Omega_o$ if it exists, such that

(3.5) $$J(\Omega_o) = \inf J(\Omega), \quad \Omega \in \mathcal{U}.$$

One difficulty here is to define a structure on $\mathcal{U}$ such that a *differential calculus* becomes possible.

A *local* theory, giving necessary conditions, can be constructed as follows.

Let us set

(3.6) $$\Gamma_o = \partial \Omega_o ,$$

assuming that a *smooth* [2] $\Omega_o$ exists satisfying (3.5).

Let us define a "double family" of neighbouring surfaces to $\Gamma_o$ as follows : we first introduce a function $\alpha$ such that

(3.7) $$\alpha \in \mathcal{C}(\Gamma_o) = \text{space of continuous functions on } \Gamma_o;$$

we introduce next

(3.8) $\nu(x)$ = unitary normal to $\Gamma_o$ at point $x \in \Gamma_o$, directed towards the exterior of $\Omega_o$

and we define $\Gamma_\lambda$ by

(3.9) $$\Gamma_\lambda = \{x + \lambda\,\alpha(x) \ \nu(x) \mid \ x \in \Gamma_o\}$$

where $\lambda$ is $>0$ and "small".

If $\lambda$ is small enough, the "topological nature" of $\Gamma_\lambda$ will be "the same" than the one of $\Gamma_o$ and we can define

(1) We choose this form to fix ideas.

(2) The existence of a solution with a *smooth* boundary is a major difficulty.

(3.10) $\Omega_\lambda$ = "interior of $\Gamma_\lambda$".

We have therefore

(3.11) $J(\Omega_o) \leqslant J(\Omega_\lambda) \quad \forall \lambda$ in the neighbourhood of 0

hence - if the derivative exists -

(3.12) $\frac{d}{d\lambda} J(\Omega_\lambda)\Big|_{\lambda=0} = 0, \quad \forall \alpha \in \mathcal{C}(\Gamma_o).$

Let us set

(3.13) $y_\lambda = y(\Omega_\lambda)$

and let us set - if the derivative exists -

(3.14) $\dot{y}(x) = \frac{d}{d\lambda} y_\lambda(x)\big|_{\lambda=0}, \quad x \in \Omega_o.$

Then

(3.15) $\frac{d}{d\lambda} J(\Omega_\lambda)\Big|_{\lambda=0} = 2\int_E (y_o - z_d)\, \dot{y}\, dx + N\int_{\Gamma_o} \alpha(x) d\Gamma.$

We now compute $\dot{y}$ ; the computations which follow are formal but can be justified (cf. for instance F. Murat and J. Simon [1]). One finds firstly that

(3.16) $\Delta\dot{y} = 0$ in $\Omega_o$.

The boundary condition for $y_\lambda$ can be written

(3.17) $y_\lambda(x + \lambda\alpha(x)\nu(x)) = 0$ for $x \in \Gamma_o$.

We expand (3.17) :

(3.18) $y_\lambda(x) + \lambda\,\alpha(x)\frac{\partial y_\lambda}{\partial \nu}(x) + \ldots = 0, \quad x \in \Gamma_o,$

where the dots indicate (formally) higher order terms in the $\lambda$-expansion near 0 ; from (3.18) and from $y_o(x) = 0$, $x \in \Gamma_o$, we deduce that

(3.19) $\frac{1}{\lambda}[y_\lambda(x) - y_o(x)] + \alpha(x)\frac{\partial y_\lambda}{\partial \nu}(x) + \ldots = 0, \quad x \in \Gamma_o;$

letting $\lambda \to 0$ in (3.19) we obtain

(3.20) $\dot{y}(x) + \alpha(x)\frac{\partial y_o}{\partial \nu}(x) = 0$ on $\Gamma_o$.

Therefore $\dot{y}$ is given by the solution of (3.16)(3.20).
We introduce now the adjoint state([1]) p given by

(3.21) $-\Delta p = 2(y_o - z_d)\chi_E$ in $\Omega_o$,

where $\chi_E$ = characteristic function of E, subject to the boundary condition

(3.22) $p = 0$ on $\Gamma_\sigma$.

Applying Green's formula, $\int_{\Omega_o} (-\Delta p)\dot{y}\, dx = 2\int_E (y_o - z_d)\dot{y}\, dx =$

$= \int_{\Gamma_o} \frac{\partial p}{\partial \nu}\, \alpha \frac{\partial y_o}{\partial \nu}\, \partial\Gamma_o$, hence it follows from (3.15) and (3.12) that

---

([1]) Compare to Section 2.

(3.23) $$\int_{\Gamma_o} \alpha \left[\frac{\partial p}{\partial \nu}\frac{\partial y_o}{\partial \nu} + N\right] d\Gamma_o = 0 \quad \forall \alpha ,$$

so that

(3.24) $$\frac{\partial p}{\partial \nu}\frac{\partial y_o}{\partial \nu} + N = 0 \text{ on } \Gamma_o .$$

We have obtained the following *optimality system* giving *necessary conditions* for $\Omega_o$ to be an *optimal control* ; if $\Omega_o$ satisfies (3.11) *and* has a smooth boundary, then the following *free boundary problem* admits a solution :

(3.25) $$\left\{\begin{array}{l} -\Delta y_o = f , \quad -\Delta p = 2(y_o - z_d)\chi_E , \text{ in } \Omega_o , \\ y_o = 0 , \quad p = 0 \quad \text{on } \Gamma_o \\ \text{and} \\ \frac{\partial p}{\partial \nu}\frac{\partial y_o}{\partial \nu} + N = 0 \quad \text{on } \Gamma_o . \end{array}\right.$$ ■

Remark 3.1. Let us comment on the aspect "*free boundary problem*" of (3.25). If we choose *arbitrarily* $\Omega_o$ (subject only to $E \subset \Omega_o$) then the equations in (3.25) with $y_o = 0$ and $p = 0$ on $\Gamma_o$ admit a unique solution but this solution *will not satisfy in general* the last condition in (3.25) (i.e. (3.24)) ; the problem "reduces" to finding the "free boundary" $\Gamma_o$ such that (3.24) holds true. ■

Remark 3.2. The *type* of problem described above appears in very many applications in many branches of Mechanics and Physics. Let us refer here to O. Pironneau [2] and to the bibliography therein.

## 4. CONTROL OF FREE SURFACES

One of the main difficulties in the optimal control of free surfaces (1) is the fact that the free surface does not depend (in general) in a regular way of the control variable.

Let us confine ourselves to a simple example, the so called "obstacle problem". We want to find a function $y$ such that

(4.1) $$\left\{\begin{array}{l} Ay - (f+v) \geq 0 \quad \text{in } \Omega , \\ y \geq 0 \quad \text{and} \\ [Ay - (f+v)]y = 0 \text{ on } \Omega \end{array}\right.$$

where $Ay = -\Delta y$, where $f$ is given in $L^2(\Omega)$ and where $v$ is the "control variable" , in $L^2(\Omega)$ ; in (4.1) $y$ is subject to the boundary condition

(4.2) $$y = 0 \text{ on } \Gamma .$$

Problem (4.1)(4.2) enters in the family of *Variational Inequalities* (cf. J.L. Lions and G. Stampacchia [1], H. Brézis and G.Stampacchia [1]). According to (4.1) there are two regions in $\Omega$ ; in $\Omega_o^v$, $y=0$ and in $\Omega \backslash \Omega_o^v$, $y$ is $> 0$ and $Ay = f+v$ ; the interface between these two regions is denoted by $S(v)$.

(1) Problems of this type arise in many applications. Cf. in particular Ch. Saguez [1] .

Both $\Omega_o^v$ and $S(v)$ *depend on* $v$ and, in general, $S(v)$ depends in a singular way on $v$.
It is assumed that

(4.3) $$v \in \mathcal{U}_{ad} = \text{a given subset of } L^2(\Omega)$$

and one wants to find $v$ such that $v \in \mathcal{U}_{ad}$ and $S(v)$ is "as close as possible" to a given surface $S_d$.
To express this last condition in quantitative manner, one introduces :

(4.4) $$\begin{cases} \chi(v) = \text{characteristic function of } \Omega_o^v \\ \chi_d = \text{characteristic function of the "good" part of } \Omega\ (^1) \text{ with boundary } S_d, \end{cases}$$

and one defines

(4.5) $$J(v) = \int_\Omega [\chi(v) - \chi_d]^2 dx. \qquad \blacksquare$$

The difficulty lies now in the *non differentiability* of (4.5).
One uses (cf. Ch.Saguez [1] ) *smooth approximations* of the above problem.

Let $\lambda \to \beta(\lambda)$ be a $C^1$ function from $\mathbb{R} \to \mathbb{R}$, such that $\beta(\lambda) > 0$ for $\lambda > 0$ , $\beta$ is convex ; we define $y_\varepsilon$ (for $\varepsilon > 0$) as the solution (which exists and is unique) of

(4.6) $$\begin{cases} Ay_\varepsilon + \frac{1}{\varepsilon}\beta(y_\varepsilon) = f + v & \text{in } \Omega, \\ y_\varepsilon = 0 \text{ on } \Gamma ; \end{cases}$$

now $y_\varepsilon = y_\varepsilon(v)$ depends in a smooth way of $v$ and a smooth approximation of $\chi(v)$ is given by $\frac{\varepsilon}{\varepsilon + y_\varepsilon(v)}$ hence the *new cost function is given by*

(4.7) $$J_\varepsilon(v) = \int_\Omega \left[\frac{\varepsilon}{\varepsilon + y_\varepsilon(v)} - \chi_d\right]^2 dx.$$

One can then write an optimality system for problem (4.6)(4.7) ; numerical results obtained by methods of this type are satisfactory (cf. Ch. Saguez, loc.cit.). ■ .

## 5. PERTURBATIONS

The general idea of using *perturbation theory* in optimal control is as follows :

assuming that we "know" how to control the system given by

(5.1) $$\mathcal{A}_o(y) = \mathcal{B}(v)$$

(using formal notations similar to those of (1.1)), how can we use

---

(1) There is in general no ambiguity in the applications.

this information in order to control a system which is "closed" to (5.1) :

$$\mathcal{A}_0(y) + \varepsilon \mathcal{A}_1(y) = \mathcal{B}(v). \tag{5.2}$$

In the applications the *"real" system* to control is given by (5.2) and we want to use the fact that ε is *"small"* (1) and that (5.1) is *"simpler"*.

This is connected to the theory of *singular perturbations*.

We refer to J.L. Lions [3] and to the bibliography therein and to [4] .

## 6. FINAL REMARKS

We confine ourselves, in these final remarks to indicate important topics not touched upon in this lecture.

Problems of *controllability* are important, in particular for *hyperbolic systems*. We refer to D. Russel [1] and to the bibliography therein.

Another important question is that of *stabilization :* "how much" control has to be applied so that we "stabilize an unstable system"? For questions of this type (in a more precise form !) we refer to J.M. Ball and M. Slemrod [1] and to the bibliography therein.

Some problems of optimal control lead to the introduction of new functional spaces, as it is shown in J.L. Lions [5][6].

Decomposition technique, parallel computations are now under study at various places and will no doubt be of more and more wide use for solving problems arising in larger and larger classes of applications.

## REFERENCES

J.M. Ball and M. Slemrod [1] *Non harmonic Fourier Series and the stabilization of distributed semi-linear control systems*. C.P.A.M. XXXII (1979), p. 555-587.

C.M. Brauner [1] *Thesis*, Paris , 1976.

H. Brézis and G. Stampacchia [1] *Sur la régularité de la solution d'inéquations elliptiques*. Bull. S.M.F. 96 (1968), p. 153-180.

(1) A situation which frequently arises in practice.

G. Chavent [1] *Identification of functional parameters in Partial Differential Equations*. A.S.M.E. New-York (1974) p. 31-48, Proc. of JACC Symposium, Austin, June 1974.

G. Chavent [2] *Identification of distributed parameter systems about the out-put of least square method : its implementation, and identifiability*. Proceedings of the V$^{th}$ IFAC Symposium on Identification and Systems Parameter Estimation, (1979).To appear, Pergamon Press, Oxford.

R. Glowinski and al. [1] *Application of optimal control of Finite Element Methods to the calculation of transonic flows and incompressible viscous flows*. IRIA-LABORIA Report Nb. 294 and in Numerical Methods in Applied Fluid Dynamics, B. Hunt, ed. Academic Press, London.

M. Hestenes [1] *Calculus of variations and optimal control theory*. Wiley, 1966.

J.P. Kernevez [1] *Control of the flux of substrate entering an enzymatic membrane by an inhibition at the boundary*. To appear.

[2] Book under preparation. North-Holland.

J.L. Lions [1] *Contrôle optimal de systèmes gouvernés par des équations aux dérivées partielles*. Dunod, Gauthier Villars 1968. English translation by S.K. Mitter, Springer, 1970.

[2] *Some aspects of the optimal control of distributed parameter systems*. SIAM Reg. Conf. Series, Vol. 6, 1972. *Remarks on the theory of optimal control of distributed systems*, in Control theory of systems governed by partial differential equations. Acad. Press, 1977.

[3] *Asymptotic methods in the optimal control of distributed systems*. I.F.A.C. Conference, Warwick, in Automatica 14 (1978), p. 199-211.

[4] *Asymptotic calculus of variations, Symposium*. Mathematics Research Centre, Madison, Spring 1980.

[5] *Sur de nouveaux espaces fonctionnels en contrôle optimal*. Note C.R. Ac. Sci., Paris, September 1979, t. 289, p. 315-319.

[6] *Function spaces and optimal control of distributed systems*. Notas de Matematicas. U.F.R.J., Rio-de-Janeiro, 1980.

J.L. Lions and E. Magenes [1] *Problèmes aux limites non homogènes et applications*. Three volumes : Dunod, 1968 (vol. 1, 2), 1970 (vol. 3). English translation by P. Kenneth, Springer, 1972.

J.L. Lions and G. Stampacchia [1] *Variational Inequalities*. C.P.A.M. 20 (1967), p. 493-519.

O. Pironneau [1] *Sur les problèmes d'optimisation de structure en Mécanique des Fluides*. Thèse, Paris, 1976.

[2] *Existence de solutions de certains problèmes d'optimum design*. Note C.R.Ac. Sci. Paris, 1974.

D. Russell [1] *Non harmonic Fourier series in the control theory of distributed parameter systems*. J.M.A.A. 18 (1967), p. 542-560.

[2] *Controllability and stabilizability theory for linear partial differential equations : recent progress and open questions*. SIAM Review 20 (4), 1978, p. 639-739.

F. Murat and J. Simon [1] To appear.

Ch. Saguez [1] *Un problème de Stefan avec source sur la frontière libre*. IRIA-LABORIA Report, no. 268, November 1977.

[2] Thesis, Paris, 1980.

J.P. Yvon [1] Thesis, Paris, 1974 : *Contrôle optimal d'un four industriel* (Part. 1 - IRIA-LABORIA Report, no. 22) ; *Contrôle optimal de systèmes gouvernés par des inéquations variationnelles* (Part II - IRIA-LABORIA Report, no. 53).

[2] Book, North-Holland, to appear.

# SECTION II

# SYSTEMS

# A MIXED TIME-FREQUENCY DOMAIN APPROACH TO MODEL REDUCTION

S.J.VAROUFAKIS

Nuclear Research Center "Democritus"
Aghia Paraskevi, Attikis
Greece

P.N.PARASKEVOPOULOS

Chair of Automatic Control Systems
School of Engineering
Democritus University of Thrace
Xanthi, Greece

The paper considers the problem of reducing the transfer function of a given system to a lower order model. The approach proposed consists of simultaneously applying both time and frequency domain fitting criteria in determining the lower order model.This approach appears to be suitable in cases where both the frequency domain as well as the time domain characteristics of the given system are to be preserved in the reduced order model.

## I. INTRODUCTION

It is usually desirable (for realization, control, computation and other purposes) to be able to represent "adequately" a high order system by a lower order model. To this end, several methods have been developed:The dominant eigenvalue and eigenvector [1], the Padè approximation [2], [3], [7], the time-moments matching [4], the continued fractions [5], the Routh approximation [6], [7], a state-space technique [8], etc. A survey of this area is given in [9].

The aforementioned order reduction techniques are either in the frequency domain or in the time domain and they are designed in such a way as to secure good fitting in their corresponding domain. This, consequently, results in a reduced order model which represents the high order system adequately in one domain, but inadequately in the other domain.

This paper presents some first results of a technique which aims in compromising the frequency domain and the time domain approaches by introducing a new method wherein an adequate approximation is sought in both domains. The approach followed is based on



*D. G. Lainiotis and N. S. Tzannes (eds.), Advances in Control, 21–29.*

the idea of expanding the transfer functions H(s) and G(s) of the given system and of the lower order model, respectively, in power series of s and their corresponding impulse responses h(t) and g(t) in power series of t. Subsequently, the coefficients of the first q terms in the s-power series of H(s) and G(s) as well as the coefficients of the first m terms in the t-power series of h(t) and g(t) are equated, where m+q is the number of the unknown parameters in G(s). These m+q equations constitute the mixed time-frequency domain fitting criteria and are solved to yield the reduced order model. This model appears to represent the original system adequately enough in both domains.

## II. PROBLEM STATEMENT

Let a given physical linear time-invariant system be described by the following transfer function

$$H(s) = \frac{b_0 + b_1 s + b_2 s^2 + \ldots + b_{n-1} s^{n-1}}{a_0 + a_1 s + a_2 s^2 + \ldots + a_{n-1} s^{n-1} + s^n} = \frac{b(s)}{a(s)} \tag{1}$$

Also let the reduced order transfer function sought be given by

$$G(s) = \frac{d_0 + d_1 s + d_2 s^2 + \ldots + d_{p-1} s^{p-1}}{e_0 + e_1 s + e_2 s^2 + \ldots + e_{p-1} s^{p-1} + s^p} = \frac{d(s)}{e(s)} \quad , \; p<n \tag{2}$$

Now, let $h(t) = \mathcal{L}^{-1}H(s)$ and $g(t) = \mathcal{L}^{-1}G(s)$, where $\mathcal{L}^{-1}$ denotes the inverse Laplace transform and h(t) and g(t) are the impulse response functions of the given system (1) and the reduced order model (2), respectively. Then, the problem considered in this paper is to develop a method for determining the 2p parameters $d_0$, $d_1, \ldots, d_{p-1}$, $e_0$, $e_1, \ldots, e_{p-1}$ of G(s) such that G(s) and g(t) simultaneously approximate H(s) and h(t), respectively.

## III. DEVELOPMENT OF THE METHOD

Let the power series of H(s) and G(s) be given by

$$H(s) = \sum_{i=0}^{\infty} H_i s^i \qquad \text{and} \qquad G(s) = \sum_{i=0}^{\infty} G_i s^i \tag{3}$$

Also let the power series of h(t) and g(t) be given by

$$h(t) = \sum_{i=0}^{\infty} h_i t^i \qquad \text{and} \qquad g(t) = \sum_{i=0}^{\infty} g_i t^i \tag{4}$$

Then the proposed method consists in determining the 2p parameters of G(s) such that the following 2p equations hold

$$H_i = G_i \;, \quad i = 0,1,\ldots,k \tag{5a}$$

$$h_i = g_i \;, \quad i = 0,1,\ldots,2p-k \tag{5b}$$

The set of equations (5a) constitute the frequency domain fitting criteria, whereas the set of equations (5b) constitute the time domain fitting criteria. Clearly, the greater the k the more "weight" is put for a better frequency domain fitting and the smaller the k the more "weight" is put for a better time domain fitting. In particular, when k=2p, then the problem reduces to the Padé-type approximation [2] , [3] in which case all the approximation is done in the frequency domain, whereas when k=0, all the approximation is done in the time domain.

In what follows we let k = p, thus imposing an equal "weight" on both the time and frequency domain by letting the time and the frequency domain fitting constraint equations to be of the same number. Using the well known results from the Padé-type approximation [2], [3], the set of equations (5a) may be written explicitly in terms of the parameters of the reduced order model transfer function G(s) and the power series coefficients of H(s) as follows

$$\mathbf{d} = \tilde{\mathbf{H}}\,\mathbf{e} \tag{6}$$

where

$$\mathbf{d} = \begin{bmatrix} d_0 \\ d_1 \\ d_2 \\ \vdots \\ d_{p-2} \\ d_{p-1} \end{bmatrix}, \quad \mathbf{e} = \begin{bmatrix} e_0 \\ e_1 \\ e_2 \\ \vdots \\ e_{p-2} \\ e_{p-1} \end{bmatrix}, \quad \tilde{\mathbf{H}} = \begin{bmatrix} H_0 & 0 & 0 & \cdots & 0 & 0 \\ H_1 & H_0 & 0 & \cdots & 0 & 0 \\ H_2 & H_1 & H_0 & \cdots & 0 & 0 \\ \vdots & \vdots & \vdots & \ddots & \vdots & \vdots \\ H_{p-2} & H_{p-3} & H_{p-4} & \cdots & H_0 & 0 \\ H_{p-1} & H_{p-2} & H_{p-3} & \cdots & H_1 & H_0 \end{bmatrix}$$

In deriving a similar expression to (6) for the set of equations (5b) the following results will be used. From the initial value theorem we have that

$$\lim_{t\to 0} g(t) = \lim_{s\to\infty} s\,\mathcal{L}[g(t)] = \lim_{s\to\infty} s\,G(s) \tag{7}$$

Using the expression (2) for G(s) we find that

$$\lim_{s\to\infty} s\,G(s) = d_{p-1}$$

But

$$\lim_{t \to 0} g(t) = g_0(0) = g_0$$

where $g_0$ is the first coefficient of the time power series of g(t) defined in (4). Thus

$$g_0 = d_{p-1} \tag{8}$$

Similarly, from the initial value theorem we have that

$$\lim_{t \to 0} g^{(1)}(t) = \lim_{s \to \infty} s\mathcal{L}[g^{(1)}(t)] = \lim_{s \to \infty} [s^2 G(s) - sg(0)] \tag{9}$$

where $g^{(k)}(t)$ denotes the kth derivative of g(t).
Using (8) and the expression (2) for G(s) we find that

$$\lim_{s \to \infty} [s^2 G(s) - sg(0)] = \lim_{s \to \infty} \frac{s^2 d(s) - sd_{p-1} e(s)}{e(s)} =$$

$$= \lim_{s \to \infty} \frac{d_0 s^2 + d_1 s^3 + \ldots + d_{p-1} s^{p+1} - d_{p-1}(e_0 s + \ldots + s^{p+1})}{e_0 + e_1 s + \ldots + e_{p-1} s^{p-1} + s^p}$$

$$= d_{p-2} - d_{p-1} e_{p-1} = d_{p-2} - g_0 e_{p-1}$$

But

$$\lim_{t \to 0} g^{(1)}(t) = g_1$$

where $g_1$ is the second coefficient of the time power series of g(t) given in (4). Thus

$$g_1 = d_{p-2} - e_{p-1} g_0 \tag{10}$$

Continuing in this manner we find the general iterative expression for the kth coefficient $g_k$ to be

$$g_k = \frac{1}{k!} \left[ d_{p-(k+1)} - \sum_{j=1}^{k} (j-1)! e_{p+j-(k+1)} h_{j-1} \right] \tag{11}$$

where use was made of (5b), i.e. of the relation $h_i = g_i$. Now, using (11), the set of equations (5b) may be written as

$$
\begin{aligned}
d_{p-1} &= h_0 \\
d_{p-2} &= h_1 + h_0 e_{p-1} \\
d_{p-3} &= 2!h_2 + h_1 e_{p-1} + h_0 e_{p-2} \\
&\vdots \\
d_1 &= (p-2)!h_{p-2} + (p-3)!h_{p-3} e_{p-1} + \ldots + h_1 e_3 + h_0 e_2 \\
d_0 &= (p-1)!h_{p-1} + (p-2)!h_{p-2} e_{p-1} + \ldots + 2!h_2 e_3 + h_1 e_2 + h_0 e_1
\end{aligned}
$$

or more compactly as

$$\mathbf{d} = \boldsymbol{\sigma} + \hat{\mathbf{H}}\,\mathbf{e} \tag{12}$$

where

$$
\boldsymbol{\sigma} = \begin{bmatrix} (p-1)!h_{p-1} \\ (p-2)!h_{p-2} \\ \vdots \\ 2!h_2 \\ h_1 \\ h_0 \end{bmatrix}, \quad
\hat{\mathbf{H}} = \begin{bmatrix}
0 & h_0 & \cdots & (p-3)!h_{p-3} & (p-2)!h_{p-2} \\
0 & 0 & \cdots & (p-4)!h_{p-4} & (p-3)!h_{p-3} \\
0 & 0 & \cdots & (p-5)!h_{p-5} & (p-4)!h_{p-4} \\
\vdots & \vdots & & \vdots & \vdots \\
0 & 0 & \cdots & 0 & h_0 \\
0 & 0 & \cdots & 0 & 0
\end{bmatrix}
$$

Heretofore, we have converted the sets of constraint equations (5a) and (5b) to the new sets of equations (6) and (12), respectively, i.e. to the new sets of equations

$$\mathbf{d} = \tilde{\mathbf{H}}\,\mathbf{e} \tag{13a}$$

$$\mathbf{d} = \boldsymbol{\sigma} + \hat{\mathbf{H}}\,\mathbf{e} \tag{13b}$$

From (13a,b) we can now readily determined **d** and **e**. Indeed, if we eliminate the vector **d** from (13a,b) we get

$$\mathbf{H}\,\mathbf{e} = \boldsymbol{\sigma}\,, \quad \mathbf{H} = \tilde{\mathbf{H}} - \hat{\mathbf{H}}$$

or

$$\mathbf{e} = \mathbf{H}^{-1}\,\boldsymbol{\sigma} \tag{14}$$

provided that the matrix **H** is invertible. Substitution of (14) in (13a) yields

$$\mathbf{d} = \tilde{\mathbf{H}}\,\mathbf{H}^{-1}\,\boldsymbol{\sigma} \tag{15}$$

Relations (14) and (15) constitute the simple expressions sought for determining the parameters of the transfer function of the reduced order model.

## IV. ILLUSTRATIVE EXAMPLES

Example 1. For comparison purposes, this example is identical to an example considered by Chidambara [8], where the transfer function H(s) of the original system is

$$H(s) = \frac{-3}{1+s} + \frac{4}{2+s} - \frac{6}{3+s} + \frac{8}{4+s} + \frac{5}{5+s} - \frac{12}{6+s} + \frac{14}{7+s} + \frac{8}{8+s}$$

and the reduced order model transfer functions G(s) is assumed of the form

$$G(s) = \frac{d_0 + d_1 s}{e_0 + e_1 s + s^2}$$

We have

$$H(s) = H_0 + H_1 s + H_2 s^2 + \dots$$

where $H_0 = 1.000000$ , $H_1 = 1.889286$ , $H_2 = -2.556336$ , ... and

$$h(t) = h_0 + h_1 t + h_2 t^2 + \dots$$

where $h_0 = 18$ , $h_1 = -134$ , $h_2 = 489$ , ...

Application of the proposed method yields

$$\mathbf{e} = \begin{bmatrix} e_0 \\ e_1 \end{bmatrix} = \begin{bmatrix} 5.42746 \\ 7.74597 \end{bmatrix} , \quad \mathbf{d} = \begin{bmatrix} d_0 \\ d_1 \end{bmatrix} = \begin{bmatrix} 5.42746 \\ 18 \end{bmatrix}$$

and thus

$$G(s) = \frac{5.42746 + 18s}{5.42746 + 7.74597s + s^2} = \frac{19.389}{6.967 + s} - \frac{1.389}{0.779 + s}$$

The reduced order model transfer function K(s) obtained by Chidambara [8] is

$$K(s) = \frac{3.4855961 + 11.02579s}{3.4856064 + 4.44044s + s^2} = \frac{14.249}{3.422 + s} - \frac{3.223}{1.102 + s}$$

In figure 1 the amplitude curves of H(s), G(s) and K(s) are shown, and in figure 2 their corresponding impulse response curves of h(t), g(t) and k(t). From these figures one readily concludes that the present method yields better results in both the time and frequency domain than that of Chidambara [8] while giving a satisfactory approximation of the original system by the reduced order model in both the time and the frequency domain.

Example 2. Here again, for comparison purposes, the example is taken from Shamash [3], where the transfer function H(s) of the original system is

$$H(s) = \frac{1}{1+s} + \frac{2}{2+s} - \frac{6}{3+s} + \frac{4}{4+s}$$

and the reduced order model transfer function G(s) is assumed of the form

$$G(s) = \frac{d_0 + d_1 s}{e_0 + e_1 s + s^2}$$

We have

$$H(s) = H_0 + H_1 s + H_2 s^2 + \ldots$$

where $H_0 = 1$ , $H_1 = -1.083$, $H_2 = 1.0903$, ... and

$$h(t) = h_0 + h_1 t + h_2 t^2 + \ldots$$

where $h_0 = 1$ , $h_1 = -3$ , $h_2 = 9.5$ , ...

Application of the proposed method yields

$$\mathbf{e} = \begin{bmatrix} e_0 \\ e_1 \end{bmatrix} = \begin{bmatrix} 24 \\ 27 \end{bmatrix} , \quad \mathbf{d} = \begin{bmatrix} d_0 \\ d_1 \end{bmatrix} = \begin{bmatrix} 24 \\ 1 \end{bmatrix} .$$

and thus

$$G(s) = \frac{24 + s}{24 + 27s + s^2} = \frac{0.9173}{0.92+s} + \frac{0.0829}{26.08+s}$$

The reduced order model transfer function R(s) derived by Shamash [3] is

$$R(s) = \frac{2 + 0.8345}{2 + 3s + s^2} = \frac{1.166}{1+s} - \frac{0.332}{2+s}$$

In figure 3 the amplitude curves of H(s),G(s) and R(s) are given, and in figure 4 their corresponding impulse response curves of h(t),g(t) and r(t). From figure 3, one may conclude that R(s) approximates H(s) better than G(s). But from figure 4, one may conclude that g(t) approximates h(t) better than r(t). Overall, the present method yields a reduced order model which approximates the original system fairly well in both the time and the frequency domain.

Comparing the two examples we see that the proposed technique was by far more successfull in example 1 than in example 2.

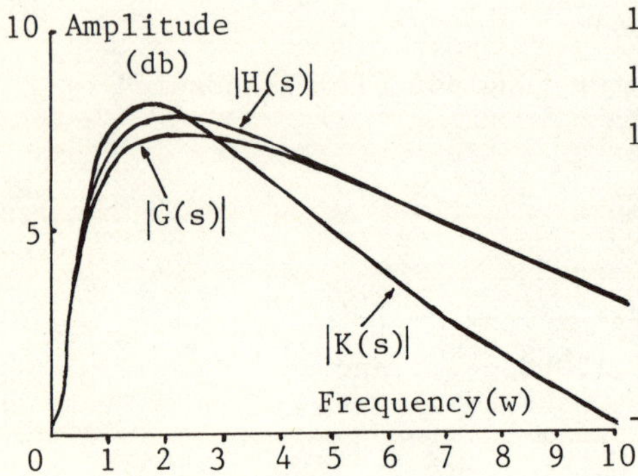


Figure 1

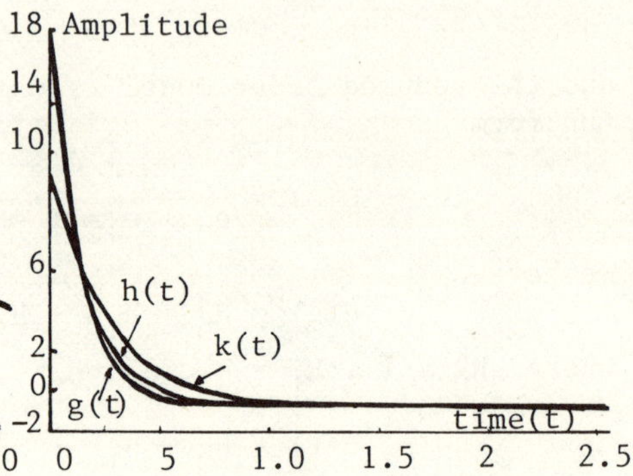


Figure 2

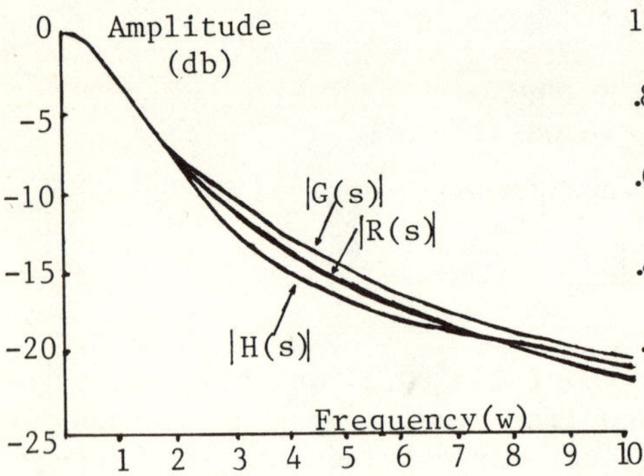


Figure 3

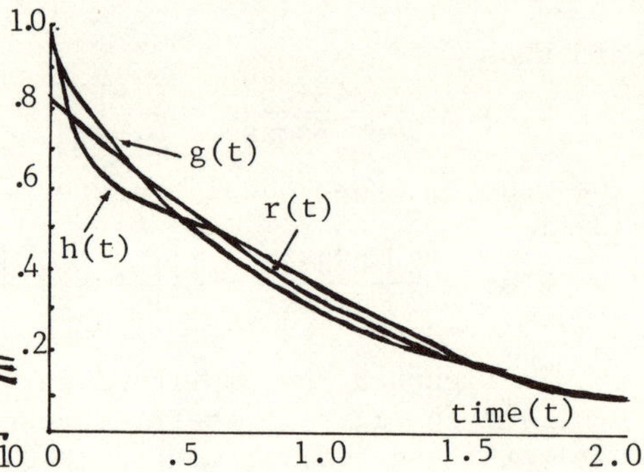


Figure 4

## V. CONCLUSIONS

A novel approach has been presented for order reduction of linear time-invariant systems. This approach yields a lower order model which preserves both time and frequency domain characteristics by introducing fitting criteria in both domains, simultaneously. In the past, the approximation was done in either the time or the frequency domain.

The proposed model reduction technique is simple since it reduces the problem to that of solving a linear system of equations.This method can readily be extended to cover the multi-input multi-output case.

## REFERENCES

[1] Davison,E.J.,"A method for simplifying linear dynamic system", IEEE Trans. Autom. Control, vol.11, pp 93-101, Jan. 1966.

[2] Shamash,Y.,"Stable reduced-order models using Padè-type approximations", IEEE Trans.Autom.Contr.,vol.19,pp.615-616,Oct.1974.

[3] Shamash,Y.,"Linear system reduction using Padè approximation to allow retention of dominant modes",Int.J.Control, vol.21, No. 2, pp. 257-272, 1975.

[4] Lal,M. and Mitra,R.,"Simplification of large system dynamics using a moment evaluation algorithm", IEEE Trans. Autom. Contr., vol. 19, pp. 602-603, Oct. 1974.

[5] Chen,C.F. and Shieh,L.S.,"A novel approach to linear model simplification", Int. J. Contr.,vol.8, pp.561-570, Dec. 1968.

[6] Hutton,M.F. and Friedland,B.,"Routh approximations for reducing order of linear, time-invariant systems", IEEE Trans.Autom.Control, vol. 20, pp. 329-337, June 1975.

[7] Shamash,Y.,"Model reduction using the Routh stability criterion and the Padè approximation technique", Int. J. Contr., vol. 21, pp. 475-484, 1975.

[8] Chidambra,M.R.,"A new canonical form of state-variable equations and its applications in the determination of a mathematical model of an unknown system", Int.J.Contr.,vol.14,No.5,pp.897-909, 1971.

[9] Bosley,M.J and Lees,F.P.,"A survey of simple transfer function derivations from higher-order state-variable models", Automatica, vol. 8, pp. 765-775, 1972.

# OPTIMAL DECOUPLING OF GENERALIZED NON-LINEAR SYSTEMS

J. Nó, J.M. Guillen

Instituto de Automática Industrial CSIC SPAIN

In this work we obtain an optimal solution of the decoupling problem when the input dimension is larger than that of the output, for generalized non-linear multivariable systems, caracterized by the application of a general linear operator, L, instead of the more usual derivative operator in the state equation.

By the usual techniques we obtain the decoupling conditions. Because of the overdimensionality of the input we can take the square control vector norm as index and transform the problem into another with a minimun norm in a Hilbert space whose constraints are the two equations.

## 1. INTRODUCTION

The problem of decoupling multivariable systems has been broadly treated in literature (1) (2) (3). The decoupling of non-linear systems, and linear systems using generalized operators instead of the more usual derivative operator, has been studied recently. (4).

The case of generalized non-linear systems is studied, extending it to group decoupling in (3).

The possibility of being able to impose a feed back law, using a new imput vector that equals the dimension of the output, and in such a way as to decouple the system, is evident when the dimension of the output vector is less than that of the input one.



D. G. Lainiotis and N. S. Tzannes (eds.), Advances in Control, 30–36.

Additionally it is obvious that the decoupling conditions will have infinite solutions in that case. That circumstance makes feasible the improvement of the solution by imposing an optimum condition, wich, if solved, would lead to an explicit and single expresion of the requested decoupling matrices.

The paper is developed acording to these lines, and we have actually arrived to a solution which is valid for a wide class of systems, subject to a series of constraints. These constraints appear in the process of getting this solution.

## 2. DEFINITIONS

Let us consider the system:

$$L_x^t = A(x,t) + B(x,t)u$$

$$Y = C(x,t) + D(x,t)u$$

Where $L^t$ is an invertible scalar operator, which is evaluated at the moment t. A(x,t) B(x,t)C(x,t) and D(x,t) are matrices whose elements can be non-linear in X and t. The dimension of x is n, that of y is p and that of u is m>p. Let us assume the appropiate dimensions on the others.

Let us define a feedback law:

$$u = F(x,t) + G(x,t)\, w$$

where dim w=p. Let us try to find F(x,t) and G(x,t) in such a way that each component of the output depends only on a single component of w. In other words let us decouple the system.

Henceforth, we will employ a more simple notation, by eliminating the arguments, keeping in mind the meaning of the simbols

Let us define the operators $L_x$ and H by means the equation:

$$L\,|C_i| = L_x\,|C_i|\cdot L_x + H|C_i|$$

Where xx is the ith element of the (px1) matrix C.

The operator $\Delta_a^j$ is defined by means of:

$$\Delta_a^j\,|C_i| = H\,|\Delta_A^{j-1}(C_i)| + \{L_x\,|\Delta_A^{j-1}(C_i)|\}.A \;//$$

with:

$$\Delta_A^o\,|C_i| = C_i$$

The decoupling indexes are defined in the following way: $d_i$ if $D_i \neq 0$

$$d_i = \min\{j \,|\, |L_x(\Lambda_A^{j-1} C_i)|B \neq 0\} \quad \text{if } D_i = 0$$

## 3. DECOUPLING

With the feedback law, the system equations are:

$$Lx = A + BF + BGw$$
$$Y = C + DF + DGw$$

If we considered the ith component in the output vector:

$$Y_i = C_i + D_i F + D_i Gw$$

By applying the L operator to this equation we obtain:

$$LY_i = (L_x C_i).Lx + HC_i = HC_i + L_x C_i A + L_x C_i A + L_x C_i BF +$$
$$+ L_x C_i BGw$$

If $d_i > 1$ then:

$$L_x C_i B = 0 \qquad LY_i = \Lambda_A^1 \, C_i$$

By assuming that:

$$L^k \, Y_i = {}^k_A \Lambda C_i$$

then:

$$L^{k+1} \, Y_i = H(\Lambda_A^k \, C_i) + (L_x \Lambda_A^k \, C_i).(A + BF + BGw)$$

and if: $k < d_i - 1$

$$L^{k+1} \, Y_i = H(\Lambda_A^k \, C_i) + (L_x \Lambda_A^k \, C_i)A + \Lambda_A^{K+1} \, C_i$$

Thus this relation has been proved by recurrence.

And for $k = d_i - i$...

$$L^{d_i} \, Y_i = \Lambda_A^{d_i} \, C_i + |L_x \Lambda_A^{d_i - 1} \, C_i|BF + |L_x \Lambda_A^{d_i - 1} \, C_i|BGw$$

Let us compose now a system by means of the definitions:

$$Y_i^* = \begin{cases} L^{d_i} Y_i & \text{if} \quad d_i \neq 0 \\ Y_i & \text{if} \quad d_i = 0 \end{cases}$$

Acoording to all this we can write,

$$Y^* = C^* + D^*F + D^*Gw$$

If the system is to be decoupled it would be written:

$$Y^* = Sw$$

where s is a diagonal operator matrix, which does not depend on x. That implies:

$$C^* = -D^*F$$
$$S = D^*G$$

If we decompose the matrices S and G in columns, we are able to write:

$$S^j = D^*G^j$$
$$C^* = -D^*F$$

where the first members are (px1) matrices and F, G* are (mx1) matrices. Consequently D* is an operator from a certain space of dimension m into one of dimension p<m:

Restricting the study to the cases where $G^j$ and F can be considered elements of a Hilbert space and the operator D* can be considered as a linear operator between both spaces, then it is posible to find and optimal solution, because the addition of a vector wich belongs to the null-space of D* to $G^j$ or F does not affect to the decoupling conditions. This yields a multiplicity of solutions. Among these it is posible to select the best one for a given performance index.

## 4. OPTIMIZATION

The chosen performance index is cuadratic in the control vector u:

$$J = u'Qu$$

This perfomance index can be considered as the inner product of a vector and itself in a Hilbert space. We can establish orthogonality conditions in this way.

As we have already seen:

$$u = F+\sum^{p} G^{j} w_{j}$$

Were w works as a scalar constant for the purpose of optimization.

By expressing F and $G^{j}$ in its unique decomposition as an addition of vectors of the null-space of D* and its orthogonal complement we obtain:

$$J = ( F+\sum^{p} G^{j} w_{j})' Q ( F+\sum^{p} G^{j} w_{j})+$$
$$+( {}^{N}F+ \sum^{p} {}^{N}G^{j} w_{j})' Q( {}^{N}F+ \sum^{p} {}^{N}G^{j} w_{j})$$

This perfomance index will be minimum when the second term is null. That is feasible, since,as we have previously seen, a variation of the components confined in the null-space of D* does not alter the decoupling conditions.

Hence, the vectors orthogonal to the null-space make minimum, the perfomance index.

As it is well know,

$$|N(D^*)| = R\ |(D^*)^{a}|$$

where x is the adjoint operator or:

$$F = (D^*)^{a}\ z$$

where z is a vector of the space on which $(D^*)^{a}$ acts.

By sustituing in the decoupling condition,

$$C^* = -D^*\ (D^*)^{a}\ z$$

If the operator $D^*\ (D^*)^{a}$ is invertible:

$$z = -|(D^*)^{a} - \ D^*\ (D^*)^{a}\ |^{-1}\ .C^*$$

$$F = -(D^*)^{a} - \ |D^*(D^*)^{a}\ |^{-1}\ .C^*$$

In a similar way we obtain,

$$G^j = (D^*)^a \, |D^*(D^*)^a|^{-1} \, S^j$$

By grouping columns.

$$^mG = (D^*)^a \, |D^*(D^*)^a|^{-1} \, S$$

where $^mG$ indicates a matrix which make the perfomance index a minimum.

The adjoint operator is obtained from its definition and from the definition of the inner product in the considered spaces. We have already defined the inner product in a space of dimension m as u'Qv. Defining that the product in the space of dimension p as z' P w, where p is some weighting matrix, it result for any couple of vectors x and y:

$$x'Q(D^*)^a \, y = x'(D^*)'Py$$

That implies:

$$Q \, (D^*)^a = (D^*)'p \; ; \; (D^*)^a = Q^{-1} \, (D^*)'p$$

By sustituing it is finally obtained.

$$^mG = Q^{-1}(D^*)' \, |D^* \, Q^{-1} \, (D^*)'|^{-1} \, \dot{S};$$

$$^mF = -Q^{-1} \, (D^*)' \, |D^* \, Q^{-1} \, (D^*)^1|^{-1} \, C^*$$

These expresions decouple and verify the minimum.

## REFERENCES

Falb, P.L - Wolovich, W.A. - dec 1967 "DECOUPLING IN THE DESIGN AND SYNTHESIS OF MULTIVARIABLE CONTROL SYSTEMS". IEEE/Trans Aut. Control AC-12 pp 651-659.

Morse, A.D.- Wonham,W.M. dec 1971 "STATUS OF NONINTERACTING CONTROL". IEEE/Trans Aut Control AC-16 pp 568-581.

Nó, J. May- 1979 "CONTRIBUCION AL CONTROL POR DESACOPLO DE SISTEMAS MULTIVARIABLES". Tesis Doctoral. Universidad de Valladolid.

Tzafestas, S.G., Paraskevopoulos, P.N.- 1977 "GENERALIZED APPROACH TO STATIC AND DYNAMIC DECOUPLING OF LINEAR CONTROL SYSTEMS". Int. J. Syst.Sci pp171-183

Luenberger, D.G. New York, 1969 "OPTIMIZATION BY VECTOR SPACE METHODS". John Wiley.

# SENSITIVITY COMPENSATION AND ADAPTIVE SAMPLING

S. Dormido*, M. de la Sen** and M. Mellado*

* Dpto. Informática y Automática, Facultad de Ciencias Físicas, Univ. Complutense, Madrid-3, SPAIN.
** Dpto. Automática e Informática, Facultad de Ciencias Universidad de Bilbao, Apt° 644, Bilbao, SPAIN.

ABSTRACT. The paper presents the possibility of using the sampling interval adjustment scheme for achieving compensation to plant parameter variation.
The behaviour of a sampled-data system is normally very depending to changes in the sampling period. Therefore, knowledge of the sampling interval sensitivity should be quite important. Sampling interval sensitivity may be used as a basis for determining the universal sampling interval or individual sampling intervals. These two points of view are significant in that they explain two different sampling interval sensitivity functions.
The first approach (global sensitivity function) is applied to the problem of compensating changes in the parameters of a sampled data system. The compensation is achieved through the variation of the sampling period.
The second approach (local sensitivity function) is applied to the problem of compensating deviations of the plant parameters from their nominal values in an adaptive sampling system by modification of the threshold value in the adaptive sampling law.

## 1. INTRODUCTION

In a sampled-data system, the sampling period may be used as a general compensator in the problem of compensating perturbations in the parameters of the system. This point of view has special relevance when a digital controller has been designed to satisfy a set of specifications given a priori. Under these conditions, a change in the parameters, keeping the sampling period constant, normally leads to design the compensation again and therefore increasing the cost.



*D. G. Lainiotis and N. S. Tzannes (eds.), Advances in Control, 37–44.*

However, it is possible to compensate small variations in the adaptive sampling law due to perturbations in its parameter vector for an adaptive sampling system, keeping the sampling instants approximately constant during the process.

## 2. SYSTEMS COMPENSATION BY CHANGING THE SAMPLING INTERVAL

The compensation methods developed so far (1) make use of the sensitivity function in the domain of the z-transform. In this paper the sensitivity function in the time domain is applied. In the following the system of the figure 1 is considered. In this analysis, all the initial conditions are assumed to be zero.

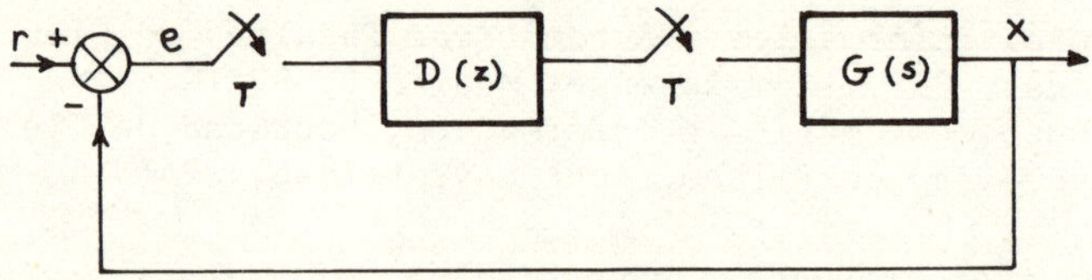


Fig. 1 Sampled Data System with Compensation Network

The time-domain matrix is a matrix expression for the convolution summation of a sampled system (2). Using time-domain matrices the following relations are obtained.

$$\underline{x} = G\underline{\ell}, \quad \underline{\ell} = D\underline{e} \quad \text{and } \underline{e} = \underline{r} - \underline{x} \tag{2-1}$$

where

$$\begin{aligned}
\underline{x} &= (x_0, x_1, \ldots, x_n)^t \text{ with } x_i = x(iT) \\
\underline{\ell} &= (\ell_0, \ell_1, \ldots, \ell_n)^t \quad " \quad \ell_i = \ell(iT) \qquad ({}^t \text{ stands for the transpose}) \\
\underline{e} &= (e_0, e_1, \ldots, e_n)^t \quad " \quad e_i = e(iT) \\
\underline{r} &= (r_0, r_1, \ldots, r_n)^t \quad " \quad r_i = r(iT)
\end{aligned}$$

$$G = \begin{pmatrix} g_0 & 0 & \cdots & 0 \\ g_1 & g_0 & \cdots & 0 \\ \cdots & \cdots & \cdots & \cdots \\ g_n & g_{n-1} & \cdots & g_0 \end{pmatrix} \qquad D = \begin{pmatrix} d_0 & 0 & \cdots & 0 \\ d_1 & d_0 & \cdots & 0 \\ \cdots & \cdots & \cdots & \cdots \\ d_n & d_{n-1} & \cdots & d_0 \end{pmatrix} \tag{2-2}$$

Therefore, the vectors $\underline{x}$, $\underline{\ell}$, $\underline{e}$ and $\underline{r}$ represent a sequence of values of the different magnitudes in successive sampling instants.

It is assumed that the reference input is a step function. $G$ and $D$ are both triangular matrices for the system and the compensator respectively. Furthermore, $g_{ij} = g(jT) = g_j$ means the impulse response of the system $G(s)$ in the sampling instant $jT$ and $d_{ij} = d(jT) = d_j$ defines the digital controller

$$D(z) = \sum_{i=1}^{N} d_i z^{-i}$$

Let us suppose that the system goes under a perturbation in the nominal vector of the plant parameter, and write with dash notation the disturbed parameter system. It is obvious that the global sensitivities of the output of both systems with respect to variations in the parameters and period are given by

$$\begin{aligned} \underline{v}^T &= G_t D \underline{e} - G D \underline{v}^T \\ \underline{v}^{T'} &= G'_t D e' - G' D \underline{v}^{T'} \\ \underline{v}^{p_i} &= G_{p_i} D \underline{e} - G D \underline{v}^{p_i} \qquad i = 1,\dots,r \end{aligned} \tag{2-3}$$

$$\begin{aligned} \underline{v}^T &= \frac{d\underline{x}}{dT} = (v_0^T, v_1^T, \dots, v_n^T)^t \\ \underline{v}^{T'} &= \frac{d\underline{\underline{x}}}{dT'} = (v_0^{T'}, v_1^{T'}, \dots v_n^{T'})^t \\ \underline{v}^{p_i} &= \frac{d\underline{x}}{dp_i} = (v_0^{p_i}, v_1^{p_i}, \dots v_n^{p_i})^t \qquad i = 1,\dots,r \end{aligned} \tag{2-4}$$

$$G_t = \begin{pmatrix} \frac{\partial g_0}{\partial t} & 0 & \cdots & 0 \\ \frac{\partial g_1}{\partial t} & \frac{\partial g_0}{\partial t} & \cdots & 0 \\ \cdots & \cdots & \cdots & \cdots \\ \frac{\partial g_n}{\partial t} & \frac{\partial g_{n-1}}{\partial t} & \cdots & \frac{\partial g_0}{\partial t} \end{pmatrix} \tag{2-5}$$

$$G_{t'} = \begin{pmatrix} \frac{\partial g'_0}{\partial t} & 0 & \cdots & 0 \\ \frac{\partial g'_1}{\partial t} & \frac{\partial g'_0}{\partial t} & \cdots & 0 \\ \cdots & \cdots & \cdots & \cdots \\ \frac{\partial g'_n}{\partial t} & \frac{\partial g'_{n-1}}{\partial t} & \cdots & \frac{\partial g'_0}{\partial t} \end{pmatrix}$$

$$G_{p_i} = \begin{pmatrix} \frac{\partial g_0}{\partial p_i} & 0 & \cdots & 0 \\ \frac{\partial g_1}{\partial p_i} & \frac{\partial g_0}{\partial p_i} & \cdots & 0 \\ \cdots & \cdots & \cdots & \cdots \\ \frac{\partial g_n}{\partial p_i} & \frac{\partial g_{n-1}}{\partial p_i} & \cdots & \frac{\partial g_0}{\partial p_i} \end{pmatrix} \qquad i = 1 \ldots r \tag{2-5}$$

being the r vector $\underline{p}$ the plant parameters.

$$\underline{p} = (p_1, p_2, \ldots, p_r)^t \tag{2-6}$$

The compensating method requires to provide an analytical expression for the output of three systems: the original one, the plant parameter perturbed system, and the perturbed system in both plant parameters and period. Using the first order approximation given by the global sensitivities function defined in (2-3), the variation of the sampling period $\Delta T$ is determined by minimizing the expression:

$$\sum_{k=1}^{N} (x'_{k+\Delta k} - x_{k+\Delta k})^2$$

where $x'_{k+\Delta k} = x'[k(T+\Delta T)]$ represents the output of the plant parameter perturbed system in the new sampling instants $j(T+\Delta T)$.

$x_{k+\Delta k} = x[k(T+\Delta T)]$ represents the output of the original system (without perturbation) in the new sampling instants $j(T+\Delta T)$.

$N$ denotes the number of sampling instants over which the compensation is achieved.

The variation in the sampling period $\Delta T$ is given by:

$$\Delta T = - \frac{\sum_{k=1}^{N} < \underline{v}_k^p . \Delta \underline{p} > . (v_k^{T'} - v_k^T)}{\sum_{k=1}^{N} (v_k^{T'} - v_k^T)^2} \qquad (2\text{-}7)$$

where

$$\underline{v}_k^p = (v_k^{p_1}, v_k^{p_2}, \dots, v_k^{p_r})^t$$

$$\Delta \underline{p} = (\Delta p_1, \Delta p_2, \dots, \Delta p_r)^t$$

Obviously, the value of $\Delta T$ in the above expression depends on the chosen value for N, which it must not be very large because the first order approximation performed in the global sensitivity functions is less valid when the number of sampling periods is increased.

In order to analyse the efficiency of the previous procedure the overshoot as a performance index has been chosen. It has been shown that the system with both, parameters plant and period perturbed, shows lower overshoot than the plant parameter perturbed system. (1)

## 3. COMPENSATION IN AN ADAPTIVE SAMPLING SYSTEM

In an adaptive sampling system the variation in the parameters of the plant can be compensated with proper changes in the parameters of the sampling law. In the following, a method with a similar philosophy as one mentioned before is used. Essentially it is based in disturbing the plant parameter perturbed system by a proper modification of the constant in the sampling law, and comparing the output of the plant parameter perturbed system with the original one (nominal values in the plant parameter). The system is perturbed only in the constants of the sampling law in such a way that the sampling instants in both systems are approximately equals. The required increment in the sampling law constants is obtained by a least-square approximation between both outputs providing that the number of sampling instants is not very large.

By using sensitivity functions and assuming a constant input in the instant $t_k$ and $g(0) = 0$, for a first order approximation the following is obtained:

$$x'_{t_k} \simeq x_{t_k} + \{ \sum_{i=1}^{r} \frac{\partial g(t)}{\partial p_i} + [\frac{\partial g(t)}{\partial p_i} \frac{\Delta t}{\Delta p_i}]_{t=t_k-t_n} (r(t_n) -$$

$$- x(t_n)) - g(t_k-t_n)\, v_{t_n}^{p_i} \} \Delta p_i + \sum_{i=1}^{s} \{ \sum_{n=0}^{k-1} [\frac{\partial g'(t)}{\partial t} \frac{\Delta t}{\Delta \alpha_i}]_{t=t_k-t_n} (r(t_n) -$$

$$- x(t_n)) - g'(t_k-t_n)\ \rho_{t_n}^{\alpha_i} \} \Delta\alpha_i \qquad (3\text{-}1)$$

where

$$\nu_{t_k}^{p_i} = \sum_{n=0}^{k-1} \left[\frac{\partial g(t)}{\partial p_i}\right]_{t=t_k-t_n} (r(t_n) - x(t_n)) + \sum_{n=0}^{k-1} \left[\frac{\partial g(t)}{\partial t} \frac{\Delta t}{\Delta p_i}\right]_{t=t_k-t_n} (r(t_n) - x(t_n)) - \sum_{n=0}^{k-1} g(t_k-t_n)\ \nu_{t_n}^{p_i} \qquad i=1,2,\ldots,r$$

$$\rho_{t_k}^{p_i} = \sum_{n=0}^{k-1} \left[\frac{\partial g(t)}{\partial t} \frac{\Delta t}{\Delta\alpha_i}\right]_{t=t_k-t_n} (r(t_n) - x(t_n)) - \sum_{n=0}^{k-1} g(t_k-t_n)\rho_{t_n}^{\alpha_i} \qquad i=1,2,\ldots,s \qquad (3\text{-}2)$$

with

$$\nu_0^{p_i} = \rho_0^{\alpha_j} = 0 \qquad i = 1,2,\ldots,r\ ; \quad j = 1,2,\ldots, s.$$

and being the plant parameter vector and the adaptive sampling law parameter vector $\underline{p}^t = (p_1, p_2, \ldots, p_r)$ and $\alpha^t = (\alpha_1, \alpha_2, \ldots, \alpha_s)$ respectively. $\nu_{t_k}^{p_i}$ and $\rho_{t_k}^{\alpha_i}$ are the sensitivity functions of the output with respect to the perturbation in the plant parameters and in the constants of the sampling law respectively.
It should be noted in expression (3-1) the effects of the double perturbation, i.e. to the perturbation in the parameters vector (which implicitely implies a variation in the sampling instant like it is shown in (3-2)) follows a perturbation in the vector of constants of the sampling law.
The output of the original system only perturbed by the second mechanism is in the instant $t_k$ in first approximation given by:

$$x''_{t_k} \simeq x_{t_k} + \sum_{i=1}^{s} \left( \sum_{n=0}^{k-1} \left[\frac{\partial g(t)}{\partial t} \frac{\Delta t}{\Delta\alpha_i}\right]_{t=t_k-t_n} (r(t_n)-x(t_n))-g(t_k-t_n) \cdot \mu_{t_k}^{\alpha_i}\right)\Delta_{\alpha_i} \qquad (3\text{-}3)$$

where

$$\mu_{t_k}^{\alpha_i} = \sum_{n=0}^{k-1} \left[\frac{\partial g(t)}{\partial t} \frac{\Delta t}{\Delta\alpha_i}\right]_{t=t_k-t_n} (r(t_n)-x(t_n)) - \sum_{n-0}^{k-1} g(t_k-t_n)\mu_{t_n}^{\alpha_i} \qquad i=1,2,\ldots,s \qquad (3\text{-}4)$$

with

$$\mu_0^{\alpha_i} = 0 \qquad i=1,2,\ldots,s$$

is the output sensitivity function with respect to variations in the adaptive sampling law parameter vector.
Following now the same procedure as in the case of periodic sampling we obtain when $\underline{\alpha}$ reduces to an element only

$$\Delta_\alpha = - \frac{\sum_{k=1}^{N} (\underline{\nu}_{t_k}^{p_i} \Delta \underline{p}) (\rho_{t_k}^{\alpha} - \mu_{t_k}^{\alpha})}{\sum_{k=1}^{N} (\rho_{t_k}^{\alpha} - \mu_{t_k}^{\alpha})^2} \qquad (3\text{-}5)$$

where

$$\nu_{t_k}^{p_i} = (\nu_{t_k}^{p_1}, \nu_{t_k}^{p_2}, \ldots, \nu_{t_k}^{p_r})^t$$

and

$$\Delta \underline{p} = (\Delta p_1, \Delta p_2, \ldots \Delta p_r)^t$$

It should be pointed out that the above approximation is only valid when the sampling instants of both systems are nearly equal. This is accomplished if the principal mechanism of change in those instants is the chosen sampling law, with the condition that the parameters are not too much different from their nominal values In the Table I some simulations for different sampling laws (3), (4), with a plant of the form:

$$G(s) = \frac{K(1-exp(-Ts))(s+a)}{s^3} \quad with\ K_0 = a_0 = \alpha_0 = 1 \text{ (nominal values)}$$

and with a unit step function as input are shown.
A comparison is made over a minimum number of sampling periods that takes a time greater than 3 sec. and performance index used has been the area enclosed by the output of the zero order sample and hold with respect to a pair of axis output-time.
The calculation of the sensitivity of the sampling instants with respect to the variation of the different parameters has been done by programming, assuming linearity for small variations.
It should be emphasized that the results in columns 4 and 6 show a better approximation than those in columns 4 and 5, as expected.

## 4. CONCLUSION

It has been shown that the sensitivity in the time-domain is a useful tool in the problem of varying the sampling period to compensate possible perturbation in the others parameters of the system, and in some cases this method may be applied to the compensation of adaptive sampling systems.

TABLE I

| Sampling law | Δk(%) | Δα | $J_{orig}$ | $J_{pert}$ | $J_{comp}$ |
|---|---|---|---|---|---|
| $T_i = \dfrac{1}{\alpha\sqrt{\lvert\dot{e}(t_i)\rvert}}$ | 5 | -0,03153 | 5,021 | 4,896 | 5,053 |
| | 10 | -0,05868 | " | 4,782 | 5,086 |
| | 15 | -0,08123 | " | 4,677 | 5,113 |
| $T_i = \dfrac{1}{\alpha\lvert\dot{e}(t_i)\rvert}$ | 5 | -0,01605 | 5,474 | 5,308 | 5,400 |
| | 10 | -0,03409 | " | 5,070 | 5,259 |
| | 15 | -0,07381 | " | 5,856 | 5,143 |
| $T_i = \dfrac{1}{\alpha\lvert\dot{e}(t_i)\rvert^{2/3}}$ | 5 | -0,02751 | 5,365 | 5,195 | 5,347 |
| | 10 | -0,05347 | " | 5,038 | 5,341 |
| | 15 | -0,07728 | " | 4,893 | 5,341 |

## References

(1) S. Dormido "Control de sistemas con el tiempo como parámetro" Tesis doctoral, Univ. de Bilbao, Sept. 1971.

(2) J. Sevely "Systemes et asservissements lineaires echantillonnés". Dunod, Paris, 1969

(3) M. de la Sen "Propiedades fundamentales de los sistemas dinámicos con elección aperiódica de los instantes de muestreo" Tesis doctoral, Univ. de Bilbao, Jun. 1979.

(4) S. Dormido, M. de la Sen y M. Mellado "Criterios generales de determinación de leyes de muestreo adaptivo". Partes I y II. Rev. de Informática y Automática, pp. 13-29, Oct. 1978

# DECOUPLING AND SENSITIVITY

Bagher Haghighian-Roodsary and Baxter F. Womack

Department of Electrical Engineering and Electronics Research Center, The University of Texas at Austin, Austin, Texas, 78712, U.S.A.

## ABSTRACT

Decoupled multiple input-multiple output systems (MI-MO) are an important subclass of MI-MO systems. Loss of decoupling because of plant variations can affect system operation significantly. This paper has as its aims: (1) The identification of classes of MI-MO systems which are decoupled; (2) Development of design methods using sensitivity functions to minimize the coupling problem which maintains the decoupling conditions to first order in parameter variations.

## I. INTRODUCTION

Multivariable systems are those systems in which one or more output variables are controlled by one or more input variables [8]. Generally, in linear multivariable control systems, a change in any input variable will usually result in changes in all output variables [6]. An important problem in the study of multivariable systems is that of synthesizing a system which has noninteracting control [6].

The decoupling problem in multivariable control systems is of importance [2,3,4,9,10] because one usually wants to design a state dependent controller so that each input will control each output independently. The advantage of such a design is that once a decoupled system is obtained the system is thus reduced to a number of single variable systems to which the well known linear design techniques can be applied [8].



*D. G. Lainiotis and N. S. Tzannes (eds.), Advances in Control, 45–55.*

The goal of sensitive decoupling theory is the reduction of the effect of system parameter variations on the actual input-output decoupling achieved by a controller designed upon the assumption that the system parameters remain undisturbed.

## II. THE INPUT-OUTPUT DECOUPLING PROBLEM AND ITS SOLUTION

There has been considerable interest in recent years in the problem of decoupling linear multivariable systems by state feedback and a transformation of the input [2,3,4,5,9].

### 2.1. Solution of the Problem

Consider a time-invariant linear system which may be described by the set of differential equations:

$$\begin{aligned} \dot{x}(t) &= A\,x(t) + B\,u(t) \\ y(t) &= C\,x(t) \end{aligned} \tag{1}$$

where x(t) is a real n-vector representing the state of the system, and y(t) and u(t) are real m-vectors representing system output and input, respectively. Then the input of the system using state variable feedback is

$$U = Fx + G\omega \qquad , \text{ G invertible} \tag{2}$$

where the m-vector $\omega$ is the reference input.

It is desired that complete input-output noninteraction between the outputs and the controlling inputs exists in the compensated system. Complete input-output noninteraction implies there is a one to one correspondance between each output and controlling input.

So the decoupling problem is: "Given the model (1) find the linear pair [F,G] in the feedback control law (2) such that the transfer function matrix of the closed-loop system is a diagonal matrix".

The equation of the overall feedback multivariable decoupled system will be as follows with the substitution of (2) into (1). i.e.

$$\begin{aligned} \dot{x} &= (A + BF)x + BG\omega \\ y &= Cx. \end{aligned} \tag{3}$$

A block diagram of the system (3) is shown in Fig. 1.

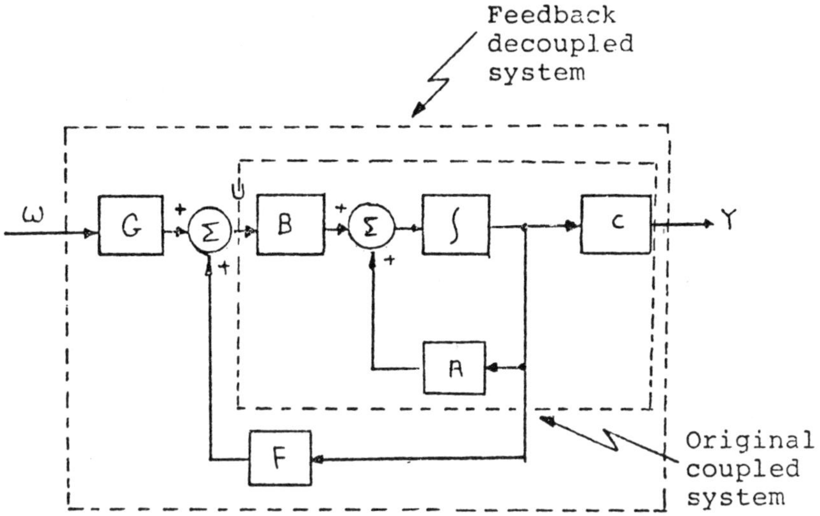


Fig. 1 Block diagram of the feedback system (3)

Define the set of positive integers $\{d_i\}$; $i = 1,2,\ldots, m$ as [3,4,5].

$$d_i = \begin{cases} \min j : C_i A^j B \neq 0 , \ j = 0,1,2,\ldots, n-1 \\ n-1, \text{ if } C_i A^j B = 0 \quad \text{for all } j \end{cases} \tag{4}$$

where $A^j = \underbrace{AA\ldots A}_{j\ \text{times}}$ and $C_i$ is the $i^{th}$ row of the matrix C [4,5].
By repeated application of the derivative operator upon $y_i$, and using $C_i(A + BF)^K = C_i A^K$ $K = 0,1,2,\ldots$ [2,3],

$$y_i^{(d_i+1)} = [C_i A^{d_i+1} + C_i A^{d_i} BF]x + C_i A^{d_i} BG\omega \quad \text{for } i=1,2,\ldots,m$$

So in matrix form

$$\begin{bmatrix} y_1^{(d_1+1)} \\ y_2^{(d_2+1)} \\ \cdot \\ \cdot \\ \cdot \\ y_m^{(d_m+1)} \end{bmatrix} = \left[ \begin{bmatrix} C_1 A^{d_1+1} \\ \cdot \\ \cdot \\ \cdot \\ \cdot \\ C_m A^{d_m+1} \end{bmatrix} + \begin{bmatrix} C_1 A^{d_1} B \\ \cdot \\ \cdot \\ \cdot \\ \cdot \\ C_m A^{d_m} B \end{bmatrix} F \right] x + \begin{bmatrix} C_1 A^{d_1} B \\ \cdot \\ \cdot \\ \cdot \\ \cdot \\ C_m A^{d_m} B \end{bmatrix} G\omega$$

or

$$y^* = (A^* + B^*F)x + B^*G\omega \tag{5}$$

Now it is clear that by choosing F and G such that [5,11]:

$$A^* + B^*F = \underline{0} \rightarrow F = (-B^*)^{-1} A^*$$
$$\text{and} \quad B^*G = I \rightarrow \underline{G} = (B^*)^{-1} \tag{6}$$

equation (5) reduces to

$$y^* = \omega$$

or

$$y_i^{(d_i+1)} = \omega_i \qquad i = 1,2,\ldots m. \tag{7}$$

Equation (7) shows that the $i^{th}$ component of $y(t)$, namely $y_i(t)$, is controlled by the $i^{th}$ component of $\omega(t)$, that is $\omega_i(t)$. Thus the solution of the above decoupling problem is provided by the following condition [2,11]:

> $B^*$ must be invertible; then the pair of matrices $[F,G,]$ given by (6) with G nonsingular decouples the system (1).

2.2 Example

The following example illustrates the application of the above theory. Consider the linear time-invariant system:

$$\dot{x} = A_0 x + B_0 U \quad \& \quad y = C_0 x \tag{8}$$

where

$$A_0 = \begin{bmatrix} 1 & 1 & 0 \\ 0 & 2 & 0 \\ 0 & 1 & 3 \end{bmatrix}, \quad B_o = \begin{bmatrix} 1 & 1 \\ -1 & 1 \\ 0 & 0 \end{bmatrix}, \quad C_0 = \begin{bmatrix} 1 & 0 & 0 \\ 0 & 0 & 1 \end{bmatrix} = \begin{bmatrix} \underline{C}_1 \\ \underline{C}_2 \end{bmatrix}.$$

the nominal (undisturbed) system will be decoupled with the following pair of matrices $[F_0, G_0]$ and a new control vector $(\underline{\omega})$ in the linear feedback control law $(U = F_0 x + G_0 \omega)$:

$$F_o = (-B_0^*)^{-1} A_0^* = -\frac{1}{2}\begin{bmatrix} 1 & -4 & -9 \\ 1 & 6 & 9 \end{bmatrix}.$$

where $B_0^* = \begin{bmatrix} 1 & 1 \\ 0 & 1 \end{bmatrix}$ and $A_0^* = \begin{bmatrix} 1 & 1 & 0 \\ 0 & 5 & 9 \end{bmatrix}$.

And from ( 6 )

$$G_o = (B_0^*)^{-1} = \frac{1}{2}\begin{bmatrix} 1 & -1 \\ 1 & 1 \end{bmatrix} = \begin{bmatrix} \frac{1}{2} & -\frac{1}{2} \\ \frac{1}{2} & \frac{1}{2} \end{bmatrix}.$$

The transfer function $(H_0)$ of the system [9] can be obtained by using (2),(8) and Laplace transforms:

$$Y(s) = C_0[SI - (A_0 + B_0 F_0)]^{-1} B_0 G_0 \omega(s) \triangleq H_0(s)\omega(s) \tag{9}$$

so

$$H_0(S) = C_0[SI - (A_0 + B_0F_0)]^{-1}B_0G_0.$$

$$H_0(s) = \begin{bmatrix} \frac{1}{S} & 0 & 0 \\ 0 & \frac{1}{S^2} & \frac{S+3}{S^2} \end{bmatrix} \begin{bmatrix} 1 & 0 \\ 0 & 1 \\ 0 & 0 \end{bmatrix} = \begin{bmatrix} \frac{1}{S} & 0 \\ 0 & \frac{1}{S^2} \end{bmatrix} . \tag{10}$$

Using equation (9) and (10):

$$\begin{aligned} \dot{y}_1(t) &= +\,\omega_2(t) \\ \ddot{y}_2(t) &= +\,\omega_2(t) \,. \end{aligned} \tag{11}$$

It is obvious that each component of the new control vector $\omega = \begin{bmatrix} \omega_1 \\ \omega_2 \end{bmatrix}$ will control one output, i.e., we have reached a perfectly decoupled system, using a new control vector $\omega$ and a pair of matrices $[F_0, G_0]$ in the feedback control law.

## III. STATEMENT OF THE SENSITIVE DECOUPLING PROBLEM

In Section 2 decouping of a nominal (undisturbed) coupled system was discussed. In the procedure of decoupling, there were no parameter variations, and the system was decoupled perfectly. The purpose of this section is to consider parameter variations in the nominal system, and then to design a controller to decouple the disturbed system to the first order of parameter variation. This controller is called a "sensitive decoupling controller" [11].

The precise statement of the sensitive decoupling controller design problem to be solved is as follows. Given the model (1) with $A=A_0 + \Delta A$, $B=B_0 + \Delta B$ and $C=C_0 + \Delta C$ where $A_0$, $B_0$ and $C_0$ are the nominal values (assumed known) and $\Delta A$, $\Delta B$ and $\Delta C$ are the parameter variations (the variations assumed to be small). The pair of matrices $\{F,G\}$ must be found in the feedback law (2) such that it decouples the system to the first order of parameter variations $\Delta A, \Delta B$, and $\Delta C$, that is, the $i^{th}$ input ($\omega_i$) affects only the $i^{th}$ output $y_i$, if second and higher powers of the parameter variations $\Delta A$, $\Delta B$ and $\Delta C$ are neglected.

Consider the decoupling condition which is repeated as follows:

$$\begin{aligned} A_0^* + B_0^* F_0 &= 0 \\ B_0^* G_0 &= I \end{aligned} \tag{11}$$

where these conditions should be satisfied to decouple the nominal (undisturbed) system. In (11) $A_0$, $B_0$ are the nominal values of the system and the pair of matrices $[F_0, G_0]$ have the nominal values which decouple the system (1). Suppose θ is the parameter vector composed of all parameters contained in A, B and C, that is:

$$\theta = [a_{11}, a_{12}, \ldots, d_{mm}, b_{11}, b_{12}, \ldots, b_{nm}, c_{11}, c_{12}, \ldots, c_{mn}]'.$$

A prime denotes transpose of the vector. Suppose θ varies from its nominal value by Δθ, then (11) cannot decouple the system; instead, the following conditions can be considered to decouple the disturbed system [11].

$$A^* + B^*F = 0$$
$$B^* G = I \tag{12}$$

where

$$A^* = A_0^* + \Delta A^*, \; B^* = B_0^* + \Delta B^*, \; F = F_0 + \Delta F,$$
$$G = G_0 + \Delta G. \qquad (13) \;\&\; (14)$$

Now it is natural that control input U will change for the disturbed system. Suppose the new control input is of the following form:

$$U = U_0 + \Delta U \tag{15}$$

where ΔU is the variation of $U_0$. The feedback control law in this case will be:

$$U = U_0 + U = FX + G\omega \tag{16}$$

where F, G and U are introduced in (14) and (15). Now ΔF and ΔG in F and G should be selected so as to correct $U_0$ by ΔU and decouple the system to the first order in ΔA, ΔB and ΔC [11]. Substitute (14) into the first equation (12), yielding

$$A^* + B^*F = A^* + B^* [F_0 + \Delta F]. \tag{17}$$

Expanding the matrices A* and B* in Taylor series about their nominal values yields:

$$A^* + B^*F = (A_0^* + B_0^*F_0) + [(A_\theta^*) \Delta\theta + (B_\theta^*)_0 \Delta\theta \; F_0 +$$
$$B_0^*(F_\theta)_0 \; \Delta\theta] + \varepsilon_1^2 \tag{18}$$

where $(A_\theta^*)_0$, $(B_\theta^*)_0$ and $(F_\theta^*)_0$ denote the partial derivatives of the matrix-valued functions A*(θ), B*(θ) and F*(θ) with respect to the parameter vector θ at the initial time [11], (and are tensors), $\varepsilon_1^2$ is a function in terms of second or higher order norms of $\Delta\theta$. It is supposed $\Delta\theta$ is small enough so that second and higher norms of $\Delta\theta$ can be neglected [11] in the above equation $\varepsilon_1^2 \cong 0$. Also the first term in (18) is zero, according to equation (11), so

$$A^* + B^*F = (A^*_\theta)\,\Delta\theta + (B^*_\theta)_0\,\Delta\theta F_0 + B^*_0\,(F_\theta)_0\,\Delta\theta. \tag{19}$$

also

$$F \triangleq F_0 + \Delta\,F = F_0 + (F_\theta)_0\,\Delta\theta + \varepsilon_1^2 \tag{20}$$

where $\varepsilon_1^2$ again is assumed to be zero so:

$$F = F_0 + \Delta F \cong F_0 + (F_\theta)_0\,\Delta\theta.$$

Hence

$$\Delta F = (F_\theta)_0\,\Delta\theta; \tag{21}$$

With the same procedures $\Delta G$ can be derived [11].

$$\begin{aligned} B^*G &= [B^*_0 + (B^*_\theta)_0\,\Delta\theta + \varepsilon_1^2]\,[G_0 + (G_\theta)_0\,\Delta\theta + \varepsilon_2^2] \\ &= B^*_0G_0 + B^*_0\,(G_\theta)_0\,\Delta\theta + (B^*_\theta)_0\,\Delta\theta G_0 + \varepsilon_3^2 \\ &\cong (B^*_0G_0) + [(B^*_\theta)_0\,\Delta\theta G_0 + B^*_0\,(G_\theta)_0\,\Delta\theta]. \end{aligned} \tag{22}$$

By (11) the first term in (22) is the unit matrix I, thus:

$$B^*G = I + [(B^*_\theta)_0\,\Delta\theta G_0 + B^*_0\,(G_\theta)_0\,\Delta\theta]. \tag{23}$$

To satisfy the decoupling condition (12),

$$(B^*_\theta)_0\,\Delta\theta G_0 + B^*_0\,(G_\theta)_0\,\Delta\theta = 0$$

or by using (11):

$$(G_\theta)_0\,\Delta\theta = -G_0\,(B^*_\theta)_0\,\Delta\theta G_0. \tag{24}$$

But

$$G = G_0 + \Delta G \cong G_0 + (G_\theta)_0\,\Delta\theta, \text{ so}$$

$$\Delta G = (G_\theta)_0\,\Delta\theta, \text{ or}$$

$$\Delta G = -G_0\,(B^*_\theta)_0\,\Delta\theta G_0. \tag{25}$$

So the correction $\Delta F$ and $\Delta G$ to the nominal values of $F_0$ and $G_0$, which guarantee the decoupling of the disturbed system are given as follows [11]:

$$\Delta F = k_f\,\Delta\theta,\ \Delta G = k_g\,\Delta\theta, \text{ where} \tag{26}$$

$$k_f = -G_0 \; [(A^*_\theta)_0 \; \Delta\theta + (B^*_\theta)_0 \; \Delta\theta F_0], \text{ and} \tag{27}$$

$$k_g = -G_0 \; (B^*_\theta)_0 \; \Delta\theta G_0. \tag{28}$$

In other words, when the system is disturbed by parameter variation, $U_0$ should be corrected by $\Delta U$ to decouple the disturbed system to the first order of $\Delta\theta$, where the correction in $U_0$ is determined by $\Delta U$:

$$\text{where } \Delta U = (\Delta F)X + (\Delta G)\omega. \tag{29}$$

Using (26) in (29) the correction will be:

$$\Delta U = (k_f \Delta\theta)X + (k_g \Delta\theta)\omega. \tag{30}$$

Example

The following example illustrates the decoupling of a disturbed system, when its state is precisely known.

Consider the linear time invariant system that we had in section 2.2. Suppose one of the parameters of matrix B, $b_{11}$, is disturbed by $\Delta b_{11}$. So the disturbed system will be:

$$\begin{aligned} \dot{X} &= AX + BU \\ y &= CX \end{aligned} \tag{31}$$

where

$$A = \begin{bmatrix} 1 & 1 & 0 \\ 0 & 2 & 0 \\ 0 & 1 & 3 \end{bmatrix}, \quad B = \begin{bmatrix} 1+\Delta b_{11} & 1 \\ -1 & 1 \\ 0 & 0 \end{bmatrix} \text{ and } C = \begin{bmatrix} 1 & 0 & 0 \\ 0 & 0 & 1 \end{bmatrix}.$$

The transfer matrix of the overall system has been derived in Section 2.1 as follows [15]:

$$H(S) = C\,[SI - (A + BF)]^{-1}BG. \tag{32}$$

Using $F_0$ and $G_0$ for the undisturbed system,

$$H(S) = \frac{1}{S^2(S + \frac{\Delta b_{11}}{2})} \begin{bmatrix} S^2(1+\frac{\Delta b_{11}}{2}) & -\frac{\Delta b_{11}}{2}(S-1)(S-3) \\ 0 & S + \frac{\Delta b_{11}}{2} \end{bmatrix} \tag{33}$$

It is obvious that the system is not decoupled anymore and there exists an off diagonal term in the system closed-loop transfer function. Comparison of (33) and (10) shows that the disturbance in $b_{11}$ also affects the diagonal terms. All terms which have been introduced, because of the disturbance $\Delta b_{11}$, are of first order in

$\Delta b_{11}$, so the system is not decoupled to first order in the parameter variation.

Now, let's apply the present theory for decoupling; that is, using the following $\tilde{F}$ and $\tilde{G}$ (in the feedback law):

$$\tilde{F} = F_o + K_f \Delta b_{11} \text{ and } \tilde{G} = G_o + K_g \Delta b_{11}, \text{ where} \tag{34}$$

$$K_f = -G_o[(A^*_{b_{11}})_o + (B^*_{b_{11}})_o F_o], \text{ and} \tag{35}$$

$$K_g = -G_o (B^*_{b_{11}})_o G_o. \tag{36}$$

$F_o$ and $G_o$ are the nominal pair in the feedback law given in section 2.2 as:

$$F_o = -\frac{1}{2}\begin{bmatrix} 1 & -4 & -9 \\ 1 & 6 & 9 \end{bmatrix} \text{ and } G_o = \frac{1}{2}\begin{bmatrix} 1 & -1 \\ 1 & 1 \end{bmatrix}. \tag{37}$$

Also

$$(A^*_{b_{11}})_o = \left(\frac{\partial A^*}{\partial b_{11}}\right)_o = \underline{0} \tag{38}$$

$$(B^*_{b_{11}})_o = \left(\frac{\partial B^*}{\partial b_{11}}\right)_o = \begin{bmatrix} 1 & 0 \\ 0 & 0 \end{bmatrix}. \tag{39}$$

$$\text{So, } K_f = -G_o[o + (B_{b_{11}})_o F_o] = -G_o (B_{b_{11}})_o F_o = \frac{1}{4}\begin{bmatrix} 1 & -4 & -9 \\ 1 & -4 & -9 \end{bmatrix} \tag{40}$$

$$\text{and, } K_g = -G_o (B^*_{b_{11}})_o G_o = -\frac{1}{4}\begin{bmatrix} 1 & -1 \\ 1 & -1 \end{bmatrix} \tag{41}$$

Using (40) and (41) in (34):

$$\tilde{F} = F_o + K_f \Delta b_{11} = \frac{1}{4}\begin{bmatrix} -2 + \Delta b_{11} & +8 - 4\Delta b_{11} & 18 - 9\Delta b_{11} \\ -2 + \Delta b_{11} & -12 - 4\Delta b_{11} & -18 - 9\Delta b_{11} \end{bmatrix} \tag{42}$$

and

$$\tilde{G} = G_o + K_g \Delta b_{11} = \frac{1}{4}\begin{bmatrix} 2 - \Delta b_{11} & -2 + \Delta b_{11} \\ 2 - \Delta b_{11} & 2 + \Delta b_{11} \end{bmatrix}. \tag{43}$$

The transfer matrix of the closed-loop system, which has been derived in section 2.1, with the new pair $\{\tilde{F}, \tilde{G}\}$ will be [4]:

$$\tilde{H} = C[SI - (A + B\tilde{F})]^{-1} B\tilde{G} \tag{44}$$

substituting $\tilde{F}$ and $\tilde{G}$ in (44):

$$\tilde{H} = \frac{1}{S^2(S - \frac{\Delta b_{11}^2}{4})} \begin{bmatrix} S^2(1 - \frac{\Delta b_{11}^2}{4}) & \frac{\Delta b_{11}^2}{4}(S-1)(S-3) \\ 0 & S - \frac{\Delta b_{11}^2}{4} \end{bmatrix}$$

There is no term in $\tilde{H}(S)$ which is of first order in $\Delta b_{11}$, and all terms in $\tilde{H}(S)$ are affected by $\Delta b_{11}^2$. If $\Delta b_{11}$ is "small enough, then $\Delta b_{11}^2$ can be ignored, and we get the following transfer function matrix:

$$\tilde{H}(S) \simeq \frac{1}{S^3} \begin{bmatrix} S^2 & 0 \\ 0 & S \end{bmatrix} = \begin{bmatrix} \frac{1}{S} & 0 \\ 0 & \frac{1}{S^2} \end{bmatrix}$$

which shows that the closed-loop transfer function is diagonal or the system is decoupled. Thus by correcting the linear feedback controllers from $F_o$ to $\tilde{F} = F_o + \Delta F$ and $G_o$ to $\tilde{G} = G_o + \Delta G$ where

$$\Delta F = K_f \Delta b_{11} \text{ and } \Delta G = K_g \Delta b_{11},$$

The system will be decoupled to first order in $\Delta b_{11}$, in other words the closed loop system will be insensitive to the first order parameter variation $b_{11}$.

## CONCLUSION

The problem of decoupling a time-invariant linear system using state variable feedback has been considered in the first section of this paper. Necessary and sufficient conditions for decoupling [2,3,11] have been determined in terms of the nonsingularity of the matrix B*. In essence, the major theoretical questions relating to decoupling a time-invariant linear system with unknown and disturbed parameters, by state variable feedback, have been resolved. The effort here was to make the system insensitive to parameter variations and decouple its inputs and outputs by a single controller. An alternate approach can be to first make the system insensitive by one controller and then decouple the resulting system by another controller.

The sensitive decoupling controller designed here is based on the first order approximation in the parameter variations, however any order approximation could be used, [11]; of course the higher order sensitive decoupling controller would usually be more complicated. It should be noted that the parameter disturbance vector $\Delta\theta$ is assumed to be known in this paper; in practice such an assumption may not hold; hence the problem of finding a way to measure or estimate $\Delta\theta$ arises.

Another problem of decoupling which has not been discussed in this paper is the derivation of the sensitivities of the off diagonal elements in the input-output transfer function matrix to the parameter variations. Finally the more general group-decoupling problem can be discussed in which a subset of inputs can control a subset of outputs rather than one input controlling one output.

REFERENCES

(1) Alanso-Concheiro, A. "A Complete Set of Independent Decoupling Invariant." Int. J. Control, 1973, Vol. 18, No. 6, pp. 1211-1220.

(2) Falb, Peter L. & William Wolovich "On the Decoupling of Multi-Variable Systems." Proc. 1967 JACC. pp. 791-796.

(3) Falb, Peter L. & William Wolovich "Decoupling in the Design and Synthesis of Multivariable Control Ssytems." IEEE Trans. on Automatic Control, Vol. AC-12, No. 6, Dec. 1967.

(4) Gilbert, Elmer G. "The Decoupling of Multivariable Systems by State Feedback." Siam J. Control, Vol. 7, Feb. 1969.

(5) Greensite, Arthur L. Elements of Modern Control Theory. Spartan-MacMilan & Co., Ltd, N. Y. 1970.

(6) Kavanagh, R.J. "Noninteracting Control in Linear Multivariable Systems," Trans. AIEE, Vol. 76, Part II, 1957.

(7) Miller, William L. & John B. Lewis. "Dynamic State Estimation in Power Systems." IEEE Trans. Automatic Control, Vol. AC-16, No. 6, Dec. 1971.

(8) Morgan Jr., Bernard S. "The Synthesis of Linear Multivariable Systems by State Variable Feedback" Proc. 1964 JACC.

(9) Ogata, Katsuhiko. State Space Analysis of Control Systems. Prentice-Hall Inc., Englewood Cliffs, N. J. 1967.

(10) Rhodes, Ian B. "A Tutorial Introduction to Estimation and Filtering." IEEE Trans. on Automatic Control, Vol. AC-16, No. 6, Dec. 1971.

(11) Tzaferta, S.G. & P.N. Paraskevopoulos. "Sensitive Decoupling Control of Linear Disturbed Parameter Systems." IEEE Trans. on Automatic Control, Feb. 1975, pp. 151-153.

# GENERAL SYSTEMS CONCEPTS FOR PROCESS CONTROL

Roger E. Cavallo and Franz Pichler

Division of Systems Theory,
Institute of Systems Science
Johannes Kepler University
A-4040 Linz/Austria

## 1. INTRODUCTION

In recent years there have been a number of discussions regarding the utility (or non-utility) of the more or less sophisticated results of "Classical Mathematical Systems Theory" for practical applications (theory of dynamical systems, automata and others). It is also the case that many of the results derived from the perspective of "General Systems Theory" are such that their direct application is not immediately apparent. In recognition of this there has been a growing emphasis on general systems methodology which focusses on the study of method and the development of method transformable into working procedures for system investigation (see e.g. Cavallo (1980), Cavallo and Pichler (1979), Klir (1980)). Thus there now exist concepts and derivable methods which are directly addressed towards empirical situations and which are of practical use.

Our paper explores certain fundamental concepts of process control from a General Systems point of view. The intention is to build a foundation from which the broadest range of methods can be applied to the widest range of control applications. We believe that it contributes to help see process control in a wider sense than usual and - by emphasizing the idea of a "stratified methodology" - indicates a top down structure for the design of process controllers.

## 2. CONTROLLABLE GENERAL PROCESSES

A process is defined here by a set P consisting of time-functions $p:T \rightarrow M$, where T is a completely ordered set (the time-set of P) with initial time 0 and M is the set of process-values (process-



D. G. Lainiotis and N. S. Tzannes (eds.), Advances in Control, 56–64.

states). A process P defined in this way is a rather strong generalization of real processes. To approximate a real process $P_{real}$ "better" one has usually to assume that P has special properties, e.g. additional mathematical structure on T,M and on the defining set of time-functions.

A process P is called an Input/Output process (I/O process, directed process) if $M = A \times B$, whereby the process-states are considered of an input component (A) and on output component (B). Then each time-function $p \in P$ is parallel-decomposed into two time-functions $u \in U$ and $y \in Y$, $p = uy$, with $p(t) = (u(t), y(t)) \in A \times B$. The function $u: T \to A$ is called the input-function, $y: T \to B$ is called the output-function.

By a (dynamical) realization of a given I/O process P we understand a corresponding dynamical system, dynP, with input and output, which is able to "generate" P. Over the last years there has been some effort spent on the problem of realization for different types of processes and related dynamical systems. The results achieved by this research are available for use in practical applications. It is of some interest to mention here that - regardless of process-type - one can always "solve" the realization problem; that is, it is always possible to make a hypothesis on the "internal dynamics" of a given process. But by the existing realization theory it might be also possible to select the "best" dynamical system from among a number of possible ones. For more details we refer the interested reader to Zadeh (1969), Kalman (1968), Mesarovic (1975) and Klir (1969).

Next we discuss the concept of controllablity of general dynamical systems (with input and output) dyn P. It is convenient to describe dyn P by two functional relations $\varphi$ and $\beta$ - the state-relation $\varphi \subset (T \times T \times Q \times U) \times Q$ and the output-relation $\beta \subset (T \times Q \times A) \times B$. In order that $(\varphi, \beta)$ constitutes a dynamical systems dyn P, $\varphi$ has to fulfill the usual "dynamical properties" ("consistency", "semi-group property", "causality") (compare e.g. Mesarovic (1975) or Pichler (1975)).

An event $(\underline{t}, q) \in T \times Q$ of dyn P is called controllable to the event $(\underline{t}, \bar{q})$ $(t < \underline{t})$ if there exists an input-function $u \in U$ such that $(t, t, q, u) \varphi q$. In the case that every event $(t, q)$ is controllable to any event $(\bar{t}, \bar{q})$ the dynamical system dyn $P = (\varphi, \beta)$ is called (completely) controllable. It is obvious - with respect to applications - the property of a dynamical system dyn P being controllable is very desirable. Since the "invention" of the concept of controllability by Kalman (1968) there has been a great deal of success in establishing criteria to ensure controllability for different types of dynamical systems. Such criteria and related theory constitute a fundamental basis of the modern approach to process control.

## 3. PROCESS-GOALS

There are many ways to introduce "goals" of a dynamical system dyn P. The goal could be "stability" or "easy to build" etc. to mention some examples. For our purposes here we restrict ourselves to goals defined for the state-trajectories of dyn P.
Let (t,q) denote an arbitrary given reached event of dyn P (which means that dyn P is assumed to have reached state q at time t). Then we define a related corresponding goal $z_{(t,q)}$ as a function of the type $z_{(t,q)}: T_t \to 2^Q \setminus \{\emptyset\}$ where $T_t$ denotes the set of all time-points t' of T with $t' \geq t$; $2^Q \setminus \{\emptyset\}$ denotes the power-set of Q minus the empty set.

We say that the goal $z_{(t,q)}$ is reachable from (t,q) if there exists an input-function $u \in U$ such that for all $t' \geq t$ we have $(t',t,q,u)\varphi \subset z_{(t,q)}(t')$. The reader may observe that in case that dyn P is completely controllable in the strong sense defined above, every event (t,q) will reach any defined corresponding goal. Figure 1 depicts an event (t,q) and corresponding goal $z_{(t,q)}$ together with a state-trajectory $(\cdot,t,q,u)\varphi$ by which the goal $z_{(t,q)}$ is reached from (t,q).

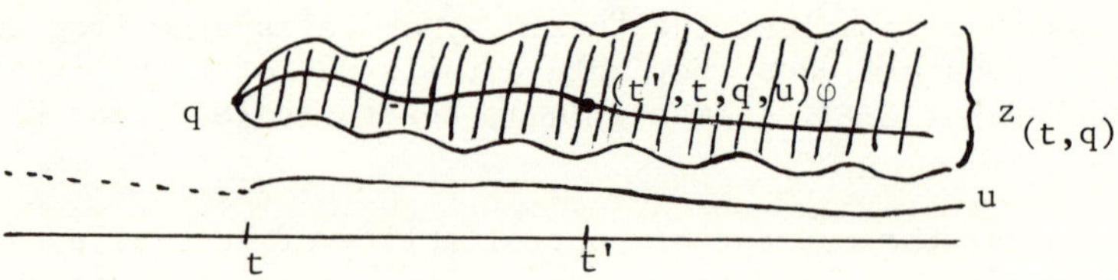


Figure 1: Example of an event (t,q) with goal $z_{(t,q)}$

In most cases of applications we have the situation that besides "reaching a goal" it is desirable to measure the "grade" to which a goal is reached (in order to be in the position to select a "control" which reaches then the goal "best"). For our purpose it is enough to measure this "grade" by a "distance function" $d : TR \times Z \to \mathbb{R}$ which assign to a pair (tr,z) consisting of a trajectory tr and a goal z a "distance" d(tr,z) between them. TR is the set of all state-trajectories of dyn P. There are several possibilities which can be meaningfully used for the metric d. Usually one would assume that the state-set Q is a metric space and induce the metric from there into $TR \cup Z$ (for example, by the method of least-squares). Another property which makes it easier to work with our concept of a goal-space Z is the following: The goal-space Z is called initial if all goals $z_{(t,q)} \in Z$ are time-restrictions of "initial goals" $z_{(0,q_o)}$ (0 the initial time in T) that is that for any goal $z_{(t,q)} \in Z$ there exists a goal $z_{(0,q_o)} \in Z$ such that $z_{(0,q_o)} | T_t = z(t,q)$.

If a goal-space Z is initial the knowledge of the set $Z_o$ of initial-goals ($Z_o := \{z_{(0,q_o)} \mid q_o \in Q_o\}$) is sufficient to know the whole goal-space Z.

Besides the introduced properties "metrical" and "initial" of goal-spaces there are many others which can be introduced and which may be important for certain classes of widespread applications (e.g. "time-invariance" oder "equi-finality").

## 4. PROCESS-CONTROL

Again, let P denote an I/O process (which models a real process $P_{real}$) and dyn P = $(\varphi, \beta)$ denote a generating dynamical system (as a hypothesis of the "internal dynamics" of $P_{real}$). $E_o \subset \{0\} \times Q$ denotes a set of given initial events $(0,q_o)$ and $Z_o$ the set of corresponding goals. Furthermore we assume the goal-space to be initial and provided with a metric d (which measures the distance between state-trajectories and goals). For this situation we want to consider two kinds of control-problems: (a) the problem of input-control and (b) the problem of structure-control.

### (a) Input-Control

We define the <u>input-control problem</u> in the following way: Determine a control-operation $c : U \to U$ such that for any I/O pair $uy \in P$ and associated initial event $(0,q_o)$ the following is true:

(1) the dynamical system dyn P answers to input-function c(u) with output-function y
(2) the goal $z_{(0,q_o)}$, corresponding to $(0,q_o)$ is reached by the input-function c(u); thus, for all t>0 we have $(t,0,q_o,c(u))\varphi \subset z_{(0,q_o)}(t)$.

Figure 2 shows a block-diagramm for input-control. The box c is called the <u>input-controller</u>.

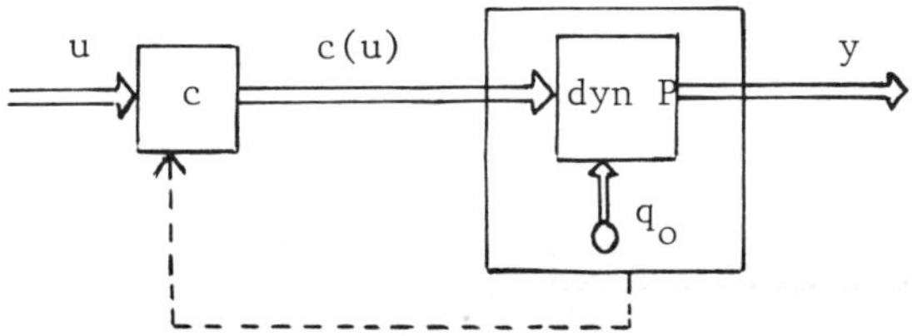


<u>Figure 2:</u> Input-Control

The dashed line indicates the "information line" between dyn P and the input-controller c which is in general necessary to design c.
If P, dyn P, and $Z_o$ relate to concrete situations, it is possible that the input-control problem has more than only one (unique) solution c. Then the optimal input control problem can be considered: Determine from the set C of solutions of the input-control problem the "optimal solution" $c_{opt}$ in the sense that for all initial events $(0,q_o)$ and all $c \in C$ we have

$$d(tr_{c_{opt}(u)}, z_{(0,q_o)}) \leq d(tr_{c(u)}, z_{(0,q_o)}).$$

Here $tr_{c_{opt}(u)}$ and $tr_{c(u)}$ denote the state-trajectories corresponding to $c_{opt}(u)$ and $c(u)$ respectively. A solution $c_{opt} \in C$ will be called an <u>optimum input-controller</u>.

(b) Structure-Control

Input-control was based on the fact that the input-function u is transformed to a "new" input function c(u) such that the output y is preserved but the goals $z_{(0,q_o)}$ corresponding to the initial events are reached.
There exists a second principle of control which is based on a transformation of the state-relation $\varphi$ of the dynamical system dyn P = $(\varphi, \beta)$. We call it structure-control and define the related <u>structure-control problem</u> in the following manner: Determine a control-operation $c : \Phi \to \Phi$ which assigns to the given state-relation $\varphi$ a state-relation $c(\varphi)$ such that for any I/O pair $uy \in P$ and associated initial event $(0,q_o)$ the following are true:

(1) the dynamical system $(c(\varphi), \beta)$ answers to input-function u with output-function y;
(2) the goal $z_{(0,q_o)}$ corresponding to $(0,q_o)$, is reached by the dynamical system $(c(\varphi), \beta)$ which means that for all $t>0$ we have $(t,0,q_o,u)c(\varphi) \subset z_{(0,q_o)}(t)$.

Figure 3 shows a block-diagram for structure-control. The box c is here called the <u>structure-controller</u>

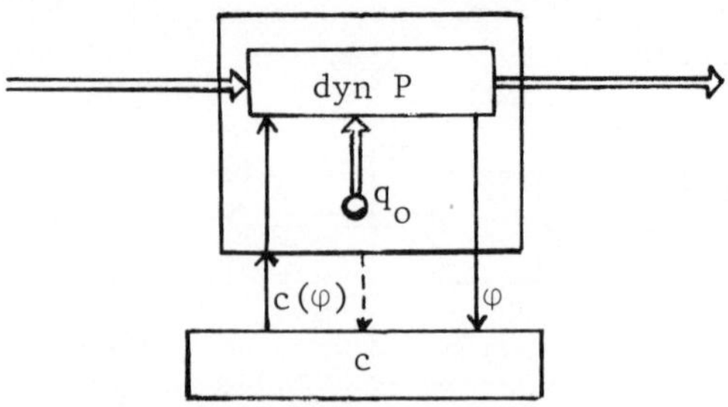


<u>Figure 3:</u> Structure-Control

Again the dashed line indicates the path of information going from dyn P to c which is necessary to design the controller.

The concept of structure-control depends strongly on the ability to be able to "create" for the given dynamical system $(\varphi,\beta)$ alternative systems $(c(\varphi),\beta)$ which generate the same process P but reach the goals "better". Realization theory and the theory of dynamical systems decomposition (represention theory, structure identification theory), which both have been strongly developed over the last years provide important tools for the implementation of this form of control (Broekstra (1979); Cavallo (1979); Cavallo and Klir (1978), (1979); Pichler-Ottendörfer (1978).

If $(\varphi,\beta)$ is derived as the "internal dynamics" of a real process $P_{real}$ it is often the case that the state-relation $\varphi$ depends on a collection of "measurable internal parameter" $p_1,p_2,\ldots,p_k$; $\varphi = \varphi(p_1,p_2,\ldots,p_k)$. The applicability of structure-control depends then on the question of wether the parameters $p_1,p_2,\ldots,p_k$ can be influenced from "outside" in order to change $\varphi$ to $c(\varphi)$. In general this question cannot be answered without information dependent on context of particular applications, but the framework defined her allows easy integration of such information (see Cavallo and Pichler (1979)).

The optimal structure-control problem can be defined in a manner analogous to that of the optimal input-control problem and it is not difficult to give concrete examples of input control-problems and of structure control-problems (see, for example, the literature related to Finite Automata and to Linear Systems).

## 5. PROCESS META-GOALS

At this point it is appropiate to address the question of goals other than our prevously defined process goals. We concentrate here on two rather important goals often desired for dynamical systems, the goals of "stability" and of "attractivity". We will see that both can be considered as properties of process-goals.

To define stability it is sufficient to assume that both the state space Q and the space TR of state-trajectories (i.e., the state process) are metric spaces. Then the dynamical system dyn P can be called stable if, for every $t\in T$ and for every $u\in U$, the function $\varphi(\cdot,t,\cdot,u) : Q \to TR$ which assigns to each state $q\in Q$ the corresponding state-trajectory $\varphi(\cdot,t,q,u)$ is continuous.

Furthermore, dyn P is called attractive if, for every pair $q,\bar{q}\in Q$ and every $t\in T$, $u\in U$ and $\varepsilon>0$, there exists a time $t(\varepsilon)$ such that for all $t'>t(\varepsilon)$ it is true that

$$d(\varphi(t',t,q,u),\varphi(t',t,\bar{q},u))< \varepsilon .$$

dyn P is called asymptotically stable if dyn P is stable and attractive.

It is not difficult to observe that for a dynamical system $(\varphi,\beta)$ to be stable, attractive or asymptotically stable depends on properties of process goals in the sense defined above. Therefore it may be justifiable to call such goals process meta-goals.

## 6. PROCESS-CONTROL DESIGN

The design of a controller c for a process P via its generating dynamical system dyn P can be performed by three steps:

(1) decision wether input- or structure-control should be used
(2) determination of the kind of "information" which is available from dyn P for the design of the controller (content of dashed lines in Figure 2 and Figure 3)
(3) determination of the type of system to be used to realize the controller c (e.g. static or dynamical)

Depending on the type of the dynamical system dyn P and of the kind of goals desired to be reached there, a solution of the stated control-problem will exist or will not exist. The general system concepts discussed here, together with general systems methodological developments, provide a basis for the determination of solutions in a general systematic way. Furthermore it is believed that by such an approach the feasability limits which are given by the natural constraints of specific situations are more easily discovered and integrated into the problem.
An example of a result of general systems theory which may help here is given by Ashby's "Law of Requisite Variety". This fact was pointed out recently by B. Porter [ ]. Concerning literatur in general systems theory which can be used to determine solutions of the control-problem we refer the reader to the work of Klir (1969), (198o), Cavallo (1978), Mesarovic (1975), Pichler (1975) and Jaron (1978).

## 7. PRACTICAL APPLICATIONS

As an example of the design of a process-controller we mention our project on the urban traffic control in the City of Linz/Austria. In that project we use the general systems concepts defined here for building up the model and take a top-down approach in the philosophy of the concept of a "stratified methodology". In a first step the traffic system is modelled by a directed graph in which the nodes model the different sources, sinks and crossings and the arrows stand for the different street-connections between the nodes. The nodes as well as the arrows of the directed graph are in a second step provided with related dynamical systems describing the dynamics of the traffic process at the nodes and arrows, respectively. Certain subgraphs (the selection depends on the local real situation) will be provided with controllers to improve the performance of the system. Both principles of control, input-control as well as structure control will be used (roughly, input-control applying mainly to the streets and structure control to the crossings). The performance criteria which define the goals of the system are many: Besides the usual criteria such as reasonable length of queuing and numbers of stoppings/km we also take traffic-noise and air-pollution into account. The controllers at the crossings will be designed as finite sequential switching networks realized in micro-processor technology.

## 8. CONCLUSION

We are aware that the considered concepts for input-control and structure-control of dynamical systems are only a first step towards the complete formulation of control problems from a general systems point of view. For an xperienced control engineer these concepts should not be surprising but they should help because of their generality to indicate the applicability of basic results and techniques to problems outside of the usual domain of control engineering.

On the other hand general systems theoreticians may learn from them some fundamentals of control engineering and should able to transfer this knowledge for applications in other fields. If our consideration should be considered to be trivial our excuse is, that it seems to us, that the general systems concepts for process control have not yet been developed so far by others. But in the believe of the importance of the subject we wanted to take a first step.

## REFERENCES

Broekstra, G. (1979): Probabilistic Constraint Analysis for Structure Identification: An Overview and Some Social Science Applications, in: B.P.Ziegler et al (eds.), Methodology in Systems Modelling and Simulation. Amsterdam: North Holland

Cavallo, R. E. (1979): The Role of Systems Methodology in Social Science Research, Bosotn: M. Nijhoff

Cavallo, R.E. (1980): Reconstructability Analysis/ Evaluation of Reconstruction Hypothses, Proceedings of the Annual Meeting San Francisco, January 7-1o, 198o

Cavallo, R.E. and F.R.Pichler (1979):General Systems Methodology: Designs for Intuition-Amplification, in: R.Ericson (ed.), Improving the Human Conditio Quality and Stability in Social Systems, New York: Springer-Verlag

Cavallo, R.E. and G. Klir (1979):The Structure of Reconstructable Relations: A Comprehensive Study, Journal of Cybernetics, Vol.9, No.4, pp.399-413

Jaron, J. (1978): Systemic prolegomena to theoretical cybernetic, Wrocław, Scientific Papers of the Institute of Technical Cyberentic of Wrocław, Technical University, Monograph Nr.8

Kalman, R.E. (1968): Lectures on Controllability and Observability, C.I.M.E. Bologna Juli 1968

Kalman, R.E., P.L.Falb, M.A. Arbib (1969): Topics in Mathematical System Theory, Mc-Graw-Hill Book Company, New York 1969

Klir, G. (1980): General Systems Concepts, in: R.Trappl (ed.), Cybernetics: A Sourcebook, Washington, D.C., Hemisphere

Klir, G. (1969): An Approach to General Systems Theory, Van Nostran Reinhald Company, New York 1969

Mesarovic, M.D. and Y. Takahara (1975): General Systems Theory: Mathematical Foundations, Academic Press, New York

Pichler, F. (1975): Mathematische Systemtheorie: Dynamische Konstru tionen, Walter de Gruyter, Berlin 1975

Pichler, F. and W. Ottendörfer (1978): Decomposition of General Dynamical Systems, University of Linz, Austria

Porter, B. (1976): Requisite Variety in the Systems and Control Sciences, Int. J.General Systems, Vol.2, pp.225-229

Zadeh, L. (1969): The concepts of systems, aggregate and state in system theory, in: System Theory (ed.L.A.Zadeh and E.Polak), Mc-Graw-Hill, New York 1969, pp.3-43

# ON ADDITIVITY AND LINEARITY IN GENERAL SYSTEMS THEORY

Paulo A. S. Veloso

Deptº Informática, Pontifícia Universidade Católica
Progr. Eng. Sistemas e Computação, COPPE - UFRJ
Rio de Janeiro, RJ ; Brazil

ABSTRACT

This paper shows that many properties of linear time-invariant systems are due to the underlying group structure, even if not commutative. Within a general time-systems framework some properties of additive time-varying systems are examined : attainability, observability, reachability,controllability, realizability and state-space construction. These are regarded from the viewpoint of mathematical foundations of general time-systems theory.

In Windeknecht's general time-systems theory (1) a system is defined as a relation between input and output signals (2). Another distinct concept is that of realization, which involves internal states (3). The problems of analysis and synthesis can be described as that of correlating a system with a realization with the same input-output behavior. Thus, non-anticipatory systems represent the behavior of deterministic realizations. This paper employs this framework extended to time-varying systems in order to investigate the role played by linearity. The aim is showing that several properties usually attributed to linearity are actually due to a group structure, which does not have to be commutative (4).

We shall fix a <u>time domain</u> $T \neq \emptyset$ ordered by $<$ together with a <u>output-observation domain</u> $\emptyset \neq J \subseteq T$ . Also fixed are the groups I and Z ( input and output alphabets ) and subgroups U of functions from T into I and Y of functions from J into Z ( input and output spaces ). We shall be dealing with ( sets of ) functions with domain included in T and using the suffixes *t and §t to denote restriction to the set of instants prior to t , resp. from t onwards ( including t ).



*D. G. Lainiotis and N. S. Tzannes (eds.), Advances in Control, 65–69.*

In order to simplify the treatment we shall assume the input space U to have a closure property : for all t in J and all u and v in U $u\underline{t}v$ = u*t ∪ v§t is in U . Because of this, we have a commutative separation : for all u and v in U $u\underline{t}0 + 0\underline{t}v = u\underline{t}v = 0\underline{t}v + u\underline{t}0$ . This will be used frequently in the sequel.

By an <u>additive t-register</u> $\mathcal{R}$ we mean a state group Q together with a pair of group homomorphisms m*t : U*t ⟶ Q ( memory map ) and n§t : QxU§t ⟶ Y§t ( output map ). Because of the separation, we get an equivalent definition of additive t-register via the representation n§t(q,u§t) = c§t(q) + d§t(u§t) , where we have two homomorphisms c§t : Q ⟶ Y§t and d§t : U§t ⟶ Y§t .

The set of t-attainable states is a subgroup , the image of m*t , whereas the set of states t-equivalent to the zero state is a normal subgroup kerc§t of Q. Since q is t-equivalent to q' iff (q-q') is in kerc§t , we can transform $\mathcal{R}$ into its reduction $\bar{\mathcal{R}}$ , which is completely t-attainable and t-observable. This reduction is minimal among all t-registers having the ( input-output) <u>behavio</u> n§t(m*t(u*t),u§t) of $\mathcal{R}$ , in the sense of being a homomorphic image of a subgroup.

A register embodies the idea of the state space being used to store information with part of the input signal setting the initial condition. ( Notice that T is not assumed to have minimal elements. The next definition starts incorporating state structures for several instants of time.

An <u>additive transition structure</u> consists of a family Q(j) of state groups together with homomorphisms M(j,i) from Q(i)xU§i*j into Q(j) for i≤j in J, satisfying the identity and semigroup composition properties. Equivalently, because of separation, M(j,i)(q,u§i*j) = A(j,i)q + B(j,i)u§i*j , where A(j,i) and B(j,i) are homomorphims such that A(i,i)q=q , B(i,i)u§i*i=0 , A(j,i)q= =A(j,t)A(t,i)q , B(j,i)u§i*j = A(j,t)B(t,i)u§i*t + B(j,t)u§t*j , for i≤t≤j in J .

Now consider the sets Rch(j,i) of states <u>reachable</u> at j from the state 0 at i and Ctr(j,i) of states at i <u>controllable</u> to 0 at j Clearly, Rch(j,i)=ImB(j,i) is a subgroup of Q(j) and Ctr(j,i) = = $A(j,i)^{-1}$[ImB(j,i)] is a subgroup of Q(i) . The next result generalizes a condition for a finite-dimensional linear time-invariant system to have all its reachable states controllable (7).

<u>Proposition</u> - Let i≤t≤t'≤j≤k in J and assume Rch(k,i) ⊆ Rch(k,j) . Then Rch(t',t) ⊆ Ctr(k,t'), thus Rch(j,i) ⊆ Ctr(k,j) .
<u>Proof</u>. Follows from q∈Ctr(k,t') iff A(k,t')q∈Rch(k,t') . QED

From an additive i-register and an additive j-i transition function one can obtain naturally an additive j-register. However, the output of non-anticipatory systems can generally be determined pointwise, which suggests the following definition.

An <u>additive machine</u> $M$ consists of an additive transition structure together with a homomorphism $N(t) : Q(t) \times U(t) \to Z$, for each t in J. Equivalently, the instantaneous output map can be represented as $N(t)(q,u(t))=C(t)q+D(t)u(t)$, with homomorphisms $C(t) : Q(t) \to Z$ and $D(t) : U(t) \to Z$. For each i in J we have an induced output map $n§i : Q(i) \times U§i \to Y§i$ given by, for j=i in J, $n§i(q,u§i)(j)=N(j)(M(j,i)(q,u§i*j),u(j))$, with similar expressions for c§i and d§i. Thus $\ker c§i = \bigcap_{j \geq i} \ker(C(j)A(j,i))$ and for $j \geq i$ $A(j,i)\ker c§i \subseteq \ker c§j$.

<u>Proposition</u> - Given $M$ there exists a unique machine $\tilde{M}$, which is completely observable on J, such that for $j \geq i$ in J, $\tilde{A}(j,i)p(i) = p(j)A(j,i)$, $\tilde{B}(j,i)=p(j)B(j,i)$, $\tilde{C}(j)p(j)=C(j)$, $\tilde{D}(j)=D(j)$; where p(t) denotes the canonical projection onto $Q(t)/\ker c§t$.

The idea of using a register together with a transition structure is taken up in the following definition.

Given t in J, an <u>additive realization</u> $P$ consists of an additive machine together with a memory homomorphism $m*t : U*t \to Q(t)$. Notice that for each $i \geq t$ in J $P$ gives rise to an additive i-register, thus being reducible to a completely observable and attainable $\tilde{P}$ with the same behavior, which is, for $j \geq t$ in J, $C(j)A(j,t)m*t(u*t) + C(j)B(j,t)u§t*j + D(j)u(j)$.

By an <u>additive system</u> we mean an additive relation S from U to Y, i.e. a subgroup of UxY, with domain U.

Given an instant t in T we define the <u>t-section</u> S§t of S to consist of those (u,y§t) with (u,y) in S and call S <u>t-functional</u> iff S§t is a function, which is then a homomorphism such that for all u in U $S§t(u) = S§t(u\underline{t}0) + S§t(0\underline{t}u)$.

Now, the behavior of an additive t-register gives a t-section of a t-functional additive system. We shall now extend Nerode's construction to obtain the converse.

<u>Theorem</u> - Any t-section of a t-functional additive system S is the behavior of an additive t-register $R$, which is minimal.
<u>Proof</u>. As $\ker S§t(.\underline{t}0)$ is a normal subgroup of U*t we can take Q to be the corresponding quotient and use the canonical projection to define m*t and n§t in a natural way. QED

Now consider for i in J the relation S£i consisting of those ((u*i,u(i)),y(i)) with (u,y) in S. We shall call S <u>t-J-causal</u> iff for each $i \geq t$ in J S£i is a function, whence a homomorphism from U*ixU(i) into Z. With these concepts we can characterize which systems have a realization.

Theorem - A t-section of an additive system S is the behavior of an additive t-J- realization iff S is t-J-causal.
Proof. If S is t-J-causal then for each i≥t in J S is i-functional and the preceding theorem gives an additive i-register with Q(i)= U*i/K(i) where K(i)=kerS§i(.t0) . So we can define , for j≥i in J, M(j,i)(u*i+K(i),v§i*j) = uiv*j + K(j) and N(j,i)(u*i+K(i),v(i)) = = S£i(u*i,v(i)) . The other direction follows from previous remarks on the behavior.QED

We have examined the role played by additivity as opposed to linearity with respect to some basic concepts and results of genera time-systems theory. These results have been derived from simple facts about groups. In contrast to the commutative additive machine of (3) we did not assume commutativity, for whatever should commute does commute and we only need quotients by kernels of homomorphisms Of course, linearity is a very useful, and frequently natural, assumption, especially in providing convenient representations. But we can at least get started simply with additivity, which is interesting from the viewpoint of foundations.

Some points deserve comments. We have used heavily the closure property assumed for the input space. Had we not assumed this all the results would still hold at the expense of considering the sets of continuations of input signals. Also, by extending Nerode's construction what we actually get is a phase space for each instant These can then be merged into a single attainable and observable state space when some extra machinery is applicable.

Finally, most definitions make sense with the word 'additive' omitted or changed to, say, 'continuous'. This translates the concepts to another ( concrete ) category, where most proofs can be suitably adapted, in general.

## ACKNOWLEDGEMENTS

This paper reports on the continuation of research conducted under L. Kerschberg's supervision. Partial financial support from IBM of Brazil is gratefully acknowledged.

## REFERENCES

1. Windeknecht, T. G. : 1971, General dynamical processes : a mathematical introduction; Academic Press.

2. Zadeh, L. A., and Polak, E. (eds.) : 1969, System theory; McGraw-Hill.

3. Kalman, R. E., Falb, P. L., and Arbib, M. A. : 1969, Topics in mathematical systems theory; McGraw-Hill.

4. MacLane, S., and Birkhoff, G. : 1968, Algebra; MacMillan.

5. Desoer, C. A. : 1970, Notes for a second course on linear systems; van Nostrand Rheinhold.

6. Harrison, M. A. : 1969, Lectures on linear sequential machines; Academic Press.

7. Kerschberg, L. : 1969, An algebraic approach to linear and multilinear systems; Case Western Reserve University, Ph. D. thesis.

# LINEAR ARRANGEMENT OF A DIGRAPH

A. Vaquero, J.M. Troya, J.M. Sánchez P.

Dept° de Informática y Automática, Facultad de Físicas,
Universidad Complutense, Madrid-3, SPAIN.

ABSTRACT. In many applications it is necessary to arrange the nodes of a digraph in a linear order. The problem is to elect a sequence that optimizes a determined cost function for each application. In general, N! sequences are possible, but this complexity can be reduced for each particular case.
When we want to go from one node to another, if it is possible, on an one-dimensional placed digraph, a problem is to place the nodes in such a manner that the sum of all path lengths be a minimum. For this problem we define the cost function and we study two cases depending on the starting node from which a path is traversed. Both cases have easy solution for particular digraph, but in general, the complexity time is $0(3^n)$ and $0(2^n)$ respectively. We obtain near-optimal solutions for both cases.

## 1. INTRODUCTION

From a digraph G=(A,R) with N nodes, there are N! linear ordering of the nodes. A linear ordering is a permutation $\Pi$ of the N nodes (1). If the length of an arc $(a_i,a_j)\in R$ in a permutation is defined as the number of visited nodes to go from $a_i$ to $a_j$, it is possible to arrange all nodes so that the sum of the lengths of all arcs is a minimum, i.e. $\sum_{(a_i,a_j)\in R} l_{ij}$ must be a minimum. In some

applications the arcs of the digraph have got a weight, i.e. each arc $(a_i,a_j)\in R$ has a weight $w_{ij}$. In these cases the problem may be to find a permutation such that $\sum_{(a_i,a_j)\in R} w_{ij}\, l_{ij}$ be a minimum.

This is a permutational problem that can be solved by a appropri-



*D. G. Lainiotis and N. S. Tzannes (eds.), Advances in Control, 70–79.*

ate dynamic programming algorithm (2)(3). However, this algorithm would have an exponential time complexity. Therefore it is necessary to search approximate solutions for great values of N.
There are many applications for which it is useful to minimize the length of the paths in the graph (4). This problem can be reduced to minimize the length of the arcs when all considered paths are known. We study this problem when all paths are single and none of them are included in the others.

## 2. PROBLEM FORMULATION

Let G=(A,R ) be a digraph such that the paths which lengths must be minimized are known. If $\ell_k$ is a path length,

$L(\Pi,R) = \sum_{k=1}^{M} \ell_k$ is the cost function. We name it length of the permutation $\Pi$ .

Definition. The length $\ell_k$ of a path $p_k$ is the sum of the length of its arcs. Then, the permutation length will be:

$$L(\Pi,R) = \sum_{k=1}^{M} \ell_k = \sum_{(a_i,a_j)\in R} \omega_{ij}\, \ell_{ij} \qquad (2.1)$$

where $\omega_{ij}$ is the number of paths which the arc $(a_i, a_j)$ belongs to.
Definition. A gap, $g$, is the space between two consecutive nodes in a permutation. An arc $(a_i, a_j)$ crosses a gap $g_k$ when $a_i$ is placed on a side of $g_k$ and $a_j$ is on the other side. The weight $\omega_k^g$ of a gap $g_k$ is the sum of the weights of all arcs that cross it.
From previous definition, it is easy to deduce that the length of an arc is the number of gaps that crosses it. And thus, the length of a permutation $\Pi$ may be written:

$$L(\Pi,R) = \sum_{k=1}^{N-1} \omega_k^g \qquad (2.2)$$

This expression is equivalent to (2.1) and it will be very useful later. In this paper we only study acyclic digraphs with a unique node of indegree zero, named initial node $a_I$ and a unique node of outdegree zero, named end node $a_F$. We consider the paths from $a_I$ to $a_F$.
Definition. The weight $\omega_i$ of a node $a_i$ is the number of paths that traverses it. If $M$ is the number of paths, it is clear that $\omega_I = \omega_F = M$
As each time that a path is traversed it is necessary to begin from the initial node, in some applications it could be useful to consider the return from $a_F$ to $a_I$ . Hence, we study two cases. In the first case, the return from $a_F$ to $a_I$ is not considered. The cost function is:

$$L(\Pi,R) = \sum_{(a_i,a_j)\in R} \omega_{ij}\, \ell_{ij} \qquad (2.3)$$

In the second case, the return from $a_E$ to $a_I$ is considered. The cost function is:

$$L(\Pi,R) = \sum_{(a_i,a_j)\in R} w_{ij}\,\ell_{ij} + M.\ell_{EI} \tag{2.4}$$

This is the same as adding the arc $(a_E,a_I)$ of weight $M$ , to the digraph $G$.

## 3. METHOD

As two symmetrical permutations have the same length, we select the N!/2 permutations such that the node $a_E$ is placed on the right of node $a_I$. In general, one of these permutations is:

$$x_r\ x_{r-1} \ldots x_1\ a_I\ y_1\ y_2 \ldots y_s\ a_E\ z_1\ z_2 \ldots z_t \tag{3.1}$$

where $X = \{x_i\}$ $1 \leqslant i \leqslant r$ is the set of nodes placed on the left of $a_I$ , $Y = \{y_i\}$ $1 < i < s$ is the set of nodes placed between $a_I$ and $a_E$, and $Z = \{z_i\}$ $1 \leqslant i \leqslant t$ is the set of nodes placed on the right of $a_E$.

The length of a permutation $\Pi$ will be

$$L(\Pi,R) = L(\Pi_x,R)+L(\Pi_y,R)+(\Pi_z,R) = \sum_{i=1}^{r} w_i^{gx} + \sum_{i=1}^{s+1} w_i^{gy} + \sum_{i=1}^{t} w_i^{gz} \tag{3.2}$$

where $\Pi_X$, $\Pi_Y$ and $\Pi_Z$ are the partial permutations of $X$, $Y$ and $Z$ sets and $gx$, $gy$, $gz$ represent the gaps on the left of $a_I$, between $a_I$ and $a_E$ and on the right of $a_E$, respectively.

Given the $X$, $Y$ and $Z$ sets the length of each partial permutation is independent of the other partial permutations. Therefore, we study the problem in two phases: firstly, we study the sorting of the nodes of each $X$, $Y$ , $Z$ set in order to minimize the length of the corresponding partial permutation and then we study which nodes must belong to each set, i.e. the distribution of the nodes $a_i \in A$ in the three sets.

## 4. SORTING CONDITIONS

In this section the sorting of $X$, $Y$ and $Z$ sets for a given distribution are studied. We obtain algorithms for finding an approximate partial permutation in each set.

### 4.1 Sorting of set $X$.

We enumerate the gaps and nodes from right to left. So, for a partial permutation $\Pi_X$, the first node $x_1$ is the nearest to $a_I$ and the first gap $gx_1$ is between $a_I$ and $x_1$. The following Lemmas and Theorems are applied to any permutation.

Lemma 4.1 The weight of a gap $gx_k$ can be computed by:

$$w^{gx}_{k+1} = w^{gx}_{k} + w^{i} \qquad (4.1)$$

where:

$$w^{i} = 2w_{i} - 2\Big(\sum_{(x_i,a_j)\in R} w_{ij} + \sum_{(a_j,x_i)\in R} w_{ij}\Big)\ \forall a_j \qquad (4.2)$$

(placed on the right of $x_i$)

And for the first gap is:

$$w^{gx}_{1} = \sum_{(x_i,a_j)\in R} w_{ij} + \sum_{(a_j,x_i)\in R} w_{ji} \qquad \forall a_j \notin X,\ \forall x_i \in X \qquad (4.3)$$

Lemma 4.2 There is at least a node $x_i \in X$ which none of its direct predecessors belong to set X.

Corollary 4.1 If $X^1 \cup X^2 = X$ and $X^1 \cap X^2 = \emptyset$ , there is at least a node $x^2_i \in X^2$ which none of its direct predecessors belong to $X^2$.

Theorem 4.1 There is always a partial permutation $\Pi_X$ such that

$$w^{gx}_{k+1} \leqslant w^{gx}_{k} \qquad 1 \leqslant k \leqslant r-1 \qquad (4.4)$$

Theorem 4.2 If $\Pi^*_X$ is a partial permutation such that for two gaps $gx_k$ and $gx_{k+1}$ is $w^{gx}_{x+1} > w^{gx}_{k}$ there is a partial permutation $\Pi_X$ which verifies (4.4) and $L(\Pi_X, R) < L(\Pi^*_X, R)$.

From these theorems we can obtain an algorithm for sorting the nodes $x_i \in X$.

4.1.1 Algorithm for Sorting X.

The algorithm $SORT(X)$ sorts the nodes $x_i \in X$ according to previous theorems. The array $X(r)$ stores the nodes. The procedure $COMPW(X(j), wh1)$ computes $w^{i}$ for each node according to Lemma 4.1. In the step $i$, a node $x_i$ is elected in such a manner that the weight of the gap $gx_{i+1}$ is a minimum. The algorithm is the following:

```
procedure SORT (X(r));
begin
  for i: = 1 to  r  do
  begin
     wh2: = 0;  j1: = i;
     for i: =j to r do
     begin
        COMPW(X(j), wh1);
        if wh2 ⩾ wh1 then if wh2 > wh1 or WEIGHT(j)>WEIGHT(j1)then
        begin
           wh2: = wh1;    j1: = j;
        end;
     end;
     k: = X(i); X(i): = X(j1); X(j1): = k;
  end;      end of SORT
```

The partial permutation obtained by applying the procedure $SORT(X)$ is $X(r), X(r-1), \ldots, X(1)$. The algorithm has a time complexity $O(r^2)$ without taking account the time of $COMPW(X(j), wh1)$. In the worst case this time is $O(N)$; so the total time is $O(N.r^2)$.

4.2 Sorting of Set Y.

The length of a partial permutation $\Pi_y$ is different depending on whether the return from $a_E$ to $a_I$ is considered or not. If $L'(\Pi_y, R)$ is the length when the return is considered and $L(\Pi_y, R)$ when not considered, then

$$L'(\Pi_y, R) = L(\Pi_y, R) + (s+1).M$$

As $(s+1).M$ is constant for all partial permutations $\Pi_y$, we only consider the value $L(\Pi_y, R)$.

Definition. In a partial permutation $\Pi_y$ an arc $(y_i, y_j) \in R$, $y_i, y_j \in Y$ is called "forward", if $y_j$ is placed on the right of $y_i$ and it is called "backward" if $y_j$ is placed on the left of $y_i$. The arcs $(y_i, x_j) \in R$ and $(z_i, y_j) \in R$, $y_i$, $y_j \in R$, $z_i \in Z$, $x_j \in X$, are called "backward crossed arcs".

Theorem 4.3 The length of a partial permutation $\Pi_y$ is

$$L(\Pi_y, R) = (s+1)M + 2L' \tag{4.5}$$

where

$$L' = \sum_{(y_i, x_j) \in R} w_{ij}\, \ell_{iI} + \sum_{(z_i, y_j) \in R} w_{ij}\, \ell_{jE} + \sum_{(y_i, y_j) \in R} w_{ij}\, \ell_{ij} \tag{4.6}$$

$\ell_{iI}$ and $\ell_{jE}$ represent the lengths from $y_i$ to $a_I$ and from $y_j$ to $a_E$ respectively. $(y_i, y_j)$ are backward arcs and $(y_i, x_j)$, $(z_i, y_j)$ are backward crossed arcs.

As $(s+1)M$ is constant, $\Pi_y$ is an optimum when $L$ is a minimum. If there are no backward crossed arcs or backwards arcs, then $\Pi_y$ is optimum but if there are backward crossed arcs, it is not possible to compute an optimum in polynomial time. A good approximation is to obtain a $\Pi_y$ such that there are no backward arcs, i.e. to do a topological sorting set $Y$.

4.2.1 <u>Algorithm for Sorting</u> It is very easy to do a topological sorting (5). We do it by using a procedure $TSORS(j, STACK)$. $S(j)$ and $P(j)$ denote each direct succesor and each direct predecessor of a node $a_j$ respectively. Initially, $INDEG(j)$ and $OUTDEG(j)$ store the number of direct predecessors and direct successors belonging to set $Y$ for all nodes $a_i \in Y$. $STACK$ is a pushdown-list. $PUSH(k)$ and $POP(k)$ are the insert and delete operations on $STACK$. If $INDEG(j) = 0$ or $OUTDEG(j) = 0$ then $a_j \notin Y$.

```
procedure TSORS(j, STACK);
begin
  for all the S(j)  do
  begin
    k: = S(j);
    if INDEG(k) ≠ 0  then
    begin
      INDEG(k): = INDEG(k) - 1;
      if INDEG(k) ≠ 0  and OUTDEG(k) = 0 then PUSH(k);
    end;
  end;
end
```

A topological sorting of nodes $a_i \in A$ with $INDEG(j) = 0$ and $OUTDEG(j) = 0$ using $TSORS(j,STACK)$ is the following:

```
procedure TSORT(Y(s));
begin
  i: = aI; TSORS(i,STACK);
  while STACK not empty do
  begin
    POP(j); insert j in Y;
    TSORS(j,STACK);
  end;
end
```

4.3 Sorting of Set $Z$.

The sorting of $Z$ is similar to sorting of $X$ because the nodes placed on the right of $a_F$ verify the same conditions as the nodes placed on the left of $a_I$ . So, it can be proved that the gap placed on the right of $a_F$ has a constant value and the nodes can be sorted in such a manner that the weight of gaps have values not increasing from left to right. Therefore, it is possible to use the same algorithm to obtain an approximate solution.

5. DISTRIBUTION OF THE NODES IN THE FIRST CASE

An algorithm would need a time $0(3^{N-2})$ to obtain the $3^{N-2}$ possible distributions of the N-2 nodes in the $X$, $Y$ and $Z$ sets. For this, we study some properties concerning the distributions in order to find an approximate solution.
We name $D_s$ the distribution $X = \emptyset$, $Z = \emptyset$, $Y = A-\{a_I, a_F\}$ . From Theorem 4.3, for this distribution, any permutation $\Pi^*$ obtained by topological sorting of the digraph is optimal and its length is $L(\Pi^*, R) = L_s = (N-1).M$.
The problem is to find a distribution $D = D_s$ such that a permutation $\Pi$ can be obtained with $L(\Pi, R) < L_s$

Theorem 5.1. If for all $a_i \in A$ is $W_i \geq M/2$ then $L_s$ is optimum.
We have made an algorithm for obtaining a distribution $D$ according

to next theorem

Theorem 5.2. A permutation $\Pi$ may be always obtained with $L(\Pi,R) < L_s$ from a distribution that performs $wg_1^x \leq M$, $wg_1^z \leq M$ without backward crossed arcs.

The nodes that must belong to $X$ and $Z$ are selected using two simila procedures $INSUC(I,k)$ and $INPRE(E,k)$ where $I$ and $E$ are the subscripts of $a_I$ and $a_E$, and $k$ is the selected node for X or Z respectively.

```
procedure INSUC(I,k);
begin
  k: = I;
  for all the S(j) do
  begin
    i: = S(I);
    if INDEG(i) = 1 and WEIGHT(i)< WEIGHT(k)  then k: = i;
  end
INDEG(k): = 0;
end
```

In the same way, the procedure $INPRE(E,k)$ would select a node and include it in set $Z$.

## 5.1 Algorithm for Distribution of the Nodes in the First Case.

The procedure $DIST1(X,Z)$ distributes the nodes in such a manner that Theorem 5.2 is verified. It selects alternately a direct successor of $a_I$ using $INSUC(I,k)$ and a direct predecessor of $a_E$ using $INPRE(E,k)$. From these nodes, all nodes that can belong to $X$ or $Z$ are selected. The weights of $gx_1$ and $gz_1$ are $wgx$ and $wgz$ respectively. The algorithm is:

```
procedure DIST1(X,Z);
begin
  wgx:=0; wgz:=0;
  while wgx < M/2 or wgz < M/2 do
  begin
    if wgx < M/2 then INSUC(I,k);
    if wgx + WEIGHT(k) < M/2 then
    begin
      wgx:=wgx + WEIGHT(k); PUSH(k);
      while STACK not empty  do
      begin
        POP(j); insert j in X;
        TSORS(j, STACK);
    end;
  end;
if wgz < M/2 then INPRE(E,k);
if wgz < WEIGHT(k) + M/2 then
begin
  wgz:=wgz + WEIGHT(k); PUSH(k);
  while  STACK not empty  do
```

```
      begin
        POP(j); insert j in Z;
        TSORP(j,STACK);
      end;
    end;
  end;
end;
```

The algorithm uses the procedures *TSORS* and *TSORP*. *TSORP* is similar to *TSORS* for set Z. The set Y contains the nodes that do not belong to X or Z. For the X, Y, and Z sets the procedures *ORDER(X)*, *TSORT(Y)* and *ORDER(Z)* obtain an approximate permutation for the nodes $a_i \in A$.

## 6. DISTRIBUTION OF THE NODES IN THE SECOND CASE

When the return from $a_F$ to $a_I$ is considered, the distribution of the nodes must be different. In this way, for the distribution $D_s$ the topological sorting of the digraph gives a permutation $\Pi'$ which length is:

$$L(\Pi',R) = (N-1)M + (N-1)M = 2(N-1)M \qquad (6.1)$$

It is always possible to obtain a distribution $D = D_s$ from which a permutation $\Pi$ with $L(\Pi,R) < L_s$ can be found as we prove below.

Lemma 6.1.- It is always possible to obtain a distribution with $Y = \emptyset$, $w_1^{gx} \leqslant 2M$, $w_1^{gz} \leqslant 2M$.

Theorem 6.1.- There are always an optimal permutation $\Pi$ such that $Y = \emptyset$, $w_1^{gx} \leqslant 2M$ and $w_1^{gz} \leqslant 2M$.
The Theorem 6.1 is not truth for digraphs with three nodes, but we only consider digraphs with a great number of nodes.
Now we need to know what is the way to obtain a distribution which satisfies these conditions. An algorithm would need a time $O(2^{N-2})$ to obtain all distributions with $Y = \emptyset$. We search an algorithm to find a distribution from which an approximate solution can be obtained.

Theorem 6.2 For a permutation $\Pi$ with $Y = \emptyset$ and without arcs $(z_i, x_j) \in R$ it is always $w_1^{gx} \leqslant 2M$ and $w_1^{gz} \leqslant 2M$.
The problem is to do a distribution such that an approximate permutation can be obtained. We have done an algorithm that uses the above procedures and finds a good distribution.

### 6.1 Algorithm for Distribution the Nodes in the Second Case.

The algorithm obtains a distribution $Y = \emptyset$ with $w_1^{gx} \leqslant 2M$ and $w_1^{gz} \leqslant 2M$ and without arcs $(z_i, x_j)$. In each step the node with greater weight is selected among the nodes that can belong to X or Z sets. Initially the direct successors of $a_I$ and the direct predecessors of $a_F$ are stored in *VNOD*. The algorithm is:

```
procedure DIST2(X,Z);
begin
  r:=0; t:=0; i:=0; TSORS(I, VNOD); TSORP(E, VNOD);
  while VNOD not empty do
  begin
    wg1:=0
    for j:=1 to i do
    begin
      COMPW(VNOD(j), wg2);
      if wg1 ≤ wg2 then
      begin
        wg1:=wg2; nod:=VNOD(j);
      end;
    end;
    delete nod in VNOD; i:=i - 1;
    if OUTDEG(nod) ≠ 0 then
    begin
      r:=r+1; X(r):=nod; TSORS(nod, VNOD(i));
    end
    else
    begin
      if INDEG(nod) ≠ 0 or r > t then
      begin
        t:=t + 1; Z(t):=nod;
        TSORP(nod,VNOD(i));
      end
      else
      begin
        r:=r+1; X(r):=nod;
        TSORS(nod,VNOD(i));
      end;
    end;
  end;
end;
```

As the algorithm selects in each step the nodes with greater weight a good approximation is to obtain the permutation with $X$ and $Z$ sets in the order that $DIST2$ obtained it. In some cases for the $X$ and $Z$ sets the procedures $ORDER(X)$ and $ORDER(Z)$ obtain a better permutation.

## 7. CONCLUSIONS

We have found algorithms to obtain approximate solution in polynomial time. An optimal permutation could be found by algorithms of exponential time complexity, but they would not be applicable for digraphs with a great number of nodes. We have studied a class of acyclic digraph, but it is possible to extend the study for othe digraphs. This can be done by transformation of digraphs. This work may be applied to placement of records on a secondary storage device in order to minimize the access time.

REFERENCES

1. C.P. Earnest, K.G. Balke and J. Anderson, "Analysis of Graphs by Ordering of Nodes", Journal of ACM, vol.19, n°1, Jan. 1972, pp. 23-42.
2. W. H. Kohler, and K. Steiglitz, "Enumerative and Iterative Computational Approaches" in "Computer and Job/Shop Scheduling Theory". Ed. by F. G. Coffman, Jr. Wiley Intersc. 1976.
3. A. Vaquero S., J.M. Troya L. and J.M. Sánchez Pérez, "Almacenamiento de registros en memorias auxiliares en función de sus probabilidades de acceso", IV Congreso de Automática e Informática, Oct. 1979, Madrid, Spain.
4. A. Vaquero S., J.M. Troya L. and J.M. Sánchez P. "Disposición lineal de un grafo para almacenamiento óptimo en memorias secuenciales" Ibid.
5. Kahn, A.B. "Topological Sorting of Large Networks" CACM 5, 1962 pp. 558-562.

# OUTPUT REPRODUCIBILITY IN LINEAR MULTIVARIABLE SYSTEMS: A POLYNOMIAL APPROACH

*S. Beghelli and R. Guidorzi*

*University of Bologna, Bologna, Italy*

## ABSTRACT

The problem of the output reproducibility in linear multivariable systems is considered with reference to canonical input-output representations of the type $P(z)y(t) = Q(z)u(t)$ that can be immediately obtained from observable block-companion state-space canonical forms. These representations lead to a simple formulation and solution of the considered problem. Necessary and sufficient conditions for the reproducibility of arbitrary output sequences, based on a simple algebraic test on the matrix $Q(z)$, are proposed. An algorithm for the synthesis of the process control is then developed and applied to a numerical example.

## I. INTRODUCTION

The problem of the reproducibility of an arbitrary output sequence by a given dynamical system under the action of a suitable control and the synthesis of this control have been treated in the literature mainly with reference to state-space descriptions. This problem, which is equivalent to the synthesis of inverse systems, has found both algebraical |1|, |2|, |3|, |4|, |5| and geometrical solutions |6|. The approach considered in this work is algebraical but differs from previous ones in that is based on the use of polynomial input-output representations. These representations can be directly obtained from state-space observable multicompanion canonical forms but give the advantage of avoiding the inclusion of the state which does not play a direct role in the output reproducibility problem. The obtained necessary and sufficient conditions are more simple than the corresponding ones



*D. G. Lainiotis and N. S. Tzannes (eds.), Advances in Control, 80–86.*

obtained for state-space models and lead in a natural way to the synthesis of inverse systems to generate the required input control.

## II. STATEMENT OF THE PROBLEM

The linear dynamical discrete-time multivariable systems that will be considered in the following are described by the usual state-space equations

$$x(t+1) = F\,x(t) + G\,u(t) \tag{2.1a}$$
$$y(t) = H\,x(t) + L\,u(t) \tag{2.1b}$$

where $t \in I$, $x \in R^n$, $u \in R^r$, $y \in R^m$ and F, G, H, L are real matrices of suitable dimensions. Since only the reachable and observable part of system (2.1) is relevant in the output reproducibility, system (2.1) will be assumed to be completely reachable and observable. A strictly equivalent input-output representation |7| for the considered system is given by the polynomial model

$$P(z)\,y(t) = Q(z)\,u(t) \tag{2.2}$$

where z is the unitary advance operator and P and Q are $(m \times m)$ and $(m \times r)$ matrices defined over the ring of the polynomials in z and $\deg \det P(z) = n > 0$. The algorithms for the transformations between these two classes of representations are described in |7| and |8|. The output reproducibility problem can be defined as follows:

*Definition 2.1* - Given an output sequence $y[0,t] = \{y(0), y(1), \ldots, y(t)\}$ find an initial condition (initial state for the models (2.1), output initial conditions for the models of the type (2.2)) and an input sequence $u[0,t] = \{u(0), u(1), \ldots, u(t)\}$ which are compatible with $y[0,t]$.

In the following sections this problem is treated with reference to a subset, $S_{oc}$, of canonical models defined in the set, $S_o$, of all the representations (2.2) with respect to the equivalence relation given by the premultiplication of P(z) and Q(z) for a nonsingular unimodular matrix M(z) |9|, |10|. Denote with $(\tilde{P},\tilde{Q})$ the elements of $S_{oc}$, where $\tilde{P}(z) = \{p_{ij}\}$, $\tilde{Q}(z) = \{q_{ij}\}$. Then it holds

$$\deg p_{ii}(z) > \deg p_{ji}(z), \quad i \neq j \tag{2.3a}$$
$$\deg p_{ii}(z) > \deg p_{ij}(z), \quad j > i \tag{2.3b}$$
$$\deg p_{ii}(z) \geq \deg p_{ij}(z), \quad j < i \tag{2.3c}$$
$$\deg p_{ii}(z) \geq \deg q_{ij}(z). \tag{2.3d}$$

Moreover the entries $p_{ii}(z)$ are monic. It can be noted that the row degrees of $\tilde{P}(z)$, i.e. the degrees of $p_{11}, \ldots, p_{mm}$ are the ordered set of Kronecker invariants $(\nu_1, \ldots, \nu_m)$ of the pair $(F^T, H^T)$ of every strictly equivalent |7| state-space realization of the pair $(\tilde{P},\tilde{Q})$.

## III. OUTPUT REPRODUCIBILITY CONDITIONS

Denote with $\nu_M = \max_i\{\nu_i\}$ the maximal element among the $\nu_i$. Then, because of condition (2.3), $\tilde{P}(z)$ and $\tilde{Q}(z)$ can be written as

$$\tilde{P}(z) = \sum_{0}^{\nu_M} z^i P_i \quad ; \quad \tilde{Q}(z) = \sum_{0}^{\nu_M} z^i Q_i \tag{3.1}$$

where $P_i$ and $Q_i$ are real matrices. Equation (2.2) can thus assume the following equivalent form

$$P_0 y(t) + P_1 y(t+1) + .. + P_{\nu_M} y(t+\nu_M) = Q_0 u(t) + .. + Q_{\nu_M} u(t+\nu_M). \tag{3.2}$$

If $y[0,t] = \{y(0),\dots,y(t)\}$ is a generic output sequence of system (2.2) then the corresponding input sequence $u[0,t] = \{u(0),\dots,u(t)\}$ must satisfy the following relation

$$\begin{bmatrix} P_0 & P_1 & \cdots & P_{\nu_M} & 0 & \cdots & 0 \\ 0 & P_0 & P_1 & \cdots & P_{\nu_M} & 0 \cdots & 0 \\ \vdots & \ddots & \ddots & \ddots & & \ddots & \vdots \\ 0 & \cdots & 0 & P_0 & P_1 & \cdots & P_{\nu_M} \end{bmatrix} \begin{bmatrix} y(0) \\ y(1) \\ \vdots \\ y(t) \end{bmatrix} =$$

$$\begin{bmatrix} Q_0 & Q_1 & \cdots & Q_{\nu_M} & 0 & \cdots & 0 \\ 0 & Q_0 & Q_1 & \cdots & Q_{\nu_M} & 0 \cdots & 0 \\ \vdots & \ddots & \ddots & \ddots & & \ddots & \vdots \\ 0 & \cdots & 0 & Q_0 & Q_1 & \cdots & Q_{\nu_M} \end{bmatrix} \begin{bmatrix} u(0) \\ u(1) \\ \vdots \\ u(t) \end{bmatrix}$$

or, in a more compact form,

$$M_P\, y^t = M_Q\, u^t \tag{3.4}$$

where $M_P$ is a $m(t-\nu_M+1) \times m(t+1)$ real matrix and $M_Q$ is a $m(t-\nu_M+1) \times r(t+1)$ real matrix.

*Remark 1* - It is very important to note that because of the structure of the canonical pair $(\tilde{P},\tilde{Q})$ due to the relations (2.3) the n output components $y_1(0),..,y_1(\nu_1-1),\dots,y_m(0),..,y_m(\nu_m-1)$ must be considered as initial conditions, i.e. are not necessarily determined by the input sequence $u^t$. Moreover the output components

$y_i(\tau)$ are taken into account by the equations (3.3) and (3.4) only till times $t+\nu_i-\nu_M$. The input sequence $u^t$ determines therefore completely the output vector between $y(\nu_M)$ and $y(t-\nu_M+\nu_{min})$ where $\nu_{min}=\min_i\{\nu_i\}$.

Given an output sequence, $y^t$, a compatible input sequence $u^t$ satisfying eq. (3.3) - (3.4) exists iff $M_P\, y^t \in \mathscr{R}(M_Q)$. The minimal norm solution can be obtained by means of the expression

$$\hat{u}^t = \tilde{M}_Q \, M_P \, y^t \tag{3.5}$$

where $\tilde{M}_Q$ denotes the pseudoinverse of $M_Q$. When the general output reproducibility problem is considered, i.e. when the existence of compatible input sequences is considered for quite general output sequences then a simple algebraic test on a matrix of reduced dimensions is possible. In fact, denote with $M_P^k$ and $M_Q^k$ the matrices reported in (3.6) and (3.7); then because of the canonical form $(\tilde{P},\tilde{Q})$ that has been considered, it follows that $\mathrm{rank}(M_P^k) = m\,k$ for every k. When matrix $M_Q^k$ is considered, from the properties of block Hankel matrices it follows that if $M_Q^{\nu_M+1}$ has maximal rank (i.e. $m(\nu_M+1)$) then for every k $M_Q^k$ has maximal rank.

$$M_P^k = \begin{bmatrix} P_0 & P_1 & \cdots & P_{\nu_M} & 0 & \cdots & 0 \\ 0 & P_0 & P_1 & \cdots & P_{\nu_M} & & 0 \\ \vdots & \ddots & \ddots & \ddots & & \ddots & \vdots \\ 0 & \cdots & 0 & P_0 & P_1 & \cdots & P_{\nu_M} \end{bmatrix} \begin{matrix} 1 \\ 2 \\ \vdots \\ k \end{matrix} \tag{3.6}$$

$$M_Q^k = \begin{bmatrix} Q_0 & Q_1 & \cdots & Q_{\nu_M} & 0 & \cdots & 0 \\ 0 & Q_0 & Q_1 & \cdots & Q_{\nu_M} & & 0 \\ \vdots & \ddots & \ddots & \ddots & & \ddots & \vdots \\ 0 & \cdots & 0 & Q_0 & Q_1 & \cdots & Q_{\nu_M} \end{bmatrix} \begin{matrix} 1 \\ 2 \\ \vdots \\ k \end{matrix} \tag{3.7}$$

The following main result has thus been proved.

*Theorem 1 (Output Reproducibility Conditions)* - Every arbitrary output sequence can be reproduced by system (2.2) iff the condition

$$\mathrm{rank}\,(M_Q^{\nu_M+1}) = m(\nu_M+1) \tag{3.8}$$

is met.

*Remark 2* - Condition (3.8) can be satisfied only if $r \geq m$; this is a well known necessary condition for output reproducibility.

## IV. OUTPUT SEQUENCE GENERATION

The synthesis of the input sequence which generates a given output sequence will now be considered. Assume now that the minimal number of necessary inputs is present, i.e. $r = m$. $\tilde{P}$ and $\tilde{Q}$ are therefore square matrices; perform now the transformation of the pair $(\tilde{Q},\tilde{P})$ to the canonical form defined by conditions (2.3) by means of the algorithm described in |9|. Note that the transformation is now performed with reference to the matrix $\tilde{Q}$ so that conditions (2.3a), (2.3b) and (2.3c) will hold for the polynomials in $\tilde{Q}(z)$ while condition (2.3d) will not be (necessarily) valid since the inverse system can be anticipative. In the obtained reversed canonical model

$$\tilde{Q}(z)\ u(t) = \tilde{P}(z)\ y(t) \tag{4.1}$$

because of the canonical structure now achieved by $\tilde{Q}(z)$, every component of $u(t)$ is expressed as a linear combination of the *past* values of the components of u and of a limited number of past and (possibly) future samples of y. This allows to compute the necessary input components by means of the inverse system (4.1) instead of solving the large set of equations (3.4) with a batch procedure. When more than m inputs are present it may or may not be possible to perform the output reproducibility by means of a subset of m inputs. The test may be easily performed by selecting the m desired columns in the matrix $\tilde{Q}$ and testing the condition (3.8).

It can also be noted |11| that when a dynamical system is externally reduced i.e. when the number of inputs and outputs is the minimal one requested for the complete reachability and observability, the functional output controllability can always be performed. The external reduction can always be performed by means of simple algebraic combinations of the input and output components |12|.

## V. A NUMERICAL EXAMPLE

Consider the non purely dynamic system described by the following canonical input-output pair

$$\tilde{P}(z) = \begin{bmatrix} z^2 - 1 & z \\ z + 1 & z \end{bmatrix} \qquad \tilde{Q}(z) = \begin{bmatrix} z^2 & 1 \\ 1 & z \end{bmatrix}$$

Then $\nu_1 = 2$, $\nu_2 = 1$, $\nu_M = 2$;

$$Q_0 = \begin{bmatrix} 0 & 1 \\ 1 & 0 \end{bmatrix} \qquad Q_1 = \begin{bmatrix} 0 & 0 \\ 0 & 1 \end{bmatrix} \qquad Q_2 = \begin{bmatrix} 1 & 0 \\ 0 & 0 \end{bmatrix}$$

and the matrix $M_Q^3$ to be tested is given by

$$M_Q^3 = \begin{bmatrix} 0 & 1 & 0 & 0 & 1 & 0 & & & & \\ 1 & 0 & 0 & 1 & 0 & 0 & & & & \\ & & 0 & 1 & 0 & 0 & 1 & 0 & & \\ & & 1 & 0 & 0 & 1 & 0 & 0 & & \\ & & & & 0 & 1 & 0 & 0 & 1 & 0 \\ & & & & 1 & 0 & 0 & 1 & 0 & 0 \end{bmatrix}$$

Since $\text{rank}(M_Q^3) = 6$ the system output can be functionally controlled.

Consider the output sequence given by $y(0)=(1,1)^T$, $y(1)=(0,0)^T$, $y(2)=(1,1)^T$, $y(3)=(0,0)^T$, $y(4)=(1,1)^T$, $y(5)=(0,0)^T$, $y(6)=(1,1)^T$, $y(7)=(0,0)^T$, $y(8)=(1,1)^T$....

Since the pair $(\tilde{Q},\tilde{P})$ is already canonical the inputs can be cyclically deduced from the output sequence. In fact from the equation

$$\tilde{Q}(z)\ u(t) = \tilde{P}(z)\ y(t) \tag{5.1}$$

it follows

$$\begin{aligned} u_1(t+2) &= y_1(t+2) - y_1(t) + y_2(t) - u_2(t) \\ u_2(t+1) &= y_2(t+1) + y_1(t) + y_1(t+1) - u_1(t). \end{aligned} \tag{5.2}$$

If zero input and zero output is assumed for $t < 0$ (every other set of compatible values could be given; their knowledge is necessary back to $t=-\nu_M$) relations (5.2) allow to determine step by step the desired input sequence which is given by $u(0)=(1,2)^T$, $u(1)=(0,0)^T$, $u(2)=(-1,2)^T$, $u(3)=(0,2)^T$, $u(4)=(-1,2)^T$, $u(5)=(-2,2)^T$ $u(6)=(-1,4)^T$, $u(7)=(-2,2)^T$, $u(8)=(-3,4)^T$,.....

Note that, in this example, the inverse system (5.1) is not asymptotically stable since its poles are given by $p_1=1$, $p_2=1$, $p_3=1$. They are however simple roots of the minimal polynomial of the dynamical matrix of every strictly equivalent realization so that the system (5.1) is not unstable.

## REFERENCES

|1| Brockett, R.W.; Mesarovic, M.D.: *The reproducibility of multivariable systems*, Journal of Mathematical Analysis and Applications, 11, 548 (1965).

|2| Falb, P.L.; Wolovich, W.A.: *Decoupling in the design and synthesis of multivariable control systems*, IEEE Trans. on Automatic Control, AC-12, 651 (1967).

|3| Dorato, P.: *On the inverse of linear dynamical systems*, IEEE Trans. on Systems Science and Cybernetics, SSC-5, 43 (1969).

|4| Sain, M.A.; Massey, J.L.: *Invertibility of linear time-invariant dynamical systems*, IEEE Trans. on Automatic Control, AC-14, 141 (1969).

|5| Silverman, L.M.: *Inversion of multivariable linear systems*, IEEE Trans. on Automatic Control, AC-14, 270 (1969).

|6| Basile, G.; Marro, G.: *A state space approach to non-interacting controls*, Ricerche di Automatica, 1, 68 (1970).

|7| Guidorzi, R.: *Canonical structures in the identification of multivariable systems*, Automatica, 11, 361 (1975).

|8| Beghelli, S.; Guidorzi, R.: *A new realization algorithm for linear multivariable systems*, Ricerche di Automatica, 7, 164 (1976).

|9| Beghelli, S.; Guidorzi, R.: *A new input-output canonical form for multivariable systems*, IEEE Trans. on Automatic Control, AC-21, 692 (1976).

|10| Guidorzi, R.: *Equivalence, invariance and dynamical systems modelling*, Res. Rept., Istituto di Automatica, Univ. of Bologna, March 1979.

|11| Guidorzi, R.: *Structural properties of externally reduced systems*, Ricerche di Automatica, 4, 68 (1973).

|12| Guidorzi, R.: *External reduction of linear multivariable systems*, Ricerche di Automatica, 3, 113 (1972).

# ON THE STRUCTURE OF MULTILINEAR SYSTEMS

Paulo A. S. Veloso

Deptº Informática, Pontifícia Universidade Católica
Progr. Eng. Sistemas e Computação, COPPE-UFRJ
Rio de Janeiro, RJ, Brazil

ABSTRACT

The structure of a class of multilinear systems is investigated from a general system-theoretic viewpoint by algebraic methods. This class encompasses not only stationary and time-varying systems but also systems with discrete or continuous time.
Multilinear systems may be regarded as generalizations of linear systems, their nonlinearity being of a special kind. They occur in such diverse areas as pattern recognition, nuclear reactor control, biological modeling.
It is shown that many multilinear systems can be decomposed into parallel linear systems interconnected by a tensor-product block followed by a linear system, thus indicating that the multilinearity can be concentrated.

## 1. INTRODUCTION

In a linear system the inputs act linearly on the output. This is no longer the case for multilinear systems, where this dependence is multilinear rather than linear. A bilinear systems has two input lines and the influence of each input on the output is linear provided the other input is held fixed. The combined effect on the output of variations on both input lines need not be linear. Thus, a multilinear system presents a non-linearity of a particular kind, being somehow almost linear.

Multilinear systems are involved in the modeling of phenomena from several areas, such as pattern recognition, nuclear reactor control, biological population growth, etc.



*D. G. Lainiotis and N. S. Tzannes (eds.), Advances in Control, 87–93.*

Imagine a potentiometer, say on an analog computer, the value of which is linearly controlled by a signal w(t). Its response to input v(t) will have the form k.w(t).v(t). If its output line is fed into a linear system $L_o$ we obtain a bilinear system

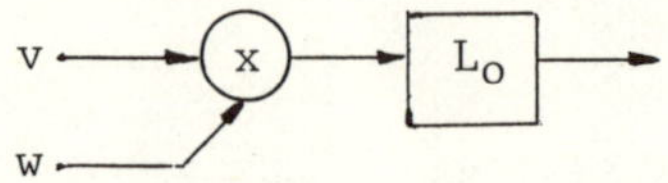


Another natural way to obtain a bilinear system consists in connecting two linear systems $L_1$ and $L_2$ in parallel via a multiplier. The result of combining both methods will be a bilinear system,

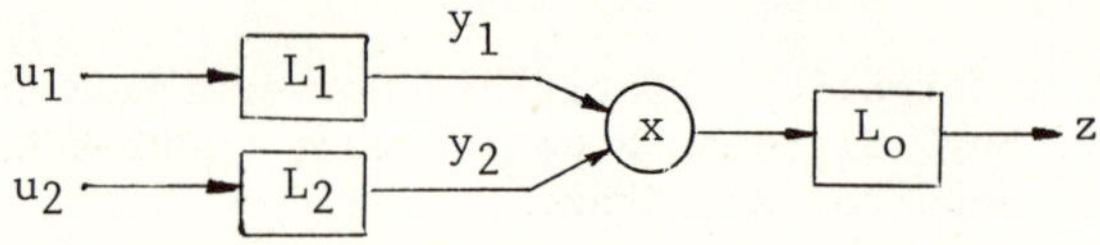


Notice that the non-linearity is concentrated in a memoryless multiplier block, all the other components being linear. It will be shown that many bilinear systems can be obtained this way.

We shall employ a generalization of Windehnecht's (1) general time-systems framework, designed to allow for more flexibility in the time domains. We shall need some standard algebraic terminology and basic results (2), (3).

## 2. SYSTEMS

A system is frequently regarded as a relation between input and output signals. Here we shall imagine that input is fed into the system at certain instants $p \in P$ and its output is observed at instants $t \in T$, considering its input-output behavior as basic.

We shall define a system as a 7-tuple

$\mathcal{S} = \langle P, T, E, Z, U, Y, S \rangle$ , where

- P and T are non-empty sets (input and output time domains);
- E and Z are non-empty sets (input and output alphabets);
- U is a non-empty subset of $E^P$ (input space);
- Y is a non-empty subset of $Z^T$ (output space);
- S is a relation from U to Y with domain U.

An important property of many physical systems is non-anticipation. For instance the output over the interval $[t_0,t_1)$ for real numbers $t_0<t_1$ does not depend on the input over $[t_1, +\infty)$ only on the part over $(-\infty, t_o)$. This motivates the next definition.

Given subsets I of P and J of T consider the relation (with '|' denoting restriction)

$${}_I S_J = \{(u|I, y|J) \ / \ (u,y) \in S\}$$

and call $S$ I-J-deterministic iff ${}_I S_J$ is a function.

From now on we fix a positive natural n and a commutative ring K with unit.

We shall a system $S$ n-linear (over K) iff

. $E = E_1 \times E_2 \times \ldots \times E_n$ , with each $E_j$ a K-module;
. $U = U_1 \times U_2 \times \ldots \times U_n$ , with each $U_j$ a K-submodule of $E_j^P$ ;
. Z is a K-module and Y a K-submodule of $Z^T$.
. the relation S is n-linear over K, i.e.,
for each $i = 1,\ldots,n$ and all $u_1 \in U_1,\ldots,u_{i-1} \in U_{i-1}$ ;
$u_{i+1} \in U_{i+1},\ldots,u_n \in U_n$ the induced relation

$$\{(u_i,y) \ / \ (u_1,\ldots,u_i,\ldots,u_n) \ Sy\}$$

is a K-submodule of $U_i \times Y$.

Notice that the usual linear systems are exactly the 1-linear ones. Thus, multilinear systems generalize the linear ones.

Clearly, a system $S$ is n-linear and I-J-deterministic iff

$${}_I S_J : (U_1 \times \ldots \times U_n)|I \rightarrow Y|J$$

is an n-linear function, which becomes a linear function in the case of linear system.

Some remarks about our concept of system may be in order. Firstly, it is possible to consider P and T as subsets of a common time domain, which may be ordered (but it is important to notice that this ordering need have a minimal element). Secondly, it is not necessary to take P and T as related to "physical time". In fact, it may be more convenient in some cases not to do so. We shall now give some examples to illustrate these points.

For a continuous-time system (say, simulated on an analog computer by means of integrators, potentiometers, multipliers, switches, etc), we may take P to be the reals and T the non-negative reals. For a discrete-time system (say, described by difference equations) we may consider P to be the integers and T the naturals. In both cases we do not care to observe the negative part, regarding it as setting the initial condition.

Now, for A/D (resp. D/A) converters it may be more convenient to take $P=\mathbb{R}$ and $T=\mathbb{N}$ (resp. $P=\mathbb{Z}$ and $T=\mathbb{R}_+$).

Finally, consider the case of Turing machines with a designate initial state. Here the initial contents of the tape(s) determine the final contents, if any. So, we regard the input as having the form $u : \mathbb{N} \to I$ (initial contents) and the output of the form $y : \mathbb{Z} \to Z$ (final contents), and take $P = \mathbb{N}$ and $T = \mathbb{Z}$. Thus we have an $\mathbb{N}$-$\mathbb{Z}$-deterministic system (for halting machines), which we would not have had we taken time related to the number of steps in the computation. For the case of one input read-only tape with $k$ working read-write tapes, we can consider $P = \mathbb{N}$ and $T = \mathbb{Z}^k$.

## 3. DECOMPOSITIONS

We shall examine the rough structure of multilinear systems. These shall be described in terms of decompositions.

Consider two systems $S$ , $R$ with output components (T,Z,Y) of $S$ coinciding with the corresponding input components of $R$. By their serial connection $S$ followed by $R$ we mean the system $S.R$ with relation

$$S.R = \{(u,w) \; / \; \text{for some } y \in Y, (u,y) \in S \text{ and } (y,w) \in R\}$$

If the input components of $S$ coincide with the corresponding ones of $R$ and likewise for the output components then we define its direct product $S \times R$ to be the system with relation.

$$S \times R = \{((u,v),(y,w)) \; / \; (u,y) \in S \text{ and } (v,w) \in R\}$$

Finally, the parallel connection of $S$ and $S'$ by $R$, assumed compatible, is the system $(S \times S').R$

In decomposing n-linear systems we shall try to isolate the non-linear, so as to, roughly speaking, diminish its dimensionality We consider first the case of deterministic systems.

Lemma - An n-linear (I-J-deterministic system $M$ can be decomposed into the parallel connection of n linear I-I-deterministic systems by an n-linear system $R$(which will be I-J-deterministic if $M$ is) so that $M = (L_1 \times \ldots \times L_n).R$.
Proof - Let $i = 1,\ldots,n$ and define

$$N_i|I = \{u_i|I \; / \; (u_1|I,\ldots,u_i|I,\ldots,u_n|I)\, {}_I M_J \, 0^J \;, \text{ for all } u_j \in U_j \text{ with } j \neq i\}$$

Notice that $N_i|I$ is a K-submodule of $U_i|I$ inducing a K-submodule $N_i = \{u_i \in U_i \; / \; u_i|I \in N_i|I\}$ of $U_i$.
The canonical projection $L_i$ of $U_i$ onto the quotient $V_i = U_i/N_i$

gives a linear I-I-deterministic system $L_i$. Now define $R$ by

$$R = \{((u_1 + N_1,\ldots,u_n + N_n) , y) / (u_1,\ldots,u_n) My\}$$

Notice that $R$ is n-linear, inheriting I-J-determinism from $M$ QED.

Notice that if system $M$ in the above lemma has input and output time domains P and T, resp., them $L_1,\ldots,L_n$ have P as both time domains. In the next proposition, $R$ will be further decomposed, yielding $L_o$ with T as both time domains.

Proposition. An n-linear I-J-deterministic system $M$ can be decomposed into a parallel interconnection of n linear I-I-deterministic systems by a P-T-deterministic tensor product $\bar{R}$ followed by a linear J-J-deterministic system $L_o$.

Proof. We have $M = (L_1 \times \ldots \times L_n).R$ with an n-linear function ${}_IR_J : (V_1 \times\ldots\times V_n)|I \to Y|J$ from the previous lemma. Consider the tensor products over K

$$\bar{R} = V_1 \times\ldots\times V_n \to W = V_1 \times \ldots \times V_n$$

$$\tilde{R} = V_1 \times\ldots\times V_n|I \to X = V_1|I \times \ldots \times V_n|I$$

and the tensor product $D : W \to X$ of the restrictions from $V_i$ to $V_i|I$ for $i = 1,\ldots,n$. There exists a linear function $L : X \to Y|J$ such that ${}_IR_J = D.L$. Define $L_o$ by

$$L_o = \{(w,y) / wD.L = y|J\}$$

Then $R = \bar{R}.L_o$ with $L_o$ as required. QED

Deterministic systems form a broad class. A particular case of special interest is that of systems where the output can be determined pointwise as in the following definition.

Call a system $S$ I-J-causal iff for each $t \in J$ there exists $I^t \subseteq I$ such that $S$ is $I^t - \{t\}$-deterministic and $I = \bigcup_{t\in J} I^t$.

For an I-J-causal system we have $y(t)$ as a function of $u|I^t$. But frequently we can obtain more information about the output from the knowledge of $u|I^t$. Indeed, given $t \in T$, consider the family of all $J' \subseteq J$ such that ${}_{I^t}S_{J'}$ is a function. By Zorn's lemma (2), this family has a maximal element $J^t$, with $t \in J^t \subseteq J$.

Theorem - An n-linear I-J-causal system $M$ can be decomposed as $M = (L_1.\hat{L}_1 \times\ldots\times L_n.\hat{L}_n).\hat{R}$ with $L_1,\ldots,L_n$ linear I-I-deterministic, $\hat{L}_1,\ldots,\hat{L}_n$ linear I-J-causal and $\hat{R}$ n-linear I-J-causal.

Proof - Let $t \in J$ and notice that $M$ is $I^t$-$J^t$-deterministic according to the above remarks. So, the previous lemma gives a decomposition $M = (L_1 \times\ldots\times L_n).R$ with R n-linear $I^t$-$J^t$-deterministic. Thus we have an n-linear function

$$(v_1^t ,\ldots,v_n^t ) R^t = (t) (v_1^t,\ldots,v_n^t) {}_{I^t}R_{J^t}$$

where $v_i^t = v_i|I^t$, for $i = 1,\ldots,n$. Now, we have a K-submodule $N_i(t)$ of $V_i^t$ with canonical projection $\hat{L}_i^t : V_i^t \to W_i^t$, for $i = 1,\ldots,n$ and an n-linear function $\hat{R}$ factoring $R^t = (\hat{L}_1^t \times\ldots\times \hat{L}_n^t).\hat{R}$. Thus we have the required systems. QED.

We can further decompose $\mathcal{R}$ in a manner analogous to the deterministic case.

<u>Corollary</u> - An n-linear I-J-causal system $M$ can be decomposed as $M = (L_1 \cdot \hat{L}_1 \times \ldots \times L_n \cdot \hat{L}_n) \cdot \tilde{R}.\hat{L}_o$ with $L_1,\ldots,L_n,\hat{L}_1,\ldots,\hat{L}_n$ as abov $\mathcal{R}$ a P-Q-causal tensor product and $\hat{L}_o$ linear J-I-causal.

A further specialization of interest is that of memoryless systems, such as adders, multipliers, where the input-output transformation is instantaneous. The next definition embodies a slight relaxation of this constraint.

Call a system $S$ I-J-static iff for each $t \in J$ there exists $p_t \in I$ such that $S$ is $\{p_t\}$ - $\{t\}$-deterministic and $I = \{p_t/t \in J\}$. Clearly an I-J-static system is an I-J-causal one where $y(t)$ is a function of the input value $u(p_t)$ at instant $p_t \in I$. The case $p_t=t$ corresponds to combinational switching circuits, whereas the more general case $p_t \leq t$ corresponds to definite machines.

Multilinear static systems have decompositions similar to the other cases.

<u>Proposition</u> - An n-linear I-J-static system $M$ can be decomposed int n linear I-I-static systems in parallel by a P-T-static tensor product followed by an J-J-static linear system, so that $M = (L_1 \times \ldots \times L_n) \cdot R \cdot L_o$

## 4. CONCLUSION

A very broad definition of time-system was given, accounting for different time behaviors on the input and output sides. Serial and parallel decomposition for such systems were taken as the natural extensions of the familiar concepts. Three increasingly restrictive subclasses were considered, deterministic, causal and static systems, which formalize various notions of non-anticipation and causality.

Multilinear systems over a commutative ring with unity were defined as encompassing the linear ones and then their overall structure described by decomposing them into parallel connections of linear systems via a tensor product followed by another linear system.

These results do not give much information about the fine structure of multilinear systems. This can sometimes be obtained in special cases, e.g. discrete stationary systems (6). However, they do have some interesting consequences. For instance, many, though not all, problems concerning multilinear systems can be attacked by studying the effect of a tensor-product connection of linear systems.

Moreover, the decompositions were obtained without explicit state-space construction, thus allowing unconstrained state assignments to the component systems. As an illustration, only multipliers have to be added to the stock of building blocks for linear systems (adders, scalers, delays and integrators) in order to construct any multilinear system described by differential/difference equations.

## ACKNOWLEDGEMENTS

Helpful discussions with L. Kerschberg and G. de La Penha as well as partial financial support from IBM of Brazil are gratefully acknowledged.

## REFERENCES

1. Windeknecht, T.G.: 1971, General dynamical processes: a mathematical introduction; Academic Press.
2. MacLane, J, and Birkhoff, G: 1968, Algebra; MacMillan Co.
3. Greub, W.: 1967, Multilinear algebra; Springer-Verlag.
4. Arbib, M.A.: 1969, Theories of abstract automata; Prentice-Hall.
5. Rink, R.E., and Mohler, R.R.: 1968, Completely controllable bilinear systems; SIAM J. on Control, pp. 477-486.
6. Arbib, M.A.: 1969, A characterization of multilinear systems; IEEE Transactions on Automatic Control, AC-11, pp. 699-704.
7. Kerschberg, L.H.: 1969, An algebraic approach to linear and multilinear systems theory; Case Western Reserve University, Ph.D. thesis.

# PREDICTION OF HARMONIC RESPONSE MEASUREMENTS OF A NONLINEAR SYSTEM VIA A TIME DOMAIN DIGITAL MODEL

C. Morgan and D.R. Towill

Dynamic Analysis Group, University of Wales Institute of Science and Technology, Cardiff, U.K.

The development of fault isolation schemes for dynamic, analogue systems using input-output information only generally requires responses to be obtained for all the expected parametric fault conditions. These responses can be measured by testing either the actual system or an analogue model of the system or, more conveniently, can be predicted by digital computer simulation methods. If the system under consideration is linear and the excitation signals of interest are deterministic, then a relatively simple digital computer model can be used to predict the steady-state response. Such models usually require little computing time. If, however, the system is significantly non-linear the digital model must simulate the real time behaviour of the system and consequently considerably more computing time is required. This paper describes the optimisation of a digital model of a system containing three nonlinear functions: deadzone, saturation and backlash, for the particular case where the excitation signal is serial mode sinusoids and the required format of the response data is amplitude ratio and phase angle of all measurable harmonic components. The model is then used to assess a fault isolation algorithm.

## 1. INTRODUCTION

The operating environment of many modern control systems is such that faults occurring in sub-assemblies which have failed routine checkout tests must be located rapidly using a minimum number of access points and measurements. These requirements have generated considerable interest in the development of automatic test routines in which the response to certain deterministic forcing



*D. G. Lainiotis and N. S. Tzannes (eds.), Advances in Control, 94–107.*

functions is used in a decision making algorithm (1),(2). In general the forcing function is chosen so that the system under test (SUT) is in a dynamic state when measurements are made.

Because of the large amount of test data required to design a comprehensive fault isolation routine, it is usually necessary to construct a simulation of the SUT (3) using either analogue or digital computing techniques. Digital simulation is the preferred method since no hardware test equipment is required and also the transfer, storage and processing of response data is relatively simple. Furthermore, any nonlinear functions which are present in the SUT can be accurately modelled by relatively simple software. One disadvantage of the digital approach, however, is the difficulty of designing a suitable numerical integrator. This means that the simulation must be optimised with respect to the SUT before reliable response predictions can be made.

In this paper an optimal digital simulation of a nonlinear, 2nd order servomechanism is developed for the particular case where the forcing function is composed of serial-mode sinusoids. Its performance is checked by comparing the predicted responses with those measured on an accurate analogue simulation of the SUT. The digital simulation is then used to assess a fault isolation algorithm using frequency response measurements.

## 2. THE SYSTEM UNDER TEST

The SUT is a second order servomechanism possessing three significant amplitude dependant nonlinear functions; deadzone, saturation and backlash. A block diagram of the system is shown in Fig.1. Several phenomena, which are attributable to the nonlinearities, are observed in the system response signal for certain operating conditions, viz. a limit cycle, jump resonances, steady state position error and higher harmonic signals (3). This behaviour indicates that the system cannot be analysed using simple linear theory, and consequently it is necessary to construct a simulation in order that the responses to deterministic forcing functions can be predicted.

## 3. NUMERICAL INTEGRATION ALGORITHM

Consider the ordinary differential equation

$$\dot{y} = f(t, y) \tag{3.1}$$

Integrating between the limits $t_n$ and $t_{n+1}$ we obtain the recurrence relation

$$y_{n+1} = y_n + \int_{t_n}^{t_{n+1}} f(t,y)\, dt \tag{3.2}$$

where $y_n$ and $y_{n+1}$ denote $y(t_n)$ and $y(t_{n+1})$ respectively and $t_{n+1}-t_n = h$, a constant for all n.

Several numerical integration methods have been developed to solve such equations: for the work described herein the Adams-Moulton predictor-corrector (4) is used. In this method an m'th order interpolating polynomial, $r_m$, is passed through the (m+1) points $(t_n,y_n)$, $(t_{n-1},y_{n-1})$,........, $(t_{n-m},y_{n-m})$ and equation (3.2) is approximated by

$$\hat{y}_{n+1} = y_n + \int_{t_n}^{t_{n+1}} r_m(t)\, dt \tag{3.3}$$

This is referred to as a predictor formula since it predicts the value of y at time $t_{n+1}$ from the values of y at $t \leq t_n$. To obtain a more accurate estimate of $y_{n+1}$ an (m+1)'th order polynomial $s_{m+1}$ is fitted to the (m+2) points $(t_{n+1},\hat{y}_{n+1})$, $(t_n,y_n)$,....., $(t_{n-m},y_{n-m})$, and is substituted in equation (3.2) to give the corrected value of $y_{n+1}$

$$y_{n+1} = y_n + \int_{t_n}^{t_{n+1}} s_{m+1}(t)\, dt \tag{3.4}$$

The corrector formula (3.4) is iterative and requires an estimate of $y_{n+1}$ to begin operation, this estimate is obtained from the predictor formula (3.3). The predictor-corrector pair are usually expressed in Lagrangean form as below

$$\hat{y}_{n+1} = y_n + \frac{h}{u'}\left\{u_0 f_n + u_1 f_{n-1} + \cdots + u_m f_{n-m}\right\} \tag{3.5}$$

$$y_{n+1} = y_n + \frac{h}{v'}\left\{v_{-1} \hat{f}_{n+1} + v_0 f_n + \cdots + v_m f_{n-m}\right\} \tag{3.6}$$

where $f_n$ denotes $f(t_n,y_n)$ and $\hat{f}_{n+1}$ is calculated using $\hat{y}_{n+1}$, i.e. $\hat{f}_{n+1}=f(t_{n+1},\hat{y}_{n+1})$. The u's and v's are functions of m and are given in Table 1 for $2 \leq m \leq 5$. Although the corrector is iterative it is usually used once only after the predictor since it is considered generally better to shorten the time increment (h) than to do repeated evaluations of equation (3.6).

Table 1 Coefficients for Lagrangean Form of Predictor-Corrector

| Predictor Coefficient | Order of Integration 4 | 5 | 6 | 7 |
|---|---|---|---|---|
| $u'$ | 12 | 24 | 720 | 1440 |
| $u_o$ | 23 | 55 | 1901 | 4277 |
| $u_1$ | -16 | -59 | -2774 | -7923 |
| $u_2$ | 5 | 37 | 2616 | 9982 |
| $u_3$ | | -9 | -1274 | -7298 |
| $u_4$ | | | 251 | 2877 |
| $u_5$ | | | | -475 |
| **Corrector Coefficient** | **4** | **5** | **6** | **7** |
| $v'$ | 24 | 720 | 1440 | 60480 |
| $v_{-1}$ | 9 | 251 | 475 | 19087 |
| $v_o$ | 19 | 646 | 1427 | 65112 |
| $v_1$ | -5 | -264 | -798 | -46461 |
| $v_2$ | 1 | 106 | 482 | 37504 |
| $v_3$ | | -19 | -173 | -20211 |
| $v_4$ | | | 27 | 6312 |
| $v_5$ | | | | -863 |

## 4. DIGITAL SIMULATION OF A DYNAMIC SYSTEM

A dynamic system containing k integrators can be represented by a block diagram of the form shown in Fig.2, where all the linear and nonlinear time independent functions have been omitted for simplicity leaving only the k integrators. The input signal to the i'th integrator, denoted $f^{(i)}$ is, in general, a function of the signals at other parts of the system, including the excitations. Such a function is referred to as a loop closure equation and can be expressed in the form

$$f^{(i)} = F_i\left(y^{(1)}, y^{(2)}, \ldots, y^{(k)}, x^{(1)}, x^{(2)}, \ldots, x^{(\ell)}\right) \tag{4.1}$$

where $F_i(\ldots)$ denotes the i'th function of $(\ldots..)$ and $x^{(1)}, \ldots,$ $x^{(\ell)}$ are the system excitations.

Consider now the state of the system at time $t_n$. The input and output signals of integrator i, estimated by the predictor-corrector, are denoted $f_n^{(i)}$ and $y_n^{(i)}$ respectively and the corresponding instantaneous value of the j'th excitation is $x_n^{(j)}$. It should be noted at this point that $f_n^{(i)}$ and $y_n^{(i)}$ will be exact if and only if $f^{(i)}(t)$ is in reality a polynomial of degree $\leqslant m+2$, which for a system containing 'hard' nonlinearities is impossible. The predictor-corrector formulae can now be used to estimate the i'th integrator output signal $y_{n+1}^{(i)}$ at time $t_{n+1}$ by first predicting the output signal using equation (3.5) then updating the estimate of the integrator input signals $f_{n+1}^{(i)}$ using equation (4.1) and finally employing equation (3.6) to obtain the corrected estimate $y_{n+1}^{(i)}$.

From the block diagram of Fig.1 it can be seen that the loop closure equations for the SUT are

$$f^{(1)} = \{K_f x^{(1)} - K_g y^{(1)} - N_2(\overset{*}{y}^{(2)}) K_f y^{(2)}\} K_a N_1(z) K_t - B_m J_m^{-1} y^{(1)} \qquad (4.2)$$

$$f^{(2)} = y^{(1)} \qquad (4.3)$$

Where $N_2(\overset{*}{y}^{(2)})$ means that the actual value of this function depends on the history of $y^{(2)}$ since the nonlinearity in question possesses memory.

## 5. OPTIMAL FREQUENCY DOMAIN DIGITAL SIMULATION

When simulating a dynamic system on the digital computer two main sources of error can be expected. Firstly there are truncation errors which are introduced (a) as a result of discretising the signals within the system so that the continuous waveforms are approximated by a series of instantaneous values and (b) by the degree of approximation in the interpolating polynomials $r_m$ and $s_{m+1}$. Secondly there are rounding errors caused by the discrete signal values and resulting computation being truncated due to the computers' finite word length. A digital simulation is therefore a compromise between these two sources of error because to minimise trunction effects it is necessary to decrease the time increment and to minimise rounding errors the time increment must be increased. Another important factor which influences the choice of increment is the total time used by the computer in running the simulation, since computer time is obviously at a premium. Comparisons made between the digital simulation and an analogue simulation of the

SUT have shown that setting m=4 in equations (3.5) and (3.6) gives good response predictions. Little improvement is obtained when m is increased to 5 or 6. Assuming m=4 then an optimal digital simulation will use the largest increment which guarantees at least the minimum required accuracy in predicting the signals of interest in the system. In general, therefore, the major source of error will be due to truncation effects introduced by discretisation.

## 5.1. Optimising the Time Increment

Consider the case where a system under analysis is excited by serial mode sinusoidal signals. If the system is linear then the steady state response at any measurement point will be sinusoidal. The integration of such signals by predictor-corrector methods; requiring a polynomial function to be fitted to a certain number of discrete points, will thus be subject to truncation errors. The magnitude of the error will obviously be proportional to the inadequacy of the fitted function. For a nonlinear system of the type discussed in this paper, however, the steady state signals of interest will, in general, be quasi-piecewise linear due to the action of the 'hard' nonlinear functions. Hence it is evident that for a given time increment the polynomial approximations and consequent integrations will provide more accurate response estimates for the nonlinear system than for the equivalent linear system, providing Gibbs phenomenon is insignificant. The increment which guarantees a certain minimum prediction error for a nonlinear simulation at a given forcing frequency can thus be obtained by comparing the equivalent linear simulation response with the exact theoretical response. Thus for the linear simulation the per-unit prediction error may be defined as

$$e(\omega,h) = \left\{ \left[A(\omega)\cos\phi(\omega) - a(\omega,h)\cos\Theta(\omega,h)\right]^2 + \left[A(\omega)\sin\phi(\omega) - a(\omega,h)\sin\Theta(\omega,h)\right]^2 \right\} A(\omega)^{-1} \quad (5.1)$$

Where $A(\omega)$ and $\phi(\omega)$ are the theoretical amplitude ratio and phase angle values at frequency $\omega$, and $a(\omega,h)$ and $\Theta(\omega, h)$ are the corresponding digital simulation estimates for a time increment h. At frequency $\omega$ the nonlinear simulation time increment which guarantees a per-unit prediction error $e^*$ in the fundamental component of the response is $h_1$, such that

$$e(\omega,h_1) \leqslant e^* \quad (5.2)$$

The error function defined in equation (5.1) is plotted against the number of sample points per cycle ($N=2\pi/\omega h$) at various

frequencies in Fig.3, for the system shown in Fig.1. It can be seen that to obtain accuracies of 1% and 0.5% the number of sample points per cycle should be 200 and 400 respectively.

The approach described above will in general provide reliable prediction of the fundamental and higher harmonic frequency responses of the nonlinear SUT; however, it is desirable to check the predictions at a selection of test frequencies by measuring the actual response on either the system or analogue simulation of the system. Then obvious errors due to, for example, limit cycling can be easily detected and the necessary corrective action taken.

### 5.2. Optimising Simulation Run-time

Frequency Response measurements are made when all transients have decayed to insignificant levels and thus considerable reduction in the total run-time of the digital simulation can be realised by using a large time increment (small N) for the initial transient region of the response, changing to the optimum accuracy value once the response has reached steady state. The initial time increment must be chosen so that the simulation is stable and the steady state response is reasonably close to the true response. Once the response has settled a polynomial function is fitted to a convenient number of sequential points at the input and output ports of each integrator and then the corresponding discrete values calculated for the optimum time increment. Only half a complete cycle need then be injected at the new time increment to deduce the Fourier components in the response waveform, since the waveforms in the SUT are symmetrical about the zero axis. The initial increment can be determined by plotting the error in harmonic component predictions for a range of values of N, taking the optimum N as datum. A suitable initial N is then chosen such that a reasonably stable and accurate response is obtained. Typical plots of error against the number of sample points per cycle for the nominal nonlinear SUT are shown in Fig.4.

## 6. COMPARISON OF RESULTS

Extensive frequency domain testing has been performed to verify the responses predicted by the digital simulation. In these tests, described fully in reference (1), data responses are measured on the hardware system and an analogue simulation of the system. The analogue simulation is more useful than the hardware system for this purpose since it allows greater control over the parameter values and hence the unique effect of each nonlinear function can be determined.

Typical responses for two sets of nonlinear parameter values are shown in Figs. 5 and 6. In Fig.5 both deadzone and backlash have been set to zero and a significant jump resonance is observed on the fundamental and third harmonic responses. This phenomenon is accurately predicted by the digital simulation. Fig.6 compares the responses of the hardware system, the analogue simulation and the digital simulation. It can be seen that the higher harmonic matching is not as good as that obtained with only the saturation function present. This is due to unavoidable inaccuracies in the analogue modelling of the non-linear parameters, particularly the backlash function. The predicted responses are, however, acceptable for system verification and fault isolation applications.

## 7. THE DIGITAL SIMULATION USED TO ASSESS A FAULT ISOLATION METHOD

In fault isolation we are interested in determining the faulty parameters in a 'sick' SUT. For the purposes of this paper faults are restricted to non-catastophic drift variations from nominal value in single parameters only. Such assumptions are justified since (a) it is usual that catastrophic faults are relatively easy to identify and (b) reasonably small deviations from nominal value in one parameter are not likely to cause large deviation in other parameters. A further requirement of a fault isolation method is that a minimum number of measurements or data points are required to correctly identify faults. One approach to fault isolation is outlined below.

Let the nominal response vector of system be

$$U = \{u_i\} \quad , \quad i = 1,2,\ldots\ldots,Q \tag{7.1}$$

where Q is the number of data points. If there are P fault cases then a Q x P recognition matrix S can be formed by firstly obtaining the response vectors for representatives of each fault case, thus

$$V_j = \{v_{ij}\} \quad , \quad \begin{array}{l} i = 1,2,\ldots\ldots,Q, \\ j = 1,2,\ \ldots\ldots,P \end{array} \tag{7.2}$$

the recognition matrix is then

$$S_j = \{s_{ij}^*\} = \{V_j - U\}^* \tag{7.3}$$

where * denotes normalisation.

The response of a system with an unknown fault is now measured and normalised giving the fault vector

$$F = \{f_i^*\} \quad , \quad i = 1,2,\ldots\ldots,Q \tag{7.4}$$

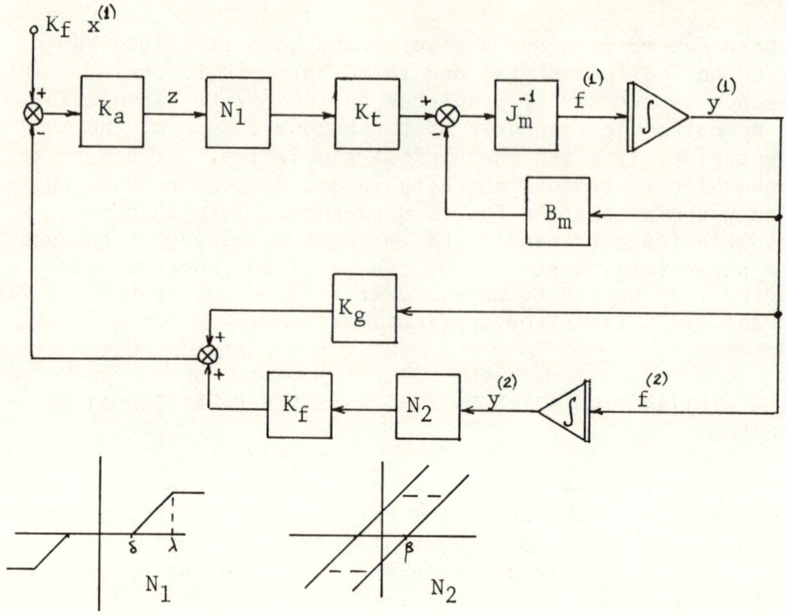


Fig.1 Block Diagram of Nonlinear System Under Test

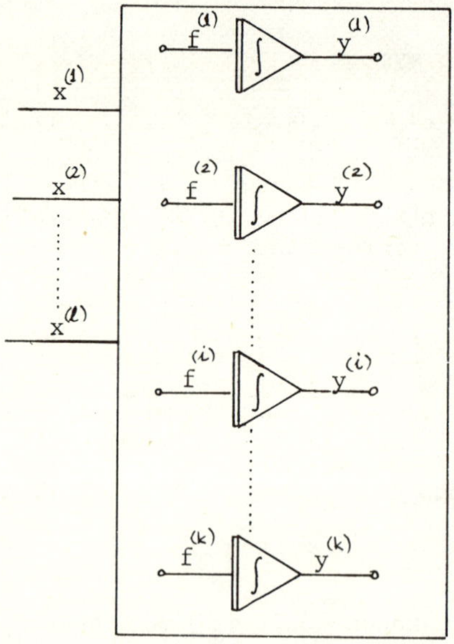


$$f = fn(X,Y)$$
$$Y = (y^{(1)}, y^{(2)}, \ldots\ldots, y^{(k)})$$
$$X = (x^{(1)}, x^{(2)}, \ldots\ldots, x^{(\ell)})$$

Fig.2 General Layout of a Dynamic System

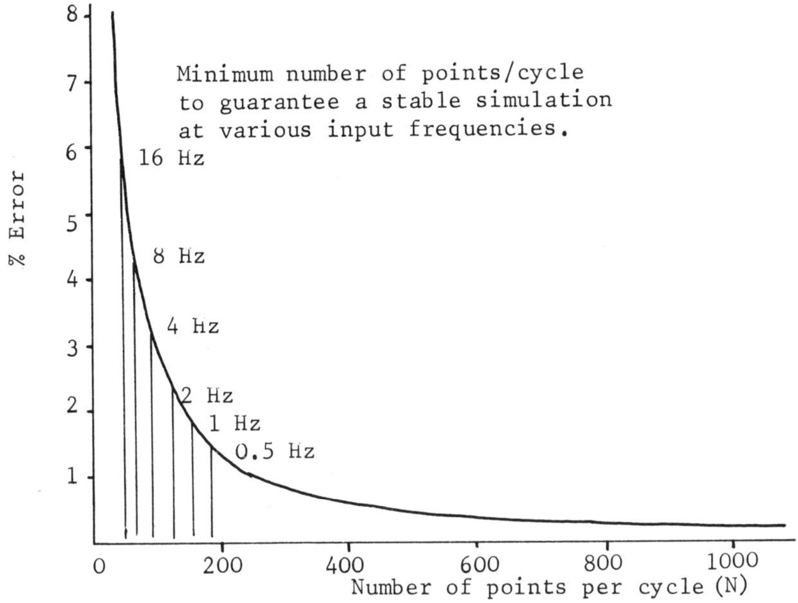


Fig.3 Accuracy of Linear Simulation as a Function of the Number of Data Points

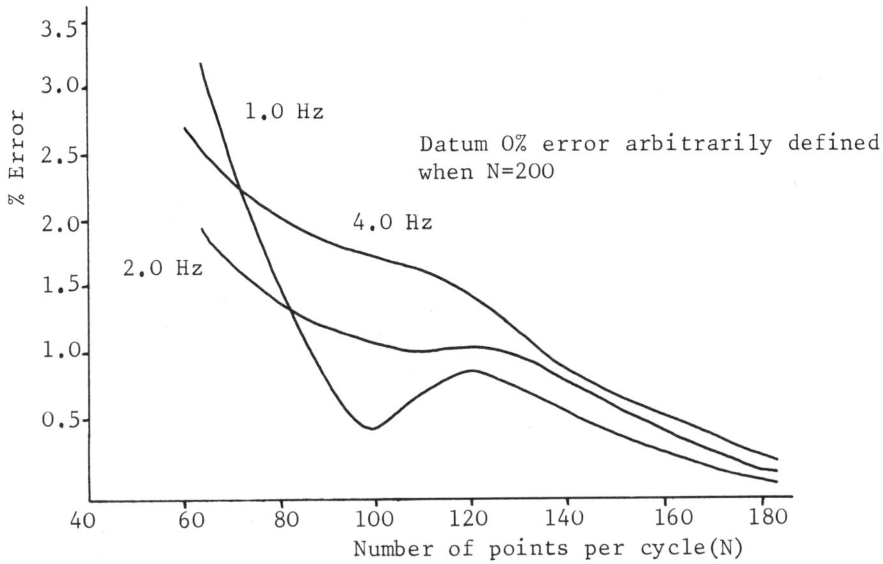


Fig.4 Accuracy of Nonlinear Simulation at Various Input Frequencies

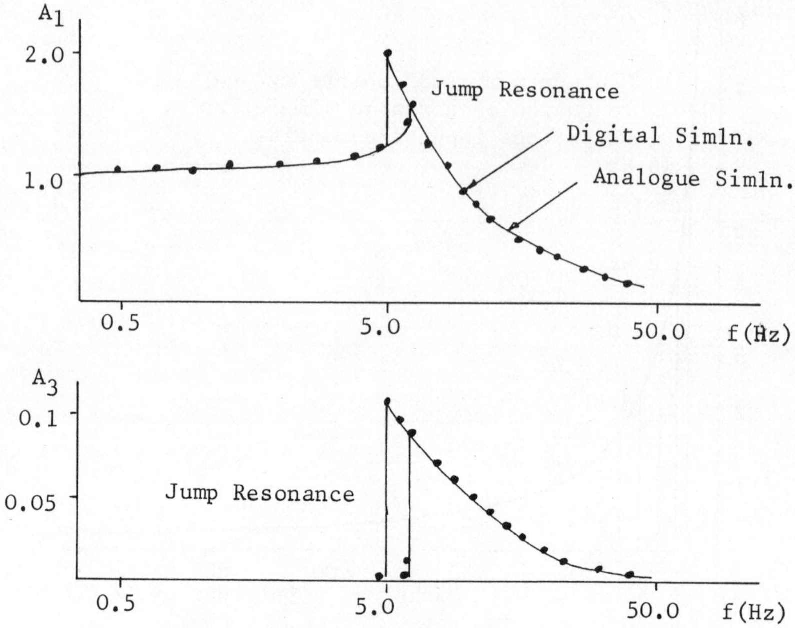


Fig.5 Fundamental and Third Harmonic Frequency Responses of the Saturated System

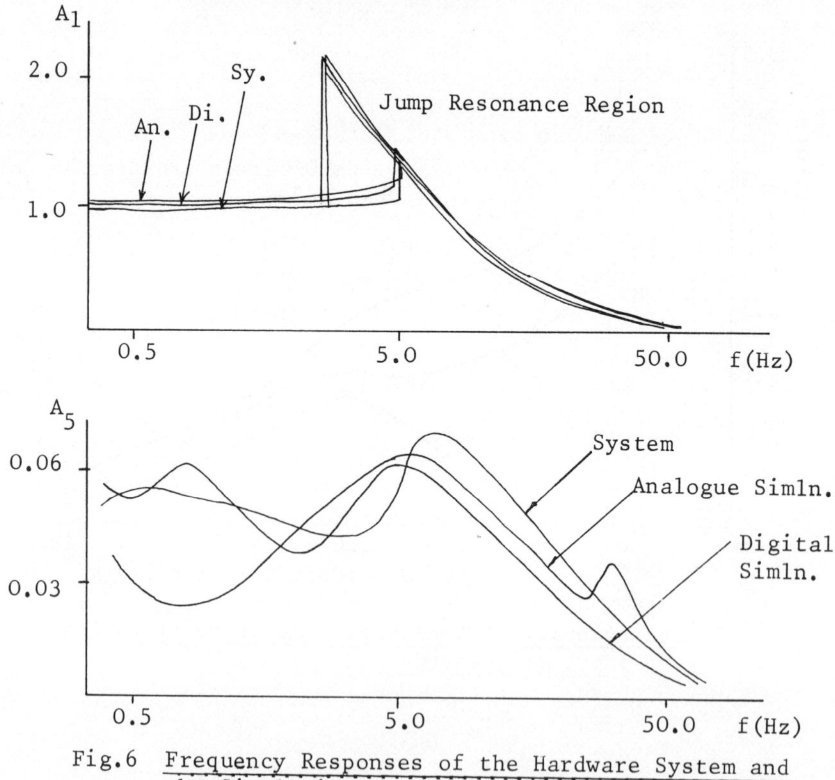


Fig.6 Frequency Responses of the Hardware System and the Simulations

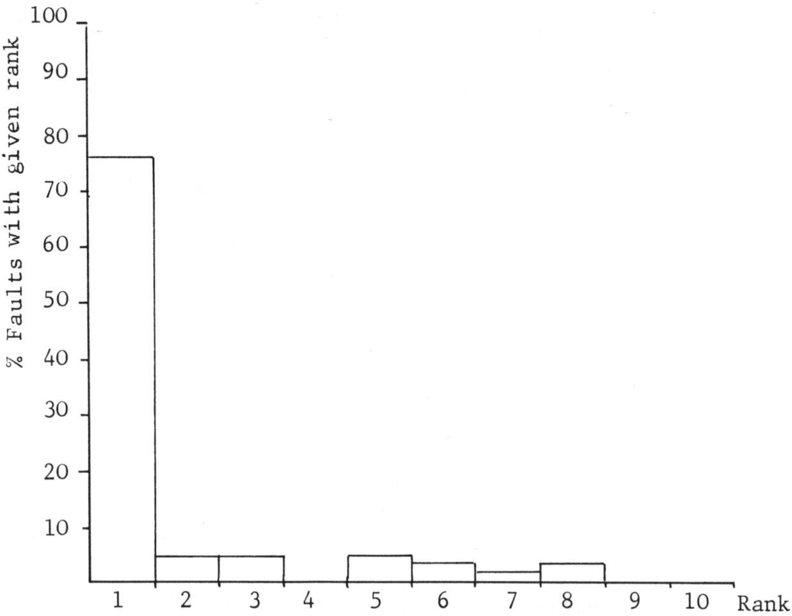


Fig.7 Performance of the NNR Used to Detect Faults in the Nonlinear System Under Investigation

In general F will not correspond exactly with any of the column vectors $S_j$ because system measurements will be affected by noise and variances in the faulty and non-faulty parameters. To choose the most likely fault a decision making algorithm such as the nearest neighbour rule (NNR)(5) is used. Thus we define a "distance" for each fault vector

$$d_j = \sum_{i=1}^{i=Q} (s_{ij}^* - f_i^*)^2 \tag{7.5}$$

The most likely fault is then fault number k where

$$d_k = \min (d_j) \quad , \; j = 1,2,\ldots,P \tag{7.6}$$

The performance of this method for the SUT, using frequency response data predicted by the digital simulation is summarised in Fig.7. Data used were estimates of the fundamental, third and fifth harmonics at the forcing frequency set {2.0,2.8,4.0, 5.6,8.0} Hz. The faulty parameters ($P_j$) were randomly chosen within the limits $0.3p_{jnom} \leqslant p_j \leqslant 0.7p_{jnom}$ and $1.3p_{jnom} \leqslant p_j \leqslant 1.7p_{jnom}$ and the non-faulty parameters were assumed to be normally distributed about nominal with standard deviation $\sigma = 0.03p_{nom}$.

In all 108 faulty systems were simulated. Fig.7 indicates that the probability that the actual fault is ranked 1 is 0.75 and the probability of the actual fault being among the three highest ranked faults is 0.87.

## 8. CONCLUSIONS

A digital computer simulation of a dynamic non-linear system can be used to predict the response of the system when it is excited by serial mode sinusoidal signals. The simulation can be optimised by adjusting the integration time increment so that a minimum amount of time is spent in calculating the transient part of the response. Such a digital simulation provides a good vehicle for the development of dynamic test schemes.

## REFERENCES

1. Sriyananda, H., Towill, D.R., and Williams, J.H.: "Voting Techniques for Fault Diagnosis from Frequency Domain Test Data", IEEE Trans. Rel., Vol.R-24, Oct. 1975, pp 260-267.
2. Morgan, C., and Towill, D.R.: "Application of the Multi-harmonic Fourier Filter to Nonlinear System Fault Location", IEEE Trans. Inst.Meas., Vol. IM-26, June 1977, pp 164-169.

3. Morgan, C., and Towill, D.R.: "Dynamic Analysis of a Non-Linear Servomechanism", UWIST, Cardiff, Tech. Note DAG 82, Sept. 1975.
4. Phillips, G.M.: "Theory and Application of Numerical Analysis", Academic Press Inc., (London), 1973, pp 304-309.
5. Batchelor, B.G.: "Pattern Recognition", Plenum Press, New York, 1978, p 77.

# REALIZATION THEORY FOR MULTIVARIATE GAUSSIAN PROCESSES II: STATE SPACE THEORY REVISITED AND DYNAMICAL REPRESENTATIONS OF FINITE DIMENSIONAL STATE SPACES

Anders Lindquist and Giorgio Picci

University of Kentucky, Lexington, Kentucky
LADSEB-CNR, Padova, Italy

ABSTRACT: The purpose of this paper is twofold. First, some of the results of Part I are generalized and clarified, reformulated in a mathematical framework which we now perceive as more natural and conceptually sounder, the basic concept being perpendicularly intersecting subspaces. By relating minimality of stochastic realizations to the factorization of certain Hankel operators, the connections to deterministic realization theory is further clarified. Furthermore, spectral domain criteria for minimality, observability and constructibility are presented. Secondly, stochastic differential equation representations are presented for all state spaces which are finite dimensional. The infinite dimensional case will be the topic of Part III of this sequence of papers.

## 1. INTRODUCTION

This is part II in a sequence of papers, the first one being [1]. We have tried to make the presentation essentially self-contained by restating certain results without proofs, but we shall frequently refer the reader to Part I for details. Part I also contains a more detailed introduction to the problem under study and an historical account of it, as well as a relevant list of references.

Given an m-dimensional stationary stochastic process $\{y(t);\ t \in R\}$, which is real, Gaussian, purely nondeterminstic, mean square continuous and centered, the basic problem is to find representations of the type



*D. G. Lainiotis and N. S. Tzannes (eds.), Advances in Control, 108–129.*

$$y(t) = Cx(t) \tag{1.1}$$

where C is a linear time-invariant operator and $\{x(t);\ t \in R\}$ is a (generally infinite dimensional) vector Markov process, for which we can obtain a differential equation representation.

This problem can be given a mathematically more precise formulation in a Hilbert space setting. To this end, define H to be the Gaussian space generated by y, i.e., the linear hull of the random variables $\{y_k(t);\ t \in R,\ k = 1,2,\dots,m\}$ closed in $L_2$ norm. The space H is a Hilbert space with inner product $\langle\xi,\eta\rangle = E\{\xi\eta\}$, where $E\{\cdot\}$ denotes mathematical expectation. To describe the dynamical aspects of the problem we need to define the past and the future of H and to introduce a shift operator on H: Let the past space $H^-$ and the future space $H^+$ be the closed linear hulls in H of $\{y_k(t);\ t \le 0,\ k = 1,2,\dots,m\}$ and $\{y_k(t);\ t \ge 0,\ k = 1,2,\dots,m\}$ respectively. Moreover, since y is stationary, there is a strongly continuous group $\{U_t;\ t \in R\}$ of unitary operators $H \to H$ such that $y_k(t) = U_{t-s}y_k(s)$ for all t,s and k[2]. Given three arbitrary subspaces A,B and X of H, we shall write $A \perp B$ when A and B are orthogonal and $A \perp B|X$ when they are conditionally orthogonal given X, i.e., $\langle E^X\alpha, E^X\beta\rangle = \langle\alpha,\beta\rangle$ for all $\alpha \in A$ and $\beta \in B$, where $E^X$ denotes orthogonal projection onto X. Finally we shall write $A \vee B$ to denote the closed linear hull of A and B.

The problem at hand can now be formulated as a geometric problem in the Hilbert space H: Find all subspaces X such that

(i) X is *Markovian*, i.e.,

$$X^- \perp X^+ | X \tag{1.2}$$

where $X^- := \bigvee_{t\le 0}(U_tX)$ and $X^+ := \bigvee_{t\ge 0}(U_tX)$.

(ii) $y_k(0) \in X$ for $k = 1,2,\dots,m$

(iii) X is *minimal*, i.e., there is no proper subspace of X satisfying (i) and (ii).

Each solution X of this problem gives rise to a representation of type (1.1). In fact, a vector process $\{x(t);\ t \in R\}$ taking values in some suitable function space can be constructed by appropriately choosing a basis in X and applying the shift $U_t$ to its components. (Note that, since y is mean-square continuous, H is separable.) In view of (i), this process must be Markov. Condition (ii) then yields a representation (1.1). Finally, (iii) insures that the *state space* X is as small as possible. In Section 7 we shall describe how this choice of basis is made in the finite dimensional case (dim $X < \infty$), postponing the infinite dimensional case to a subsequent paper, Part III. (The geometric theory of Sections 2-6 does not, however, require any restriction of the dimension of X.)

This paper constitutes not merely a sequel of Part I; it contains many extensions, generalizations and other improvements of the results presented in Part I. We have found that the geometric theory is most easily understood and explained in terms of *perpendicularly intersecting subspaces*, and we devote Sections 2 and 3 to this reformulation. In Section 4, we show how the concepts of deterministic realization theory [3,4] can be used to provide further insight into the stochastic theory, and Section 5 and 6 contain extensions of the spectral domain theory presented in [1]. Section 7, finally, is devoted to dynamical representations of finite dimensional state spaces.

## 2. PERPENDICULARLY INTERSECTING SUBSPACES

Let A and B be two (closed) subspaces of the Hilbert space H. We shall say that the subspace $X \subset H$ is an *(A,B)-splitting subspace* if

$$A \perp B | X . \tag{2.1}$$

If there is no *proper* subspace of X which also satisfies (2.1), X is said to be *minimal*. The purpose of this section is to determine under what conditions on A and B there is only *one* minimal (A,B)-splitting subspace.

LEMMA 2.1 *There is one and only one minimal (A,B)-splitting subspace contained in* A *(in* B*), namely* $X = \bar{E}^A B$ $(X = \bar{E}^B A)$. (The bar over the E denotes closure.)

Proof. The useful decomposition formula

$$A = \bar{E}^A B \oplus (A \cap B^{\perp}) \tag{2.2}$$

implies that $\bar{E}^A B = A \ominus (A \cap B^{\perp})$ and that $A \ominus \bar{E}^A B \perp B$, the latter of which is equivalent to (2.1) with $X = \bar{E}^A B$. On the other hand, any (A,B)-splitting subspace X containing A satisfies $A \ominus X \perp B$, i.e., $A \ominus X \subset A \cap B^{\perp}$, or equivalently $X \supset A \ominus (A \cap B^{\perp})$, i.e., $X \supset \bar{E}^B A$. □

LEMMA 2.2 *All* (A,B)*-splitting subspaces contain* $A \cap B$.

Proof. Let $\eta \in A \cap B$. Then $\eta \perp \eta | X$, i.e., $\eta \in X$. □

Consequently, if there is only one (A,B)-splitting subspace, we have $\bar{E}^A B = \bar{E}^B A = A \cap B$. Conversely, if

$$\bar{E}^A B = A \cap B, \tag{2.3}$$

$\bar{E}^A B$ is also the minimal (A,B)-splitting subspace contained in B, i.e.,

$$\bar{E}^B A = A \cap B, \tag{2.4}$$

and it must be the only minimal (A,B)-splitting subspace, since all such subspaces contain $A \cap B$ (Lemma 2.2). We shall say that A and B *intersect perpendicularly* if one of the two equivalent conditions (2.3) or (2.4) holds.

PROPOSITION 2.1 *There is a unique minimal (A,B)-splitting subspace if and only if A and B intersect perpendicularly and it is given by (2.3) and (2.4).*

The following proposition provides an alternative characterization of perpendicular intersection of subspaces in the case that they span the whole space.

PROPOSITION 2.2 *Let $A \vee B = H$. Then A and B intersect perpendicularly if and only if $B^\perp \subset A$, or equivalently, $A^\perp \subset B$.* (Here $A^\perp$ denotes the orthogonal complement of A in H.)

Proof. (if): Assume that $B^\perp \subset A$ holds. Set $X = \bar{E}^A B$. Then (2.2) yields $A = X \oplus B^\perp$, from which it follows that $X \subset B$ and that $A \cap B = X$.

(only if): Set $X = \bar{E}^A B$ and assume that $X = A \cap B$. Then it follows from (2.2) that $(A \ominus X) \perp B$. Therefore, since $A \vee B = H$ and $B \supset X$, $B^\perp \subset A$. □

In the sequel, we shall need the following simple observation, the proof of which is trivial.

LEMMA 2.3 *Let A, B, A' and B' be four subspaces such that $A \subset A'$ and $B \subset B'$. Then any (A',B')-splitting subspace is also an (A,B)-splitting subspace.*

## 3. THE GEOMETRY OF SPLITTING SUBSPACES REVISITED

Our basic problem is to determine the set of all minimal $(H^-,H^+)$-splitting subspaces, where $H^-$ and $H^+$ are the past and future spaces of the given stochastic process $\{y(t);\ t \in R\}$; we shall usually drop the prefix $(H^-,H^+)$, plain "splitting subspace" referring to the pair $(H^-,H^+)$. If $H^-$ and $H^+$ intersect perpendicularly, there is only one minimal splitting subspace, namely the *present space*

$$H^o = H^- \cap H^+ \tag{3.1}$$

(Proposition 2.1). When y has a rational spectral density, this corresponds to the case where y can be realized by a purely autoregressive scheme. However, in general this is not the case, and there is a whole family of minimal splitting subspaces, two of which are the forward and backward predictor spaces $X_- := \bar{E}^{H^-}H^+$ and $X_+ := \bar{E}^{H^+}H^-$ respectively (Lemma 2.1). Defining $N^- := H^- \ominus X_-$ and $N^+ := H^+ \ominus X_+$ we obtain the orthogonal decomposition

$$H = N^- \oplus H^{\square} \oplus N^+ , \tag{3.2}$$

where H is the Gaussian space of the process y and $H^{\square}$ is the *frame space* $H^{\square} := X_- \vee X_+$. In Part I, we showed that all minimal splitting subspaces X satisfy

$$H^{\circ} \subset X \subset H^{\square}, \tag{3.3}$$

i.e., the frame space is the closed linear hull of all minimal splitting subspaces, hence containing all pertinent information about y. We shall call $N^-$ and $N^+$ the *junk spaces* since all information in them may be discarded. By applying (2.2), it is immediately seen that $N^- = H^- \cap (H^+)^{\perp}$ and $N^+ = H^+ \cap (H^-)^{\perp}$. In the special case that $H^-$ and $H^+$ intersect perpendicularly $H^{\square}$ and $H^{\circ}$ coincide.

The following theorem expresses the fact that any (minimal or nonminimal) splitting subspace can be regarded as a *minimal* splitting subspace if the past and future spaces are extended so that they intersect perpendicularly.

THEOREM 3.1 *The space* $X \subset H$ *is a splitting subspace if and only if it is the (unique) minimal* $(S,\bar{S})$*-splitting subspace*

$$X = S \cap \bar{S} \tag{3.4}$$

*for some perpendicularly intersecting subspaces* S *and* $\bar{S}$ *such that* $S \supset H^-$ *and* $\bar{S} \supset H^+$. *The correspondence* $X \leftrightarrow (S,\bar{S})$ *is one-one, the pair* $(S,\bar{S})$ *being uniquely determined by the relations* $S = H^- \vee X$ *and* $\bar{S} = H^+ \vee X$.

COROLLARY 3.1 *Representation (3.4) may also be written in any of the following four equivalent ways:*

$$X = S \ominus \bar{S}^{\perp} \tag{3.5}$$

$$X = \bar{S} \ominus S^{\perp} \tag{3.6}$$

$$X = \bar{E}^{S}\bar{S} \tag{3.7}$$

$$X = \bar{E}^{\bar{S}}S \tag{3.8}$$

Proof. Theorem 3.1 in Part I states that X is a splitting subspace if and only if (3.5) or (3.6) holds for some $S \supset H^-$ and $\bar{S} \supset H^+$, and that X and $(S,\bar{S})$ are related as in the last sentence of the present theorem. But then $S^{\perp} \subset \bar{S}$, and therefore S and $\bar{S}$

intersect perpendicularly (Proposition 2.2) and X is given by (3.4) (Proposition 2.1). Relations (3.7) and (3.8) follow by comparing (3.5) and (3.6) to (2.2). □

A subspace containing the past space $H^-$ will be called an *extended past space* and one containing the future $H^+$ an *extended future space*. We shall say that an extended past space S (extended future space $\bar{S}$) is *minimal* if there is a minimal splitting subspace X such that $S = H^- \vee X$ ($\bar{S} = H^+ \vee X$). This definition coincides with that in [5] for regular Markovian splitting subspaces (to be defined shortly).

PROPOSITION 3.1 *An extended past space* S *is minimal if and only if*

$$S \subset (N^+)^\perp \tag{3.9}$$

*and an extended future space* $\bar{S}$ *is minimal if and only if*

$$\bar{S} \subset (N^-)^\perp . \tag{3.10}$$

Proof. (if): In the first case, choose $\bar{S} = S^\perp \vee H^+$ and in the second case, $S = \bar{S}^\perp \vee H^-$. Then it follows from Theorem 3.1 in Part I that $X := S \cap \bar{S}$ is the required X.

(only if): Since a minimal splitting subspace X is contained in $H^\square$, we have $S = H^- \vee X \subset (N^+)^\perp$ and $\bar{S} = H^+ \vee X \subset (N^-)^\perp$. □

The reason for this definition of minimality will become clear in Section 5. Note that an arbitrary pair $(S,\bar{S})$ of perpendicularly intersecting minimal extended past and future spaces will not necessarily define a *minimal* splitting subspace X, although any such X will be contained in the frame space $H^\square$. For X to be minimal, the pairing of S and $\bar{S}$ must be done in a more precise manner. This leads to the concepts of observability and constructibility as defined by Ruckebusch.

A splitting subspace is said to be *observable* if $\bar{E}^X H^+ = X$ and *constructible* if $\bar{E}^X H^- = X$. In Part I, we showed that $X = S \cap \bar{S}$ is observable if and only if

$$\bar{S} = S^\perp \vee H^+ \tag{3.11}$$

and constructible if and only if

$$S = \bar{S}^\perp \vee H^- . \tag{3.12}$$

PROPOSITION 3.2 *Let* X *be a splitting subspace, and let* S *and* $\bar{S}$ *be the corresponding extended past and future spaces. Then the following conditions are equivalent:*

(i) X *is minimal*

(ii) X *is observable and constructible*

(iii) X *is observable and* S *is minimal*

(iv) X *is constructible and* $\bar{S}$ *is minimal*

For a proof of the equivalence of (i) and (ii), see [6] or Proposition 4.2 below. Keeping Proposition 3.1 in mind, the rest follows from Theorem 3.1 of Part I. Note that minimality of X imposes a definite pairing of S and $\bar{S}$. For example, $X_-$ corresponds to $S_- := H_-$ and $\bar{S}_- := (N^-)^\perp$ and $X_+$ to $S_+ := (N^+)^\perp$ and $\bar{S}_+ := H^+$.

Among all minimal splitting subspaces X we shall eventually only be able to use those which are *Markovian*, i.e., those which satisfy condition (1.2). As pointed out in Part I (the original reference is [6]), X is Markovian if and only if the corresponding extended spaces S and $\bar{S}$ satisfy the following two conditions:

$$U_t S \subset S \quad \text{for all} \quad t \le 0 \tag{3.13}$$

$$U_t \bar{S} \subset \bar{S} \quad \text{for all} \quad t \ge 0 \quad . \tag{3.14}$$

i.e., S must be left invariant and $\bar{S}$ right invariant (under the shift $U_t$). Note that, if X is observable, we need only impose condition (3.13); then, in view of (3.11), (3.14) will follow automatically. In the same way, only (3.14) is needed when X is constructible. It is easy to see that for a Markovian splitting subspace $S = X^-$ and $\bar{S} = X^+$, i.e., $X^-$ and $X^+$ intersect perpendicularly. In general, S and $\bar{S}$ are given by $S = S_1 \oplus S_2$ and $\bar{S} = \bar{S}_1 \oplus \bar{S}_2$ where $S_2$ and $\bar{S}_2$ are *doubly invariant*, i.e., they satisfy both (3.13) and (3.14), whereas $S_1$ and $\bar{S}_1$ are only simply invariant. Then $\cap_{t\in R} U_t S = S_2$ and $\cap_{t\in R} U_t \bar{S} = \bar{S}_2$, i.e., $S_2$ and $\bar{S}_2$ are the purely deterministic parts of S and $\bar{S}$ respectively. If $S_2 = 0$ ($\bar{S}_2=0$), we shall say that S($\bar{S}$) is *purely nondeterministic*. A splitting subspace for which both extended spaces S and $\bar{S}$ are purely nondeterministic will be called *regular*. In view of Lemma 2.1 of Part I and the fact that $S \subset \bar{S}$, both S and $\bar{S}$ are full range whenever X is regular. The usefulness of this concept is that the class of regular Markovian splitting subspaces is in one-one correspondence with a certain class of pairs $(W,\bar{W})$ of full-rank factors of y, W being stable and $\bar{W}$ strictly unstable. We shall return to this in Section 5.

To insure that all minimal splitting subspaces are regular, we need to introduce the concept of noncyclicity. We say that the given process y is *noncyclic* if it has nontrivial junk spaces, i.e. $N^- \ne 0$ and $N^+ \ne 0$, and *strictly noncyclic* if both $N^-$ and $N^+$ are full range.

PROPOSITION 3.3 *Let* y *be strictly noncyclic. Then all minimal splitting subspaces are regular.*

Proof. If $N^-$ and $N^+$ are full range, $(N^-)^\perp$ and $(N^+)^\perp$ are purely nondeterministic (Part I; Lemma 2.1). Hence the proposition follows from (3.9) and (3.10). □

## 4. HANKEL OPERATORS AND MINIMALITY OF SPLITTING SUBSPACES

Given an arbitrary splitting subspace X, there is a unique pair $(S,\bar{S})$ of perpendicularly intersecting subspaces such that $S \supset H^-$, $\bar{S} \supset H^+$ and $X = S \cap \bar{S}$ (Theorem 3.1). We shall try to gain some further insight into the conditions under which the state space X is minimal by applying the basic concepts of deterministic realization theory [3,4]. Let $G : \bar{S} \to S$ and $G^* : S \to \bar{S}$ be the *Hankel operators* $G = E^S|_{\bar{S}}$ and $G^* = E^{\bar{S}}|_S$, where $A|_B$ denotes the restriction of the operator A to the domain B. Furthermore, let $\bar{R}(A)$ denote the closure of the range and $N(A)$ the null space of A.

PROPOSITION 4.1 *The operators* G *and* G* *are adjoints. Moreover,* $N(G) = S^\perp$, $N(G^*) = \bar{S}^\perp$ *and*

$$\bar{R}(G) = \bar{R}(G^*) = X , \tag{4.1}$$

*where* $X = S \cap \bar{S}$.

Proof. To see that the first statement is true, note that for all $s \in S$ and $\bar{s} \in \bar{S}$

$$\langle G\bar{s}, s\rangle_S = \langle \bar{s}, s\rangle = \langle \bar{s}, G^* s\rangle_{\bar{S}} ,$$

where $\langle\cdot,\cdot\rangle_S$ is the inner product restricted to S. Relations (4.1) follow from (3.7) and (3.8), and, since $S^\perp \subset \bar{S}$ (Proposition 2.2), $N(G) = S^\perp$. In the same way, it is seen that $N(G^*) = \bar{S}^\perp$. □

By Corollary 3.1, $X = S \ominus \bar{S}^\perp = \bar{S} \ominus S^\perp$. Consequently, X is isomorphic to the quotient spaces $S/\bar{S}^\perp$ and $\bar{S}/S^\perp$, i.e.,

$$X \simeq \bar{S}/N(G) \quad \text{and} \quad X \simeq S/N(G^*) \tag{4.2}$$

(Proposition 4.1). Actually, by choosing representations in the equivalence classes properly, we may identify X directly with the quotient spaces in (4.2).

The formulas (4.1) show that the factorization of G and G* through X described by the commutative diagrams

(4.3)

(where $J : X \to S$ and $\bar{J} : X \to \bar{S}$ are the insertion maps $Jx = \bar{J}x = x$) are *canonical* in the sense that $E^X|_{\bar{S}}$ ($E^X|_S$) maps *onto* a dense subset of X and $J(\bar{J})$ is *one-one*. This canonical property, as well as (4.2), illustrates the fact that X is a minimal $(S,\bar{S})$-splitting subspace. (Cf[3; p.259].)

We shall now relate this approach to the concepts of observability and constructibility defined in Section 3 by investigating under what conditions X is a *minimal* splitting subspace [with respect to $(H^-,H^+)$]. The following lemma provides us with two splitting subspaces which are potentially smaller.

LEMMA 4.1 *Consider the restricted Hankel operators* $G^+ := G|_{H^+}$ *and* $G^- := G^*|_{H^-}$. *Then* $\bar{R}(G^+)$ *and* $\bar{R}(G^-)$ *are splitting subspaces contained in* X.

Proof. It suffices to prove the statement concerning $G^+$; then the one about $G^-$ follows by symmetry. Trivially $R(G^+) \subset R(G)$, and hence $\bar{R}(G^+) \subset X$ (Proposition 4.1). Obviously, the predictor space $\bar{R}(G^+) = \bar{E}^S H^+$ is an $(S,H^+)$-splitting subspace, and hence also a splitting subspace (Lemma 2.3). □

It is therefore natural to consider the factorizations of $G^+$ and $G^-$ through X induced by those in (4.3), i.e.,

(4.4)

where $0 := E^X|_{H^+}$ is the *observability operator* and $C := E^X|_{H^-}$ the *constructibility operator*. The first factorization is canonical if and only if 0 maps onto a dense subset of X, i.e., X is observable, in which case $X = \bar{R}(G^+)$. The same statement holds for the second factorization (4.4) if we exchange 0 for C, $G^+$ for $G^-$ and observable for constructible. Consequently, minimality of X is equivalent to both factorizations being canonical (Proposition 3.2). Note that, unlike the situation in deterministic realization theory [3,4], *two* factorizations are needed. In this respect, constructibility is a form of dual observability and does not correspond to reachability. The "reachability operators" J and its dual $\bar{J}$ are always one-one and play no important role here. (Since in effect we are working with spaces of functionals of inputs and outputs, both diagrams are dualized as compared to the deterministic setting in the sense that all arrows and the order between the observability and reachability operators have been reversed.)

The previous discussion suggests that a splitting subspace X is observable if and only if, for a given S, $\bar{S}$ is as small as possible, and that, dually, the same holds for constructibility exchanging S for $\bar{S}$, and vice versa. To clarify this point, first consider the following lemma.

LEMMA 4.2 *Let* X *and* X' *be two splitting subspaces and let* $(S,\bar{S})$ *and* $(S',\bar{S}')$ *be the corresponding extended past and future spaces. Then* $X \subset X'$ *if and only if* $S \subset S'$ *and* $\bar{S} \subset \bar{S}'$.

Proof. In view of (3.4), the (if)-part is trivial. It remains to prove (only if). Since $S = H^- \vee X$ (Theorem 3.1), $X' \subset X$ implies that $S' \subset S$. The rest follows by symmetry. □

Consequently, in order to obtain a minimal $X = S \cap \bar{S}$, we must reduce S and $\bar{S}$ as much as possible. Now, keeping S fixed, it is easy to see that the smallest $\bar{S}$ which both contains $H^+$ and intersects S perpendicularly is $\bar{S} = H^+ \vee S^\perp$, i.e., exactly the condition (3.11) for observability. Likewise, keeping $\bar{S}$ fixed, $S = H^- \vee \bar{S}^\perp$ is the smallest subspace containing $H^-$ and intersecting $\bar{S}$ perpendicularly, yielding the constructibility condition (3.12). We shall now discuss some consequences of these observations.

First, this provides us with an alternative proof of the equivalence of conditions (i) and (ii) in Proposition 3.2, which we restate here in a somewhat stronger form.

PROPOSITION 4.2 *Let* $S \supset H^-$ *and* $\bar{S} \supset H^+$ *be two arbitrary subspaces, and set* $X = S \cap \bar{S}$. *Then* X *is a minimal splitting subspace if and only if both conditions* (3.11) *and* (3.12) *hold.*

Proof. (if): Assume that (3.11) and (3.12) hold. The $S^\perp \subset \bar{S}$, i.e., S and $\bar{S}$ intersect perpendicularly (Proposition 2.2). Hence, X is a splitting subspace (Theorem 3.1). It remains to show that X is minimal. In order to reduce X either S or $\bar{S}$ or both must be reduced while neither can be enlarged (Lemma 4.2). But this is inconsistent with (3.11) and (3.12), since S and $\bar{S}$ will not be reduced. Hence, X is minimal.

(only if): Suppose that X is a minimal splitting subspace. Then the pair $(S,\bar{S})$ is uniquely determined; we must have $S = H^- \vee X$ and $\bar{S} = H^+ \vee X$ as in Theorem 3.1. In fact, under the given conditions, neither S nor $\bar{S}$ can be smaller, and enlarging (say) S (as given by Theorem 3.1), to $S \oplus Z$ will yield $(S \oplus Z) \cap \bar{S} = X \oplus Z \neq X$, since $Z \subset S^\perp \subset \bar{S}$ (Proposition 2.2). Now assume that (say) (3.11) does not hold. Since $\bar{S}$ contains $H^+$ and intersects S perpendicularly, $\bar{S} \supset H^+ \vee S^\perp$. Then, by the discussion above, $\bar{S}$ can be reduced to

$H^+ \vee S^\perp$. Since $\bar{S} = H^+ \vee X$, this would reduce X too, contradicting minimality. □

Secondly, given any splitting subspace, the above procedure defines an algorithm by which a minimal splitting subspace can be constructed.

PROPOSITION 4.3 *Let* X' *be an arbitrary splitting subspace and define* $\bar{S} := H^+ \vee (H^- \vee X')^\perp$ *and* $S := H^- \vee \bar{S}^\perp$. *Let* $X = S \cap \bar{S}$. *Then* X *is a minimal splitting subspace. If* X' *is Markovian, then so is* X.

Proof. (i) To show that X is a minimal splitting subspace, it only remains to prove that $\bar{S} = H^+ \vee S^\perp$ (Proposition 4.2). The given condition $S = H^- \vee \bar{S}^\perp$ implies that $S^\perp \subset \bar{S}$, and consequently $H^+ \vee S^\perp \subset \bar{S}$. Set $S' = H^- \vee X'$. Then, since $(S')^\perp \subset \bar{S}$, S' contains $H^-$ and intersects $\bar{S}$ perpendicularly. But S is the smallest subspace with this property. Hence, $S \subset S'$, i.e., $(S')^\perp \subset S$ , which implies that $\bar{S} \subset H^+ \vee S^\perp$. Hence, $\bar{S} = H^+ \vee S^\perp$ as required.

(ii) To prove that X is Markovian, we must show that (3.13) and (3.14) hold. Since X' is Markovian, S' satisfies (3.13). Then $(S')^\perp$ satisfies (3.14) and, since the same is true for $H^+$, $\bar{S}$ also satisfies (3.14). Now, by a symmetric argument, it is seen that $S := H^- \vee \bar{S}^\perp$ satisfies (3.13). □

## 5. SPECTRAL REPRESENTATION OF REGULAR MARKOVIAN SPLITTING SUBSPACES

Since the given process y is stationary, mean-square continuous and purely nondeterministic, it has the spectral representation

$$y(t) = \int e^{st}\, d\hat{y}(s) \tag{5.1}$$

where integration is over the imaginary axis and $d\hat{y}$ is an orthogonal stochastic vector measure such that

$$E\{d\hat{y}(i\omega)d\hat{y}(i\omega)^*\} = \Phi(i\omega)d\omega, \tag{5.2}$$

$\Phi$ being the $m\times m$ spectral density. Assume that $\Phi$ has rank $p \le m$. Then a *full-rank spectral factor* is any $m\times p$-matrix solution of

$$W(s)W(-s)' = \Phi(s) \tag{5.3}$$

such that rank $W = p$. To any such spectral factor, we may associate a p-dimensional Wiener process

$$u(t) = \int \frac{e^{st}-1}{s}\, d\hat{u}(s) \ ; \ d\hat{u} = W^{-L} d\hat{y} \ , \tag{5.4}$$

which spans the whole space H (see Part I). ($W^{-L}$ is the left inverse of W.) Let $\mathcal{U}$ denote the class of all such Wiener processes. Let $H^-(du)$ and $H^+(du)$ be the closed linear hulls in H of respectively the past and future of the process u. We shall say that W is *stable* if its rows belong to the Hardy space $H_2^+$ of p-dimensional functions f which are analytic in the open right half-plane and such that $\int \|f(\sigma + i\omega)\|^2 d\omega$ is uniformly bounded for all $\sigma > 0$, and *strictly unstable* if its rows belong to the conjugate Hardy space $H_2^-$. Let $\mathcal{U}^+$ and $\mathcal{U}^-$ be the subclasses of $\mathcal{U}$ corresponding to stable and strictly stable spectral factors respectively.

LEMMA 5.1 *There is a one-one correspondence between stable full-rank spectral factors* W *(determined modulo multiplication with a constant unitary matrix) and left invariant [i.e., satisfying* (3.13)*] and purely nondeterministic subspaces* $S \supset H^-$. *The subspace* S *is related to* W *by*

$$S = H^-(du) \tag{5.5}$$

*where* $u \in \mathcal{U}^+$ *is the Wiener process corresponding to* W.

LEMMA 5.2 *There is a one-one correspondence between strictly unstable full-rank spectral factors* $\bar{W}$ *and right invariant, [i.e., satisfying* (3.14)*] and purely nondeterministic subspaces* $\bar{S} \supset H^+$. *The subspace* $\bar{S}$ *is related to* $\bar{W}$ *by*

$$\bar{S} = H^+(d\bar{u}) \tag{5.6}$$

*where* $u \in \mathcal{U}^-$ *is the Wiener process corresponding to* $\bar{W}$.

The proofs of these lemmas can be found in Part I, where somewhat differently stated but equivalent results are given.

It was shown in Part I that, to each $u \in \mathcal{U}$,

$$\int f\, d\hat{u} \xmapsto{Q_u} f \tag{5.7}$$

defines an isomorphism between H and the space $L_2^p(I)$ of all p-dimensional row vector functions which are square-integrable on the imaginary axis. By the Paley-Wiener Theorem [7,8,11], the isomorphic image of $H^-(du)$ under $Q_u$ is precisely the Hardy space $H_2^+$.

Now, in view of Theorem 3.1 and Lemmas 5.1 and 5.2, X is a regular Markovian splitting subspace if and only if

$$X = H^-(du) \cap H^+(d\bar{u}) \tag{5.8}$$

for some $u \in \mathcal{U}^+$ and $\bar{u} \in \mathcal{U}^-$ such that $H^-(du)$ and $H^+(d\bar{u})$ intersect perpendicularly.

LEMMA 5.3 *Let* $u \in \mathcal{U}^+$ *and* $u \in \mathcal{U}^-$, *and let* $W$ *and* $\overline{W}$ *be the corresponding spectral factors. Set*

$$K = \overline{W}^{-L}W. \tag{5.9}$$

*Then* $S := H^-(du)$ *and* $\bar{S} := H^+(d\bar{u})$ *intersect perpendicularly if and only if* $K$ *is an inner function.* (See Part I for definition.)

Proof. The subspaces $S$ and $\bar{S}$ intersect perpendicularly if and only if $\bar{S}^\perp \subset S$ (Proposition 2.2), i.e., $H^-(d\bar{u}) \subset H^-(du)$, the isomorphic image of which (under $Q_u$) is $H_2^+K \subset H_2^+$; to see this use Lemma 4.3 of Part I. Since $\bar{S}^\perp$ is full range and left invariant, Lemma 4.1 of Part I can be used to show that $Q_u\bar{S}^\perp = H_2^+K$ is a full-range invariant subspace of $L_2^p(I)$ [7]. Hence, in view of the Beurling-Lax-Helson Theorem [7], $H_2^+K \subset H_2^+$ if and only if $K$ is inner. □

By Corollary 3.1, (5.8) can be written $X = H^-(du) \ominus H^-(d\bar{u})$, the isomorphic image of which (under $Q_u$) is $Q_uX = H_2^+ \ominus (H_2^+K)$. Consequently

$$X = \int (H_2^+K)^\perp d\hat{u}, \tag{5.10}$$

where $u \in \mathcal{U}^+$ is the Wiener process corresponding to $W$, and the superscript $\perp$ denotes orthogonal complement in $H_2^+$. We collect these observations in the following theorem, which is a slight generalization of Theorem 4.1 in Part I.

THEOREM 5.1 *The subspace* $X$ *is a regular Markovian splitting subspace if and only if (5.10) holds for some pair* $(W,\overline{W})$ *of full rank spectral factors such that* $W$ *is stable,* $\overline{W}$ *is strictly unstable, and* $K := \overline{W}^{-L}W$ *is inner.*

In particular, if $y$ is strictly noncyclic, all minimal Markovian splitting subspaces are given by Theorem 5.1 (Proposition 3.3).

## 6. SPECTRAL DOMAIN CRITERIA FOR OBSERVABILITY, CONSTRUCTIBILITY AND MINIMALITY

Theorem 5.1 provides us with a procedure to find all regular Markovian splitting subspaces: All possible pairs $(W,\overline{W})$ of full-rank spectral factors with $W$ stable, $\overline{W}$ strictly unstable, and $K = \overline{W}^{-L}W$ inner, inserted into (5.9) and (5.10), generate the whole family of such splitting subspaces. But how can we decide whether such a pair will provide an observable, or a constructible or a minimal splitting subspace? We need to translate the geometric criteria of Section 3 into spectral domain language.

To this end, first note that W is a stable full-rank spectral factor if and only if it can be written

$$W = W_-\theta, \tag{6.1}$$

where $\theta$ is an inner function (for $H_2^+$) and $W_-$ is the unique outer spectral factor, corresponding to $S = S_- := H^-$. Similarly, $\bar{W}$ is a strictly unstable full-rank spectral factor if and only if it has the representation

$$\bar{W} = \bar{W}_+ \bar{\theta}, \tag{6.2}$$

where $\bar{\theta}$ is conjugate inner (inner for $H_2^-$) and $\bar{W}_+$ is the unique conjugate outer spectral factor, which corresponds to $\bar{S} = \bar{S}_+ := H^+$.

PROPOSITION 6.1 *Let X be a regular Markovian splitting subspace, let (6.1) and (6.2) be the corresponding spectral factors, and let K be defined by (5.9). Then X is observable if and only if $K^{-1}$ and $\bar{\theta}$ are right coprime and constructible if and only if K and $\theta$ are right coprime.*

Proof. Let S and $\bar{S}$ be respectively the extended past and future spaces of X, and let u and $\bar{u}$ be the Wiener processes defined through the representations (5.5) and (5.6). Using Lemma 4.3 in Part I, it is seen that the isomorphic image of the constructibility condition (3.12) under the map $Q_u$ is $H_2^+ = (H_2^+K) \vee (H_2^+\theta)$, which holds if and only if K and $\theta$ are right coprime. In the same way, using the map $Q_{\bar{u}}$, the observability condition (3.12) is seen to be equivalent to $K^{-1}$ and $\bar{\theta}$ being right coprime. □

In order to apply conditions (iii) and (iv) of Proposition 3.2, we also need to characterize minimality of S and $\bar{S}$ in the spectral domain. We shall say that a stable (strictly unstable) full-rank spectral factor is *minimal* if it corresponds to a minimal extended past (future) space. Assume that y is strictly noncyclic. Then $X_-$ and $X_+$ are regular splitting subspaces (Proposition 3.3), and consequently there are associated pairs $(W_-,\bar{W}_-)$ and $(W_+,\bar{W}_+)$ of full-rank spectral factors (Theorem 5.1). Then define $K_-$ and $K_+$ to be the corresponding functions (5.9). We showed in Part I that in the scalar case, all minimal Markovian splitting subspaces have the same K, so that in particular $K_- = K_+$, but it is still unclear to us what happens in the general case. Moreover, let $\theta_+$ be the $\theta$ in (6.1) corresponding to $W_+$ and $\bar{\theta}_-$ the $\bar{\theta}$ in (6.2) corresponding to $\bar{W}$.

PROPOSITION 6.2 *Suppose that y is strictly noncyclic. Let W be a stable full-rank spectral factor, and define $J = W^{-L}W_+$. Then J is inner and the following conditions are equivalent:*

(i) W *is minimal*

(ii) $\theta$ *is a left inner divisor of* $\theta_+$, *i.e., there is an inner function* $\phi$ *such that* $\theta\phi = \theta_+$.

(iii) J *and* $K_+$ *are right coprime.*

Proof. The equivalence between (i) and (ii) follows from Proposition 4.1 in Part I. The equivalence of (i) and (iii) follows from Theorem 5.1 in Part I. □

The fact that (i) and (ii) are equivalent was first proven in [5] in the scalar case. The dual version of Proposition 6.2 goes as follows.

PROPOSITION 6.3 *Suppose that* y *is strictly noncyclic. Let* $\overline{W}$ *be a strictly unstable full-rank spectral factor, and define* $\overline{J} = \overline{W}^{-L}\overline{W}_-$. *Then* $\overline{J}$ *is conjugate inner and the following conditions are equivalent:*

(i) $\overline{W}$ *is minimal*

(ii) $\bar{\theta}$ *is a left inner divisor of* $\bar{\theta}_-$

(iii) $\overline{J}$ *and* $K_-^{-1}$ *are right coprime.*

## 7. STATE EQUATIONS FOR FINITE DIMENSIONAL MARKOVIAN SPLITTING SUBSPACES

If the matrix function K is rational, the representation (5.10) is particularly simple to realize.

PROPOSITION 7.1 *Let* X *be a regular Markovian splitting subspace. Then* X *is finite dimensional if and only if* K, *defined by* (5.9), *is rational.*

Proof. As pointed out in Section 4, $X = \bar{R}(G)$, where $G = E^S|_{\bar{S}}$. Let u be the Wiener process associated to S through relation (5.5). Then, under $Q_u$, $S \simeq H_2^+$ and $\bar{S} \simeq H_2^- K$ (Part I: Lemma 4.3), and consequently, using the notations of Section 5 of Part I, $G$ is isomorphic to the Hankel operator $H_K$. Since therefore $X \simeq \bar{R}(H_K)$, dim $X < \infty$ if and only if K is rational [4; p.395, Theorem 3.8]. (To prove the if-part, follow the lines of the last part of the proof of Lemma 7.1.) □

In general, there are no finite dimensional splitting subspaces. One very important exception, however, is the case when the spectral density of y is rational.

PROPOSITION 7.2 *Let the spectral density* $\Phi$ *be rational. Then all minimal splitting subspaces are finite dimensional.*

Proof. It is a well-known fact from Wiener filtering theory [8] that, under the stated condition, the closed linear hull in H of the predictors $\{E^{H^-}y(t);\ t \geq 0\}$ is finite-dimensional. But this is precisely $X_- := \bar{E}^{H^-}H^+$, and hence $\dim X_- < \infty$. Then, by symmetry, $X_+ := \bar{E}^{H^+}H^-$ is also finite dimensional, and consequently so is the frame space $H^\square$. Since all minimal splitting subspaces are contained in $H^\square$ [see (3.3)], the desired result follows. □

COROLLARY 7.1 *Let the spectral density* $\Phi$ *be rational. Then all splitting subspaces contained in the frame space are finite dimensional.*

Now consider a finite dimensional regular Markovian splitting subspace with corresponding spectral factors $(W,\bar{W})$. Then, by Proposition 7.1, $K = \bar{W}^{-1}W$ is rational. Following Forney [9], we consider the vector space $V_K$ of 2p-tuples over the field of rational functions, generated by the columns of the $2p \times p$ matrix $\tilde{K} = [I,K']'$. Then use the algorithm presented in [9] to find a minimal polynomial basis in $V_K$. Let $n_i$ be the degree of the i:th element in this basis, i.e., the greatest degree of the 2p-polynomials in the i:th 2p-tuple. Then the minimality of the basis means that $n := n_1 + n_2 + \ldots + n_p$ is as small as possible. Let $\tilde{Q} := [Q',\bar{Q}']'$ be the $2p \times p$ polynomial matrix formed by this basis. Then

$$K = \bar{Q}Q^{-1}, \tag{7.1}$$

where $Q$ and $\bar{Q}$ are right coprime polynomial $p \times p$ matrices which are *column proper* [10], i.e., the high order coefficient matrices $[Q]_h$ and $[\bar{Q}]_h$ are full rank ($[Q]_h$ is the constant matrix whose i:th column consists of the coefficients of $s^{n_i}$ in the i:th column of Q.) The numbers $n_1, n_2, \ldots, n_p$ are called the *indices* of Q (or of $\bar{Q}$) [9; Main Theorem and Theorem 5].

LEMMA 7.1 *Suppose that* K *is given by (7.1). Let* $f \in (H_2^+K)^\perp$ *and define* $g = fQ$. *Then* $g \in R^p[s]$, *where* $R^p[s]$ *is the class of p-dimensional (row) vectors of polynomials with real coefficients.*

Proof. In view of (7.1), the representation

$$(H_2^+K)^\perp = \{f \in H_2^+ \mid fK^{-1} \in H_2^-\} \tag{7.2}$$

[7, p.75] can be written in the form

$$(H_2^+K)^\perp = \{gQ^{-1} \in H_2^+ \mid g\bar{Q}^{-1} \in H_2^-\}. \tag{7.3}$$

Set $h := g\bar{Q}^{-1}$. Let $f \in (H_2^+K)^\perp$. Then, since $f \in H_2^+$, $g = fQ$ is analytic in the open right half plane. Similarly, since $h \in H_2^-$, $g = h\bar{Q}$ is analytic in the open left half plane. Also $f \in (H_2^+K)^\perp$ implies that f can be analytically extended to the imaginary axis

[4; p.270, Lemma 13.6], so therefore $g = fQ$ is analytic there too. Hence, $g$ is an entire function. We want to prove that $g \in \mathbb{R}^p[s]$. To this end, first note that $k := \det K$ is a Blasche product, i.e., $k$ is the product of relatively prime functions $k_i(s) := (s - s_i)^{n_i}/(s + \bar{s}_i)^{n_i}$ (where, for each function $k_i$, $s_i$ is a complex number and $n_i$ an integer), and that $K$ is an inner divisor of $kI$ [7; p.70]. Then $H_2^+ k \subset H_2^+ K$ [7; p.69] and $H_2^+ k = \cap_i (H_2^+ k_i)$ so that

$$(H_2^+ K)^\perp \subset (H_2^+ k)^\perp = \vee_i (H_2^+ k_i)^\perp. \tag{7.4}$$

Therefore, if we could prove that, for each $i$, $(H_2^+ k_i)^\perp$ is a space of rational vector functions, the same holds true for $(H_2^+ K)^\perp$, and consequently the entire function $g$ must be a vector polynomial. Since $k_i$ is a scalar inner function, it suffices to prove that an arbitrary component of any vector function in $(H_2^+ k_i)^\perp$ is rational; hence, for the rest of the proof, it is no restriction to assume that $H_2^+$ is a Hardy space of *scalar* functions. Now, using Cauchy's formula, it can be shown [8; p.34] that

$$e_j(s) = \frac{1}{s + \bar{s}_i} \left( \frac{s - s_i}{s + \bar{s}_i} \right)^j \; ; \; j = 0, 1, 2, \ldots$$

is an orthogonal basis of $H_2^+$. However, $e_j k_i = e_{j+n_i}$, and consequently $H_2^+ k_i$ is the closed linear hull of $\{e_{n_i}, e_{n_i+1}, \ldots\}$. Hence, $(H_2^+ k_i)^\perp$ is the linear span of $\{e_0, e_1, \ldots, e_{n_i-1}\}$, which is a space of rational functions as required. □

PROPOSITION 7.3 *Let* $X$ *be a finite dimensional regular Markovian splitting subspace and let* $(W, \bar{W})$ *be the corresponding pair of spectral factors. Then there is an* $m \times p$ *polynomial matrix* $P$ *such that*

$$W = PQ^{-1} \text{ and } \bar{W} = P\bar{Q}^{-1} \tag{7.5}$$

*where* $Q$ *and* $\bar{Q}$ *are defined by the (right coprime and column proper) factorization* (7.1) *of* (5.9). *The function* $P$ *has a left inverse.*

Proof. Equations (5.9) and (7.1) together yield $WQ = \bar{W}\bar{Q}$. Call this matrix function $P$; then equations (7.5) hold. Since $W$ has a left inverse, so does $P$. It remains to show that $P$ is a matrix of polynomials. Since $WK^{-1} = \bar{W}$, (7.2) implies that the rows of $W$ belong to $(H_2^+ K)^\perp$. Hence, by Lemma 7.1, $P$ is a polynomial matrix. □

LEMMA 7.2 *For* $i,j = 1, 2, \ldots, p$, *let* $p_{ij}$ *and* $q_{ij}$ *be polynomials such that* $p_{ij}/q_{ij} = (Q^{-1})_{ij}$. *Then, for each* $i$, $\deg p_{ij} - \deg q_{ij} \leq -n_i$ *with equality for some* $j$.

Proof. By Cramer's rule, $[Q^{-1}(s)]_{ij} = (-1)^{i+j}\Delta_{ji}(s)/\Delta(s)$, where $\Delta := \det Q$ and $\Delta_{ji}$ is the determinant of the matrix obtained by deleting row j and column i in Q. Hence, $\Delta_{ji}$ is a sum of products of one element from each of all columns of Q except the i:th, and consequently $\deg \Delta_{ji} \le n - n_i$. Since $[Q]_h$ is full rank, for each i there is a j such that equality holds. In fact, in each column i of Q there is a row j such that deleting row j and column i in $[Q]_h$ leaves a nonsingular matrix. Hence in forming $\Delta_{ji}$ there is at least one product that contains only factors of highest degree. Since $\deg \Delta = n$, the lemma follows. □

THEOREM 7.1 *Let* K *be given by (7.1), and let* $n_1, n_2, \dots, n_p$ *be the indices of* Q. *Then*

$$(H_2^+K)^\perp = \Gamma_{(n_1, n_2, \dots, n_p)}Q^{-1} \tag{7.6}$$

*where* $\Gamma_{(n_1, n_2, \dots, n_p)}$ *is the set of all* $g \in R^p[s]$ *such that, for each* $i = 1, 2, \dots, p$, $\deg g_i < n_i$.

Proof. From (7.3) it follows that the functions in $(H_2^+K)^\perp$ are of the form $gQ^{-1}$ where g is a row vector of polynomials (Lemma 7.1). It just remains to show that the functions g satisfying $gQ^{-1} \in H_2^+$ and $g\bar{Q}^{-1} \in H_2^-$ are precisely those in $\Gamma_{(n_1, n_2, \dots, n_p)}$. But, since Q and $\bar{Q}$ have the same indices, this follows from Lemma 7.2. □

COROLLARY 7.2 *Let* X *be a finite dimensional regular Markovian splitting subspace, let* $(W,\bar{W})$ *be the corresponding pair of spectral factors, and let* Q *and* $\bar{Q}$ *be column proper right coprime polynomial matrices satisfying*

$$\bar{Q}Q^{-1} = \bar{W}^{-L}W. \tag{7.7}$$

*Then*

$$X = \int_{-\infty}^{\infty}\Gamma_{(n_1, n_2, \dots, n_p)}(i\omega)P^{-L}(i\omega)d\hat{y}(i\omega), \tag{7.8}$$

*where* P *is the polynomial matrix* $P = WQ$, $n_1, n_2, \dots, n_p$ *are the indices of* Q, *and* $\Gamma$ *is defined as in Theorem 7.1. The dimension of* X *is equal to* $n := n_1 + n_2 + \dots + n_p$.

Proof. Theorems 5.2 and 7.1 imply that

$$X = \int_{-\infty}^{\infty}\Gamma_{(n_1, n_2, \dots, n_p)}(i\omega)Q^{-1}(i\omega)d\hat{u}(i\omega). \tag{7.9}$$

Since $W = PQ^{-1}$, $d\hat{u} = QP^{-L}d\hat{y}$, and consequently (7.8) follows. By Proposition 7.3, P is a polynomial matrix. The statement concerning dim X is immediate. □

Next we proceed to find a basis in X. To this end, first define the $n \times p$ polynomial matrix

$$\Pi(s) = \text{diag}\{\pi_{n_1}(s), \pi_{n_2}(s), \ldots, \pi_{n_p}(s)\} \tag{7.10}$$

where $\pi_k$ is the k-dimensional column vector $\pi_k(s) := (1, s, s^2, \ldots, s^{k-1})'$ of powers of s, and $n_1, n_2, \ldots, n_p$ are the indices of Q. Then, since $[Q]_h$ is nonsingular, Q may be written

$$Q(s) = [Q]_h\{\text{diag}(s^{n_1}, s^{n_2}, \ldots, s^{n_p}) + \tilde{A}\Pi(s)\}, \tag{7.11}$$

where $\tilde{A}$ is the $p \times n$-dimensional constant matrix of polynomial coefficients. Now, in view of (7.8), the stochastic vector x defined by

$$x = \int_{-\infty}^{\infty} \Pi(i\omega)P^{-L}(i\omega)d\hat{y}(i\omega) \tag{7.12}$$

is a basis in X.

To obtain a dynamical representation for x, define the shift matrix

$$J = \text{diag}\{J_{n_1}, J_{n_2}, \ldots, J_{n_p}\}, \tag{7.13}$$

where $J_k$ is the companion matrix with characteristic polynomial $\chi_{J_k}(s) = s^k$ [i.e., $J_k$ is a $k \times k$ matrix with $(J_k)_{ij} = 1$ whenever $j = i + 1$ and zero otherwise], and the $p \times n$ matrix E with ones in position $(\Sigma_{i=1}^{j} n_i, j)$ and zeros elsewhere.

LEMMA 7.3 [10]. *Let the* $n \times n$ *matrix* A *and the* $n \times p$ *matrix* B *be defined by*

$$\begin{cases} A = J - E\tilde{A} \\ B = E[Q]_h^{-1} \end{cases} \tag{7.14}$$

*Then*

$$\Pi(s)Q(s)^{-1} = (sI - A)^{-1}B. \tag{7.15}$$

Proof. By definition,

$$(sI - A)\Pi(s) = (sI - J)\Pi(s) + E\tilde{A}\Pi(s).$$

But it is not hard to see that

$$(sI - J)\Pi(s) = E\,\mathrm{diag}\{s^{n_1}, s^{n_2}, \ldots, s^{n_p}\}$$

and consequently $(sI - A)\Pi(s) = BQ(s)$. □

The pair (A,B) defined by (7.14) is the Luenburger *controllable canonical form*, and $n_1, n_2, \ldots, n_p$ are the controllability indices [9,10].

THEOREM 7.2 *Let* rank $\Phi = p$. *Then to each regular Markovian splitting subspace* X *of dimension* $n < \infty$ *there corresponds a stochastic realization*

$$\begin{cases} dx = Axdt + Bdu & (7.16a) \\ y = Cx, & (7.16b) \end{cases}$$

*where* A, B *and* C *are constant matrices of dimensions* $n \times n$, $n \times p$ *and* $m \times n$ *respectively, such that, for all* $t \in R$, x(t) *is a basis in* $U_tX$, $\{u(t)\ ;\ t \in R\}$ *is a Wiener process satisfying* $H^-(du) = S = H^- \vee X$, A *is a stability matrix,* (C,A) *is observable, and* (A,B) *is in controllable canonical form with controllability indices* $n_1, n_2, \ldots, n_p$. *The triplet* (A,B,C) *is determined in the following way: Let* (7.1) *be a column proper and right coprime factorization of the inner function* K *corresponding to* X, *and let* $n_1, n_2, \ldots, n_p$ *be the indices of* Q. *Then* A *and* B *are defined by* (7.14) *and* C *as the coefficient matrix of* P, *i.e.,*

$$C\Pi(s) = P(s). \tag{7.17}$$

Proof. Define the n-dimensional process $\{x(t)\ ;\ t \in R\}$ with components $x_k(t) := U_t x_k$, $k = 1, 2, \ldots, n$, x being the stochastic vector (7.12). Then x(t) is a basis in $X_t := U_tX$ for all $t \in R$, and, since $P^{-L}d\hat{y} = Q^{-1}d\hat{u}$,

$$x(t) = \int_{-\infty}^{\infty} e^{i\omega t}\Pi(i\omega)Q^{-1}(i\omega)d\hat{u}(i\omega), \tag{7.18}$$

where the Wiener process $\{u(t)\ ;\ t \in R\}$ is determined by (5.5) and Theorem 3.1. Now, inserting (7.15) into this and transforming to the time domain, we obtain

$$x(t) = \int_{-\infty}^{t} e^{A(t-\tau)}Bdu(\tau), \tag{7.19}$$

which is the integrated version of (7.16a). In fact, since det Q is the characteristic polynomial of A [9], all eigenvalues of A lie in the open left half plane, and, since $\tilde{Q}$ is a minimal basis in $V_K$ (i.e., n is as small as possible), the pair (C,A) is observable. To establish (7.16b), first note that, in view of (3.3), $y(o) \in X$; hence, there is a matrix C such that $y(o) = Cx$. Applying the operator $U_t$ to this, (7.16b) follows. It is clear from (7.12) that C is given by (7.17). □

Note that we can obtain a whole class of equivalent realizations (7.16) through transformations of the type

$$(A,B,C) \to (T^{-1}AT,\ T^{-1}B,\ CT),$$

where T is any nonsingular constant matrix.

The complete analysis of this section can be repeated with time reversed. Indeed by replacing Q by $\bar{Q}$ in the derivation above we obtain a realization (7.16) in which the matrix A has all its eigenvalues in the open right half plane. To justify the exchange of Q and $\bar{Q}$, observe that (5.11) may also be written

$$X = \int (H_2^- K)^{\perp} d\hat{u},$$

where now $\perp$ denotes the orthogonal complement in $H_2^-$. Then the situation is completely symmetric.

ACKNOWLEDGEMENTS. This work was supported partially by the National Science Foundation under grant ENG-7903731 and partially by the Air Force Office of Scientific Research under grant AFOSR-78-3519.

The authors would like to thank Professor Sanjoy Mitter for bringing [9] to our attention and Professor Paul Fuhrman for suggestions on improving the proof of Lemma 7.1.

## REFERENCES

1. A. Lindquist and G. Picci, Realization theory for multivariate stationary Gaussian processes I: State space construction, *Proc. 4th Intern. Symp. Math. Theory of Networks and Systems*, July 1979, Delft, Holland, pp. 140-148.

2. Yu. A. Rozanov, *Stationary Random Processes*, Holden Day, 1967.

3. R.E. Kalman, P.L. Falb and M.A. Arbib, *Topics in Mathematical Systems Theory*, McGraw Hill, 1969.

4. P.A. Fuhrman, *Linear Operators and Systems in Hilbert Space*, McGraw Hill (to be published).

5. G. Ruckebusch, Factorisations minimales de densités spectrales et représentations markoviennes, *Proc. 1re Colloque AFCET-SMF*, Palaiseau, France, 1978.

6. A. Lindquist, G. Picci and G. Ruckebusch, On minimal splitting subspaces and Markovian representations, *Mathematical Systems Theory*, 12(1979), 271-279.

7. H. Helson, *Lecture Notes on Invariant Subspaces*, Academic Press, 1964.

8. H. Dym and H.P. McKean, *Gaussian Processes, Function Theory, and the Inverse Spectral Problem*, Academic Press, 1976.

9. G.D. Forney, Jr., Minimal bases of rational vector spaces, with applications to multivariable linear systems, *SIAM J. Control*, 13 (May 1975), 453-520.

10. W.A. Wolovich, *Linear Multivariable Systems*, Springer-Verlag, 1974.

# ON THE MINIMAL STATE SPACE DIMENSION OF NON-STATIONARY GAUSSIAN PROCESSES IN DISCRETE TIME

L. E. Zachrisson

Math. Dept., Royal Inst. Techn., S-100 44 Stockholm, Sweden

ABSTRACT

After the problem formulation in Section II a preliminary problem reduction is made in Section III. Its result is summed up in Prop. 1 of Section IV. It turns out that the key problem in answering the minimality question is the following: Suppose that the <u>subdiagonal</u> elements $m_{ts}$ $(t \geq s \geq 1)$ of a matrix $M$ are equal to $\sum_{i=1}^{n} c_i(t)d_i(s) = (c(t),d(s))_n$ respectively. What is the minimal $n$ for such a representation to be possible? Answer: The Semirank of $M$ as defined in Section V.

## 1. INTRODUCTION

The problem to model an input - output process as something evolving with discrete or continuous time in a vector-space with a minimal number of dimensions (the minimal realization problem) has almost exclusively been treated under the assumption that the evolution operator is time-invariant. There do exist notable exceptions, for instance the well-known book by R.W. Brockett [1] and work using the concept of Resolution Hilbert Space, see R. Saeks [2]. However, mainly deterministic problems have been in focus. The present contribution gives a discussion of a rather general but elementary stochastic case, involving Gauss-Markov processes in discrete time. (Cf. Section 5 in [3]). For that purpose the concept of <u>semi-rank</u> is introduced, another variant of the rank concept than that of <u>displacement rank</u> [4]. A very recent report on the possibilities to beautiful mathematics that the



*D. G. Lainiotis and N. S. Tzannes (eds.), Advances in Control, 130–138.*

time-invariant stochastic problem offers has been given earlier in this session [5].

## 2. FORMULATION OF THE PROBLEM

Let $\{y_t\}$, where $y(t,\omega) \in R^m$, be a Gaussian centered process $(t = 0,1,2,\ldots,T, \leq \infty)$ which can be looked upon as the output of a Gauss-Markov vector process:

$$\left.\begin{aligned} &x(t+1,\omega) = A(t)x(t,\omega) + B(t+1)v(t+1,\omega) \\ &y(t,\omega) = C(t)x(t,\omega) \\ &(t = 0,1,2,\ldots,T, \leq \infty) \\ &x(0,\omega) = 0 \end{aligned}\right\} \qquad (1)$$

Here $x(t,\omega) \in R^n$ ; $y(t,\omega) \in R^m$ ; $v(t,\omega) \in R^p$

$A(t)$ is an $(n\times n)$-matrix, $C(t)$ is $(m\times n)$, and $B(t)$ is $(n\times p)$

$v$ is Gaussian with $Ev(t) = 0$; $Ev(t)v^*(s) = I_p\delta_{ts}$

An extra assumption that we make is that <u>$A(t)$ is non-singular</u> $(t = 1,2,\ldots)$.

We ask the following questions:

α) What is the smallest $n$ making such a representation possible with intact covariances of $y$: $Q(t,s) = E(y(t)y^*(s))$? ($n$ is the dimension of the state space)

β) Give N.a.S.C. for $(A(t), B(t), C(t))$ to be minimal in the sense above.

γ) Give an algorithm for the determination of such a representation.

## 3. PRELIMINARY REDUCTIONS

There is too much arbitrariness in the model (1) to make it easy to handle, but it turns out that no real restriction is really made by making $A(t)$ equal to the identity in $R^n$ and by making $y$ and $v$ scalar (that is m=p=1). We'll indicate how this can be achieved.

<u>Step 1</u>. Transform $A$ to the identity $(n\times n)$ by the adjointness trick. With the definitions ($A^{-1}_{(t)}$ exists by assumption)

$Z(1) = \Gamma$ (invertible)

$$Z(t+1) = \Gamma A^{-1}(t)\ldots A^{-1}(t) \quad (t = 1,2,\ldots)$$

we get

$$\begin{cases} x_{new}(t+1) = x_{new}(t) + B_{new}(t+1)v(t+1) \\ y(t) = C_{new}(t)x_{new}(t) \end{cases}$$

with

$$\begin{cases} x_{new}(t) = Z(t)\, x(t) \\ B_{new}(t) = Z(t)\, B(t) \\ C_{new}(t) = C(t)\, Z^{-1}(t) \end{cases}$$

Step 2. Make $y$ scalar, by the introduction of extra points of time.

For each integer time $t$ is substituted $m$ points of time $(t,1); (t,2);\ldots,(t,m)$, all having the vector $y(t)$ in the memory (just copy from one time to the next one, that is make $B_{new} = 0$) and pick the first component of $y$ at the new time $(t,1)$, the next component of $y$ at the new time $(t,2)$ and so on. This transforms the matrix $C(t)$ to a row-vector $c^*(t)$ of format $1\times n$. The equations of the model get the form

$$\begin{cases} x(t+1) = x(t) + B(t+1)v(t+1) \\ y(t) = c^*(t)x(t) = \sum_{i=1}^{n} c_i(t)x_i(t) = (c(t),x(t)) \end{cases}$$

and the elements $q(t,s)$ of y:s covariance matrix become

$$q(t,s) = E(y(t)y(s)) = c^*(t)d(s) = (c(t),d(s)) \; (t \geq s) \tag{2}$$

with

$$d(s) = \sum_{\sigma=1}^{s} B(\sigma)B^*(\sigma)c(s) \; , \tag{3}$$

Here $d$ is, of course, a column vector of format $n\times 1$ in the same way as $c$. For scalar products in $R^n$ we use interchangeably $(c,d)$ and $c^*d$, (in rare cases $(c,d)_n$).

Step 3. Make $v$ scalar, by introducing innovations $\nu(t+1) = y(t+1) - E^t\{y(t+1)\}$ (Here $E^t$ is the operator of conditional expectations with respect to $\sigma\,(y(1),\ldots,y(t))$.)

Evidently $\nu(t+1)$ is scalar in the same way as $y(t+1)$.

Here

$$y(t) = E^t y(t) = c^*(t)E^t x(t) = c^*(t)\hat{x}(t)$$

using the Kalman notation $\hat{}$.

Furthermore, $E^t \nu(t+1) = 0$ as $\nu(t+1)$ is independent of $\sigma\{y(1),..,y(t)\}$, and we have, using rules of conditional calculus, that $\hat{x}(t) = E^t x(t+1) = E^t E^{t+1} x(t+1) = E^t \hat{x}(t+1)$ so every component of the stochastic vector

$$\hat{x}(t+1) - \hat{x}(t) = \hat{x}(t+1) - E^t \hat{x}(t+1)$$

is in the innovation space spanned by the Gaussian stochastic variable $\nu(t+1)$, that is, there exists a deterministic vector $e(t+1)$ such that

$$\left.\begin{aligned} \hat{x}(t+1) - \hat{x}(t) &= e(t+1)\nu(t+1) \\ y(t) &= c^*(t)\hat{x}(t) \end{aligned}\right\}$$

Here $\nu$ is scalar white noise, and by admitting $\hat{x}$ to the earlier role of $x$ (at the end of step 2) we get a model with scalar although not normalized white noise. However, a normalization is easy to obtain also. If $\nu(t+1) \neq 0$ (a.s.), then put $v(t+1) = \nu(t+1)/\sqrt{E\nu^2(t+1)}$ and $b(t+1) = e(t+1)\sqrt{E\nu^2}$; If $\nu(t+1) = 0$ (a.s.), then the sequence of observations $\{y(\tau)\}$ do not change if we put $e(t+1) = 0$, and having done this they do not change if we substitute $v(t+1)$ for $\nu(t+1)$ with $\|v(t+1)\| = 1$. Now we have achieved our goal and $y(t)$ is represented as

$$y(t) = c^*(t) \sum_{\tau=1}^{t} b(\tau)v(\tau) = \sum_{\tau=1}^{t} (c(t),b(\tau))v(\tau) \tag{4}$$

We see that now the innovation is $\nu(t+1) = (c(t+1),b(t+1))v(t+1)$, so it is equal to zero if $c(t+1) \perp b(t+1)$. It is natural and admissible to require $b(t+1) \perp c(t+1) \Rightarrow b(t+1) = 0$.

In equ. (4) $c(t)$ is the same as at the end of step 2, but the computation of $b$ includes determination of projections on successively generated subspaces (Kalman-Bucy filtering).

It is interesting to note that the $b$:s and $c$:s furnish a Cholesky-factorization (Lagrange factorization would be a more appropriate name) of the covariance matrix of $y$ in lower and upper triangular matrices: $Q = \mathcal{L}\mathcal{L}^*$

For equ. (4) gives

$$\sum_{t=1}^{T} u(t)y(t) = \sum_{s=1}^{T}\left[\sum_{\tau=s}^{T} u(t)(c(t),b(s))\right]v(s) = \sum_{s=1}^{T}\left(\sum_{t=s}^{T} u_t \ell_{ts}\right)v(s)$$

where

$$\ell_{ts} = \begin{cases} (c(t),b(s)) & \text{if } t \geq s \\ 0 & \text{if } t < s \end{cases}$$

so

$$u^*Qu = E\left(\sum_{t=1}^{T} u(t)y(t)\right)^2 = \sum_{s=1}^{T}\left(\sum_{\tau=1}^{T} u(t)\ell_{ts}\right)^2 = \|u^*\mathcal{L}\|^2 = u^*\mathcal{L}\mathcal{L}^*u \quad (5)$$

The Cholesky factorization is unique modulo signs of the columns of $\mathcal{L}$. If a diagonal term $\ell_{ss}$ vanishes, then the whole column (all $\ell_{ts}$) vanishes.

## 4. INSIGHTS GAINED FROM THE PRELIMINARY REDUCTIONS

From reduction step 2 above is clear that necessary for $\{y(t)\}_1^T$ to be the output of a state space process of formal dimension n is that there exist two sequences $\{c(t)\}_1^T$ and $\{d(t)\}_1^T$ of vectors in $R^n$ such that the covariances q(t,s) of y have the representation

$$E\{y(t)y(s)\} = (c(t),d(s))_n \quad (t \geq s) \tag{2}$$

From reduction step 3 is clear that it is also necessary that there exists a third sequence of n-vectors $\{b(t)\}_1^T$ such that the lower Cholesky-factor $\mathcal{L}$ of the covariance matrix Q has the form

$$\ell(t,s) = (c(t),b(s))_n \quad (t \geq s) \tag{5}$$

It is easy to show that semipositivity of Q plus condition (2) is sufficient for the existence of the vectors b such that (5) us true, so that (2) is really <u>necessary and sufficient for the existence of an "n-dimensional state representation"</u>.

Here is a sketch of a sufficiency proof by induction. The first induction step is already the typical one (apart from notations) and is finished when it is shown firstly that b(1) can be determined such that (c(t),b(1)) with t = 1,2,... form the first column of the Cholesky factor $\mathcal{L}$, and secondly that the element $q^{(2)}(t,s)$ $(t \geq s \geq 2)$ of the matrix $Q^{(2)}$ of the quadratic form

$$u^*Q^{(2)}u \triangleq u^*Qu - \left(\sum_{t=1}^{T} u(t)(c(t),b(t)\right)^2$$ can be written as

$(c(t),d^{(2)}(s))$ with an appropriate sequence of vectors $\{d^{(2)}(s)\}_2^T$. The determination of $Q^{(2)}$ is such that $u^*Q^{(2)}u$ shall not contain any term with an u(1) in it.

The value of b(1) becomes $\pm \frac{1}{\sqrt{(c(1),d(1))}}\cdot d(1)$ (if $(c(1),d(1)) > 0$; if $(c(1),d(1)) = 0$ b(1) becomes zero), and that of $d^{(2)}(s) = d(s) - d(1)\frac{(c(s),d(1))}{(c(1),d(1))}$. Now of course b(2) becomes $d^{(2)}(2)/\sqrt{(c(2),d^{(2)}(2))}$ and so on. The fact that Q is the positive semidefinite is a guarantee that $(c(t), d^{(t)}(t)) \geq 0$. The b:s become unique apart from the sign, the plus and the minus being equally good for representation.

Identifying elements in Q and $\mathcal{L}\mathcal{L}^*$ gives the simple equations

$$\left.\begin{aligned} (b(t),c(t))^2 &= (c(t),d(t)) - \sum_{s=1}^{t-1} (c(t),b(s))^2 \\ b(t) &= (d(t) - \sum_{s=1}^{t-1} b(s)\cdot(b(s),c(t))/(b(t),c(t)) \\ &\text{if } (b(t),c(t) \neq 0, \text{ else } b(t) = 0. \quad (t = 1,2,\ldots) \end{aligned}\right\} \quad (6)$$

which are enough to determine b(1), b(2),... recursively when the c:s and d:s are given.

Summing up we have

Proposition 1

Necessary and sufficient for the existence of an "n-dimensional state representation" is the existence of sequences $\{c(t)\}_1^T$ and $\{d(t)\}_1^T$ such that

$$E\{y(\bar{t})\, y(s)\} = (c(t),d(s))_n \quad (T \geq t \geq s \geq 1) \text{ (T may be infinite)}$$

Another n.a.s.c. is the existence of sequences $\{c(t)\}_0^T$ (the same as above) and $\{b(t)\}_0^T$ such that the element $\ell(t,s)$ of the lower Cholesky-factor of the covariance is

$$\ell(t,s) = (c(t),b(s))_n \quad (T \geq t \geq s \geq 1)$$

The relations between {c},{d} and {b} are given by (6). {c} and {b} give immediately a state representation of "scalar type".

## 5. THE MINIMALITY PROBLEM. SEMI-RANK

In the formulation of Proposition 1 the restriction of the equality between scalar products and matrix-elements (in the covariance matrix Q or the Cholesky-factor $\mathcal{L}$) to positions on and below the diagonal is very important. As Q is symmetric and all elements of $\mathcal{L}$ above the diagonal are zero, Q and $\mathcal{L}$ are perfectly known in spite of the restriction. Denote with U a matrix with non-zero elements strictly above the diagonal only. Then there evidently exists a U such that

$$Q + U = C\cdot D \quad \text{(equivalenly for } \mathcal{L}\text{)}$$

if below and on the diagonal $(t \geq s)$ we have

$$q(t,s) = c^*(t)ds.$$

Here $C$ is a matrix with the t:th row equal to $c^*(t)$ and $D$ is matrix with the s:th column equal to $d(s)$.

Let us make the

Definition. The semi-rank (shortened SRank) of a matrix $M$ is defined as

$$\text{SRank } M = \text{Min}\{\text{Rank}(M + U)\}$$

We remark that, as $\text{Rank}\{M + U\}$ is a non-negative integer, there exists a $U_o$ such that $\text{SRank } M = \text{Rank}(M + U_o)$

Then we have

Proposition 2:

The minimal dimension $n$ of a state representation of $y$ is equal to the semirank of the covariance matrix $Q$ and also to the semirank of its lower Cholesky-factor $\mathcal{L}$.

Proof:

If $Q + U$ has rank $n$, then there is, according to well-known theorems, a factorization of $Q + U$ in two factors $C$ and $D$, where $C$ has $n$ columns, independent of each other, and $D$ has $n$ rows, independent of each other, and this gives $q(t,s) = c^*(t)d(s)$ $(t \geq s)$; So $\text{rank}(Q + U)$ is a possible $n$ in Prop 1 (call it $n_{repr}$): $\text{Min } n_{repr} \leq \text{Min}_U \text{ Rank}(Q + U) = \text{Rank}(Q + U) = \text{Rank}(Q + U_o) = n_o$

On the other hand, the minimal $n$ (call it $n^o_{repr}$) for which sequences $\{c\}$ and $\{d\}$ (call them $\{c^o\}$ and $\{d^o\}$) exist in Proposition 1, gives a $u$ according to

$$(Q + U)_{ts} = (c^o(t),d^o(s))_{n^o_{repr}}, \text{ so } n^o_{repr} \geq \text{Min}_U \text{ Rank}(Q + U) = n_o . \quad \text{Hence } n^o_{repr} = n_o$$

This proves the first part of the proposition. As representations by scalar products of $Q$ and $\mathcal{L}$ imply each other according to Prop. 1, the proof is finished.

The definition of semi-rank does not seem promising from the algorithmic point of view. However, remembering that the rank of $Q + U_o$ is equal to the dimensionality of the space, spanned by the columns of $Q + U_o$ and that a basis of that space can be found among the column-vectors themselves, especially a "first" basis, gives an indication of a possible algorithm. Let us use $\tilde{Q}$ for $Q + U_o$, and let column $s$ of $\tilde{Q}$ be called $\tilde{q}_s$. Suppose first that $U_o$ is known. Then we get the "first" basis by using the following questionaire to be gone through sequentially:

Quest. 1. Is $\tilde{q}_1$ equal to zero?
If No, put $\tilde{q}_1$ in the basis. If Yes, put the lin comb $d_{(1)} = 0$

⋮

Quest. s. Is $\tilde{q}_s$ a lin comb of $(\tilde{q}_1,\ldots,\tilde{q}_{s-1})$?
If No, put $\tilde{q}_s$ in the basis. If Yes, determine the lin comb $d_{1s},d_{2s},\ldots)^* = d(s)$ such that $\tilde{q}_s = \sum_\nu d_{\nu s}c_\nu$ ($c_\nu$ are the columns in the present basis).

⋮

The total number of noes will give the size of the basis, that is the rank of $Q + U_o$. The (column) vectors in the basis, placed side by side, will form the matrix $C$ (column nr $\nu$ is called $c_\nu$), and the combination coefficients will form the columns of the matrix $D$.

Now, let us form a modified version of our questionnaire which will not need any knowledge of $U_o$ but will turn out to be equally effective. For the formulation of the questions we shall use the notation $P^s v$ for the "projection on the s-future of the column-vector v": $P^s(v_1,v_2,\ldots,v_{s-1},v_s,v_{s+1},\ldots) = (0,0,\ldots,0,v_s,v_{s+1},\ldots)$. (Here $v$ is written as row vector). We will call the columns in the matrix $C$, which will be constructed, the "semi-basis". Now:

Mod.quest.1. Is $P^1 q_1 (= q_1)$ equal to zero?
If No, put the $q_1$ in the semi-basis;
If Yes, put the lin comb $d_{(1)} = 0$.

⋮

Mod.quest.s. Is $P^s q_s$ a linear combination of $(P^s q_1, P^s q_2,\ldots,P^s q_{s-1})$?
If No, put $P^s q_s$ in the semi-basis. If Yes, determine the combination column $d_{(s)} = (d_{1s},d_{2s},\ldots)^*$ such that $P^s q_s = \sum_\nu d_{\nu s} P^s c_\nu$

⋮

We observe that each time we have a yes in the first questionnaire, then there is a yes also in the modified questinnaire, because $\tilde{q}_s = \sum_\nu d_{\nu s} c_\nu$ implies $P^s q_s = P^s \tilde{q}_s = \sum_\nu d_{\nu s} P^s c_\nu$.

Hence the modifications make the number of yeses tend to grow and hence the number of noes to diminish. But again, if the answer is yes in Mod.quest. s, then from the equation $P^s q_s = \sum d_{\nu s} P^s c_\nu$ we determine $u_s$ by $q_s + u_s = \sum d_{\nu s} c_\nu$ and we get $\text{Rank}(Q^\nu + U)$ really less than $\text{Rank}(Q + U_o)$ if the number of noes does diminish. So this can not be the case, the noes occur at the same places (i.e. for the same s-set) in both questionnaires, and even the same d:s can be used.

This gives the principle of an algorithm for determining the semi-rank of $Q$ and of optimal "semifactors" $C^o$ and $D^o$. One bad thing about it, however, is the very big number of multiplications that must be performed to compute the elements of Q if these are given only through the original semifactors C and D. It is, however, possible to devise an algorithm avoiding this kind of difficulty if the n of C and D is considerably smaller than T. (The number of operations of diminishes, hopefully, by a factor ~ n/

Of the questions $\alpha,\beta,\gamma$ put in Section 2 we have given answer only to $\alpha$ and $\gamma$. Lack of space and time makes us postpone an answer to $\gamma$. Another important question is: is it possible to give in special cases (constant coefficients for instance), sufficient conditions in order to make an easy choice of an optimal U?

## REFERENCES

1. R.W. Brockett; Finite Dimensional Linear Systems, Wiley 1970

2. R. Saeks; State in Hilbert Space, SIAM Review, Vol. 15, 1973, pp. 283-308

3. L.-E. Zachrisson; Remarks on Wide Sense Equivalents of Continuous Gauss-Markov Processes in $R^n$. Math. Progr. Study 5, 1976, pp. 103-114

4. T. Kailath, S.-Y. Kung, M. Morf; Displacement Ranks of a Matrix. Bulletin (New Series) Amer. Math. Soc., Vol. 1, No. 5, Sept 1979, pp. 769-773

5. A. Lindquist, G. Picci; Realization Theory for Multivariate Stationary Gaussian Processes, (I and) II. This Volume.

# A DECENTRALIZED STOCHASTIC OPTIMAL CONTROL IN HILBERT SPACE

Romano M. De Santis
Département de Génie Electrique
Ecole Polytechnique de Montréal
Montréal, CANADA

Richard Saeks
Electrical Eng. Department
Texas Tech University
Lubbock, Texas

ABSTRACT

A decentralized optimal control problem is formulated and solved in the mathematical framework of Hilbert resolution space valued random processes. The solution procedure follows a pattern similar to that adopted by Bode and Shannon in their derivation of the classical Wiener filter.

## 1. MATHEMATICAL PRELIMINARIES

A bounded linear operator P on a Hilbert space H is called an orthoprojector if for all pairs $x, y \in H$ one has $\langle Px, y\rangle = \langle x, Py\rangle$ and $P(Px) = Px$. Given a linearly ordered set $\upsilon$ with minimum and maximum elements $t_o$ and $t_\infty$, a family of orthoprojectors, $R = P^t : t \in \upsilon$, is a resolution of the identity, [Sa.1], if it enjoys the following two properties:

i) $P^{t_o}H = 0$, $P^{t_\infty}H = H$ and $P^k H \supset P^\ell H$ whenever $k > \ell$;

ii) if $P^i$ is a sequence of orthoprojectors in R and there exists and orthoprojector P such that $P^i x \to Px$ for each $x \in H$, then $P \in R$.

A Hilbert space H equipped with a resolution of the identity, $R = P^t : t \in \upsilon$, is called a Hilbert resolution space and is denoted by the symbol $[H, P^t]$. Given an orthoprojector $P^t \in R$, the orthoprojector $I-P^t$ is denoted by the symbol $P_t$.

The triple $(H, \Sigma, P)$ will denote a probability space of Hilbert space valued random processes with Hilbert splace, H, family of Borel sets, $\Sigma$, and probability measure, P. Recall that [Ba 1,



D. G. Lainiotis and N. S. Tzannes (eds.), Advances in Control, 139–144.

ch 6], with each random process, $\rho\epsilon(H, \Sigma, P)$, with a finite first and second moment, one finds associated an element, $m\epsilon H$, (the mean value of $\rho$) and an operator, $Q_\rho : H \rightarrow H$, (the covariance of $\rho$), such that

$$E(x, \rho) = (x, m_\rho), \forall x \epsilon H$$

$$\text{and } E(x, \rho - m_\rho)(y, \rho - m_\rho) = (x, Q_\rho y), \forall x, y \epsilon H$$

where the symbol "E f($\rho$)" denotes the expected value of the scalar valued random variable f($\rho$) with respect to the probability space underlying $\rho$. Similarly, with each pair $\rho$, $\pi\epsilon(H, \Sigma, P)$ one finds associated an operator $Q_{\rho\pi}$, (the cross covariance) with the property that

$$E(x, \rho - m_\rho)(y, \Pi - m_\Pi) = (x, Q_{\rho\pi} y), \forall x, y \epsilon H$$

Two random processes $\rho$ and $\Pi\epsilon(H, \Sigma, P)$ are uncorreclated if $Q_{\rho\pi}=0$. Clearly if $\rho$ and $\Pi$ are uncorrelated then $Q_{\rho+\pi}$. Observe also that linear transformations of uncorrelated random processes are uncorrelated.

## 2. THE PROBLEM

Given the multi input - multi output system represented in Figure 1, assume that $D_i$, $G_{ij}$ and $T_{ij}$, $i = 1, \ldots, n$, $j = 1, \ldots, m$, are linear causal mappings on a stochastic Hilbert resolution space, while $z_i$, are Hilbert resolution space valued random processes. Each output $e_i$ of this system can be viewed as given by the sum of two contributions: $e_i = e_{i1} + e_{i2}$, where $e_{i1}$ represents the effect of the perturbations $z_i$ acting via their assigned paths to the output $T_i z$, $e_{i1} = \sum_i T^i_{ij} z_j$, and $e_{i2}$ represents the effect of the same perturbations acting via the feed forward controllers paths $G_{ij}$, $D_i$, i.e., $e_{i2} = \sum_j G_{ij} D_i z$ assuming the perturbations $z_i$ to be statistically independent and letting $T_{ij}$ and $G_{ij}$ be assigned, the problem considered in the present paper is to choose the causal controllers $D_i$ in such a way as to minimize the expected value of the sum of the energies associated with each output of the system. In other words we want to determine m linear causal and bounded operators $D_{jo}$, $j = 1, \ldots, m$, such that

$$J(D_{1o}, \ldots, D_{mo}) = \min_{D_1, \ldots, D_m} J(D_1, \ldots, D_m)$$

where

$$J(D_1, \ldots, D_m) = \sum_{j=1}^{m} E\,||\epsilon_i||^2$$

$$\text{with } \epsilon_i = \sum_{j=1}^{m} (G_{ij} D_j + T_{ij}) \rho_j, \ i = 1, \ldots, m$$

$$= \sum_{j=1}^{m} T_{ij} \rho_j, \ i = m+1, \ldots, n.$$

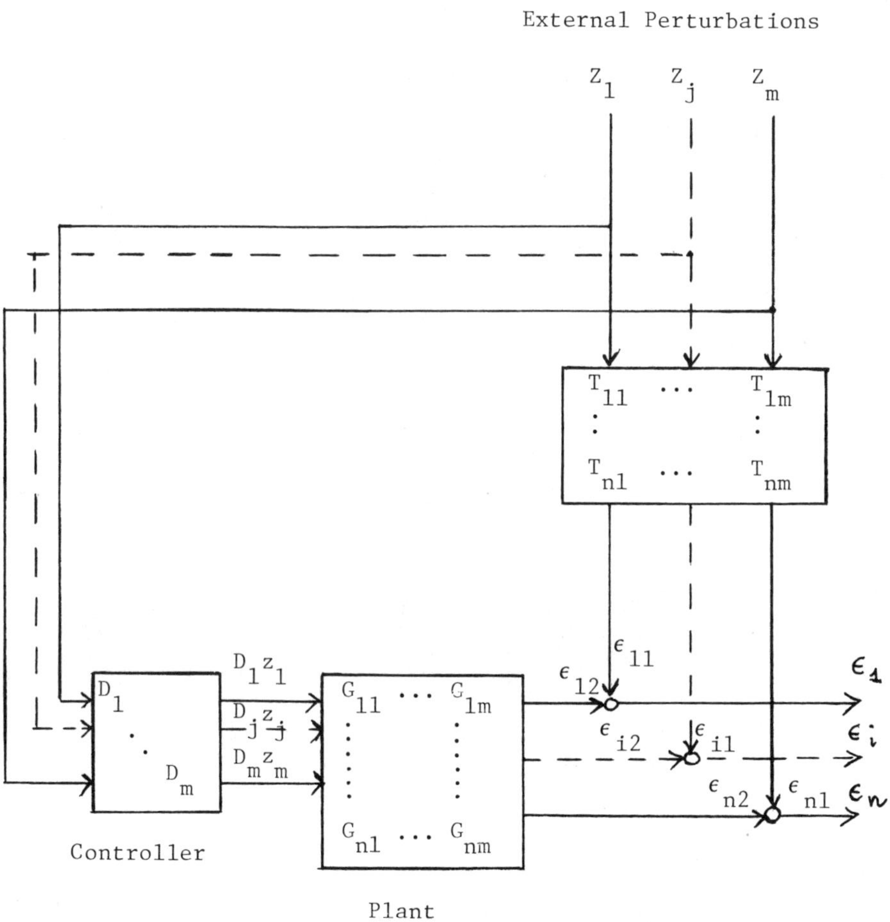


Figure 1: Block Diagram of the Decentralized Feed forward Stochastic Control.

## 3. THE SOLUTION

Denoting with $Q_j$ the covariance operator associates with $z_j$, suppose that there exist:

i) a multiplicative causality decomposition of $Q_j$

$$Q_j = \Omega_j \Omega_j^* , \quad j = 1, \ldots, m$$

with $\Omega_j$ causal and causally invertable;

ii) a multiplicative causality decomposition of

$$W_j = \sum_{i=1}^{n} G_{ij}^* G_{ij} = \Lambda_j^* \Lambda_j, \quad j = 1, \ldots, m$$

with $\Lambda_j$ causal and causally invertable;

iii) an additive causality decomposition of

$$\Pi_j = \Lambda_j^{*^{-1}} \left[ \sum_{i=1}^{n} T_{ij}^* G_{ij} \right] \Omega_j$$

$$\Pi_j = [\Pi_j]_C + [\Pi_j]_{\bar{A}}$$

where $[\Pi_j]_C$ is causal and $[\Pi_j]_{\bar{A}}$ is strictly anticausal;

Then: a solution to P2S is given by

$$D_{jo} = - \Lambda_j^{-1} [\Pi_j]_C \, \Omega_j^{-1}$$

## 4. THE PROOF

In view of the stochastic independence of the $z_j$, one has

$$E||\epsilon_i||^2 = \text{tr} \left\{ \sum_{j=1}^{m} [T_{ij} Q_j T_{ij}^* + G_{ij} D_j Q_j D_j^* G_{ij}^* + G_{ij} D_j Q_j T_{ij}^* + T_{ij} Q_j D_j^* G_{ij}^*] \right\}$$

and from here

$$\sum_{i=1}^{n} E||\epsilon_i||^2 = \left\{ \sum_{j=1}^{m} \text{tr} \left[ \sum_{j=1}^{m} (T_{ij} Q_j T_{ij}^* + G_{ij} D_j Q_j D_j^* G_{ij}^* + G_{ij} D_j Q_j T_{ij}^* + T_{ij} Q_j D_j^* G_{ij}^*] \right\}$$

This implies that

$$\min_{D_1,\dots,D_m} J(D_1,\dots,D_m) = \sum_{j=1}^{m} \min_{D_j} J(D_j)$$

where

$$J(D_j) = \sum_{i=1}^{n} (T_{ij} Q_j T_{ij}^* + G_{ij} D_j Q_j D_j^* G_{ij}^*$$
$$+ G_{ij} D_j Q_j T_{ij}^* + T_{ij} Q_j D_j^* G_{ij}^*)$$

that is

$$J(D_j) = \mathrm{tr}\,[Y_j + W_j D_j Q_j D_j^* + Z_j D_j Q_j + Q_j D_j^* Z_j^*]$$

where

$$Y_j = \sum_{i=1}^{n} T_{ij} Q_j T_{ij}^*$$

$$Z_j = \sum_{i=1}^{n} T_{ij}^* G_{ij}$$

Assuming now the existence of the following multiplicative causality decompositions

$$Q_j = \Omega_j \Omega_j^*$$
$$W_j = \Lambda_j^* \Lambda_j$$

with $\Omega_j$ and $\Lambda_j$ causal and causally invertable, it follows that

$$J(D_j) = \mathrm{tr}\,[Y_j + \Lambda_j D_j \Omega_j \Omega_j^* D_j^* \Lambda_j$$
$$+ Z_j D_j \Omega_j \Omega_j^* + \Omega_j \Omega_j^* D_j^* Z_j^*]$$

and from here (Bode's completion of the square argument)

$$J(D_j) = \mathrm{tr}\ (\Lambda_j D_j \Omega_j + \Pi_j)(\Lambda_j D_j \Omega_j + \Pi_j)^*$$
$$-\Pi_j \Pi_j^* + Y_j$$

where

$$\Pi_j = \Lambda_j^{*-1} Z_j^* \Omega_j .$$

It follows:

$$J(D_j) = \operatorname{tr} \left\{ (\Lambda_j D_j \Omega_j + [\Pi_j]_C (\Lambda_j D_j \Omega_j + [\Pi_j]_C)^* \right\}$$

$$+ \operatorname{tr} \left\{ [\Pi_j]_{\bar{A}} [\Pi_j]_{\bar{A}}^* - \Pi_j \Pi_j^* + Y_j \right\}$$

Since the second member of this expression is independent from $D_j$ it follows that a sufficient condition for $D_j$ to minimize $J^j(D_j)$ is that $\Lambda_j D_j \Omega_j + [\Pi_j]_C = 0$ hence $D_j = -\Lambda_j^{-1} [\Pi_j]_C \Omega_j^{-1}$.

References:

[Ba 1] Balakrishnan, A.V., Applied Functional Analysis, Springer-Verlag, N.Y. 1976.

[Sa 1] Saeks, R., Resolution Space Operators and Systems, Springer-Verlag, N.Y. 1973.

[De 1] De Santis, R.M., Saeks, R., Tung, L., Basic Optimal Estimation and Control Problems in Hilbert Space, Math Systems Theory Journal, December 1978.

# DECENTRALIZED LQG PROBLEM: SUBOPTIMALITY AND STRUCTURE

R.Krtolica[+] and D.D.Šiljak[++]

[+]Mathematical Institute, Serbian Academy of Sciences and Arts, 11000 Beograd,Yugoslavia
[++]School of Engineering, University of Santa Clara, Santa Clara, California 95053

Sufficient conditions for suboptimality of control and state estimation of linear time-varying stochastic systems with quadratic cost are extended to the general case with nonzero terminal cost and random initial state.

## INTRODUCTION

A design methodology for suboptimal estimation and control in the presence of model uncertainties and nonclassical constraints on control and information patterns is frequently needed in feedback design of large-scale engineering systems composed of interconnected subsystems.

Generalizing the previous deterministic work on performance deterioration and suboptimality of decentralized control of weakly coupled systems, /1-4/, the authors, /5/, have provided characterizations of suboptimality of linear feedback for systems with deterministic initial state and zero terminal cost. The purpose of the actual paper is to show that the sufficiency part of these characterization theorems extends to the case with random initial conditions and nonzero terminal cost.



*D. G. Lainiotis and N. S. Tzannes (eds.), Advances in Control, 145–152.*

## SUBOPTIMAL DECENTRALIZED CONTROL AND ESTIMATION

Let us consider a set of s interconnected subsystems described by stochastic differential equations

$$dx_i = A_i x_i dt + B_i u_i dt + \sum_{j=1}^{s} A_{ij} x_j dt + dv_i, \quad i=1,2,\ldots,s \tag{1}$$

where $x_i(t)$, $u_i(t)$, $v_i(t)$ are $n_i, p_i, n_i$ vectors, respectively, which are the state, input and noise of the i-th subsystem such that $n_1+n_2+\ldots+n_s=n$, $p_1+p_2+\ldots+p_s=p$, $(p\leq n)$, and $A_i(t), B_i(t), A_{ij}(t)$ are matrices of appropriate dimensions, whose elements are continuous functions of time on the interval $(t_o, t_f)$, $t_o \geq 0$ is the initial time, and $t_f > t_o$ is the finite, terminal time. The input noise in (1) is described by an n vector $v^T(t)=[v_1^T(t), v_2^T(t), \ldots, v_s^T(t)]$, which is a Wiener process independent of the initial state $x_o^T = [x_1^T(t_o), x_2^T(t_o), \ldots, x_s^T(t_o)]$, with $E(dv \cdot dv^T) = R_v dt$, where $E$ denotes mathematical expectation, $R_v(t)$ is the incremental covariance matrix, whose elements are continuous functions of time, and $E(dv)=0$. It is also assumed that the Gaussian statistics $Ex_o = m_o$, $\mathrm{cov}(x_o, x_o) = R_o$, and $R_v$, for the system (1) is given *a priori*.

For convenience, we introduce the notation $A_D = \mathrm{diag}\{A_1, A_2, \ldots, A_s\}$, $A_C = (A_{ij})$, $(i,j=1,2,\ldots,s)$, $A = A_D + A_C$, and $B = \mathrm{diag}\{B_1, B_2, \ldots, B_s\}$, where the subscripts D and C stand for "decoupled" and "coupled" subsystems. With this notation, the system (1) can be represented as

$$dx = A_D x dt + B u dt + A_C x dt + dv, \tag{2}$$

where $x^T(t) = [x_1^T(t), x_2^T(t), \ldots, x_s^T(t)]$ and $u^T(t) = [u_1^T(t), u_2^T(t), \ldots, u_s^T(t)]$.

We associate with the i-th subsystem in (1) a quadratic cost

$$J_i = x_{fi}^T S_{fi} x_{fi} + \int_{t_o}^{t_f} (x_i^T Q_{xi} x_i + u_i^T Q_{ui} u_i) dt \tag{3}$$

and consider the expected value $EJ_i$ of the cost $J_i$, as a measure of system performance. We denote by $x_{fi}=x_i(t_f)$ and $x_{fi}^T S_{fi} x_{fi}$ the terminal (final) state and the terminal cost of the i-th subsystem, respectively. The matrices $Q_{xi}(t)$ are symmetric positive semidefinite of dimension $n_i \times n_i$, the matrices $Q_{ui}(t)$ are symmetric positive definite of dimension $p_i \times p_i$ and the matrices $S_{fi}$ are positive semidefinite of dimension $n_i \times n_i$. It is assumed that the block diagonal matrices $Q_x=\text{diag}\{Q_{x1}, Q_{x2}, \ldots, Q_{xs}\}$ and $Q_u=\text{diag}\{Q_{u1}, Q_{u2}, \ldots, Q_{us}\}$ both have continuous elements in time. Hence, the quadratic cost associated to the interconnected system (2) is

$$J = x_f^T S_f x_f + \int_{t_o}^{t_f} (x^T Q_x x + u^T Q_u u)\,dt, \tag{4}$$

where $x_f = x(t_f)$, and $S_f=\text{diag}\{S_{f1}, S_{f2}, \ldots, S_{fs}\}$.

A way to account for the incomplete knowledge of the interconnection matix $A_C$ is to compare the expected cost of the system (2) to the expected cost of some referent system. This is performed by assuming some referent pattern of interconnections with respect to which bounded but otherwise uknown perturbations are allowed. In our case extreme perturbation values of the elements of the interconnection matrix $A_C$ are defined with respect to zero, $(A_C=0)$. Another possibility is to determine the referent interconnection matrix by some nonzero values of its elements (-extreme, average, or else). The additional interest of the referent interconnection pattern $A_C=0$ resides in the fact that the control design for the referent system (2) provides in that case automatically a decentralized control law which often proves to be a technological and computational benefit.

Let us denote the control law that corresponds to the referent system (2) in which $A_C=0$ by

$$u^o(t) = -L(t)\cdot x(t), \tag{5}$$

where the elements of the p x n matrix $L(t)$ are continuous functions of time on the interval $(t_o, t_f)$. When this control is applied to the interconnected system(2) with an arbitrary matrix A, the value of the expected cost $EJ^+$ is generally different form that of $EJ^o$ which is calculated for the referent system, that is when $A=A_D$. We propose to bound $EJ^+$ from above by $\mu^{-1}\cdot EJ^o$, where $\mu$ is a positive number. In order to extend the definition of degree of suboptimality, /5/, to the control of the systems with random initial states, we assume that (5) is an optimal control law and introduce the following

(6) DEFINITION. The control law (5) is *suboptimal with degree* $\mu$ for the system (2) and the cost (4) if there exists a positive number $\mu$ such that the inequality

$$E[J^+|x_o] \leq \mu^{-1}\cdot E[J^o|x_o] \tag{7}$$

holds for all initial states $x_o$, their expectations $m_o$, and all $t_o\varepsilon[0,+\infty)$, $t_f\varepsilon(t_o,+\infty)$.

We note that Definition (6) is designed to single out a class of nonstationary suboptimal control problems which are in a sense "quasi-stationary", that is, independent of choice of the time interval $(t_o,t_f)$. Otherwise, it is possible to define the degree of suboptimality with respect to some specific time interval, e.g. $(0,+\infty)$ according to /6/.

In order to delineate the class of interconnection matrices $A_C$ that are allowed by the degree of suboptimality of the control law (5), we introduce the matrices $N(t,t_f)$, $A_D(t)$ and $W(t)$ defined by

$$\frac{d}{dt}N(t,t_f)+\hat{A}_D^T(t)N(t,t_f)+N(t,t_f)\hat{A}_D(t)+W(t)=0,$$

$$N(t_f,t_f)=S_f,$$

$$\hat{A}_D(t) = A_D(t) - B(t)L(t),$$

$$W(t) = L^T(t)\cdot Q_u(t)\cdot L(t) + Q_x(t), \tag{8}$$

and state the following

(9) THEOREM. The control law (5) is suboptimal with degree $\mu\leq 1$ for the system (2) and the cost (4) if the matrix

$$F(t_o,t_f)=A_C^T(t_o)N(t_o,t_f)+N(t_o,t_f)A_C(t_o)-(1-\mu)W(t_o) \tag{10}$$

is negative semidefinite for all $t_o\varepsilon[0,+\infty)$ and all $t_f\varepsilon(t_o,+\infty)$.

We note that for nonzero terminal cost the degree of suboptimality must by necessity satisfy $\mu \leq 1$. Furthemore, we remark that when the control law (5) is optimal, the differential equation in (8) becomes a Riccati equation, but we do not use this fact explicitly.

By duality, the results obtained for suboptimal control can be applied to the problem of suboptimal state estimation. To show this, let us consider the state estimation problem for the system (2) with noisy observations

$$dx = A_D x dt + Budt + A_C x dt + dv,$$

$$dy = Cxdt + dw, \tag{11}$$

where $C(t) = \text{diag}\{C_1(t), C_2(t), \ldots, C_s(t)\}$ is $q \times n$ matrix, and $C_i(t)$ is $q_i \times n_i$ matrix with elements which are continuous functions of time. In (11), $w(t)$ is a $q$ vector, the observation noise, which is a Wiener process independent of the initial state $x_o$ and the input noise $v(t)$. The dimension $q$ of the vector $w$ is determined as $q=q_1+q_2+\ldots+q_s$. Furthemore, $Edw=0$, $E(dwdw^T)=R_w dt$, and $R_w(t)$ is assumed to be positive definite on $(t_o, t_f)$ with elements which are continuous functions of time. To solve the state estimation problem, we must find a state estimate $\hat{x}(t)$ as a functional of $y(\tau), \tau \varepsilon (t_o, t)$, such that the weighted mean square estimation error

$$\sigma^2(t) = E\{a^T \cdot [x(t) - \hat{x}(t)]\}^2 \tag{12}$$

is minimized for all weighting vectors $a$.

When $A \equiv A_D$, the state estimate $\hat{x}(t)$ that satisfies the stochastic differential equation

$$d\hat{x} = A\hat{x}dt + Budt + K(dy - C\hat{x}dt), \tag{13}$$

where $\hat{x}(t_o) = m_o$, is assumed to be the optimal solution $\hat{x}^o(t)$ to this problem. The elements of the $q \times n$ gain matrix $K(t)$ are continuous functions of time, /7/. The appropriate mean square error $[\sigma^o(t)]^2$ of the state estimation is generally different from $[\sigma^+(t)]^2$, which is a mean square error produced by the state estimator (13) applied to the system (11) when $A \neq A_D$. According to /5/, we use the following

(14) DEFINITION. The state estimator (13) for the system (11) has a *degree of suboptimality* $\rho$ if there exists a positive number $\rho$ such that the inequality

$$[\sigma^+]^2 \leq \rho^{-1}\cdot[\sigma^o]^2 \tag{15}$$

holds for all weighting vectors a, and all $t_o \varepsilon [0,+\infty)$, $t_f \varepsilon [t_o,+\infty)$.

Let us denote by $N_e(t,t_o)$ the solution to the matrix differential equation

$$\frac{d}{dt}N_e(t,t_o) = [A_D(t)-K(t)C(t)]N_e(t,t_o)+ N_e(t,t_o)\cdot[A_D(t)-K(t)C(t)]^T+ R_v(t) + K(t)R_w(t)K^T(t), \tag{16}$$

where $N_e(t_o,t_o)=R_o$, and state the following

(17) COROLLARY. The estimator (13) for the system (11) has a degree of suboptimality $\rho\leq 1$ if the matrix

$$H(t_f,t_o) = A_C(t_f)N_e(t_f,t_o) + N_e(t_f,t_o)A_C^T(t_f) -(1-\rho)[R_v(t_f) + K(t_f)R_w(t_f)K^T(t_f)] \tag{18}$$

is negative semidefinite for all $t_o \varepsilon [0,+\infty)$ and all $t_f \varepsilon (t_o,+\infty)$.

This Corollary is an immediate consequence of Theorem (9) and the duality principle.

It follows from Theorem (9) and Corollary (17) that the matrices A-BL and A-KC are stable, (i.e. they have all eigenvalues with negative real parts). That means that if both conditions of Theorem (9) and Corollary (17) are fulfilled, the mean square estimation error converges. The estimator (13) may or may not be a minimum variance estimator but its mean square error is bounded from above by $\rho^{-1}\cdot[\sigma^o(t)]^2$.

## CONCLUSION

The degree of suboptimality of a linear control and linear state estimator is defined for the general case of linear time-varying stochastic systems with quadratic cost. The sufficient suboptimality conditions are derived for completely decentralized control and partially decentralized estimation.

## ACKNOWLEDGEMENT

The research reported herein was partially supported by the Community for Scientific Research in Beograd, Yugoslavia, under the Project on System Theory, Control and Numerical Mathematics, No 5354/1978 at the Mathematical Institute, and by the U.S.Department of Energy under the Contract EC-77-S-03-1493.

## APPENDIX

Proof of Theorem (9)

It is easy to show, (e.g./7/), that

$$E[J^+|x_o] = x_o^T \cdot M(t_o,t_f) \cdot x_o + \int_{t_o}^{t_f} tr[M(t,t_f) \cdot R_v(t)]dt, \quad (A.1)$$

$$E[J^o|x_o] = x_o^T \cdot N(t_o,t_f) \cdot x_o + \int_{t_o}^{t_f} tr[N(t,t_f) \cdot R_v(t)]dt, \quad (A.2)$$

where

$$\frac{d}{dt}M(t,t_f) + \hat{A}^T(t) \cdot M(t,t_f) + M(t,t_f) \cdot \hat{A}(t) + W(t) = 0,$$

$$M(t_f,t_f) = S_f,$$

$$\hat{A}(t) = A(t) - B(t) \cdot L(t). \quad (A.3)$$

Multiplying (A.3) by $\mu$ and subtracting from (8), we obtain

$$\frac{d}{dt}(\mu \cdot M - N) + \hat{A}^T \cdot (\mu \cdot M - N) + (\mu \cdot M - N) \cdot \hat{A} + F = 0,$$

$$\mu \cdot M(t_f,t_f) - N(t_f,t_f) = -(1-\mu) \cdot S_f. \quad (A.4)$$

By differentiation with respect to t, it can be verified that

$$\mu \cdot M(t,t_f) - N(t,t_f) = \int_{t}^{t_f} \emptyset^T(\tau,t) \cdot F(\tau,t_f) \cdot \emptyset(\tau,t) d\tau - (1-\mu) \cdot S_f, \quad (A.5)$$

where

$$\frac{d}{dt}\emptyset(t,\tau) = \hat{A}(t)\emptyset(t,\tau),$$
$$\emptyset(\tau,\tau) = I. \qquad (A.6)$$

It follows from (A.1) and (A.2) that

$$\mu E[J^+|x_0] - E[J^o|x_0] = x_0^T[\mu M(t_0,t_f)-N(t_0,t_f)]x_0$$
$$+\int_{t_0}^{t_f} tr[\mu M(t,t_f)-N(t,t_f)]\cdot R_v(t)dt, \qquad (A.7)$$

which, together with (A.5), proves the theorem.

REFERENCES

/1/ Popov,V.M.:(1960),Proc.of the first IFAC Congress, Moscow, U.S.S.R., pp.173-176.

/2/ Rissanen,J.J.:(1966), IEEE Trans., AC-11, pp.530-532.

/3/ Bailey,F.N., and Ramapriyan,H.K.:(1973), IEEE Trans., AC-18, pp.532-534.

/4/ Šiljak,D.D., and Sundareshan,M.K.:(1976),IEEE Trans., AC-21,pp.79-84.

/5/ Krtolica,R., and Šiljak,D.D.:(1978), Proc.of the 12-th Asilomar Conf.on Circ.,Syst.and Comp., Pacific Grove, Calif., Nov.6-8.

/6/ Ikeda,M., and Šiljak,D.D.:(1979),Proc. of the 13-th Asilomar Conf.on Circ., Syst. and Comp., Pacific Grove, Calif., Nov. 5-7.

/7/ Aström,K.J.:(1970), *Introduction to Stochastic Control Theory,* Academic Press, New York.

# A BINDING MODEL FOR LARGE-SCALE SYSTEM DESIGN

Demetre D. Kouvatsos

Computing Laboratory, University of Bradford,
West Yorkshire, England.

One of the main objectives in the design of a large-scale system is to identify the pattern of interactions between many functional system requirements and to provide a heuristically based estimation process to predict their relative strengths. In this paper the complexity of the design problem is expressed in terms of a system "binding" model which involves limit and structural design parameters and satisfies the necessary heuristic and regularity conditions. The model is presented as a large system of non-linear equations and an iterative procedure is employed to determine the magnitude of the weight to be assigned to each interaction between pairs of interrelated system requirements. This model has already been used, in conjunction with a general decomposition criterion, in the modular design of a computer operating system and in the logical design of a structured multifield record.

## INTRODUCTION

The design of a large-scale system is a very complex process involving many interrelated requirements. A designer of such a system is not only confronted with the initial tasks of choosing the functional requirements and identifying the pattern of their interaction but he must also provide an objective estimation process to predict their relative strengths.

Let M denote a set of requirements associated with any prospective design at which a possible solution might fail. One cannot make a prior exhaustive list of such contingencies; the choice of a good and useful set belongs to the experience and the



*D. G. Lainiotis and N. S. Tzannes (eds.), Advances in Control, 153–162.*

talent of the designer. Two types of dependency may exist between any pair of design requirements. Any solution that tends to satisfy one of the pair may also tend to satisfy the other or it may make the other more difficult to accommodate. Such pairs are thus considered to concur, conflict or to be independent of each other. The potential requirements can be thought to interact only if the designer is aware of a good reason why they should be so correlated and as such the compilation of a useful set again depends on the skill of the designer himself. Let L denote this set of interactions between pairs of design requirements. The design problem may therefore be represented visually as a linear undirected graph G(M,L) whose vertices correspond to the set M of m say requirements and whose signed links correspond to the set L of $\ell$ say interactions. It is possible for different designers to agree about the contents of the sets M and L whether or not they agree about their relative importance; because a simple inclusion of a requirement in M or an interaction in L, as such, attaches no weight at all. What designers disagree about is the relative importance of various vertices and links. Thus an objective estimation of the relative strengths of both vertices and links of G(M,L) is the most difficult to achieve.

Let D denote the domain of all the different solutions of the design problem. One is interested in those links between vertices which hold for all solutions he can conceive. Alexander [1] has pointed out that any sample based on those possible solutions which happen to have been constructed is heavily biased towards the past. To avoid bias one needs either to examine all solutions of D exhaustively or to obtain an unbiased sample from D, neither of which is practical. To overcome the bias Alexander [1] attempts to find causal relations between any two variables and assigns as relative strengths of the interactions +1 for concurrence and −1 for conflict. Furthermore his reasoning is based on somewhat restrictive assumptions; for example all $q_i$'s must be equal to $\frac{1}{2}$ and thus does not provide an objective description of the design problem (see below).

In Kouvatsos [2] some basic ideas about the role of a mathematical model as a tool in the estimation process were briefly mentioned. It is pertinent at this stage to fully describe the development of a mathematical model which will satisfy all heuristic and regularity conditions and could be used as an objective measure of the relative strengths of the interactions.

## GENERAL MODEL CONSIDERATIONS

Let us uniquely associate with each requirement i; i=1,2,.., m, a binary random variable $y_i$ which splits D into two; into a set of those solutions that satisfy i where $y_i$ takes the value 0 with probability $q_i$ and a set of those which do not where $y_i$ takes the value 1 with probability $p_i$ i.e.

$$P_r\,(y_i = x_i) = q_i^{\,1-x_i}\, p_i^{\,x_i} \qquad (1)$$

such that

$$q_i + p_i = 1 \qquad (2)$$

where $x_i = 0$ or $1$, $i = 1, 2, \ldots, m$.

Similarly all the joint probabilities may be defined e.g. $P_r(y_i = 0,\ y_j = 0)$ etc. Further the dependencies between pairs of requirements can be formally described by means of the correlation coefficients $\rho_{ij}$; $0 \leq |\rho_{ij}| \leq 1$. Clearly, because of the binary random variables $y_i$ and $y_j$, $\rho_{ij}$ must satisfy for all i and j

$$\rho_{ij} = [P_r(y_i{=}0, y_j{=}0) - q_i\, q_j]/(p_i\, p_j\, q_i\, q_j)^{\frac{1}{2}} \qquad (3)$$

The estimation process of the relative strengths of the interactions $\{\rho_{ij}\}$ must provide the appropriate expressions for the absolute and joint probabilities in relation (3). Clearly the creation of a mathematical model for $\{\rho_{ij}\}$ necessitates the construction of m real functions $P_r(y_i{=}x_i), i{=}1,2,\ldots,m$ and at the most $[m(m-1)]/2$ real functions $P_r(y_i{=}x_i, y_j{=}x_j)$.

## THE UNIVARIATE CASE

Considering the absolute probability $q_i$ w.l.o.g., one must take into account the following three important points.

i) The probability $q_i$ must be a decreasing function as the number of links (complexity) associated with vertex i increases.

ii) One must 'differentiate' between positive and negative linkages on vertex i since negative connections reduce the size of $q_i$ more than do positive connections.

iii) The event of a vertex i being connected positively with some vertices in G is statistically independent from the event that i is related negatively with other vertices in G.

Thus $q_i$ may be presented as

$$q_i = q_{i+}\, q_{i-} \qquad (4)$$

where $q_{i+}$ and $q_{i-}$ are the absolute probabilities for the purely positive and negative implications of vertex i respectively. The probability $q_{i+}$ is expressed by

$$q_{i+} = \alpha_{i+} + \beta_{i+}\, e^{-\gamma_{i+}} \qquad (5)$$

where $\gamma_{i+}$ is the sum of all positive $\rho_{ij}$'s; $i \neq j=1,2,\ldots,m$ and $\alpha_{i+}$, $\beta_{i+}$ are real numbers between 0 and 1 such that

$$\alpha_{i+} = \beta_{i+} = 1 \qquad (6)$$

Similarly

$$q_{i-} = \alpha_{i-} + \beta_{i-}\, e^{\gamma_{i-}} \qquad (7)$$

where $\gamma_{i-}$ is the sum of all negative $\rho_{ij}$'s and $\alpha_{i-}$, $\beta_{i-}$ are real numbers between 0 and 1 such that

$$\alpha_{i-} + \beta_{i-} = 1 \qquad (8)$$

The parameters $\alpha_{i+}$ and $\alpha_{i-}$ can be interpreted as the lower limits for the absolute probabilities $q_{i+}$ and $q_{i-}$ respectively which can be chosen by the designer. Naturally $\alpha_{i-}$ is expected in general, to be less than $\alpha_{i+}$. The decrease of $q_i$ as the 'complexity' $(\gamma_{i+}, \gamma_{i-})$ on vertex i increases can be seen in Figure 1.

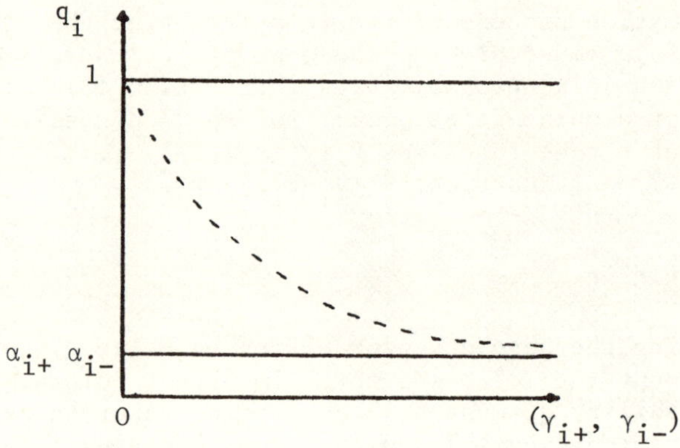


Fig. 1. Decrease in $q_i$ as 'complexity' on i increases

In general, relation (1) is written as

$$P_r\ (y_i = x_i) = x_i + (-1)^{x_i} q_i \qquad (9)$$

where $q_i$ is given by (4) and is defined as the weight of vertex i. Furthermore $q_i$ satisfies all the above conditions.

## THE BIVARIATE CASE

For all $i \neq j = 1,2,\ldots,m$ the joint probability must

i) satisfy the relation of independence i.e.

$$P_r(y_i=x_i, y_j=x_j) = P_r(y_i=x_i)\ P_r(y_j=x_j) \qquad (10)$$

and ii) provide a probabilistically consistent model i.e.

$$0 \leq P_r(y_i=x_i,\ y_j=x_j) \leq \min\{P_r(y_i=x_i),\ P(y_j=x_j)\} \quad (11)$$

Thus, in the independent case, the joint probability is, clearly given by

$$P_r(y_i=x_i, y_j=x_j) = x_i x_j + x_i(-1)^{x_j} q_j + x_j(-1)^{x_i} q_i + (-1)^{x_i+x_j} q_i q_j \quad (12)$$

In the dependent case by

$$P_r(y_i=x_i, y_j=x_j) = x_i x_j + x_i(-1)^{x_i} q_j + x_j(-1)^{x_i} q_i + (-1)^{x_i+x_j} q_{ij} \quad (13)$$

where $q_{ij} = P_r(y_i=0,\ y_j=0)$. (14)

In this case $q_{ij}$ must be defined to be dependent on the correlation coefficient $\rho_{ij}$ and on the absolute probabilities $q_i$ and $q_j$ and to be such that the property ii) is maintained. Therefore if $\rho_{ij} \geq 0$, define

$$q_{ij} = q_i\, q_j + A \quad (15)$$

where $A \geq 0$ is the increase in the joint probability due to positive correlation, such that the following relation holds

$$q_{ij} \leq \min(q_i, q_j) = q_i \quad (16)$$

if $q_i \leq q_j$ or from (15)

$$q_i\, q_j + A \leq q_i \text{ or } q_i\, p_j \geq A \quad (17)$$

To satisfy relation (17), the parameter A must depend on $q_i$, $p_j$ and also it is desirable that A is dependent on $\rho_{ij}$. Thus on the basis of these observations and after several experimental attempts, let us define

$$A = 4\, q_i\, p_j\, (e^{-\rho_{ij}} - e^{-2\rho_{ij}}) \quad (18)$$

where the relation (17) is satisfied since $4(e^{-\rho_{ij}} - e^{-2\rho_{ij}}) \leq 1$. Similar results hold if $q_i \leq q_j$. Note that relation (10) is satisfied when $\rho_{ij}=0$.

Thus, let us define setting $\rho_{ij} = \rho \geq 0$

$$0 \leq q_{ij} = \begin{cases} q_i\, q_j + 4\, q_i\, p_j\, (e^{-\rho} - e^{-2\rho}) & \text{if } q_i \leq q_j \\ q_i\, q_j + 4\, q_j\, p_i\, (e^{-\rho} - e^{-2\rho}) & \text{if } q_j \leq q_i \end{cases} \quad (19)$$

Similar reasoning to the case $\rho_{ij} \geq 0$ produces the following formula for $\rho_{ij} = \rho \leq 0$

$$0 \leq q_{ij} = \begin{cases} q_i\, q_j - 4\, q_i\, q_j\, (e^{\rho} - e^{2\rho}) \text{ if } q_i \leq p_j \\ q_i\, q_j - 4\, p_i\, p_j\, (e^{\rho} - e^{2\rho}) \text{ if } p_j \leq q_i \end{cases} \qquad (20)$$

By using expressions (13), (19) and (20) all other joint probabilities can be defined and it can be easily seen that relation (11) holds for all $x_i$ and $x_j$ together with

$$\sum_i \sum_j P_r\, (y_i = x_i,\, y_j = x_j) = 1 \qquad (21)$$

Note that the joint probability can be extended to any higher order joint distribution (if required) by making use of recurrence relations for multivariate probabilities.

## PHYSICAL INTERPRETATION

For $\rho \geq 0$, let $q_i \leq q_j$ and then, since $q_{ij} \leq \min(q_i, q_j)$, it is true that $\max q_{ij} = q_j$ and so the maximum increase of the joint distribution due to correlation is equal to $q_j - q_i q_j = q_j(1-q_i)$ and $\max\{q_j(1-q_i)\}$ is given when $q_i = q_j = \frac{1}{2}$ i.e. $\max\{q_j(1-q_i)\} = 0.25$. Therefore the maximum increase of the joint distribution due to positive correlation is equal to 0.25. In the model presented in the previous section this increase is given by A in (18) i.e. $A = 4q_i p_j (e^{-\rho} - e^{-2\rho})$. Thus the maximum which can be attained by A is 0.25 and is achieved when $q_i = p_j = \frac{1}{2}$ and $\rho = 0.7$ (See Figure 2).

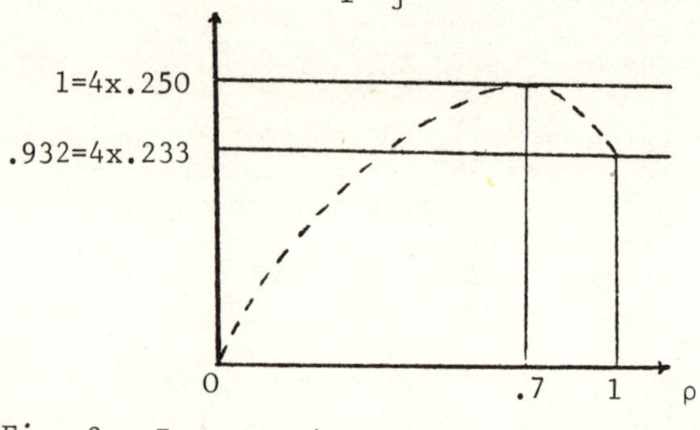


Fig. 2. Increase in A due ρ

From these observations, it is clear that the increase in A depends not only on $q_i$ and $p_j$ but also on $\rho$. Ideally the maximum should be attained when $\rho = 1$ but as can be seen in Table 1 the increase in the joint distribution may have a wide range of values, anything from 0 to 0.25. On inspection it may be noted that the increase is exactly equal to $q_i p_j$ or equivalently $q_j p_i$. Further, if $\rho < 1$, the increase is expected to be $\leq q_i p_j$. Hence the function representing the increase in the joint probability distribution should be given by $q_i p_j\, h(\rho)$ when $q_i \leq p_j$

where $h(\rho)$ is a real function having the properties i) $h(0)=0$, ii) $h(1)=1$ and iii) $h(\rho)$ is monotonic increasing between those limits.

Table 1

| $q_i$ | $q_j$ | $q_{ij}$ | Increase $q_{ij}-q_iq_j$ |
|---|---|---|---|
| .5 | .5 | .5 | .25 |
| .4 | .4 | .4 | .24 |
| .3 | .3 | .3 | .21 |
| .2 | .2 | .2 | .16 |
| .1 | .1 | .1 | .09 |

The quantity A has all these properties except that max $h(\rho)$=max $4(e^{-\rho}-e^{-2\rho})$ is attained at $\rho=0.7$. Despite this deficiency however, the function $4(e^{-\rho}-e^{-2\rho})$ is retained in A, since not only 0.233 is fairly close to 0.25 but also this function contributes towards a unique determination of $\{\rho_{ij}\}$ as can be seen in the following section. Similar comments apply when $\rho_{ij}\leq 0$.

## CONVERGENCE

The expression (3) for all i<j may be rewritten as

$$\{\rho_{ij} = f(\rho_{ij})\} \tag{22}$$

where $f(\rho_{ij}) = (q_{ij}-q_iq_j)/(q_iq_jp_ip_j)^{\frac{1}{2}}$ (23)

For $\rho_{ij}=\rho\geq 0$, using results (9) and (19) expression (23) is given by

$$f(\rho) = \begin{cases} 4q_ip_j(e^{-\rho}-e^{-2\rho})/(q_iq_jp_ip_j)^{\frac{1}{2}} & \text{if } q_i\leq q_j \\ 4q_jp_i(e^{-\rho}-e^{-2\rho})/(q_iq_jp_ip_j)^{\frac{1}{2}} & \text{if } q_j\leq q_i \end{cases} \tag{24}$$

Similarly for $\rho_{ij}=\rho\leq 0$,

$$f(\rho) = \begin{cases} 4_{qi}q_j(e^{2\rho}-e^{\rho})/(q_iq_jp_ip_j)^{\frac{1}{2}} & \text{if } q_i\leq p_j \\ 4p_ip_j(e^{2\rho}-e^{\rho})/(q_iq_jp_ip_j)^{\frac{1}{2}} & \text{if } p_j\leq q_i \end{cases} \tag{25}$$

It is important at this stage to investigate the validity and practical utility of the model by examining the convergence of the system of non-linear equations (22) to a unique non-zero solution. It has been conjectured that a necessary condition for the convergence of (22) is that $f(\rho_{ij})$; $\rho_{ij}=\rho\geq 0$, taken as a function of $\rho_{ij}$ only by keeping the remaining coefficients $\{\rho_{ik}, \rho_{jl}\}$ $k\neq i, j\neq l$ as constants, posseses the following properties

i) $f(\rho)$ is continuous and such that $0\leq f(\rho)\leq 1$

ii) $f(0)=0$
iii) $f(\rho)$ has only one maximum in the region $[0,1]$.

It has been found experimentally by using a program that the model function $f(\rho)$ given by expressions (24) and (25) satisfies conditions i) to iii). However, the function $f(\rho)$ can take two forms, for which the convergence properties are essentially different as can be seen in Figures 3 and 4.

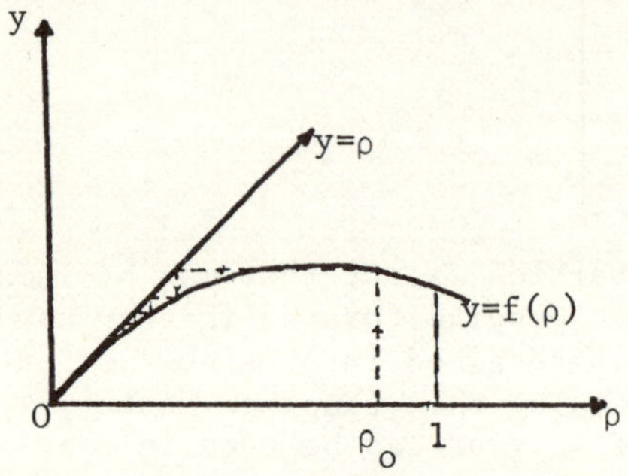


Figure 3a. $(df/d\rho)_{\rho=0} \leq 1$

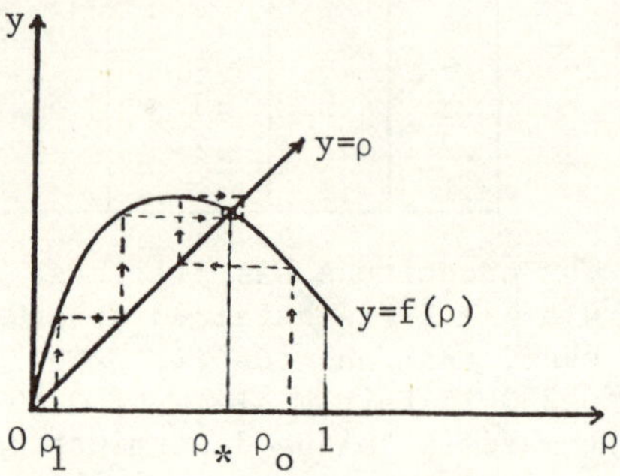


Figure 4b. $(df/d\rho)_{\rho=0} > 1$

It is clear from Figure 3 that the equation $\rho=f(\rho)$ converges to the solution $\rho=0$ for any $\rho=\rho_o>0$ as initial value. From Figure 4 it can be seen that there are two solutions, one being $\rho=0$ and the other $\rho=\rho_*>0$. However, it can be seen, that provided the starting value is non-zero, the equation $\rho=f(\rho)$ converges to a unique solution $\rho_*$ in the range $[0,1]$. At this point an examination is necessary to establish the range of values for the parameters $\{\alpha_{i+}, \alpha_{i-}\}$ satisfying the relation b. in Figure 4 and so indicate the freedom of the designer to choose the lower limit of the absolute probabilities.

Let $\rho \geq 0$. Then (24) may be rewritten as

$$f(\rho)= [4q_i p_j (e^{-\rho} - e^{-2\rho})]/g(\rho)^{\frac{1}{2}} \text{ if } q_i \leq q_j$$

where $g(\rho)=q_i q_j p_i p_j$.

Set $\zeta=e^{-\rho}$, then

$$f(\zeta)= [4q_i p_j (\zeta-\zeta^2)]/g(\zeta)^{\frac{1}{2}} = 4q_i p_j f_1(\zeta)$$

where $f_1(\zeta) = (\zeta-\zeta^2)/g(\zeta)^{\frac{1}{2}}$, $f_1(1)=0$

and $f_1'(\zeta) = [2(1-2\zeta)g(\zeta) - (\zeta-\zeta^2)\, g'(\zeta)]/[2g(2)^{\frac{3}{4}}]$

Therefore $f_1'(1) = -1/g(1)^{\frac{1}{2}}$. Now

$$f'(\zeta) = 4\, [q_i p_j f_1'(\zeta)+f_1(\zeta)(q_i p_j)']$$ and so

$$f'(1) = 4\, q_i p_j f_1'(1) = -(4\, q_i p_j)/g(1)^{\frac{1}{2}}$$

Thus $\frac{df(\rho)}{d\rho}\Big|_{\rho=0} = -\frac{df(\zeta)}{d\zeta}\Big|_{\zeta=1} = -(4q_i p_j)/g(1)^{\frac{1}{2}}$ (26)

Expression (26) is required for convergence to satisfy relation b. i.e.

$$4\, q_i\, p_j > g(1)^{\frac{1}{2}} \qquad (27)$$

Since $\max g(1)^{\frac{1}{2}} = \max (q_i q_j p_i p_j)^{\frac{1}{2}} = \frac{1}{4}$, the necessary conditions for (27) to hold are given by

$$\begin{cases} q_i\, p_j > .0625 & \text{if } q_i \leqslant q_j \\ q_j\, p_i > .0625 & \text{if } q_j \leqslant q_i \end{cases} \qquad (28)$$

Following a similar analysis when $\rho \leqslant 0$ it is found that the corresponding conditions are

$$\begin{cases} q_i\, q_j > .0625 & \text{if } q_i \leqslant p_j \\ p_i\, p_j > .0625 & \text{if } q_j \geqslant p_i \end{cases} \qquad (29)$$

It can be shown in general that condition (28) and (29) are satisfied when the model parameter $\alpha_{i+}$ lies within the following limits

$$(.5-e^{-N_{i+}})/(1-e^{-N_{i+}}) \text{ and } (.866-e^{-.2N_i})/(1-e^{-.2N_i})$$

where $N_{i+}$ is the number of positive links connected to vertex i. The same limits apply for $\alpha_{i-}$ but $N_{i+}$ must be substituted with $N_{i-}$, the number of negative links of i. If one of $N_{i+}$, $N_{i-}$ is zero, then the limits, if for example $N_{i-}=0$ are given by be

$$(.25-e^{-N_{i+}})/(1-e^{-N_{i+}}) \text{ and } (.75-e^{-.2N_{i+}})/(1-e^{-.2N_{i+}})$$

## APPLICATIONS

The 'binding' model described above has been implemented computationally by an algorithm which employs an iterative procedure, operates on initial 0, 1, and -1 link weights and generates the final weights which represent the unique solution of the system of non-linear equations (22). This model in conjunction with a general criterion of system decomposition [2] i.e.

$$\min \sum_{\pi} \rho_{ij}^{2} \qquad (30)$$

where $\rho$ is an arbitrary partition of the set M, has successfully been used in the modular design of a computer operating system [3] and in the construction of multifield structured bibliographical records resulting in the design of an optimal storage allocation scheme [4].

## REFERENCES

[1] Alexander, C., 1967, *Notes on the Synthesis of Form*, Harvard Univ. Press.

[2] Kouvatsos, D. D., 1977, Proceedings of the First International Conference on Information Science and Systems. Hemisphere Publishing Corp., Vol. 2, pp. 575-579.

[3] Kouvatsos, D. D., 1977, Int. J. of Comp. & OR, Vol. 4, No. 1, pp. 55-63.

[4] Kouvatsos, D. D. and Yannakoudakis, E. J., 1978, Papers of the Seventh Comp. Sci. Conference, Lancaster Univ., England.

# BIFURCATIONS IN NONNEGOTIABLE GAMES: A PARADIGM FOR SELF-ORGANIZATION IN COGNITIVE SYSTEMS

John S. Nicolis,
Dpt. of Electrical Eng.
University of Patras,
Greece

E.N. Protonotarios
Dpt. of Electrical Eng.
National Technical University of Athens, Greece

## Synopsis

Artificial cognitive systems perform signal processing by cross-correlating or, in general "metabolizing" markovian strings of given discrete symbols-thereby sustaining between themselves and the environment a (built in) communication language.

"Self"-organization in cognitive systems is meant to imply the emergence of constraints for the establishment of a meta- language (aiming e.g. at enrichment of perception at the expense perhaps of ambiguity resolution).

In contradistinction to natural languages (which are somehow generating their own symbols) formal ones, (i.e computer and finite state automata languages) are usually defined by their intrinsic inability to represent themselves or forward any statement referring to itself.

Still, observations, plausible isomorphisms and theories coming from natural languages as exercised between (two) human conversing partners may serve as valuable hints in inspiring ideas for self-transcending algorithms in man-made systems.

In this paper a detailed model concerning nonresolvable ambiguity of the communication dynamics(envisaged as bidirectional information transfer) between two



*D. G. Lainiotis and N. S. Tzannes (eds.), Advances in Control, 163–173.*

hierarchical structures is dealt with. The scope is to design algorithms leading to limited resolution of intrapersonal conflict via interpersonal communication. Each system possess two irreducible hierarchical levels Q,W and Q´,W´ respectively. The dynamical prosesses at each level (Q,Q´),(W,W´) are described correspondiggly as Markov and semi-Markov chains. The transition probabilities at the lower levels Q,Q´stem from the parameters of underlying nonnegotiable games-the prisoner´s dilemma (PD) and the chicken Game (CG)-which emulate all activity below Q,Q´ and are played inductively with conditional mixed strategies or "propensities".

The markovian kinetics at the levels Q,Q´ do not evolve however under fixed propensities: Since the games are played in iteration, learning takes place which means that the time evolution of the propensities is governed by a system of two first order non-linear differential equations; the time derivatives of the propensities are proportional to the gradient of the expected pay off(s) with respect to that propensity.

These dynamics give rise after a certain transient time to bifurcations: The steady states of the propensities involved are unstable or marginally stable (saddle points or centers) and as a result the games will afterwards "lock in" in the cooperative (cc) or "homeostatic" state with a probability which depend on the "threshold" values of the propensities. So the markovian chains at the levels Q,Q´ after the transient period may either disintegrate to memoryless processes i.e unconnected and mutually exclusive states or, to settle into marginally stable regimes. In the latter case it is possible to maintain beyond the instability a full markov chain with four states and fixed propensities or periodically modulated propensities a "mata-language" as it were; the establishment of this chain (in lieu of isolated states) provides some sort of "long-range coherence"amongst all available states $S_i$(i 1,...,4).

The transition probabilities at the higher levels W,W´ are parametrized on collective properties at Q,Q´ which have to do with (a) the probability of occupancy u of the "synergetic" state (cc) and (b) the crosscorrelation r between sequences of the chains (Q,W´) and (Q´,W) respectively.

The semi-Markov chains at the higher levels W,W´ are endowed with mean holding times which are inversely

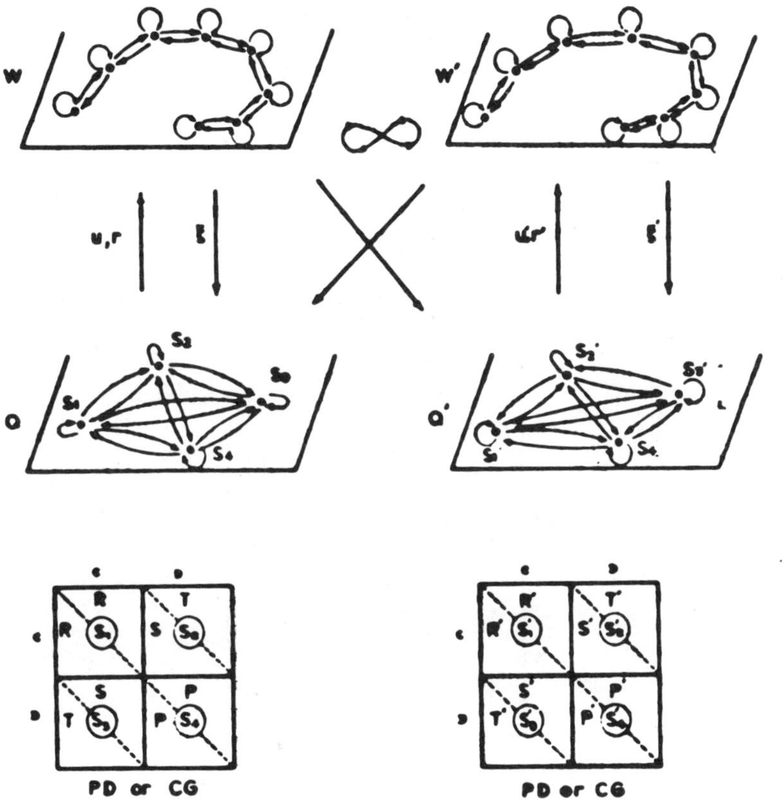


Fig. 1. The general layout of the communication process between 2 hierarchical systems.

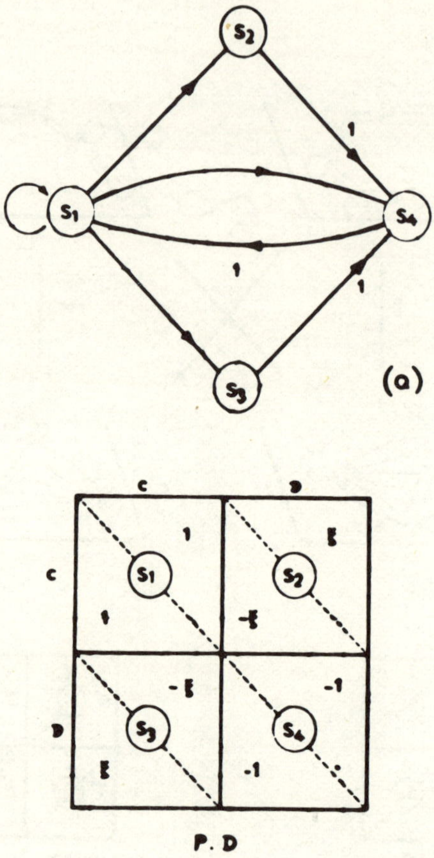


Fig. 2. The Prisoner's Dilemma played inductively and the resulting Markov chain.

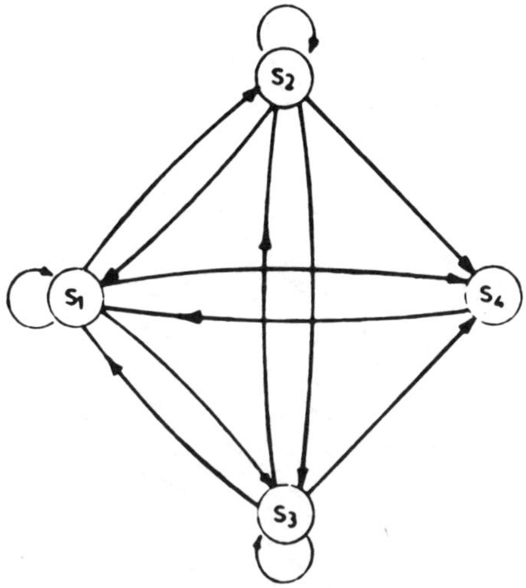


(a)

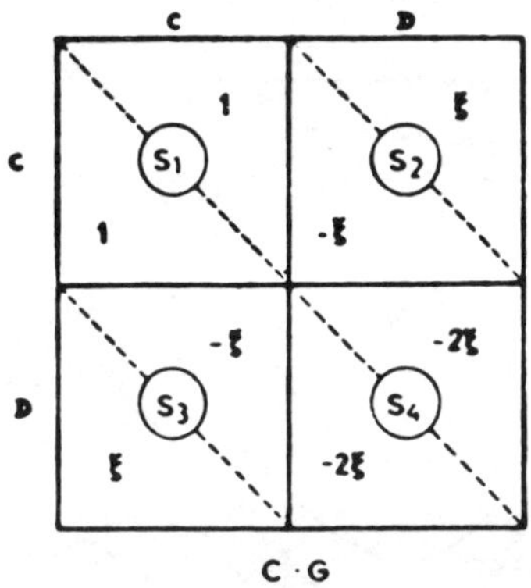


C · G

Fig. 3. The Chicken's Game and the resulting Markov chain.

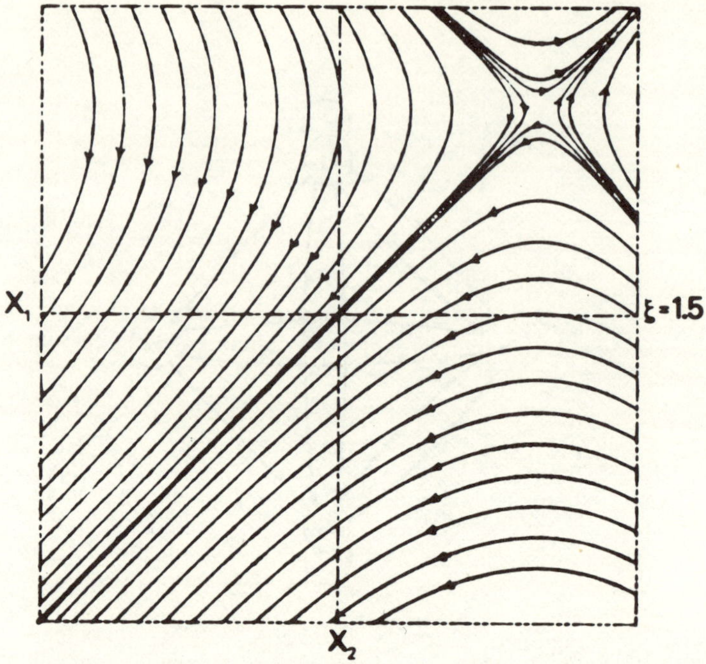


Fig. 4. Phase-space description of the Prisoner's Dilemma

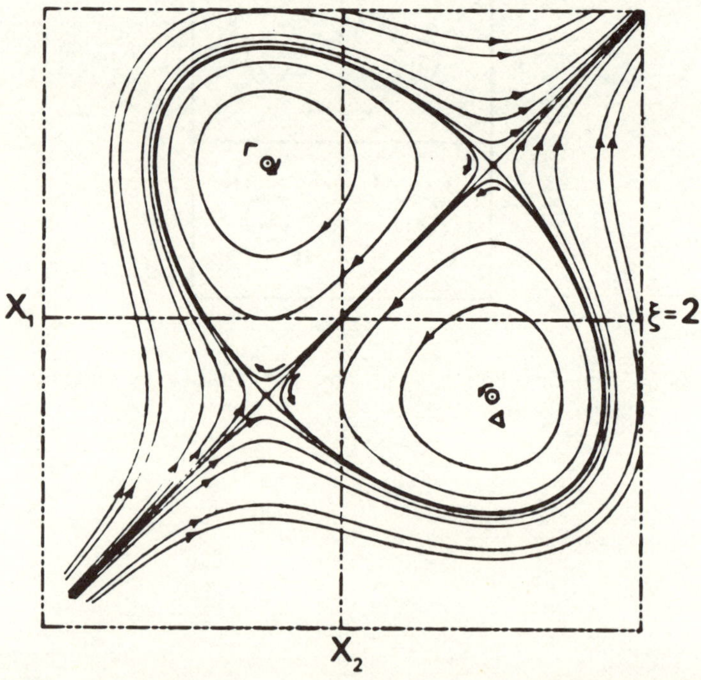


Fig. 5. (a) Phase-space description of the Chicken Game.

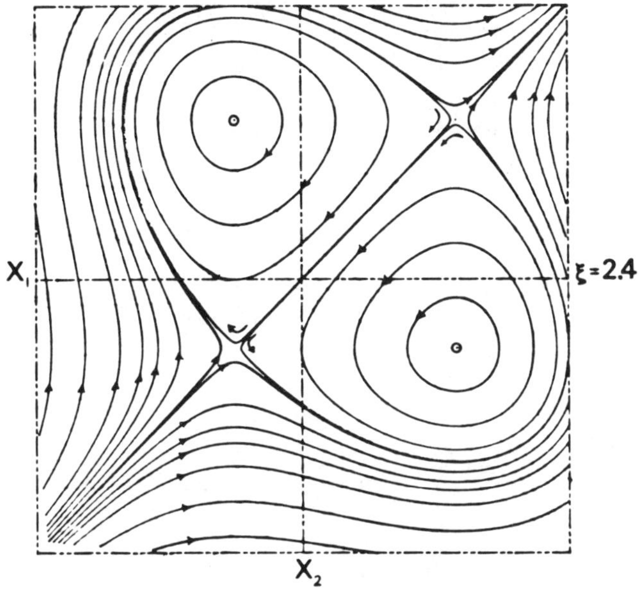


Fig. 5 (b). Phase-space description of the Chicken Game.

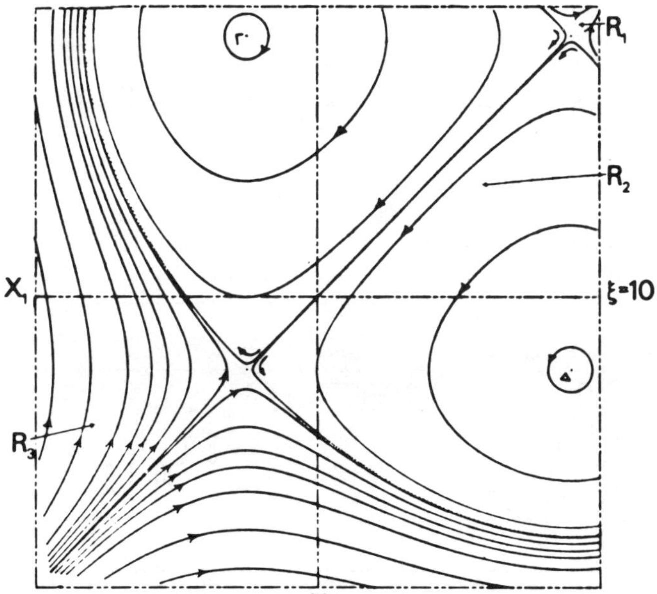


Fig. 5 (c). Phase-space description of the Chicken Game.

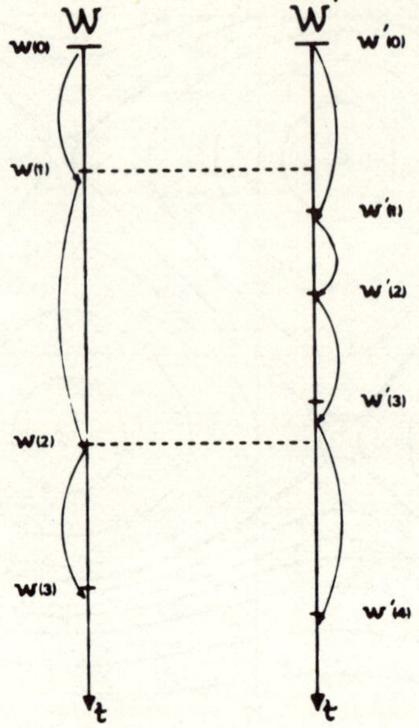


Fig. 6. Computer simulation: state transitions at the levels $W$ and $W'$.

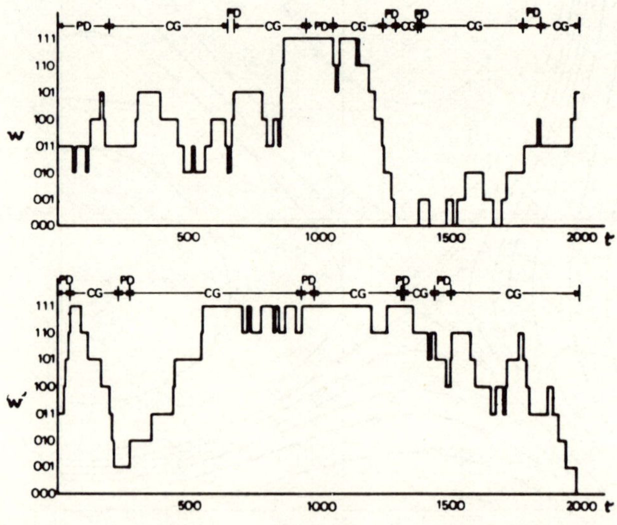


Fig. 7. Sequence of states at the higher levels $W$ and $W'$, with transition probability parameters at these levels: $\pi_k=0.1$, $g_k=0.2$, $c_k=d_k=0.8$, $\pi'_k=0.2$, $q'_k=0.1$, $c'_k=d'_k=0.8$, and $\lambda_k=\mu_k=\nu_k=\xi_k=\lambda'_k=\mu'_\kappa=\nu'_\kappa=\xi'_\kappa=2$, for $k=1, 2, \ldots, 8$. The initial value of the parameter $\xi$ in the underlying games are $\xi(0)=\xi'(0)=2$; the figure of merit is given as initial value $\mathscr{F}(0)=\mathscr{F}'(0)=0.2$. In the computer simulation $\varepsilon=0.10$ in eqn. (29) and $m=2$; also each partner switches games if $\mathscr{F}(n)<10^{-4}$.

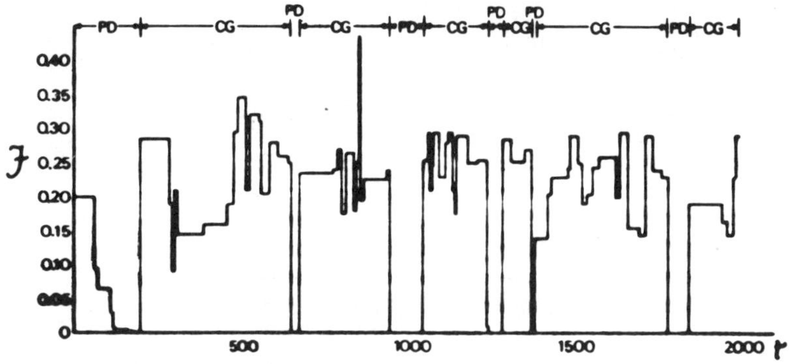


Fig. 8(a)

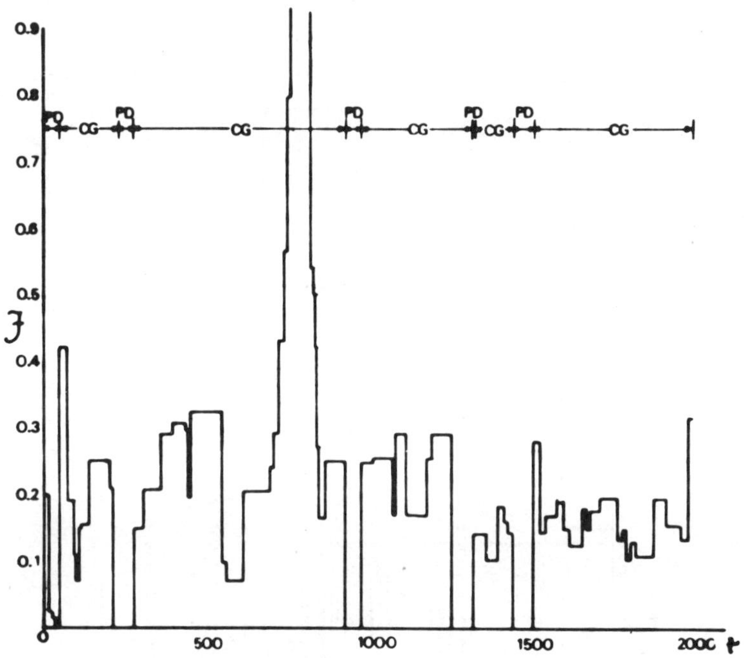


Fig. 8(b)

Fig. 8. Evolution of the instantaneous figure of merit derived from the computer simulation with parameter values as in Figure 7.

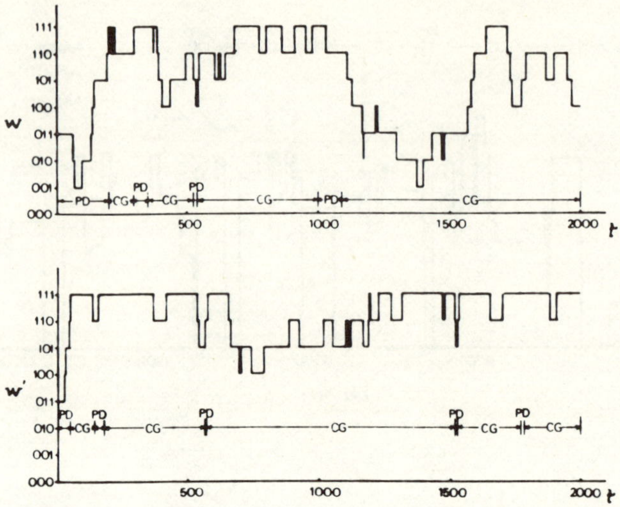


Fig. 9. Sequence of states at the higher levels $W$ and $W'$, with transition probability parameters at these levels: $\pi_k=0.1$, $g_k=0.2$, $c_k=d_k=0.8$, $\pi'_k=0.2$, $q'_k=0.1$, $c'_k=d'_k=0.8$, and $\lambda_k=\mu_k=\nu_k=\xi_k=\lambda'_k=\mu'_k=\nu'_k=\xi'_k=2$, for $k=1, 2, \ldots, 8$. The initial value of the parameter $\xi$ in the underlying games are $\xi(0)=1.1$ and $\xi'(0)=1.8$; the figure of merit is given as initial value $\mathscr{F}(0)=\mathscr{F}'(0)=0.2$. In the computer simulation $\varepsilon=0.10$ in eqn. (29) and $m=2$; also each partner switches games if $\mathscr{F}(n)<10^{-4}$.

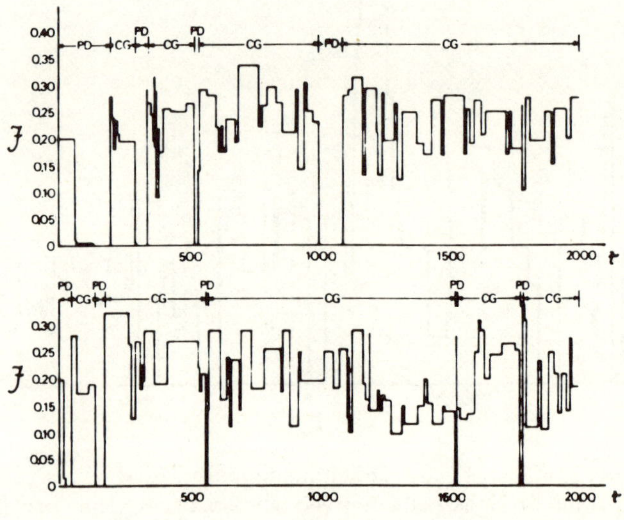


Fig. 10. Evolution of the instantaneous figure of merit derived from the computer simulation with parameter values as in Fig. 9.

proportional to the redundancy of the levels Q,Q'. The underlying games alternate between (PD) and(CG). Complete control scenarios are provided for compromizing the "intrapresonal conflict" by maximizing for each individual partner a certain functional, the "Figure of merit" which weighs in a multiplicative fashion the two basic factors: Homeostasis(i.e probability $\bar{u}$ or reaching the (cc) state) and crosscorrelation $\bar{r}$ with the partner.

The changeover between the games (PD) and (CG) takes place within each individual when his figure of merit with the previous game seems to be reaching local plateaus i.e when stationarity is established.

## REFERENCES


1. Haken H. "Synergetics" Springer verlag (1977)
2. Mac Lean.P "The Triune Brain in conflict" Psychother. Psychosm 28 pp 207-220 (1977).
3. Mac Lean P. "The Imitative-Creative Interplay of our three Mentalities" in"Astride the two Cultures"(Ed) H.Harris,Random Hause 1975.
4. Mac Lean P. "The paranoid streak in Man" in "Beyond Reductionism" (Eds) Koestler A and Smythies I.R. Hutchinson (London) 1969.
5. Nicolis J.S. Protonotarios E.N. Voulodemou I. "Control Markov chain model for Biological hierarchies" J. of theor. Biology 68 pp 563-581 (1977).
6. Nicolis J.S. Protonotarios E.N. Theologou M. "Communication between two self-organizing systems modelled by controlled Markov chains" Int. Journal of Man-Machine studies vol 10 pp 343, 1978.
7. Rapoport A Chammah A. "Prisoners Dilemma" Ann Arbor Paperbacks 1965.
8. Nicolis J.S. Protonotarios E.N "Bifurcations in non constant sum Games" to be published.

# A RANDOM WALK BETWEEN TWO RANDOM ABSORBING BARRIERS

P.-C.G. Vassiliou

Statistics Unit, University of Ioannina

## ABSTRACT

The classical two player gambler's ruin problem is well known and has attracted a great deal of attention in the literature in various forms. This paper is concerned with various extensions of the two-player gambler's ruin problem the study of which started with Vassiliou and Karakostas in (1). In the present we have two gamblers with possibilities of increasing their capital for example by borrowing or to decrease their capital for example by investing it to something else. In the picturesque language of the particle performing a random walk we have a particle performing a simple random walk with the presence of two random absorbing barriers performing also independent random walks. The probability distributions of absorptions and other related questions are being answered both by analytical methods and with the use of a computer.

## 1. INTRODUCTION

The classical two player gambler's ruin problem is well known and has attracted a great deal of attention in the literature in various forms, see (3). Extensions of these two player problems to three players have also been discussed, see (1). The present paper is concerned with various extensions of the two player gambler's ruin problem the study of which started with (6). In the present we have two gamblers with possibilities of increasing at each step their capital for example by borrowing or to decrease their capital for example by investing it to something else.



*D. G. Lainiotis and N. S. Tzannes (eds.), Advances in Control, 174–181.*

In presenting the problem we will use the picturesque language of the particle performing a simple random walk in the (x,y) axis in discrete time, i.e. unit steps at unit intervals;this is done to simplify and enliven the formulations. It should be noted however that this convextion merely lends color to the description and that a continuous time stochastic process is easily obtained from it see (3). In this context our problem is that of a particle performing a simple random walk with the presence of two random absorbing barriers performing also independent simple random walks. The treatment of such problems is of some significance among others because of the fundamental role played by the simple random walk in the theory of diffusion processes. Stochastic models in many areas such as economics, population genetics, particle physics, biology and demography give rise to random walk processes.

In section 2 of the present paper the formal description of the problem and the notation used is given. In section 3 the probabilities of the various states of the particle absorbing and non-absorbing are given. Thus the probability distribution of the particle to be absorbed by the random barrier is being given in two ways (i) by numerically solving, with the use of a described in the present algorithm a suitable difference equation and (ii) by providing a theorem through which an analytic expression for the required probability could be obtained. Both ways have advantages of their own and are useful and important. To complete the description of the process in probability terms, in the same section the probability of the particle to be at point k at time n while both the random barriers are at given distances is being calculated, both numerically and analytically. Thus important questions concerning the above stochastic process are being answered.

## 2. FORMAL DESCRIPTION OF THE PROBLEM AND NOTATION USED

Let us suppose with no loss of generality that a particle is initially at point $j>0$ and that it undergoes a simple random walk with independent increments $Z_1, Z_2,...$ where for $i=1,2,...$

$$\text{prob}\{Z_i=1\}=p_{1s}, \quad \text{prob}\{Z_i=0\}=p_{os}, \quad \text{prob}\{Z_i=-1\}=p_{-1s} \tag{1}$$

Let $X_n$ be the state of the particle at time n then obviously

$$X_n=j+\sum_{i=1}^{n} Z_i$$

We define an absorbing barrier as follows

<u>Definition 1.</u> A barrier at state $a>0$ is called an absorbing barrier if all the states in $[a,\infty)$ are absorbing. A barrier at state $a<0$

is called an absorbing barrier if all the states $(-\infty,a]$ are absorbing.

Suppose also that at point $a>0$ there is a random absorbing barrier at time 0, which we call the random barrier A, and that it undergoes an independent simple random walk with independent increments $R_1,R_2,\ldots$ where for $i=1,2,\ldots$

$$\text{prob}(R_i=1)=p_{1a}, \quad \text{prob}(R_i=0)=p_{oa}, \quad \text{prob}(R_i=-1)=p_{-1a} \tag{2}$$

Finally let us suppose that at point $-b>0$ there is a random absorbing barrier at time 0 which we call the random barrier B and that it undergoes an independent simple random walk with independent increments $W_1, W_2,\ldots$ where for $i=1,2,\ldots$

$$\text{prob}(W_i=1)=p_{1b}, \quad \text{prob}(W_i=0)=p_{ob}, \quad \text{prob}(W_i=-1)=p_{-1b} \tag{3}$$

and that the random variables $R_i$, $Z_i$ and $W_i$ are independent. Let $Y_n$ be the state of the random barrier A at time n and also. Let us define by

$$D_n^A=Y_n-X_n, \quad n=1,2,\ldots \tag{4}$$

to be the distance between the random barrier A and the particle at time n. Let S be the state of the random barrier B at time n and also let us define by

$$D_n^B=X_n-S_n, \quad n=1,2,\ldots \tag{5}$$

to be the distance between the particle and the random barrier B at time n.

## 3. PROBABILITIES OF THE VARIOUS STATES OF THE SYSTEM

It is of interest to calculate the probabilities of the various states of the system at time n. In this respect we will firstly estimate the probability of absorption of the particle by the random barrier A at point k at time n. Thus we are looking for the probability

$$\begin{aligned}\text{prob}\{&D_1^A>0,\ D_2^A>0,\ldots,D_{n-1}^A>0,\ D_n^A\leq 0,\ D_1^B>0,\ D_2^B>0,\ldots D_n^B>0,\\ &X_n=k\,|\,X_0=j,\ \ Y_0=a,\ S_0=-b\}\end{aligned} \tag{6}$$

Define the following

$$\begin{aligned}f^{(n)}(r,0,d,d_n,j,k)=\text{prob}\{&D_1^A>0,D_2^A>0,\ldots,D_{n-1}^A>0,\ D_n^A=0,\\ &D_1^B>0,D_2^B>0,\ldots,D_n^B=d_n>0,\ X_n=k\,|\,D_0^A=r,\ D_0^B=d,\ X_0=j\}\end{aligned} \tag{7}$$

Let $E_n$ denote the event "the distance between the random barrier A and the particle to be 0 for the first time at time n, the distance between the random barrier B and the particle to be $d_n$ at time n and the particle to be at point k". Then

$$f^{(n)}(r,0,d,d_n,j,k)=\text{prob}\{E_n|D_0^A=r,\ D_0^B=d,\ X_0=j\}.$$

If the first step of the particle is +1 and the first step of the random barrier A is -1 then the distance between the two becomes r-2 and the particle is at point j+1;if the first step of the random barrier B is -1 then the distance between the particle and the barrier becomes d+2. For $E_n$ now to occur we must have $E_{n-1}$ occuring conditional on (r-2,d+2, j+1). The probability of the first steps to be as described above and $E_n$ to occur is therefore

$$pq_1q_2\ \text{prob}\{E_{n-1}|D_0^A=r-2,\ D_0^B=d+2,\ X_0=j+1\}=pq_1q_2f^{(n-1)}(r-2,\ 0,d,d_nj+1,k) \tag{8}$$

All the possibilities for the first steps and the corresponding probabilities are mutually exclusive and exhaustive possibilities so we have that

$$f^{(n)}(r,0,d,d_n,j,k)=\sum_{x=-1}^{1}\sum_{y=-1}^{1}\sum_{w=-1}^{1}p_{xs}p_{ya}p_{wb}f^{(n-1)}(r+y-x,\ 0,d+x-w,d_n,j+x,k)$$

With initial and boundary conditions

$$f^{(0)}(r,0,d,d_n,j,k)=\begin{cases}1 & r=0,\quad d=d_n \quad\text{and}\quad j=k\\ 0 & \text{otherwise}\end{cases} \tag{10}$$

$$f^{(n)}(0,0,d,d_nj,k)=0,\quad f^{(n)}(r,0,0,d_n,j,k)=0,\quad f^{(n)}(-1,0,d,\ d_n,j,k)=0,\ f^{(n)}(r,0,-1,d_n,j,k)=0\quad \text{for } n=1,2,\dots;$$

$$f^{(n)}(r,0,d,d_n,j,k)=0\quad \text{for } n<|k-j|\text{or}|j-k+r|\text{or } r/2\text{ or}|k+b-d_n|$$

The difference equation (9) together with the initial and boundary conditions (10), (11) could not be solved by the usual method with the use of probability generating functions see Cox and Miller in (2) due to the fact that (9) is of the 4th degree in a and d and moreover we have three variables. Also methods as that used by Weesakul in (8), Neuts, M.F. in (5) and Hardin and Sweet in (4) could not be applied in this case for the same reasons. We solved equation (9) numerically and also we were able to obtain analytic solutions. Both ways of solving the equation presents an interest of its own. We will firstly describe the numerical solution.

Let us suppose that we want to calculate the probability $f^{(n)}(r,0,d,d_n,j,k)$ for fixed values of the parameters $n,r,d,d_n,j,k$. From equation (9) it is obvious that for the calculation of this probability we will need to calculate the probabilities of thi form for various values of the parameters n,r,d, and j while $d_n$ and k will be fixed. Let n take some value $n-n_0$ $(n_0>0)$ then consider in the space $R^3$ the parallepiped with verteses $(r-2n_0,d,j+n_0)$, $(r-2n_0, d+2n_0,j+n_0)$, $(r,d+2n_0,j+n_0)$, $(r,d,j+n_0)$ and $(r,d,j-n_0)$. Consider also the mesh of unit size with lines parallel to the vectors $\vec{x}_0$, $\vec{y}_0,\vec{z}_0$ (which constitutes a base in $R^3$) on this parallepiped. Then the coordinates of the points of this mesh determines the corresponding probabilities $f^{(n-n_0)}(x,0,y,d_n,z,k)$ needed for the calculation of the probability $f^{(n)}(r,0,d,d_n,j,k)$. Based on the above our algorithm for numerically solving equation (9) is the following.

Algorithm RA2

Step 0 (initialization). Set the variables $d_n$ and k to the prescribed values $d_n^0$ and $k^0$. Provide the computer memory with the initial conditions (10) for n=0 and for the values of (r,d,j) such that the points $(r,d,j)\in R^3$ to be the points of the mesh of unit size on the space $S_0$ which is the intersection of the parallepiped with verteces $(r-2n,d,j+n)$, $(r-2n,d+2n,j+n)$, $(r,d+2n,j+n)$, $(r,d,j+$ and $(r,d,j-n)$, $(r+2n,d,j-n)$, $(r+2n,d-2n,j-n)$, $(r,d-2n,j-n)$ and the space $A=\{(x,y,z)\in R: x\geq -1,\ y\geq -1\}$.

Step 1. Provide the computer memory with the boundary conditi for n=1 and for the values of (r,d,j) such that the points $(r,d,j,)\in R^3$ to be points of the mesh of unit size on the space $S_1$ which is the intersection of the parallepiped with verteces $(r-2(n-1),d,j+n-1)$, $(r-2(n-1),d+2(n-1),j+n-1)$, $(r,d+2(n-1),j+n-1)$, $(r,d,j+n-1)$ and $(r,d,j-n+1)$, $(r+2(n-1),d,j-n+1)$, $(r,d-2(n-1),j-n+1$ and the space A. For r=1,2,...;for d=1,2,...;and j=1,2,... (in tha order) use equation (9) to calculate the probabilities $f^{(1)}(r,0,d,d$ j,k) for the rest of the points of the mesh.

. . . . . . . . . . . . . . . . . . . .

Step n. Provide the computer memory with the boundary conditi (10) for n=n-1 and for the values of (r,d,j) such that the points $(r,d,j)\in R^3$ to be the points of the mesh of unit size on the space $S_{n-1}$ which is the intersection of the parallepiped with verteces $(r-2,d,j+1)$ $(r-2,d+2,j+1)$, $(r,d+2,j+1)$, $(r,d,j+1)$ and $(r,d,j-1)$, $(r+2,d,j-1)$, $(r,d-2,j-1)$ and the space A. Use equation (9) to calculate the probability $f^{(n)}(r,0,d,d_n,j,k)$.

Step n+1. Return to step 0 and replace the values of $K_0$ or $d_r^C$ with others if desired.

Let $P^0(s,d^*,k^*)=f^{(s)}(r,0,d,d^*,j,k^*)p^{(n-s)}(d^*,d_n,k^*,k)p^{(n-s)}(k^*,$

$$k)+f^{(s)}(r,-1,d,d^*,j,k^*)p^{(n-s)}(d^*,d_n,k^*,k)p_{**}^{(n-s)}(k^*-1,k) \qquad (12)$$

where $p^{(n-s)}(d^*,d_n,k^*,k)$ is the propability- in a simple random walk with the presence of a random absorbing barrier at point -b of the particle to be at point k at time n-s while the barrier is at a distance $d_n$ given the initial states to be k* and d*; $p_{**}^{(n-s)}(k^*,k)$ is the probability - in an unrestricted simple random walk- for the random barrier A to be at state k at time n-s given its initial state to be k*. Also let

$$[\tfrac{1}{2}(a+n-k)]=n_1 \text{ where by [a] we mean the integer part of a;}$$
$$[\tfrac{1}{2}(n+k-j)]=n_2;\ k_1=r-s;\ k_2=s+j;\ k_3=n+k-s;\ k_4=n+k-s; \qquad (13)$$
$$d_1=\max\{1,k-n+b\};\ d_2=k+n+b$$

then the following theorem holds.

Theorem 1.

Let $n^*=r/2$ if r even and $n^*=(r+1)/2$ if r odd then

a) If $k\geq n^*+j$ then

$$f^{(n)}(r,0,d,d_n,j,k)=p^{(n)}(d,d_n,j,k)p_{**}^{(n)}(j+r,k)-\sum_{s=n^*}^{n_1}\sum_{k^*=k_1}^{k_2}\sum_{d^*=d_1}^{d_2}P^0(s,d^*,k^*)-\sum_{s=n_1+1}^{n_2}\sum_{k^*=k_3}^{k_2}\sum_{d^*=d_1}^{d_2}P^0(s,d^*,k^*)-\sum_{s=n_2+1}^{n-1}\sum_{k=k_3}^{k_4}\sum_{d^*=d_1}^{d_2}P^0(s,d^*,k^*)$$

(b) If $k<n^*+j$ then

$$f^{(n)}(r,0,d,d_n,j,k)=p^{(n)}(d,d_n,j,k)p_{**}^{(n)}(j+r,k)-\sum_{s=n^*}^{n_2}\sum_{k^*=k_1}^{k_2}\sum_{d^*=d_1}^{d_2}P^0(s,d^*,k^*)-\sum_{s=n_2+1}^{n_1}\sum_{k^*=k_1}^{k_2}\sum_{d^*=d_1}^{d_2}P^0(s,d^*,k^*)-\sum_{s=n_1+1}^{n-1}\sum_{k^*=k_3}^{k_4}\sum_{d^*=d_1}^{d_2}P^0(s,d^*,k^*)$$

Proof.

Let us suppose for the moment that the random barrier A does not absorb the barrier when they meet then define

$$p^{(n)}(r,0,d,d_n,j,k)=\text{prob}\{D_1^B>0,\ D_2^B>0,\ldots,D_n^B=d_n>0,\ X_n=k,\ Y_n=k\,|\,D_0^A=r,\ D_0^B=d,\ X_0=j\} \qquad (16)$$

then since the two simple random walks are independent one could proove that

$$p^{(n)}(r,0,d,d_n,j,k)=p^{(n)}(d,d_n,j,k)p_{**}^{(n)}(r+j,k) \tag{17}$$

If we compare the probabilities $p^{(n)}(r,0,d,d_n,j,k)$ and $f^{(n)}(r,0,d,d_n j,k)$ in the former have been added the probabilities of the mutually exclusive events that the particle and the random barrier A have been met, i.e. their distance was 0 or -1, for the first time before n and they meet again irrespectively which time that is at point k at time n while the distance of the paricle and the random barrier B is positive and equal with $d_n$ at time n given the initial states to be r,d, and j respectively. The probability that the distance between the particle and the random barrier B wil be positive given their initial distances to be r and d respectivel and the initial position of the particle to be j is

$$\sum_{d^*=j+b}^{j+b+2n^*} f^{(n^*)}(r,0,d,d^*,j,j+n^*) \quad \text{where } n^*=r/2$$

Given that one of the previous mutually exclussive events have take place then the probability that at time n the particle and the ranc barrier A will meet again and that the distance of the particle anc the random barrier B will be d is $p^{(n-n^*)}(0,0,d^*,d_n,j+n^*,k)=p^{(n-n^*)}(d^*,d_n,j+n^*,k)p_{**}^{n-n}(j+n^*,k)$. Obviously the probability that both events will take place is

$$\sum_{d^*=j+b}^{j+b+2n^*} f^{(n^*)}(r,0,d,d^*,j,j+n^*)p^{(n-n^*)}(d^*,d_n,j+n^*,k)p_{**}^{(n-n^*)}(j+n^*,k)$$

If we take all the possible meetings at time less than n which are mutually exclusive events we get (14) and (15).

In a similar way we could find the probability distribution $f^{(n)}(r,-1,d,d_n,j,k)$. Letting successively $n=n^*$, $n^*+1$,.. and using theorems 1 and 2 we get the required probability distributions in analytic forms. This is so since the probabilities $p^{(n)}(d,d_n,j,k)$ could be obtained in analytic forms from (7). Now we are looking for the following probability distribution

$$\begin{aligned} p^{(n)}(r,r_n,d,d_n,j,k)=\text{prob}\{&D_1^A>0,D_2^A>0,\dots,D_n^A=r_n>0,D_1^B>0,D_2^B>0,\\ &\dots,D_n^B=d_n>0,\ X_n=k\,|\,D_0^A=r,D_0^B=d_n,\ X_0=k\}. \end{aligned} \tag{18}$$

Let

$$\begin{aligned} R(s,d^*,k^*)=&f^{(s)}(r,0,d,d^*,j,k^*)p^{(n-s)}(d^*,d_n,k^*,k)p_{**}^{(n-s)}(k^*,\\ &k+r_n)+f^{(s)}(r,-1,d,d^*,j,k^*)p^{(n-s)}(d^*,d_n,k^*,k)p^{(n-s)}(k^*-1,k+r_n) \end{aligned}$$

$$[\tfrac{1}{2}(a+n-k-r_n]=n_4, \quad |n-r_n/2|=n_5, \quad k_4=k+s-n+r_n$$

Then the following theorem could be prooved.

<u>Theorem 2.</u> If $k+r_n \geq n^*+j$ then

$$p^{(n)}(r,r_n d,d_n,j,k)=p^{(n)}(d,d_n,j,k)p^{(n)}_{**}(j+r,k+r_n)-\sum_{s=n^*}^{n_2}\sum_{k^*=k_1}^{k_2}\sum_{d^*=d_1}^{d_2}R(s,d^*,k^*)-\sum_{s=n_4+1}^{n_2}\sum_{k^*=k_4}^{k_2}\sum_{d^*=d_1}^{d_2}R(s,d^*,k^*)-\sum_{s=n_2+1}^{n_5}\sum_{k^*=k_1}^{k_4}\sum_{d^*=d_1}^{d_2}R(s,d^*,k^*)$$

If $k+r_n<n^*+j$ then

$$p^{(n)}(r,r_n,d,d_n,j,k)=p^{(n)}(d,d_nj,k)p^{(n)}_{**}(j+r,k+r_n)-\sum_{s=n^*}^{n_2}\sum_{k^*=k_1}^{k_2}\sum_{d^*=d_1}^{d_2}R(s,d^*,k^*)-\sum_{s=n_2+1}^{n_5}\sum_{k^*=k_4}^{k_3}\sum_{d^*=d_1}^{d_2}R(s,d^*,k^*)-\sum_{s=n_4+1}^{n_5}\sum_{k^*=k_4}^{k_2}\sum_{d^*=d_1}^{d_2}R(s,d^*,k^*).$$

## REFERENCES


(1). Barnett, V.D. (1964). *A three player extension of the gamblers ruin problem.* J. Appl. Prob. 1, 321-334.

(2). Cox, D.R. and Miller (1968). *The theory of stochastic processes.* Methuen, London.

(3). Feller (1971). *An Introduction to Probability theory and its Applications,* Vol. II second Edition. John Wiley.

(4). Hardin, J.C. and Sweet, A.L. (1969). *A note on absorption probabilities for a random walk between a reflecting and an absorbing barrier.* J. Appl. Prob. 6, 224-336.

(5). Neuts, M.F. (1963). *Absorption probabilities for a random walk between a reflecting and an absorbing barrier.* Bull. Soc. Math. Belg. 15, 253-258.

(6). Vassiliou, P-C.G. and Karakostas, K.X. (1978). *A simple random walk with one absorbing barrier performing a simple random walk.* Trans. of the Eighth Prague Conference, Prague, Academia Prague, Vol.B, pp. 345-351.

(7). Vassiliou, P-C.G. and Karakostas, K.X. (1979). *On certain aspects of a simple random walk with one absorbing barrier performing a simple random walk Part I, A particle and a random absorbing barrier.* To appear.

(8). Weesakul, B. (1961). *The random walk between a reflecting and an absorbing barrier,* Ann. Math. Statist. 32, 765-9.

# SECTION III

# CONTROL THEORY

# ORTHOGONAL COLLOCATION USING MODAL REDUCTION

William E. Crockett

West Virginia College of Graduate Studies
Institute, West Virginia 25112 USA

The solution of parabolic differential equations by orthogonal collocation (OC) methods is much faster than by finite difference methods. Unfortunately, OC methods produce a large set of first order constant coefficient differential equations. This high order set is easily and accurately reduced with the use of modal methods and the aggregation matrix.

## INTRODUCTION

The works of Finlayson (2) and Villadsen (5,6) have shown the utility of OC for the solution of parabolic partial differential equations. Finlayson (2) compared OC with a Crank-Nicholson finite difference solution to a packed bed reactor analysis. He found OC to be four times faster and five times as accurate.

Methods other than OC, though, are still used to solve parabolic partial differential equations. OC reduces the partial differential equation to a large set of first order constant coefficient differential equations. To solve these, modal matrices and inverses must be determined. For this reason, many people prefer the slower, easier finite difference methods.

The use of the modal reduction methods presented by Lamba-Vittal Rao (4), Georgakis (3), and Crockett (1) provides a quick accurate solution to the problem by reducing the large set of differential equations to a much smaller set of equations.



*D. G. Lainiotis and N. S. Tzannes (eds.), Advances in Control, 185–191.*

## ORTHOGONAL COLLOCATION

The collocation method for the solution of differential equations is one of a general class of approximate methods known as the Method of Weighted Residuals which includes the well-known Galerkin and moments methods as special cases. The dependent variable y is expanded in a series

$$\sum a_i(t)P_i(x)$$

of known functions of x, $P_i$, multiplied by unknown functions of t, $a_i$. The trial functions are substituted into the original differential equation, which is satisfied at discrete collocation points x. This gives a set of ordinary differential equations governing $a_i(t)$. In the orthogonal collocation method developed by Villadsen and Stewart (6), the known expansions $P_i$ are orthogonal polynomials. The collocations points are denoted by $x_i$ and are the roots of $P_n(x)=0$.

Villadsen and Stewart also point out that the differential equations can be solved in terms of the solution at the collocation points, $y(x_i)$, instead of the coefficients $a_i(t)$. This is more convenient. Villadsen and Michelsen (5) present Fortran computer programs based upon a modification of these ideas. Their formulas for y and its derivatives are

$$y(x_i) = \sum_{k=1}^{N+2} \ell_k(x_i)y_k$$

$$\frac{dy(x_i)}{dx_i} = \sum_{k=1}^{N+2} \ell_k'(x_i)y_k$$

$$\frac{d^2y(x_i)}{dx^2} = \sum_{k=1}^{N+2} \ell_k''(x_i)y_k$$

where

$$\ell_i(x) = \frac{p_N(x)}{(x-x_i)p_N'(x_i)}$$

$$p_N(x) = x(x-1)P_N(x)$$

$$P_N(x) = \text{Legendre polynomials,}$$

$$P_0(x) = 1$$

$$P_1(x) = 2x-1$$

$$P_2(x) = 6x^2-6x+1$$

etc.

and

$$y_k = y(x_k)$$

The Gaussian quadrature formula for

$$q = \int_0^1 y(x,t)dx$$

becomes

$$q = \sum_{k=1}^{N+2} w_k y_k$$

with $w_k = \int_0^1 \ell_k(x)dx.$

A SAMPLE PROBLEM

Consider unsteady-state diffusion in a slab defined by:

$$\frac{\partial^2 y}{\partial x^2} = \frac{\partial y}{\partial t}$$

$$\frac{\partial y}{\partial x}(0,t) = 0$$

$$y(1,t) = 0$$
$$y(x,0) = -1$$

with a mean concentration defined by

$$q = \int_0^1 y(x,t)dx.$$

The classical solution is

$$q = -2 \sum_{n=1}^{\infty} \frac{4}{(2n-1)^2\Pi^2} e^{-\frac{(2n-1)^2\Pi^2 t}{4}}$$

A Simple, Approximate Solution

We use the simplest orthogonal collocation, one interior collocation point determined from:

$$P_1(x) = 2x-1 = 0$$

or

$$x_i = 0.5.$$

But we use three points, this one, plus the two end points so as to satisfy the boundary conditions on x. Therefore we use

$$x_1=0 \quad ; \quad x_2=0.5 \quad ; \quad x_3=1$$

Now

$$p_{N+2}=x(x-1)(2x-1)=2x^3-3x^2+x$$

Then

$$\ell_1(x)=2x^2-3x-1; \quad \ell_2(x)=-4x(x-1); \quad \ell_3(x)=x(2x-1)$$

$$\ell_1'(x)=4x-3 \quad ; \quad \ell_2'(x)=-4(2x-1); \quad \ell_3'(x)=4x-1$$

$$\ell_1''(x)=4 \quad ; \quad \ell_2''(x)=-8 \quad ; \quad \ell_3''(x)=4$$

And the partial differential equation at the interior collocation point becomes

$$\frac{dy(x_i)}{dt} = y_{xx}(x_i) = \sum_{k=1}^{3} \ell_k''(x_i)y_k \qquad i=2$$

or

$$\frac{dy(.5)}{dt} = \dot{y}_2 = 4y_1-8y_2+4y_3$$

But notice that the two boundary conditions on x

$$y(1,t)=0 \text{ becomes } y(1)=0 \text{ becomes } y_3=0$$

and

$$\frac{dy}{\partial x}(0,t)=0 \text{ becomes } y_x(0)= \sum_{k=1}^{3} \ell'_k(0)yk=0$$

$$\text{becomes } -3y_1+4y_2-y_3=0$$

Thus all we must solve is:

$$y_3 = 0$$

$$-3y_1+4y_2-y_3=0$$

$$\dot{y}_2=4y_1-8y_2+4y_3$$

or

$$y_1 = (4/3)y_2$$

$$\dot{y}_2 = 4(4/3)y_2-8y_2= -(8/3)y_2$$

or

$$y_2= -e^{-(8/3)t}= -e^{-2.667t}$$

and the simple, approximate q is

$$q=0y_1+y_2+0y_3= -e^{-2.667t}$$

The exact q is

$$q = -.8105695e^{-2.4674011t}-.090063274e^{-22.2066094t}$$

$$-.0324228e^{-61.68502751t}-.01654223e^{-120.9026539t}$$

$$-\dots$$

so we've obtained the first eigenvalue to within 8% but the first coefficient to within only 23%. We can do better

## A Difficult, Excellent Solution

To obtain better accuracy, we need only use more interior collocation points. With six interior points (and the two end points, $x_1$=0, $x_8$=0), obtained from

$$P_6(x) = 0,$$

we have

$$\dot{y} = A\ y \qquad A = 6x6$$
$$q = D\ y \qquad D = 1x6$$
$$y(0)= (-1\ -1\ -1\ \dots)^T$$

After laboriously solving this set, we obtain

$$q = -.810569469e^{-2.46740110t} - .090069326e^{-22.2066484t}$$
$$- .031792382e^{-61.6950457t} - .022846989e^{-120.656339t}$$
$$- .001299956e^{-225.835784t} - .043421877e^{-782.208549t}$$

When compared, we find that the first four eigenvalues are all within 0.3% of the exact and the first two coefficients are accurate to at least five significant figures. The accuracy has increased tremendously, but the work has also increased. We now decrease the work.

## A Simple, Good Solution

Lamba and Vittal Rao (4) show that a reduced order model, using the aggregation matrix will provide a good representation of a high order model. They show that if the high order model is represented by:

$$\dot{y} = A\ y$$
$$q = D\ y$$
$$y(0) \text{ given}$$

then the reduced-order modal model, of order r,

$$\dot{z} = F\ z$$

$$q_m = D_m z$$

can be found from

$$F = CAC'\ (CC')^{-1}$$
$$D_m = DC'\ (CC')^{-1}$$

if the aggregation matrix C is found from

$$C = \overline{W}TW^{-1}$$

where:

W = modal matrix of A with its columns arranged from left to right in the order of increasing negative magnitude.

T = $[I_r:0]$; the identity matrix is same order as reduced model needed.
$\overline{W}$ = the first r columns and rows of W

For instance, if we want only a first order model for $q_m$, by these manipulations we find

$$q_m \cong -.832471258e^{-2.4674011t} ,$$

the second order reduced modal model is

$$q_m \cong -.785139818e^{-2.4674011t} -.08726787e^{-22.2066484t} ;$$

the third order modal model is

$$q_m \cong -.814495686e^{-2.4679011t} -.082058028e^{-22.2066484t}$$

$$-.028888285e^{-61.6950457t}$$

The approximate results are good:

| t | exact q | ortho 6 term q | approx 1 term q |
|---|---|---|---|
| .05 | -.747686 | -.727797 | -.735852 |
| .1 | -.643177 | -.643176 | -.650446 |
| .5 | -.236050 | -.236050 | -.242426 |
| 1. | -.068740 | -.068740 | -.070598 |

## CONCLUSIONS

Orthogonal collocation is a good method for solving parabolic partial differential equations. To improve the accuracy, we must increase the number of collocation points. This increases the order of the resulting set of first order ordinary differential equation. The use of the Lamba-Vittal Rao aggregation matrix gives us reduced order modal representations of very good accuracy. This method is especially useful if one wishes a simple model for use, not just a table of answers.

## REFERENCES

1. Crockett, W.E.: 1979, Proc. JACC-Denver, p. 207.
2. Finlayson, B.A.: 1971, Chem. Eng. Sci. 26, p. 1081.
3. Georgakis, C.: 1977, Proc. JACC, San Francisco, p. 639.
4. Lamba, S.S., Vittal Rao, S.: 1978, IEEE Trans. Auto. Cont. AC23, p. 81.
5. Villadsen, J., Michelson, M.L.: 1978, Solution of Differential Equation Models by Polynomial Approximations, Prentice-Hall.
6. Villadsen, J., Stewart, W.E.: 1967, Chem. Eng. Sci. 22, p. 1483.

# ORTHOGONAL TRANSFORMATIONS AND SQUARE-ROOT ALGORITHMS IN LINEAR-QUADRATIC OPTIMAL CONTROL OF DISCRETE-TIME SYSTEMS

R. Gorez

Lab. of Automatic Control and Systems Analysis
University of Louvain-La-Neuve
Bâtiment Maxwell - Place du Levant 3
B - 1348 Louvain-La-Neuve

Orthogonal transformations and square-root algorithms result in a straightforward solution of the linear quadratic optimal control of discrete-time systems. These algorithms are computationnally efficient in case of time-invariant problems. Moreover, the case of reference model-following systems can be treated, with a small increase of the computational task.

An algorithm using Gram-Schmidt procedure is given for single input-single output systems; memory requirements are very low, allowing the use of programmable pocket calculators for low-order systems. For the "cheap-control" problem, a closed form of the resulting dead-beat control law can be easily derived.

## 1. INTRODUCTION

Fast algorithms using orthogonal transformations techniques and factorisations of matrices are well known in the field of estimation and filtering (1-4), but little use of them has been done by control engineers (5-6).

However, these techniques result in a straightforward and beautifully simple derivation of the control law and Riccati equations for the linear quadratic optimal control of discrete-time systems. Moreover, in case of time-invariant problems with zero terminal cost, they yield efficient algorithms, with a computational task which is only linearly related to the dimension and the number of inputs and outputs of the system. Then, the case of reference model-following systems can be treated very easily, with a linear increase of the computational burden. Furthermore, these



*D. G. Lainiotis and N. S. Tzannes (eds.), Advances in Control, 192–200.*

techniques can be used for singular control problems as well as for the regular ones, allowing dead-beat control laws to be derived in closed form.

## 2. LINEAR QUADRATIC OPTIMAL CONTROL

We consider the discrete-time linear system :

$$x(k+1) = A(k)\,x(k) + B(k)\,u(k), \tag{1}$$

$$y(k) = C(k)\,x(k) + D(k)\,u(k), \tag{2}$$

and the quadratic performance criterion :

$$J = \Sigma_{i=0}^{N-1}\left[y^T(i)Q(i)y(i) + u^T(i)R(i)u(i)\right] + x^T(N)P_N x(N) \tag{3}$$

where $x, u, y$ are respectively state, control and output vectors, of dimensions $n$, $m$, $p$, $Q$ is a positive-definite symmetric matrix, $R$ and $P_N$ are non negative-definite symmetric matrices, of appropriate dimensions, $N$ is a given, possibly infinite, horizon.

Using a dynamic programming argument (7) results in

$$x^T Px = \underset{u}{Min}(y^T Qy + u^T Ru + x_+^T P_+ x_+)\ , \tag{4}$$

where $x$, $u$, $y$, $P$, $Q$, $R$, stand for $x(k)$, $u(k)$, $y(k)$, $P(k)$, $Q(k)$, $R(k)$, and $x_+$, $P_+$, for $x(k+1)$, $P(k+1)$, respectively. Clearly, $P$ must be a nonnegative-definite symmetric matrix, with final value $P(N)=P_N$, and $x^T(0)\ P(0)\ x(0)$ will give the optimal value of the performance criterion (3).

The quadratic form in the right hand side of (4) may be written as follows :

$$\left\|\begin{bmatrix} R^{T/2}u \\ P_+^{T/2}x_+ \\ Q^{T/2}y \end{bmatrix}\right\|^2 = \left\|\begin{bmatrix} R^{T/2} & 0 \\ P_+^{T/2}B & P_+^{T/2}A \\ Q^{T/2}D & Q^{T/2}C \end{bmatrix}\begin{bmatrix} u \\ x \end{bmatrix}\right\|^2 , \tag{5}$$

where $R^{T/2} \doteq (R^{1/2})^T$, $P_+^{T/2} = (P_+^{1/2})^T$, $Q^{T/2} = (Q^{1/2})^T$ are the square-roots of nonnegative definite matrices $R$, $P$, $Q$, and can be obtained by Cholesky decomposition (1-4).

It is well known (4,8) that there are orthogonal transformations $T$ resulting in

$$T \begin{bmatrix} R^{T/2} & 0 \\ P_+^{T/2} B & P_+^{T/2} A \\ Q^{T/2} D & Q^{T/2} C \end{bmatrix} = \begin{bmatrix} R_c^{T/2} & \bar{K} \\ 0 & P^{T/2} \\ 0 & 0 \end{bmatrix} , \tag{6}$$

where $R_c^{T/2}$ and $P^{T/2}$ are upper triangular matrices, of order $m$ and $n$ respectively. The transformation can be done via Householder transformations or modified Gram-Schmidt algorithms (4,8), even if $R$ is not of full rank (in that case, the same rows of $R_c^{T/2}$ and $\bar{K}$ are null).

Then, (5) may be replaced by

$$\| R^{T/2} u + \bar{K} x \|^2 + \| P^{T/2} x \|^2 , \tag{7}$$

the minimum of which clearly occurs for :

$$u = u_o - (R_c^{T/2})^+ \bar{K} x , \tag{8}$$

where $u_o$ is any vector in the null space of $R_c^{T/2}$ and $(R_c^{T/2})^+$ is the Moore-Penrose pseudo-inverse (8) ; if $R_c$ is of full rank, $u_o$ is obviously null and $(R_c^{T/2})^+ = (R_c^{T/2})^{-1}$ .

Therefore, we have simultaneously obtained the solution $P$ of (4) and the optimal regulator gain

$$K = (R_c^{T/2})^+ \bar{K} . \tag{9}$$

Moreover, premultiplying both sides of (6) by their transposes results in

$$R_c = R + B^T P_+ B + D^T Q D , \tag{10}$$

$$R_c K = R_c^{1/2} \bar{K} = B^T P_+ A + D^T Q C , \tag{11}$$

$$P = A^T P_+ A + C^T Q C - K^T R_c K , \tag{12}$$

which are the usual equations of the discrete linear quadratic optimal control problem.

## 3. TIME - INVARIANT ZERO TERMINAL COST PROBLEM

If $A$, $B$, $C$, $D$, $Q$, $R$, are time-invariant matrices and $P_N = 0$, it can be shown (2) that

$$P - P_+ = LL^T \tag{13}$$

where $L^T$ is a $p$ by $n$ upper trapezoidal matrix, of nonincreasing rank.

Then, using (6) and (13) in the right-hand side of (5) yields

$$\left\| \begin{bmatrix} R^{T/2}u \\ P_+^{T/2}x_+ \\ Q^{T/2}y \end{bmatrix} \right\|^2 = \left\| \begin{bmatrix} R_c^{T/2} & \bar{K} \\ 0 & L^T \\ 0 & P_+^{T/2} \\ 0 & 0 \end{bmatrix} \begin{bmatrix} u \\ x \end{bmatrix} \right\|^2 , \qquad (14)$$

but, because $A$, $B$, $C$, $D$, $Q$, $R$ are time-invariant matrices, it may also be written as

$$\left\| \begin{bmatrix} R_+^{T/2} & 0 \\ P_{++}^{T/2}B_+ & P_{++}^{T/2}A_+ \\ Q_+^{T/2}D_+ & Q_+^{T/2}C_+ \\ L_+^T B_+ & L_+^T A_+ \end{bmatrix} \begin{bmatrix} u \\ x \end{bmatrix} \right\|^2$$

where $P_{++}$ and $L_+$ stand for $P(k+2)$ and $L(k+1)$, respectively.

Then, using an orthogonal transformation as in transformation (6) for the first three block-rows, results in another expression for (14) :

$$\left\| \begin{bmatrix} R_{c+}^{T/2} & \bar{K}_+ \\ 0 & P_+^{T/2} \\ 0 & 0 \\ L_+^T B & L_+^T A \end{bmatrix} \begin{bmatrix} u \\ x \end{bmatrix} \right\|^2 = \left\| \begin{bmatrix} R_{c+}^{T/2} & \bar{K}_+ \\ L_+^T B & L_+^T A \\ 0 & P_+^{T/2} \\ 0 & 0 \end{bmatrix} \begin{bmatrix} u \\ x \end{bmatrix} \right\|^2 \qquad (15)$$

Thus, it is now obvious that the matrices in right-hand sides of (14) and (15) are related by an orthogonal transformation, yielding :

$$T \begin{bmatrix} R_{c+}^{T/2} & \bar{K}_+ \\ L_+^T B & L_+^T A \end{bmatrix} = \begin{bmatrix} R_c^{T/2} & \bar{K} \\ 0 & L^T \end{bmatrix} \qquad (16)$$

This transformation can be done in at most $r < m$ sequential Householder or Gram-Schmidt transformations, on $m+n$ vectors of dimension $m+p$, $r$ being the rank of $R_c$. Computational advantages are obvious, the total number of arithmetic operations being proportional to $r(m+p)\,(m+n)$ instead of $n^3$.

Transformation (16) is used recursively in a backward iterative procedure, starting with values of $R_{c+}$, $K_+$, $L_+$ obtained from equations (10 - 13) for $k = N-1$ ;
it can be shown that, for $k = N-1$ :

$$\bar{K} = FC \quad , \qquad F = (R_c^{1/2})^+ D^T Q \quad ,$$
$$L^T = Q_o^{T/2} \quad , \qquad Q_o = Q - FF^T \quad , \qquad (17)$$

and $Q_o$ is a non negative symmetric matrix of order $p$.

The iterative procedure is stopped when $k = 0$, if $N$ is finite, and when $L = 0$ if $N \to \infty$ .

Equation (9) is still used for calculating the optimal regulator gain. In case of control with receding or infinite horizon, only the last value of the gain will be needed.

## 4. REFERENCE MODEL FOLLOWING SYSTEM

Optimal control theory may be used in view of constraining the system output to follow the output of a reference model (9). Equations of the model are :

$$x_{r+} = A_r x_r \quad , \qquad (18)$$

$$y_r = C_r x_r \quad , \qquad (19)$$

where $x_r$ and $y_r$ are respectively state and output vectors, of dimensions $n_r$ and $p$.

The following performance criterion is used:

$$J = \Sigma\left[(y-y_r)^T Q(y-y_r) + u^T Ru\right] \qquad (20)$$

Then, we may consider an augmented system of dimension $n_a = n + n_r$ , with $m$ inputs and $p$ outputs. The state and output equations of this system are :

$$\begin{bmatrix} x_+ \\ x_{r+} \end{bmatrix} = \begin{bmatrix} A & 0 \\ 0 & A_r \end{bmatrix} \begin{bmatrix} x \\ x_r \end{bmatrix} + \begin{bmatrix} B \\ 0 \end{bmatrix} u \quad , \qquad (21)$$

$$e = \begin{bmatrix} C & -C_r \end{bmatrix} \begin{bmatrix} x \\ x_r \end{bmatrix} + Du \quad . \qquad (22)$$

Then, the optimal control law is

$$u = - Kx - K_r x_r \ , \tag{23}$$

where
$$\begin{bmatrix} K & K_r \end{bmatrix} = (R_c^{T/2})^+ \begin{bmatrix} \bar{K} & \bar{K}_r \end{bmatrix} \ , \tag{24}$$

and $\bar{K}$ , $\bar{K}_r$ are obtained by applying transformation (16) to the augmented system :

$$T \begin{bmatrix} R_{c+}^{T/2} & \bar{K}_+ & \bar{K}_{r+} \\ L_+^T B & L_+^T A & L_{r+}^T A_r \end{bmatrix} = \begin{bmatrix} R_c^{T/2} & \bar{K} & \bar{K}_r \\ 0 & L^T & L_r^T \end{bmatrix} \tag{25}$$

Moreover, at the start-up of the procedure, equations (17) yield for the augmented system :

$$\begin{aligned} \begin{bmatrix} \bar{K} & \bar{K}_r \end{bmatrix} &= F \begin{bmatrix} C & -C_r \end{bmatrix} \ , \\ \begin{bmatrix} L^T & L_r^T \end{bmatrix} &= Q_o^{T/2} \begin{bmatrix} C & -C_r \end{bmatrix} \ , \end{aligned} \tag{26}$$

Then, it is obvious that the sequence of orthogonal transformations and consequently, values of $K$ , are not altered by the augmentation of the system, and the computational burden is increased by a factor $(n + n_r + m) \ / \ (n + m)$.

## 5. SINGLE INPUT - SINGLE OUTPUT SYSTEM.

If $m = p = 1$ , equations (1-3) are written as follows :

$$x_+ = Ax + bu \ , \tag{27}$$

$$y = c^T + du \ , \tag{28}$$

$$J = \Sigma \ (y^2 + \rho u^2) \ , \tag{29}$$

where $Q$ has been normalized to unity.

Using a Gram-Schmidt technique for transformation (16) results in the algorithm given on the next page.

The computational task is very low; for each of the last $n$ columns, only $2+2$ multiplications and $2$ additions are required. Then, the algorithm can be implemented on a programmable pocket calculator, $n$ being limited by the number of memory registers : a second order system can be treated on a 8 registers calculator.

Algorithm for single input - single output systems.

1) Initialize $i = 1$

- if $d \neq 0$, $r_c = \rho + d^2$, $s = \sqrt{r_c}$

$$\bar{k}^T = c^T d/s, \qquad q_o = \rho/r_c$$

- if $d = 0$, $s = \sqrt{\rho}$, $\bar{k}^T = 0^T, q_o = 1$

$$l^T = \sqrt{q_o}\, c^T$$

2) Compute the optimal regulator gain

- if $s \neq 0$, $k^T = k^T/s$

3) Stop if $i = N$ or $l^T = 0^T$

4) Compute $\beta = l^T b$ and $a^T = l^T A$

Construct (2 x $(1+n)$ matrix) $M = \begin{bmatrix} s & \bar{k}^T \\ \beta & a^T \end{bmatrix}$

5) Increment $i \leftarrow i+1$ and transform $M$ by the following procedure :

- Compute $v = \sqrt{s^2 + \beta^2}$

- if $v \neq 0$, compute $u_1 = s/v$, $u_2 = \beta/v$

The transformation matrix is $T = \begin{bmatrix} u_1 & u_2 \\ -u_2 & u_1 \end{bmatrix}$

and $TM = \begin{bmatrix} v & u_1\bar{k}^T + u_2 a^T \\ 0 & -u_2\bar{k}^T + u_1 a^T \end{bmatrix}$

Then, substitute $v \rightarrow s$

$$u_1\bar{k}^T + u_2 a^T \rightarrow \bar{k}^T$$

$$-u_2\bar{k}^T + u_1 a^T \rightarrow l^T$$

- if $v = 0$, $T = I$ whence $0 \rightarrow s$

$k^T$ and $l^T$ are unchanged

6) Go to 2.

## 6. CHEAP CONTROL PROBLEMS.

The case $\rho = 0$ is sometimes refered to as a cheap-control problem (10).

Then, if $d \neq 0$ ,

$$M = \begin{pmatrix} d & c^T \\ 0 & 0^T \end{pmatrix}$$

is in the required form from the very beginning of the procedure; it results in an instantaneous control action with gain $k^T = c^T/d$.

If $d = 0$, and if there is a pure delay of $m$ steps in the system ( $c^T A^{i-1} b=0$, $i=1$ to $m$ ; $c^T A^m b \neq 0$), for $i=1$ to $m+1$ :

$$M = \begin{bmatrix} 0 & \underline{0}^T \\ c^T A^{i-1} b & c^T A^i \end{bmatrix} ,$$

the transformation (16) being identity for $i=1$ to $m$ ,and a row permutation for $i=m+1$. The resulting dead-beat regulator gain is $k^T = c^T A^{m+1}/c^T A^m b$.

Thus, results of (10) are obtained by the given algorithm.

## 7. CONCLUSIONS

The use of orthogonal transformations techniques has resulted in a straightforward derivation of the equations of the discrete linear quadratic optimal control problem and in an algorithm for the calculation of the optimal regulator gain. This algorithm is far more efficient than conventional Riccati equations in case of time-invariant problems with few inputs and outputs.

Cheap control problems and regular problems can be treated in the same algorithm; the first are even easier and faster to deal with.

At last, transformation techniques have separation properties which allow to treat reference model-following control systems, with a small increase of the computational task.

REFERENCES

(1) Kaminsky, P.G., Bryson, A.E., and Schmidt, S.F.: 1971, *IEEE Trans. Automat.Contr.*,AC-16, pp 727-736.

(2) Morf, M., and Kailath, T.: 1975, *IEEE Trans. Automat. Contr.*, AC-20, pp 487-497

(3) Lainiotis, D.G.: 1976, *IEEE Trans.Automat.Contr.*, AC-20, pp 555-556

(4) Bierman, G.J.: 1977, *Factorization Methods for Discrete Sequential Estimation*, New York, Academic Press

(5) Payne, H.J., and Silverman, L.M.: 1973, *Proc. IEEE Conf. Decision and Control*, pp 558-563

(6) Silverman, L.M.: 1976, in *Control and Dynamic Systems*, 12, New York, Academic Press

(7) Kwakernaak, H., and Sivan, R.: 1972, *Linear Optimal Control Systems*, New - York, Wiley

(8) Stewart, G.W.: 1973, *Introduction to Matrix Computations*, New-York, Academic Press

(9) Tyler, J.S.: 1966, *IEEE Trans. Automat. Contr.*,AC-9, pp 485-498

(10)Leden, B.: 1976, *IEEE Trans. Automat. Contr.*,AC-21, pp 791-792

AN ALGORITHM FOR POLE PLACEMENT BY A CONSTANT OUTPUT FEEDBACK

S. P. Bingulac

Institut Mihailo Pupin, Volgina 15, Beograd, Yugoslavia

## ABSTRACT

An iterative computational procedure for pole placement by a constant output feedback regulator matrix is presented. In the regulator matrix only elements in a single row and column are adjusted, while other elements may have fixed preassigned values. The pole placement procedure is equally applicable in both the state feedback and the partial pole placement. In either case the procedure is computationally more convenient than existing methods.

## INTRODUCTION

Recently published papers treating the problem of pole placement by a constant output feedback /1/ - /4/ have been devoted mainly to the formulation of both necessary and sufficient conditions. Little attention, however, has been paid to a design of readily applicable computational algorithms. The purpose of this paper is to describe an iterative algorithm for pole placement by a constant output feedback regulator matrix. According to previously published results, it is assumed that integers n, m and k, representing dimensions of the state, input and output vectors, respectively, satisfy the following /1/, /3/ pole placement condition

$$m + k - 1 = n.$$

Under this condition it is possible to adjust only elements in a single row and column of the regulator matrix, while all other elements may have fixed preassigned values. The adjustable elements are calculated by solving a set of n-1 algebraic equations



D. G. Lainiotis and N. S. Tzannes (eds.), Advances in Control, 201–209.

of the quadratic form. The pole placement algorithm requires a particular realization of the open-loop system. This realization, referred to as "output feedback" realization, is characterized by a specific form of both input and output matrices, B and C, respectively. In B there are only m rows having nonzero elements, while in C only k columns may have elements different from zero. The realization is described in the next section, while the algorithm is presented in subsequent sections.

## DEFINITION OF THE OUTPUT FEEDBACK REALIZATION

Consider a controllable and observable realization

$$\{A,\ B,\ C\} \tag{1}$$

describing a linear multivariable dynamic system

$$\begin{aligned} \dot{x}(t) &= A\ x(t) + B\ u(t) \\ y(t) &= C\ x(t) \end{aligned} \tag{2}$$

where $x(t)\varepsilon R^n$, $u(t)\varepsilon R^m$, and $y(t)\varepsilon R^k$, $1 \le m \le n$, $1 \le k \le n$, $m + k - 1 = n$ are state, control, and output vectors, respectively, while A, B and C are constant matrices of compatible dimensions.

$$\text{Let}\quad \rho|B| = m,\ \rho|C| = k \quad\text{and}\quad T \triangleq |B \vdots X| \tag{3}$$

where X is an arbitrary $nx(n-m)$ matrix assuring the regularity of T, i.e. $\rho|T| = n$.

Then, using the matrix T, the realization (1) may be transformed into

$$\{\tilde{A},\tilde{B},\tilde{C}\} = \{T^{-1}A\ T,\ T^{-1}B, C\ T\} \tag{4}$$

where, according to (3), the matrix B has the form

$$\tilde{B} = \begin{vmatrix} I_m \\ \hdashline 0 \end{vmatrix}.$$

Let $v_i$, $i\varepsilon[1,k]$, denote the indices of linearly independent colums $\tilde{c}^i$ of the matrix $\tilde{C} \triangleq |\tilde{c}^1 \ \ldots \ \tilde{c}^n|$. The test for linear independence should start from the first column of $\tilde{C}$. The linearly dependent columns $\tilde{c}^j$, $j \neq v_i$, (see the example below) may be expressed by linear combinations of linearly dependent ones, i.e.

$$\tilde{c}^j = \sum_h \bar{p}_{hj}\ \tilde{c}^h \quad \text{for} \quad h = [v_1,\ldots,v_s]\ ;\ \ s < j.$$

Let $P = \{p_{ij}\}$ be a regular triangular (nxn) matrix with

$$P_{ij} = \left\{ \begin{array}{ll} 1 & \text{for } i = j \\ -p_{hj} & \text{for } i = h \\ 0 & \text{otherwise} \end{array} \right\} .$$

Example. In the case of $n = 7$ and $v_i = [1,4,6]$ the columns $\tilde{c}^2, \tilde{c}^3$, $\tilde{c}^5$ and $\tilde{c}^7$ are related to $\tilde{c}^1$, $\tilde{c}^4$ and $\tilde{c}^6$ by
$\tilde{c}^2 = \bar{p}_{12}\tilde{c}^1$ ; $\tilde{c}^3 = \bar{p}_{13}\tilde{c}^1$ ; $\tilde{c}^5 = \bar{p}_{15}\tilde{c}^1 + \bar{p}_{45}\tilde{c}^4$ ;
$\tilde{c}^7 = \bar{p}_{17}\tilde{c}^1 + \bar{p}_{47}\tilde{c}^4 + \bar{p}_{67}\tilde{c}^6$, while the matrix P becomes

$$P = \left| \begin{array}{ccccccc} 1 & p_{12} & p_{13} & 0 & p_{15} & 0 & p_{17} \\ & 1 & 0 & 0 & 0 & 0 & 0 \\ & & 1 & 0 & 0 & 0 & 0 \\ & & & 1 & p_{45} & 0 & p_{47} \\ & & & & 1 & 0 & 0 \\ & & 0 & & & 1 & p_{67} \\ & & & & & & 1 \end{array} \right| \tag{5}$$

Using the matrix P, the realization (4) may be further transformed into

$$\{\hat{A},\hat{B},\hat{C}\} = \{P^{-1}\tilde{A}\,P,\ P^{-1}\tilde{B},\ \tilde{C}\,P\} \tag{6}$$

where, according to (5), the matrices $\hat{B}$ and $\hat{C}$ are of the form

$$\hat{B} = \left| \begin{array}{c} B_1 \\ \hline 0 \end{array} \right| ; \quad \hat{C} = |0 \,\vdots\, .. \,\vdots\, \tilde{c}^{v_1} \,\vdots\, 0 \,\vdots\, .. \,\vdots\, 0 \,\vdots\, \tilde{c}^{v_k} \,\vdots\, 0| \quad .$$

The realization (6) is referred to as "output feedback" realization, since it facilitates the solution of the pole placement problem by a constant output feedback. In a particular case, which often occurs, indices $v_i$ are given by

$$v_i = i \ ; \qquad i \varepsilon [1,k] \quad . \tag{7}$$

Eq.(7) implies that the first k columns of $\tilde{C} \triangleq |C_1 \,\vdots\, \tilde{C}_2|$ are linearly independent, i.e. $\rho|\tilde{C}_1| = k$. In this case matrices P and $\hat{C}$ are of the form

$$P = \left| \begin{array}{c|c} I_k & P_{12} \\ \hline 0 & I_{n-k} \end{array} \right| ; \quad \hat{C} = |\tilde{C}_1 \,\vdots\, \tilde{C}_2| \cdot P = |\tilde{C}_1 \,\vdots\, 0|, \text{ where } P_{12} = -\tilde{C}_1^{-1}\tilde{C}_2 .$$

## FORMULATION OF THE POLE PLACEMENT PROBLEM

Assume that an open-loop system is represented by the output feedback realization

$$\{A,B,C\} \tag{8}$$

described in the previous section. Without loss of generality and in order to simplify further discussion the particular case given by (7) will be considered. Thus, the realization (8) has the following properties

$$C = \left| C_1 \,\vdots\, 0 \right| ; \quad B = \left| \begin{array}{c} B_1 \\ \hline 0 \end{array} \right| ; \quad \rho|C_1| = k \; ; \quad \rho|B_1| = m. \tag{9}$$

The problem of pole placement by a constant output feedback may be formulated in the following way.

Given (i) a realization (8) satisfying (9), and
(ii) a set of n closed-loop eigenvalues $\lambda_i$, $i\varepsilon[1,n]$,
find a constant (mxk) output feedback regulator matrix

$$F^* = \{f^*_{ij}\} ,$$

assuring that n eigenvalues of the closed-loop system $H = A+B\ F^*C$ are equal to specified $\lambda_i$, i.e.

$$\det(Is-H) = \prod_{i=1}^{n} (s-\lambda_i). \tag{10}$$

Let all eigenvalues $\lambda_i$ be distinct. This assumption is made purely for the convenience of simplifying further discussion. The case of multiple eigenvalues will be discussed later.

## AN ALTERNATE PROBLEM FORMULATION

Using (9), the closed-loop system matrix H becomes

$$H = A + \left| \begin{array}{c|c} B_1F^*C_1 & 0 \\ \hline 0 & 0 \end{array} \right| , \tag{11}$$

which, with the partion $A \triangleq \left| \begin{array}{c|c} A_{11} & A_{12} \\ \hline A_{21} & A_{22} \end{array} \right|$ (with $A_{11}$ of dimension m × k)

and the substitution $F \triangleq A_{11} + B_1F^*C_1$, yields

$$H = \left| \begin{array}{c|c} F & A_{12} \\ \hline A_{21} & A_{22} \end{array} \right| \tag{12}$$

Thus, according to (12) the pole placement condition (10) may be expressed in the following way.

Find a (mxk) constant matrix $F = \{f_{ij}\}$ so that the Jordan form $H_j$ of the closed-loop matrix H has eigenvalues $\lambda_i$, i.e.

$$H_j = Q^{-1}\, H\, Q = \mathrm{diag}\{\lambda_i\}\ . \qquad (13)$$

In (13) Q is the modal matrix transforming H into the Jordan form,

$$H\, Q = Q\, H_j \qquad (14)$$

Using $Q \triangleq |q^1 \vdots \ldots \vdots q^n|$, $q^i$ being the eigenvector associated with the eigenvalue $\lambda_i$, eq. (14) may be expressed by

$$H\, q^i = q^i \lambda_i , \quad \text{or} \quad (H - I_n \lambda_i) q^i = 0\ ; \quad i \varepsilon [1,n], \qquad (15)$$

Eq. (15) respesents the familiar condition for calculating the eigenvalues and eigenvectors of a given matrix. In the case of pole placement, however, in eq. (15) the eigenvalues $\lambda_i$ are given, while the upper left part of the matrix H, i.e. the (mxk) block F, should be determined so that (15) is satisfied for $i \varepsilon [1,n]$.

## SELECTION OF ADJUSTABLE ELEMENTS $f_{ij}$ OF THE MATRIX F

Since it is assumed that

$$m + k - 1 = n \qquad (16)$$

it is sufficient in the matrix F to adjust only the m + k - 1 elements. As it will be shown, it is most convenient to adjust the following elements

$$f_{hj} \quad \text{and} \quad f_{ih}\ ; \quad j \varepsilon [1,k]\ , \quad i \varepsilon [1,m], \qquad (17)$$

i.e. the elements in the h-th row and the h-th column. The integer h may be selected as $1 \le h \le \min(m,k)$. To simplify further notation it will be assumed that

$$h = 1.$$

Thus, to emphasize the adjustable elements, the matrix F will be represented by

$$F = \begin{array}{c}\left[\begin{array}{c|c} f_o & f^1 \\ \hline f^2 & F_2 \end{array}\right]\begin{array}{l}\}\,1\\ \}\,m_1\end{array}\\ \begin{array}{cc}\underbrace{\quad}_{1} & \underbrace{\qquad\qquad}_{k_1}\end{array}\end{array}\ ; \qquad \begin{array}{l} m_1 = m-1 \\ k_1 = k-1 \end{array} \qquad (18)$$

According to (18), matrix H, given by (12), may be expressed by

$$H = \left|\begin{array}{c|c|c} f_o & f^1 & a^1 \\ \hline f^2 & & \\ \hline a^2 & & H_2 \end{array}\right| \tag{19}$$

which, using (15), leads to

$$\left|\begin{array}{c|c|c} f_o - \lambda_i & f^1 & a^1 \\ \hline f^2 & & \\ \hline a^2 & & H_2 - I_{n_1}\lambda_i \end{array}\right| \cdot q^i \triangleq B^i \cdot q^i = 0, \tag{20}$$

where $n_1 = n - 1$.

Note that the vectors $a^1$ and $a^2$, eq. (19), represent the first row and the first column of submatrices $A_{12}$ and $A_{21}$, eq, (11), respectively. As it has been emphasized earlier, the $(m_1 x k_1)$ block $F_2$, eq. (18), have fixed arbitrary values.

## ALGORITHM DESCRIPTION

Assume that the first element in the eigenvector $q^i$, $i\varepsilon[1,n]$, has unity value. Then eq. (20) yields

$$\begin{array}{l} f_o - \lambda_i = -\left|\begin{array}{c|c} f^1 & a^1 \end{array}\right| \tilde{q}^i \\ \left|\begin{array}{c} f^2 \\ \hline a^2 \end{array}\right| = -\left|B_2^i\right| \tilde{q}^i \end{array} \quad ; \text{ where } \quad q^i \triangleq \left|\begin{array}{c} 1 \\ \hline \tilde{q}^i \end{array}\right| ; B_2^i \triangleq (H_2 - I_{n_1}\lambda_i). \tag{21}$$

If the matrix $B_2^i$ is regular, the vector $\tilde{q}^i$ may be eliminated from (21) leading to an expression relating the adjustable elements, i.e.

$$f_o = \lambda_i + \left|\begin{array}{c|c} f^1 & a^1 \end{array}\right| (B_2^i)^{-1} \left|\begin{array}{c} f^2 \\ \hline a^2 \end{array}\right| . \tag{22}$$

The regularity of $B_2^i$ may be satisfied by an appropriate choice of the fixed elements in the submatrix $F_2$, eq.(18). Introducing further

$$(B_2^i)^{-1} = \left|\begin{array}{c|c} \overbrace{P_{11}^i}^{k_1} & P_{12}^i \\ \hline P_{21}^i & P_{22}^i \end{array}\right| \Big\} m_1 \qquad (n_1 \times n_1) \tag{23}$$

eq. (22) may be transformed into

$$f_o = \lambda_i + a^1 P_{22}^i a^2 + f^1 P_{12}^i a^2 + a^1 P_{21}^i f^2 + f^1 P_{11}^i f^2, i\varepsilon[1,n]. \tag{24}$$

Note that in (24) the following quantities are unknown

$$\begin{aligned} &f_o \text{ - scalar} \\ &f^1 \text{ - } k_1 \text{ dimensional row vector} \\ &f^2 \text{ - } m_1 \text{ dimensional column vector} \end{aligned} \tag{25}$$

leading to $1 + k_1 + m_1 = n$ unknowns.

Since eq. (24) has been derived from (20), it may be concluded that quantities (25) satisfying (24) for a given $\lambda_i$ guarantee the desired eigenvalue in the closed-loop matrix H.
Thus, the problem of pole placement by a constant output feedback regulator has been reduced to the solution of n equations, eq.(24) with n unknowns (25).

## AN ALTERNATE FORM OF EQ. (24)

Using the known fact that the trace of a given (nxn) matrix is equal to the sum of its eigenvalues, an alternate from of the pole placement condition, eq. (24), may be obtained. Thus, from (19) the element $f_o$ may be expressed as

$$f_o = \sum_{i=1}^{n} \lambda_i - \text{trace } (H_2) \tag{26}$$

In this way, a following system of $n_1$ equations with $n_1$ unknowns may be formulated

$$f^1\alpha^i + \beta^i f^2 + f^1 C\, f^2 + \gamma_i \overset{\Delta}{=} e_i(f^1, f^2) = 0$$

$$\text{where} \quad \alpha^i = P^i_{12} a^2 \;;\; \beta^i = a^1 P^i_{21} \;;\; C^i = P^i_{11} \tag{27}$$

$$\gamma_i = a^1 P^i_{22} a^2 - f_o + \lambda_i$$

Since both sets of equations (24) and (27) are of the quadratic form, a simple iterative algorithm for the adjustment of vectors $f^1$ and $f^2$ may be designed. Using the Newton method, the increments $\Delta f^1_h$ and $\Delta f^2_h$ in the h-th iteration may be calculated by solving the following system of $n_1$ linear equations

$$\begin{vmatrix} g^1 \\ \vdots \\ g^{n_1} \end{vmatrix} \begin{vmatrix} \Delta f_h^{1T} \\ \Delta f_h^2 \end{vmatrix} = -e^h \tag{28}$$

$$\text{where } g^i = \Big|\underbrace{(\alpha^i + c^i f_h^2)^T}_{k_1} \,\vdots\, \underbrace{\beta^i + f_h^1 c^i}_{m_1}\Big| \qquad (1\times n_1)$$

$$e^h \triangleq \{e_i^h\} \; ; \quad e_i^h \triangleq e_i(f_h^1, f_h^2), \quad i\varepsilon[1,n_1].$$

## COMMENTS AND GENERALIZATIONS

### Choice of the adjustable elements in F

Form (24) and (27) it may be concluded that adjusting the elements $f_{ij}$ belonging to only one row and one column of F offers computational simplifications. In this case the pole placement problem is reduced to the solution of the set of equations of the quadratic form, which is rather easy to handle. Computational experience has shown that it is not always necessary to adjust elements in the first row and column of F. The criterion for selecting a satisfactory row and column index is the following. It is possible to select any row and column index, say h-th, $1 \leq h \leq \min(m,k)$, where at least one element among

$$a_{hj} \quad \text{or} \quad a_{ih}, \qquad j\varepsilon[k+1,n] \; ; \quad i\varepsilon[m+1,n] \tag{29}$$

has nonzero value. In the case of $h = 1$, $eq_1$ (18), elements (29) correspond to the elements of the vectors $a^1$ and $a^2$, eq. (19). It may be easily verified that conditions of controllability and observability of (8) guarantee the existence of such index h.

### The case of multiple eigenvalues

The case of multiple eigenvalues, eq. (10), may be treated using the concept of generalized eigenvectors, namely considering the equation

$$(H - I_n\lambda_i)q^{i(v_j)} = \lambda_i q^{i(v_j - 1)}$$

$v_j\varepsilon[1,r_i]$; $q^{1(0)} \equiv 0$, $r_i$ being the multiplicity of $\lambda_i$. The case of multiple eigenvalues will be discussed in detail in a separate paper.

### State feedback

The approach to pole placement presented in this paper permits simplifications in solving the following related problems:

- Pole placement by the state feedback, and
- Partial pole placement.

The case of the state feedback leads to $k = n$, while the dimension of the input vector may be taken to be unity. This yields

$$m = 1, \quad m_1 = 0, \quad k = n \quad \text{and} \quad k_1 = n_1 \tag{30}$$

Thus, in the case of (30), vectors $f^2$ and $a^1$, eq.(24), vanish, i.e. they become zero dimensional, while the vectors $f^1$ and $a^2$ become $n_1$ dimenzional. In other words, eq.(24) or (22) leads to

$$f_o = \lambda_i + f^1 (B_2^i)^{-1} a^2, \qquad i\varepsilon[1,n], \tag{31}$$

which is linear in $f_o$ and $f^1$ permitting in this way direct, noniterative solution of the state feedback regulator problem. Note that calculation of the controllable canonical form is not necessary.

Partial pole placement

The problem of partial pole placement /2/ may be defined as assigning only v closed-loop eigenvalues, where

$$v = \max(m,k). \tag{32}$$

To be specific let $k \geq m$. In this case the problem of partial pole placement may be solved satisfying eq. (24) for $i\varepsilon[1,k]$ only. These k equations may be satisfied taking for the elements of the vector $f^2$ fixed arbitrary values, i.e. $f^2 = f^*$. In this case eq. (22) becomes

$$f_o = \lambda_i + |f^1 \,\vdots\, a^1| \; (B_2^i)^{-1} \begin{vmatrix} f^* \\ \hdashline a^2 \end{vmatrix}, \quad i\varepsilon[1,k], \tag{33}$$

which is, similarly to eq.(31), linear in $f_o$ and $f^1$ and may be solved in the straightforward manner.

REFERENCES

/1/ J.C.Willems & W.H.Hesselink, "General properties of the pole placement problem", Proc. IFAC Congress, Helsinki, 1978, pp. 1725-1729.

/2/ E.J. Davison & S.G.Chow, "An algorithm for the assignment of closed loop poles using output feedback in large linear multivariable systems", IEEE Trans. Automatic Control, Vol.AC-18, pp. 74-75, 1973.

/3/ E.J. Davison & S.H.Wang, "On pole assignment in linear multivariable systems using output feedback", IEEE Trans. Automatic Control, Vol. AC-20, pp.516-518, 1975.

/4/ H.Kimura, "A further result on the problem of pole placement by output feedback", IEEE Trans. Automatic Control, Vol.AC-22, pp. 458-462, 1977.

# EXACT MODEL-MATCHING OF 2-D SYSTEMS USING PID CONTROLLERS

P. N. Paraskevopoulos, O. I. Kosmidou

Chair of Automatic Control Systems
School of Engineering
Democritus University of Thrace
Xanthi, Greece

The paper considers the problem of matching exactly the transfer function matrix of a given 2-D (two-dimensional) system to that of a desired 2-D model using output feedback, where in the feedbac loop proportional (P), integral (I) and derivative (D) terms are present. The approach followed reduces the problem to that of solving a linear system of algebraic equations. Furthermore, necessar and sufficient conditions are established for exact matching.

## I. INTRODUCTION

This paper considers the problem of exact model-matching of linear, time-invariant, multivariable, 2-D systems using PID output feedback. Recently, certain results concerning this problem have been reported [1]-[3]. In [1] some first theoretical results, whic partially solve the problem,are given. [2] and [3] refer to exact model-matching via state and output feedback, respectively, where the structure of the controller is of the proportional type. This restricted form of the controller has the advantage in that it is very simple and cheap to impliment, but it has the disadvantage in that it offers a limited number of free parameters which may be used to satisfy the design requirements. This disadvantage very often results in a situation where the proportional controller is insufficient to satisfy the problem.

To circumvent the above shortcomings, we are introducing in th paper a controller of a more general structure resulting in a grea number of degrees of freedom in the design. In particular, the present paper constitutes an extension of the static output feedback case [3] to the dynamic (PID) output feedback case.



*D. G. Lainiotis and N. S. Tzannes (eds.), Advances in Control, 210–217.*

## II. FORMULATION OF THE PROBLEM

Consider a 2-D linear time-invariant multivariable system described by its transfer function matrix [4]

$$\mathbf{T}(w,z) = \frac{\mathbf{A}(w,z)}{b(w,z)} = \frac{\sum_{i=0}^{\bar{m}} \sum_{j=0}^{\bar{n}} \mathbf{A}_{ij} w^i z^j}{\sum_{i=0}^{m} \sum_{j=0}^{n} b_{ij} w^i z^j} \quad ; \ \bar{n} \leqslant n, \ \bar{m} \leqslant m \tag{1}$$

where $\mathbf{T}(w,z)$, $\mathbf{A}(w,z)$ and $\mathbf{A}_{ij}$, $i = 0,1,\ldots,\bar{m}$, $j = 0,1,\ldots,\bar{n}$ are $\ell x r$ matrices. Consider, also, a 2-D linear, time-invariant, multivariable model described by its transfer function matrix

$$\mathbf{T}_m(w,z) = \frac{\mathbf{C}(w,z)}{d(w,z)} = \frac{\sum_{i=0}^{\bar{q}} \sum_{j=0}^{\bar{p}} \mathbf{C}_{ij} w^i z^j}{\sum_{i=0}^{q} \sum_{j=0}^{p} d_{ij} w^i z^j} \quad ; \ \bar{q} \leqslant q, \ \bar{p} \leqslant p \tag{2}$$

where $\mathbf{T}_m(w,z)$, $\mathbf{C}(w,z)$ and $\mathbf{C}_{ij}$, $i = 0,1,\ldots,\bar{q}$, $j = 0,1,\ldots,\bar{p}$ are $\ell x r_m$ matrices. The control law is assumed of the form (see figure 1)

$$\mathbf{U}(w,z) = \mathbf{G}(w,z)\,\mathbf{R}(w,z) + \mathbf{K}(w,z)\,\mathbf{Y}(w,z) \tag{3}$$

where $\mathbf{G}(w,z)$ and $\mathbf{K}(w,z)$ are the dynamic controller matrices having dimensions $rxr_m$ and $rx\ell$, respectively, where $\mathbf{R}(w,z)$ and $\mathbf{Y}(w,z)$ are the input and output vectors of the overall system. Clearly, the multivariable controller matrices may have complicated structures. In this paper, we assume that $\mathbf{G}(w,z)$ and $\mathbf{K}(w,z)$ have the following special structure (see figure 2)

$$\begin{aligned} \mathbf{G}(w,z) &= \mathbf{G} \\ \mathbf{K}(w,z) &= \frac{\mathbf{K}_1}{w} + \frac{\mathbf{K}_2}{z} + \mathbf{K}_3 + w\mathbf{K}_4 + z\mathbf{K}_5 \end{aligned} \tag{4}$$

That is, we assume that $\mathbf{G}(w,z)$ involves only a proportional term, whereas $\mathbf{K}(w,z)$ involves proportional plus integral plus derivative terms. The particular case where $\mathbf{K}(w,z)$ involves only the proportional term has already been reported [3].

The closed-loop transfer function matrix $\mathbf{T}_c(w,z)$ will then be

$$\mathbf{T}_c(w,z) = \left[\mathbf{I}_\ell - \mathbf{T}(w,z)\left[\frac{\mathbf{K}_1}{w} + \frac{\mathbf{K}_2}{z} + \mathbf{K}_3 + w\mathbf{K}_4 + z\mathbf{K}_5\right]\right]^{-1} \mathbf{T}(w,z)\mathbf{G} \tag{5}$$

The precise definition of the exact transfer function matrix synthesis problem treated in this paper can be stated as follows: Given $\mathbf{T}(w,z)$ and $\mathbf{T}_m(w,z)$, find the controller matrices $\mathbf{G}, \mathbf{K}_1, \mathbf{K}_2, \mathbf{K}_3, \mathbf{K}_4$ and $\mathbf{K}_5$ such that

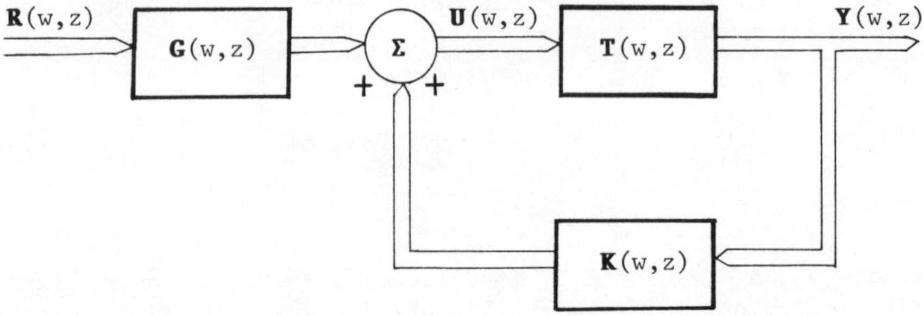


Figure 1. Closed-loop system with general Type controller

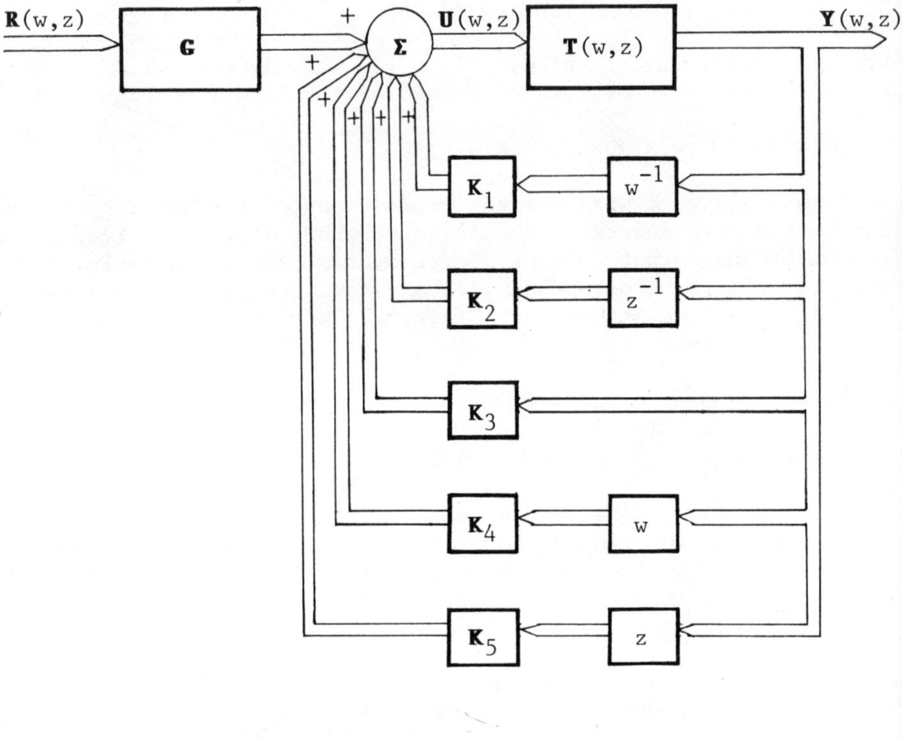


Figure 2. Closed-loop system with PID controller

$$\mathbf{T}_c(w,z) = \mathbf{T}_m(w,z) \tag{6}$$

## III. SOLUTION OF THE PROBLEM

Combining equations (5) and (6) we have

$$\left[\mathbf{I}_\ell - \mathbf{T}(w,z)\left[\frac{\mathbf{K}_1}{w} + \frac{\mathbf{K}_2}{z} + \mathbf{K}_3 + w\mathbf{K}_4 + z\mathbf{K}_5\right]\right]^{-1} \mathbf{T}(w,z)\mathbf{G} = \mathbf{T}_m(w,z) \tag{7}$$

If (1) and (2) are substituted in (7) and after some algebraic manipulations we get

$$\begin{aligned} & wzd(w,z)\mathbf{A}(w,z)\mathbf{G} + z\mathbf{A}(w,z)\mathbf{K}_1\mathbf{C}(w,z) + w\mathbf{A}(w,z)\mathbf{K}_2\mathbf{C}(w,z) + \\ & + wz\mathbf{A}(w,z)\mathbf{K}_3\mathbf{C}(w,z) + w^2 z\mathbf{A}(w,z)\mathbf{K}_4\mathbf{C}(w,z) + wz^2\mathbf{A}(w,z)\mathbf{K}_5\mathbf{C}(w,z) = \\ & = wzb(w,z)\mathbf{C}(w,z) \end{aligned} \tag{8}$$

Clearly, equation (8) is a two-variable polynomial identity which implies that (8) holds <u>if and only if</u> the sum of the coefficients of the terms $w^i z^j$ with the same i and j in the left-hand side is equal to the coefficient of the corresponding $w^i z^j$ term in the right-hand side, for all i and j. Application of this approach to (8) will lead to a linear system of algebraic equations of the form

$$\mathbf{P}\,\boldsymbol{\vartheta} = \mathbf{h} \tag{9}$$

where $\mathbf{P}$ and $\mathbf{h}$ are constant matrices involving only known parameters, namely, the parameters of $\mathbf{T}(w,z)$ and $\mathbf{T}_m(w,z)$ and $\boldsymbol{\vartheta}$ is the unknown vector involving the unknown parameters of the controller matrices $\mathbf{G}$, $\mathbf{K}_1$, $\mathbf{K}_2$, $\mathbf{K}_3$, $\mathbf{K}_4$ and $\mathbf{K}_5$.

A procedure will next be developed which determines the matrices $\mathbf{P}$ and $\mathbf{h}$. To this end equation (8) is written first in a suitable form by way of the Kronecker product as follows [2], [5].

$$\begin{aligned} & wz\mathbf{M}(w,z)\mathbf{g} + z\mathbf{N}(w,z)\mathbf{f}_1 + w\mathbf{N}(w,z)\mathbf{f}_2 + wz\mathbf{N}(w,z)\mathbf{f}_3 + w^2 z\mathbf{N}(w,z)\mathbf{f}_4 + \\ & + wz^2\mathbf{N}(w,z)\mathbf{f}_5 = wz\mathbf{r}(w,z) \end{aligned} \tag{10}$$

where

$$\mathbf{M}(w,z) = \mathbf{I}_{r_m}^T \otimes d(w,z)\mathbf{A}(w,z) = \sum_{i=0}^{\bar{m}+q} \sum_{j=0}^{\bar{n}+p} \mathbf{M}_{ij} w^i z^j \tag{11}$$

$$\mathbf{N}(w,z) = \mathbf{C}^T(w,z) \otimes \mathbf{A}(w,z) = \sum_{i=0}^{\bar{m}+\bar{q}} \sum_{j=0}^{\bar{n}+\bar{p}} \mathbf{N}_{ij} w^i z^j \tag{12}$$

$$\mathbf{r}(w,z) = b(w,z)\begin{bmatrix} \mathbf{c}_1^T(w,z) \\ \vdots \\ \mathbf{c}_\ell^T(w,z) \end{bmatrix} = \sum_{i=0}^{m+\bar{q}} \sum_{j=0}^{n+\bar{p}} \mathbf{r}_{ij} w^i z^j \tag{13}$$

$$\mathbf{g} = \begin{bmatrix} \mathbf{g}_1^T \\ \vdots \\ \mathbf{g}_r^T \end{bmatrix} , \quad \mathbf{f}_i = \begin{bmatrix} (\mathbf{k}_i)_1^T \\ \vdots \\ (\mathbf{k}_i)_r^T \end{bmatrix} \quad ; \; i = 1,2,\ldots,5$$

where $\otimes$ denotes the Kronecker product and $a_{ij}(w,z)$ is the (i,j)-element of $\mathbf{A}(w,z)$, $\mathbf{g}_i$ is the ith row of $\mathbf{G}$, $(\mathbf{k}_i)_j$ is the jth row of $\mathbf{K}_i$ and $\mathbf{c}_i(w,z)$ is the ith row of $\mathbf{C}(w,z)$. The dimensions of $\mathbf{M}(w,z)$, $\mathbf{N}(w,z)$, $\mathbf{r}(w,z)$, $\mathbf{g}$ and $\mathbf{f}_i$ are $\ell r_m \times r r_m$, $\ell r_m \times \ell r$, $\ell r_m \times 1$, $r r_m \times 1$ and $\ell r \times 1$, respectively.

The coefficient matrices $\mathbf{M}_{ij}$, $\mathbf{N}_{ij}$ and $\mathbf{r}_{ij}$ of $\mathbf{M}(w,z)$, $\mathbf{N}(w,z)$ and $\mathbf{r}(w,z)$, respectively, may be computed using an algorithm developed in [2]. Furthermore, equation (10) may be simplified by using the following: Let

$$f(w,z) = \sum_{i=0}^{p} \sum_{j=0}^{m} f_{ij} w^i z^j = \mathbf{w}_p^T \mathbf{F} \mathbf{z}_m$$

Then, we have

$$q(w,z) = w^k z^s f(w,z) = w^k z^s \mathbf{w}_p^T \mathbf{F} \mathbf{z}_m = \mathbf{w}_{p+k}^T \mathbf{Q} \mathbf{z}_{m+s} \tag{14}$$

For the case of matrix polynomials we have: Let $\mathbf{F}(w,z)$ be an $\alpha \times \beta$ matrix having the form

$$\mathbf{F}(w,z) = \sum_{i=0}^{p} \sum_{j=0}^{m} \mathbf{F}_{ij} w^i z^j = \mathbf{\Omega}_p \hat{\mathbf{F}} \mathbf{Z}_m \tag{15}$$

where

$$\mathbf{\Omega}_p = \begin{bmatrix} \mathbf{I}_\alpha & w\mathbf{I}_\alpha \ldots w^p \mathbf{I}_\alpha \end{bmatrix}, \quad \mathbf{Z}_m = \begin{bmatrix} \mathbf{I}_\beta \\ z\mathbf{I}_\beta \\ \vdots \\ z^m \mathbf{I}_\beta \end{bmatrix} \quad \text{and} \quad \hat{\mathbf{F}} = \begin{bmatrix} \mathbf{F}_{0,0} & \mathbf{F}_{0,1} \ldots \mathbf{F}_{0,m} \\ \mathbf{F}_{1,0} & \mathbf{F}_{1,1} \ldots \mathbf{F}_{1,m} \\ \vdots & \vdots & \vdots \\ \mathbf{F}_{p,0} & \mathbf{F}_{p,1} \ldots \mathbf{F}_{p,m} \end{bmatrix}$$

Then we have

$$P(w,z) = w^k z^s \mathbf{F}(w,z) = \sum_{i=0}^{p+k} \sum_{j=0}^{m+s} \mathbf{P}_{ij} w^i z^j = w^k z^s \mathbf{\Omega}_p \hat{\mathbf{F}} \mathbf{Z}_m = \mathbf{\Omega}_{p+k} \hat{\mathbf{P}} \mathbf{Z}_{m+s} \tag{16}$$

Now the matrix $\mathbf{M}(w,z)$ may be written according to (15) as follows

$$\mathbf{M}(w,z) = \sum_{i=0}^{\bar{m}+q} \sum_{j=0}^{\bar{n}+p} \mathbf{M}_{ij} w^i z^j = \mathbf{\Omega}_{\bar{m}+q} \hat{\mathbf{M}} \mathbf{Z}_{\bar{n}+p} \tag{17}$$

Then, according to (16) we have

$$wz\mathbf{M}(w,z) = \mathbf{E}(w,z) = \sum_{i=0}^{\bar{m}+q+1} \sum_{j=0}^{\bar{n}+p+1} \mathbf{E}_{ij} w^i z^j = \mathbf{\Omega}_{\bar{m}+q+1} \hat{\mathbf{E}} \mathbf{Z}_{\bar{n}+p+1} \tag{18}$$

Similarly, $\mathbf{N}(w,z)$ may be written as

$$\mathbf{N}(w,z) = \sum_{i=0}^{\bar{m}+q} \sum_{j=0}^{\bar{n}+p} \mathbf{N}_{ij} w^i z^j = \mathbf{\Omega}_{\bar{m}+\bar{q}} \hat{\mathbf{N}} \mathbf{Z}_{\bar{n}+\bar{p}} \tag{19}$$

Then, according to (16) we have

$$w^k z^\ell \mathbf{N}(w,z) = \mathbf{Q}_{k\ell}(w,z) = \sum_{i=0}^{\bar{m}+\bar{q}+k} \sum_{j=0}^{\bar{n}+\bar{p}+\ell} (\mathbf{Q}_{k\ell})_{ij} w^i z^j = \mathbf{\Omega}_{\bar{m}+\bar{q}+k} \hat{\mathbf{Q}}_{k\ell} \mathbf{Z}_{\bar{n}+\bar{p}+\ell} \tag{20}$$

The indices k and $\ell$ are introduced in the matrix $\mathbf{Q}_{k\ell}$ to distinguish for the different powers of $w^k z^\ell$ involved in (10). Finally, $\mathbf{r}(w,z)$ may be written as

$$\mathbf{r}(w,z) = \sum_{i=0}^{m+\bar{q}} \sum_{j=0}^{n+\bar{p}} \mathbf{r}_{ij} w^i z^j = \mathbf{\Omega}_{m+\bar{q}} \hat{\mathbf{r}} \mathbf{Z}_{n+\bar{p}} \tag{21}$$

Then, according to (16) we have

$$wz\mathbf{r}(w,z) = \mathbf{R}(w,z) = \sum_{i=0}^{m+\bar{q}+1} \sum_{j=0}^{n+\bar{p}+1} \mathbf{R}_{ij} w^i z^j = \mathbf{\Omega}_{m+\bar{q}+1} \hat{\mathbf{R}} \mathbf{Z}_{n+\bar{p}+1} \tag{22}$$

Using relations (17) to (22), equation (10) may be written as

$$\left[\sum_{i=0}^{\bar{m}+q+1} \sum_{j=0}^{\bar{n}+p+1} \mathbf{E}_{ij} w^i z^j\right] \mathbf{g} + \left[\sum_{i=0}^{\bar{m}+\bar{q}} \sum_{j=0}^{\bar{n}+\bar{p}+1} (\mathbf{Q}_{01})_{ij} w^i z^j\right] \mathbf{f}_1 +$$

$$\left[\sum_{i=0}^{\bar{m}+\bar{q}+1} \sum_{j=0}^{\bar{n}+\bar{p}} (\mathbf{Q}_{10})_{ij} w^i z^j\right] \mathbf{f}_2 + \left[\sum_{i=0}^{\bar{m}+\bar{q}+1} \sum_{j=0}^{\bar{n}+\bar{p}+1} (\mathbf{Q}_{11})_{ij} w^i z^j\right] \mathbf{f}_3 +$$

$$\left[\sum_{i=0}^{\bar{m}+\bar{q}+2} \sum_{j=0}^{\bar{n}+\bar{p}+1} (\mathbf{Q}_{21})_{ij} w^i z^j\right] \mathbf{f}_4 + \left[\sum_{i=0}^{\bar{m}+\bar{q}+1} \sum_{j=0}^{\bar{n}+\bar{p}+2} (\mathbf{Q}_{12})_{ij} w^i z^j\right] \mathbf{f}_5 =$$

$$= \sum_{i=0}^{m+\bar{q}+1} \sum_{j=0}^{n+\bar{p}+1} \mathbf{R}_{ij} w^i z^j \tag{23}$$

Define $\tau_1 = \max(\bar{m}+q+1, \bar{m}+\bar{q}+2, m+\bar{q}+1)$ and $\tau_2 = \max(\bar{n}+p+1, \bar{n}+\bar{p}+2, n+\bar{p}+1)$. Then the following must hold; $\mathbf{E}_{ij}=\mathbf{0}$, $i=\bar{m}+q+2,\dots,\tau_1$, $j=\bar{n}+p+2,\dots,\tau_2$ ; $(\mathbf{Q}_{01})_{ij}=\mathbf{0}$, $i=\bar{m}+\bar{q}+1,\dots,\tau_1$, $j=\bar{n}+\bar{p}+2,\dots,\tau_2$ ; $(\mathbf{Q}_{10})_{ij}=\mathbf{0}$, $i=\bar{m}+\bar{q}+2,\dots,\tau_1$, $j=\bar{n}+\bar{p}+1,\dots,\tau_2$ ; $(\mathbf{Q}_{11})_{ij}=\mathbf{0}$, $i=\bar{m}+\bar{q}+2,\dots,\tau_1$, $j=\bar{n}+\bar{p}+2,\dots,\tau_2$; $(\mathbf{Q}_{21})_{ij}=\mathbf{0}$, $i=\bar{m}+\bar{q}+3,\dots,\tau_1$, $j=\bar{n}+\bar{p}+2,\dots,\tau_2$; $(\mathbf{Q}_{12})_{ij}=\mathbf{0}$, $i=\bar{m}+\bar{q}+2,\dots,\tau_1$, $j=\bar{n}+\bar{p}+3,\dots,\tau_2$; $\mathbf{R}_{ij}=\mathbf{0}$, $i=m+\bar{q}+2,\dots,\tau_1$ , $j = n+\bar{p}+2,\dots,\tau_2$.

Using the above, equation (23) may be written as

$$\left[\sum_{i=0}^{\tau_1}\sum_{j=0}^{\tau_2}\mathbf{E}_{ij}w^iz^j\right]\mathbf{g}+\left[\sum_{i=0}^{\tau_1}\sum_{j=0}^{\tau_2}(\mathbf{Q}_{01})_{ij}w^iz^j\right]\mathbf{f}_1+\left[\sum_{i=0}^{\tau_1}\sum_{j=0}^{\tau_2}(\mathbf{Q}_{10})_{ij}w^iz^j\right]\mathbf{f}_2+$$

$$\left[\sum_{i=0}^{\tau_1}\sum_{j=0}^{\tau_2}(\mathbf{Q}_{11})_{ij}w^iz^j\right]\mathbf{f}_3+\left[\sum_{i=0}^{\tau_1}\sum_{j=0}^{\tau_2}(\mathbf{Q}_{21})_{ij}w^iz^j\right]\mathbf{f}_4+\left[\sum_{i=0}^{\tau_1}\sum_{j=0}^{\tau_2}(\mathbf{Q}_{12})_{ij}w^iz^j\right]\mathbf{f}_5=$$

$$=\sum_{i=0}^{\tau_1}\sum_{j=0}^{\tau_2}\mathbf{R}_{ij}w^iz^j \tag{24}$$

Then, upon equating coefficients of like $w^iz^j$ terms for $i=0,1,\ldots,\tau_1$, $j=0,1,\ldots,\tau_2$ equation (24) yields

$$\mathbf{E}_{ij}\mathbf{g}+(\mathbf{Q}_{01})_{ij}\mathbf{f}_1+(\mathbf{Q}_{10})_{ij}\mathbf{f}_2+(\mathbf{Q}_{11})_{ij}\mathbf{f}_3+(\mathbf{Q}_{21})_{ij}\mathbf{f}_4+(\mathbf{Q}_{12})_{ij}\mathbf{f}_5=\mathbf{R}_{ij}$$

$$i=0,1,\ldots,\tau_1\ ,\ j=0,1,\ldots,\tau_2 \tag{25}$$

Equation (25) may be written more compactly, as follows

$$\mathbf{P}\,\boldsymbol{\vartheta}=\mathbf{h} \tag{26}$$

where

$$\mathbf{P}=\begin{bmatrix}\mathbf{E}_{00} & (\mathbf{Q}_{01})_{00} & (\mathbf{Q}_{10})_{00} & (\mathbf{Q}_{11})_{00} & (\mathbf{Q}_{21})_{00} & (\mathbf{Q}_{12})_{00}\\ \mathbf{E}_{01} & (\mathbf{Q}_{01})_{01} & (\mathbf{Q}_{10})_{01} & (\mathbf{Q}_{11})_{01} & (\mathbf{Q}_{21})_{01} & (\mathbf{Q}_{12})_{01}\\ \vdots & \vdots & \vdots & \vdots & \vdots & \vdots\\ \mathbf{E}_{\tau_1\tau_2} & (\mathbf{Q}_{01})_{\tau_1\tau_2} & (\mathbf{Q}_{10})_{\tau_1\tau_2} & (\mathbf{Q}_{11})_{\tau_1\tau_2} & (\mathbf{Q}_{21})_{\tau_1\tau_2} & (\mathbf{Q}_{12})_{\tau_1\tau_2}\end{bmatrix},\ \mathbf{h}=\begin{bmatrix}\mathbf{R}_{00}\\ \mathbf{R}_{01}\\ \vdots\\ \mathbf{R}_{\tau_1\tau_2}\end{bmatrix},\quad \boldsymbol{\vartheta}=\begin{bmatrix}\mathbf{g}\\ \mathbf{f}_1\\ \mathbf{f}_2\\ \mathbf{f}_3\\ \mathbf{f}_4\\ \mathbf{f}_5\end{bmatrix}$$

where the dimensions of $\mathbf{P}$, $\boldsymbol{\vartheta}$ and $\mathbf{h}$ are $\tau\ell r_m x(rr_m+5\ell r)$, $(rr_m+5\ell r)x1$ and $r\ell r_m x1$, respectively and $\tau=(\tau_1+1)(\tau_2+1)$. Clearly, (26) is the relation (9) sought.

The procedure outlined above has reduced the problem of determining $\mathbf{G}$, $\mathbf{K}_1$, $\mathbf{K}_2$, $\mathbf{K}_3$, $\mathbf{K}_4$ and $\mathbf{K}_5$ such that (7) holds, to that of solving the overdetermined linear system of equations (26). Thus, for the problem to have a solution, equation (26) must be a consistent system of equations, that is $\mathbf{h}\in R[\mathbf{P}]$.

We have therefore established the following theorem.

*Theorem*

*Given two transfer function matrices $\mathbf{T}(w,z)$ and $\mathbf{T}_m(w,z)$, defined in equation (1) and (2) respectively, there exist constant matrices $\mathbf{G}$, $\mathbf{K}_1$, $\mathbf{K}_2$, $\mathbf{K}_3$, $\mathbf{K}_4$ and $\mathbf{K}_5$ of the controller (4) such that equation (7) holds if and only if $\mathbf{h} \in R[\mathbf{P}]$.*

Assuming that $\mathbf{h} \in R[\mathbf{P}]$, then equation (26) yields the following value for $\boldsymbol{\vartheta}$:

$$\boldsymbol{\vartheta} = (\mathbf{P}^T\mathbf{P})^{-1}\mathbf{P}^T\mathbf{h} \tag{27}$$

It is remarked that in cases where (26) is an underdetermined system of equations, the problem has always a solution, that is, the PID controller is always sufficient to satisfy the problem requirements. Furthermore, certain controller parameters may remain free and may be chosen to satisfy other design requirements such as sensitivity.

## IV. CONCLUSIONS

A method has been presented for the synthesis of transfer functions of 2-D systems using PID output feedback controllers. This method introduces more degrees of freedom in the design of 2-D systems than those reported in [3], thus offering a technique with greater potential to satisfy the design requirements. These results may have practical applications in the design of 2-D systems, particularly in the design of 2-D filters which have been used recently in a variety of fields [6],[7].

Since even a PID controller may not suffice to satisfy the design requirements, it follows that an interesting area for further research is to assume the controller matrices $\mathbf{G}(w,z)$ and $\mathbf{K}(w,z)$ of a more general form and seek for the minimum order compensator for exact matching.

## REFERENCES

[1] Eising,R. and Erme,E.,"Exact model matching of 2-D systems", IEEE Trans. of Autom. Contr., vol.24, pp 134-135, Febr. 1979.
[2] Paraskevopoulos,P.N.,"Exact model matching of 2-D systems via state feedback", Journal of the Franklin Institute (accepted).
[3] Paraskevopoulos,P.N.,"Transfer function matrix synthesis of 2-D systems", IEEE Trans. Autom. Control (accepted).
[4] Kung,S.Y.,Levy,B.,Morf,M. and Kailath,T.,"New results in 2-D systems theory-Part II", Proc.IEEE,vol. 65, pp.945-961,June 1977.
[5] Brewer,J.W.,"Kronecker products and matrix calculus in system theory", IEEE Trans. on Circuits and Systems, vol. 25, pp. 772-781, 1978.
[6] Special issue on digital filters, Proceedings of IEEE, vol. 63, No. 4, April 1975.
[7] Special issue on multidimensional systems, Proceedings of IEEE, vol. 65, No. 6, June 1977.

# 2-D SYSTEMS THEORY - A PROGRESS REPORT

E. Fornasini, G. Marchesini (^)

Dept. of Electrical Eng., Univ. of Padova, Italy

ABSTRACT. The stability problem of two-dimensional filters is receiving wide attention [1-10].

Here a preliminary account of an "internal stability" theory is given which takes into account the dynamical model point of view.

The 2-D recursive model presented in [6] is considered and a Lyapunov equation is derived.

## 1. INTRODUCTION AND PRELIMINARY DEFINITIONS

From now on a 2-D system $\Sigma = (A_1, A_2, B_1, B_2, C)$ is identified by the following pair of equations:

$$\begin{aligned} x(h+1,k+1) &= A_1 x(h,k+1)+A_2 x(h+1,k)+B_1 u(h,k+1)+B_2 u(h+1,k) \\ y(h,k) &= C\, x(h,k) \end{aligned} \tag{1}$$

where

- $(h,k)$ are elements of $\mathbb{Z}\times\mathbb{Z}$ ("time set"), partially ordered by the product of the orderings
- $x(\cdot)$: $\mathbb{Z}\times\mathbb{Z}\to\mathbb{R}^n$ is a map whose value at time $(h,k)$ is called the "local state at time $(h,k)$"
- $u(\cdot)$: $\mathbb{Z}\times\mathbb{Z}\to\mathbb{R}$ and $y(\cdot)$: $\mathbb{Z}\times\mathbb{Z}\to\mathbb{R}$ are the input and the output maps respectively, and $u(h,k)$, $y(h,k)$ are the input and the output values at time $(h,k)$
- $A_1, A_2 \in \mathbb{R}^{n\times n}$, $B_1$, $B_2 \in \mathbb{R}^{n\times 1}$, $C \in \mathbb{R}^{1\times n}$ are suitable matrices which completely characterize the 2-D system $\Sigma$.

---

(^) This work has been supported by CNR-GNAS



*D. G. Lainiotis and N. S. Tzannes (eds.), Advances in Control, 218–226.*

When an input function u(·) is given, the solution of (1) requires a complete information about a suitable set of local states called "initial global state".

Considered in $\mathbb{Z}\times\mathbb{Z}$ a non empty set $\mathcal{C}$, ("separation set") which satisfies the following characteristic properties:

(i) if h>i, k>j, (h,k) and (i,j) cannot simultaneously belong to $\mathcal{C}$.

(ii) if (h,k) belongs to $\mathcal{C}$, then $\mathcal{C}$ intersects the sets {(h-1,k), (h,k+1), (h-1,k+1)} and {(h+1,k), (h+1,k-1)} and does not contain the set {(h+1,k), (h,k+1)}

(iii) for any (i,j) in $\mathbb{Z}\times\mathbb{Z}$, the relation $(h,k)\leq(i,j)$ cannot be satisfied by infinitely many elements $\overline{(h,k)}$ in $\mathcal{C}$.

The "future of $\mathcal{C}$" is the set

$$\mathfrak{F}_{\mathcal{C}}=\{(i,j):\ (h,k)\leq(i,j)\quad\text{for some}\quad(h,k)\quad\text{in }\mathcal{C}\}$$

The "global state" $\mathcal{X}_{\mathcal{C}}$ on the separation set $\mathcal{C}$ is defined as

$$\mathcal{X}_{\mathcal{C}}\triangleq\{x(h,k):\ (h,k)\in\mathcal{C}\}$$

and the computation of a local state x(i,j),(i,j) in $\mathfrak{F}_{\mathcal{C}}$, can be performed starting from $\mathcal{X}_{\mathcal{C}}$, whenever u is known in $\mathfrak{F}_{\mathcal{C}}$. In particular if u is identically zero on $\mathfrak{F}_{\mathcal{C}}$ the local states in $\mathfrak{F}_{\mathcal{C}}$ depend only on $\mathcal{X}_{\mathcal{C}}$.

## 2. POLYNOMIAL CONDITIONS FOR INTERNAL STABILITY

The notion of internal stability of a 2-D system is related to the behaviour of the free evolution of local states resulting from a bounded global state assignment on a separation set $\mathcal{C}$. Let assume in $\mathbb{Z}\times\mathbb{Z}$ the distance function $d((i,j),(h,k))\triangleq|i-h|+|j-k|$ and denote by

$$d((i,j),\mathcal{C})=\min_{(h,k)\in\mathcal{C}}d((i,j),(h,k))$$

the distance between (i,j) and the set $\mathcal{C}$. Introduce the following notation

$$\|\mathcal{X}_{\mathcal{C}}\|=\sup_{x\in\mathcal{X}_{\mathcal{C}}}\|x\|\quad,\qquad\|x\|\ \text{euclidean norm of } x$$

Definition 1: *Let $\mathcal{C}$ be a separation set in $\mathbb{Z}\times\mathbb{Z}$, and assume* u = 0. *The 2-D system (1) is internally stable with respect to $\mathcal{C}$ if given* $\varepsilon>0$, *for every $\mathcal{X}_{\mathcal{C}}$ with $\|\mathcal{X}_{\mathcal{C}}\|<\infty$, there exists a positive integer* m *such that* $\|x(i,j)\|<\varepsilon$ *when* (i,j) *is in the future of $\mathcal{C}$ and* $d((i,j))>m$.

The internal stability depends on the pair $(A_1,A_2)$, and does not depend on the separation set .

Theorem 1 [7] *Internal stability respect to any separation $\mathcal{C}$ implies internal stability w.r. to every separation set.*

At this point the construction of internal stability criteria is perhaps the most natural topic of investigation. Is is well known that internal stability of a discrete linear system

$$x(h+1) = Ax(h) + Bu(h) \qquad y(h) = Cx(h) \tag{3}$$

can be checked (i) by evaluating the roots distribution of the characteristic polynomial of A with respect to the unit circle in the Gauss plane or (ii) by solving the Lyapunov equation.

Up to now the corresponding picture for 2-D systems is incomplete. In fact a 2-D version of Lyapunov equation is not available, despite some steps have already been done [9]. The following theorem provides the 2-D counterpart of the 1-D stability criterion based on the characteristic polynomial.

Theorem 2 [6] *A 2-D system* $\Sigma = (A_1, A_2, B_1, B_2, C)$ *is internally stable if and only if the polynomial* $\det(1 - A_1z_1 - A_2z_2)$ *is devoid of zeros in the closed polydisc* $\mathcal{P}_1 = \{(z_1, z_2) \in \mathbb{C} \times \mathbb{C}: |z_1| \leq 1, |z_2| \leq 1\}$

Several tests have been proposed in the literature [Shanks, Huang, Jury, Anderson etc.] to check if the unit polydisc $\mathcal{P}_1$ intersects the variety of a polynomial $q \in \mathbb{C}[z_1, z_2]$. The original field of application of these tests has been the external (BIBO) stability analysis, since Shanks theorem [1] states that the external stability of a two dimensional filter with transfer function p/q, p and q coprime, depends only on the zeros distribution of q.

Actually in 1978 Goodman [5] showed that a two dimensional filter can be BIBO stable even when the denominator q vanishes in some points of the torus $T_1 = \{(z_1, z_2) \in \mathbb{C} \times \mathbb{C}: |z_1| = |z_2| = 1\}$. As a consequence Shanks theorem has to be restated in the following form

Theorem 3 [5] *Let* $G(z_1, z_2) = p(z_1, z_2)/q(z_1, z_2)$ *denote the transfer function of a* BIBO *stable filter. Then* $G(z_1, z_2)$ *has no poles in the closed unit polydisc* $\mathcal{P}_1$ *and no nonessential singularities of the second kind on* $\mathcal{P}_1$ *except possibly on* $T_1$.

Then the mutual implications between roots distribution of q and BIBO stability are as follows:

$$\text{BIBO stability} \Rightarrow q(z_1, z_2) \neq 0 \quad \text{in } \mathcal{P}_1 - T_1$$
$$\text{BIBO stability} \Leftarrow q(z_1, z_2) \neq 0 \quad \text{in } \mathcal{P}_1$$

This shows that the absence of intersections between $\mathcal{P}_1$ and the variety of q, which can be checked by the above mentioned tests, ensures BIBO stability. Nevertheless we can have BIBO stability even when the intersection is nonempty, that is when the tests would give a negative result.

On the other hand by Theorem 2 these tests fit perfectly to the internal stability analysis. In particular the following Corollary of Theorem 2, derived from Huang's criterion [2], will provide (see Section 4) a frequency dependent Lyapunov equation.

Corollary 1 [9] *A 2-D system* $\Sigma = (A_1, A_2, -, -, -)$ *is internally sta-*

*ble if and only if the complex matrix* $A_1 + e^{j\omega}A_2$ *is stable (i.e. the magnitudes of its eigenvalues are less than 1) for any real* $\omega$

Remark. An obvious necessary 2-D stability condition resulting from Corollary 1 is that $A_1 \pm A_2$ have to be stable.

## 3. MATRIX CONDITIONS FOR INTERNAL STABILITY

In this section we are concerned with the extension to the two-dimensional case of some properties characterizing internally stable 1-D systems. These properties are recalled in Theorem 4.

Theorem 4. *The following propositions are equivalent:* i) *the system* (3) *is internally stable;* ii) *the series* $\sum_{i=0}^{\infty} \|A^i\|$ *converges;* iii) $\|A^k\| < 1$ *for some integer* $k > 0$*;* iv) *the series* $\sum_{i=0}^{\infty} (A^i)^T(A^i)$ *converges.*

Note that $P = \sum_{i=0}^{\infty} A^{iT}A^i$ is the solution of the Lyapunov equation

$$X = I + A^T XA$$

As we shall see in Theorem 5, the family of matrices $\{A_1^r \sqcup\!\sqcup^s A_2, r,s \in \mathbb{N}\}$, defined as

$$\begin{array}{ll} A_1^r \sqcup\!\sqcup^o A_2 \triangleq A_1^r & A_1^o \sqcup\!\sqcup^s A_2 \triangleq A_2^s \\ A_1^r \sqcup\!\sqcup^s A_2 \triangleq A_1(A_1^{r-1} \sqcup\!\sqcup^s A_2) + A_2(A_1^r \sqcup\!\sqcup^{s-1} A_2) & \text{if } r,s > 0 \end{array} \tag{4}$$

plays an essential role in extending points i)-iv) of Theorem 4.

Theorem 5 [9] *The following propositions are equivalent:* i) *the 2-D system (1) is internally stable;* ii) *the series* $\sum_{r,s=0}^{\infty} \|A_1^r \sqcup\!\sqcup^s A_2\|$ *converges;* iii) $\sum_{r+s=k} \|A_1^r \sqcup\!\sqcup^s A_2\| < 1$ *for some positive integer* $k$*;* iv) *the series* $\sum_{r,s=0}^{\infty} (A_1^r \sqcup\!\sqcup^s A_2)^T(A_1^r \sqcup\!\sqcup^s A_2)$ *converges.*

The problem of relating the series $\sum_{r,s} (A^r \sqcup\!\sqcup^s A_2)^T(A^r \sqcup\!\sqcup^s A_2)$ to the solution of an algebraic matrix equation is still unsolved, except for n=1 (see section 4).

A way to obtain a Lyapunov equation consists in reducing a two dimensional dynamics to a one dimensional, by assuming a periodic initial global state, and then in applying 1-D theory. As we shall see in the sequel this procedure is not fruitful since the existence of stability criteria under periodic initial conditions does not imply a stable behaviour under generic initial conditions.

From now on, $\mathcal{C}_i$, $i \in \mathbb{Z}$, will denote the following separation sets

$$\mathcal{C}_i = \{(h,k)\colon h+k = i\}$$

Let call a global state $\mathcal{X}_{\mathcal{C}_i}$ "H periodical" if $x(h,k) \in \mathcal{X}_{\mathcal{C}_i} \Rightarrow x(h,k) = x(h+H,k-H)$ for any $(h,k)$ in $\mathcal{C}_i$.

Clearly the stability check for a H-periodical global initial state reduces to solve a standard Lyapunov equation in dimension $nH \times nH$.

The 2-D system considered in the following example is unstable. However it shows a stable behaviour corresponding to any periodic global initial state.

Example. The 2-D system $\Sigma = (A_1, A_2, -, -, -)$ with

$$A_1 = 0.5 \begin{bmatrix} \cos\sqrt{0.5} & \sin\sqrt{0.5} \\ -\sin\sqrt{0.5} & \cos\sqrt{0.5} \end{bmatrix}, \quad A_2 = 0.5 \begin{bmatrix} \cos\sqrt{0.3} & \sin\sqrt{0.3} \\ -\sin\sqrt{0.3} & \cos\sqrt{0.3} \end{bmatrix}$$

is internally unstable. In fact the eigenvalues of $A_1 + e^{j\omega}A_2$ are $\lambda_{1,2}(\omega) = 0.5(e^{\pm j\sqrt{0.5}} + e^{j(\omega \pm \sqrt{0.3})})$ and the magnitude of one of them is 1 for $\omega = \omega_{1,2} = \pm(\sqrt{0.5} - \sqrt{0.3})$. Any other choice of $\omega$ leads to a stable matrix $A_1 + e^{j\omega}A_2$.

Let now change the basis in $X = \mathbb{C}^2$ (local state space), and refer to the complex basis

$$v_1 = \begin{bmatrix} -j \\ 1 \end{bmatrix}, \quad v_2 = \begin{bmatrix} j \\ 1 \end{bmatrix}$$

which constitutes a spectral basis for $A_1, A_2$ and $A_1 + e^{j\omega}A_2, \forall\omega$. Correspondingly, matrices $A_1$ and $A_2$ assume the diagonal form

$$\hat{A}_1 = \frac{e^{j\sqrt{0.5}}}{0.5} \begin{bmatrix} 1 & 0 \\ 0 & 1 \end{bmatrix} \triangleq \begin{bmatrix} \nu_1 & 0 \\ 0 & \nu_2 \end{bmatrix}; \quad \hat{A}_2 = \frac{e^{j\sqrt{0.3}}}{0.5} \begin{bmatrix} 1 & 0 \\ 0 & 1 \end{bmatrix} \triangleq \begin{bmatrix} \mu_1 & 0 \\ 0 & \mu_2 \end{bmatrix}$$

and the free evolution of local states in given by

$$\hat{x}(h+1,k+1) \triangleq \begin{bmatrix} \hat{x}_1(h+1,k+1) \\ \hat{x}_2(h+1,k+1) \end{bmatrix} = \begin{bmatrix} \nu_1 \hat{x}_1(h,k+1) + \mu_1 \hat{x}_1(h+1,k) \\ \nu_2 \hat{x}_2(h,k+1) + \mu_2 \hat{x}_2(h+1,k) \end{bmatrix}$$

Assume the local states in $\mathcal{X}_{\mathcal{C}_0}$ have a periodical shape, say $\hat{x}(h,-h) = \hat{x}(h+H,-h-H)$ for some $H > 0$ and for every $h$ in $\mathbb{Z}$. Then

$$\hat{A}_1^r \sqcup\!\sqcup\, ^s\hat{A}_2 = \begin{bmatrix} \binom{r+s}{s} \nu_1^r \mu_1^s & 0 \\ 0 & \binom{r+s}{s} \nu_2^r \mu_2^s \end{bmatrix}$$

gives

$$\begin{bmatrix} \hat{x}_1(h,-h+H) \\ \hat{x}_1(h+1,-h-1+H) \\ \vdots \\ \hat{x}_1(h+H-1,-h) \end{bmatrix} = \overbrace{\begin{bmatrix} \binom{H-1}{0}\nu_1^{H-1} & \binom{H-1}{1}\nu^{H-2}\mu_1 & \cdots & \binom{H-1}{H-1}\mu^{H-1} \\ \binom{H-1}{1}\nu_1^{H-2}\mu_1 & \binom{H-1}{2}\nu^{H-3}\mu_1^2 & \cdots & \binom{H-1}{0}\nu^{H-1} \\ \cdots & \cdots & \cdots & \cdots \\ \cdots & \cdots & \cdots & \cdots \\ \binom{H-1}{H-1}\mu_1^{H-1} & \binom{H-1}{0}\nu_1^{H-1} & \cdots & \binom{H-1}{H-2}\nu_1\mu_1^{H-2} \end{bmatrix}}^{G} \begin{bmatrix} \hat{x}_1(h,-h) \\ \hat{x}_1(h+1,-h-1 \\ \cdots \\ \cdots \\ \hat{x}_1(h+H-1,-h-H+1 \end{bmatrix}$$

Since G is a circulant matrix [11], its eigenvalues are

$$\chi_\ell = (\nu_1 + \mu_1 r_\ell)^{H-1} = 0.5(e^{j\sqrt{0.5}} \pm r_\ell e^{j\sqrt{0.3}})^{H-1}$$

where $r_\ell$, $\ell = 1,2,\ldots H$ denote the H-th roots of 1. Due to the fact that $|\nu_1+\mu_1 e^{j\omega}|$ is less than 1 for any real $\omega$, except $\omega = \sqrt{0.5}-\sqrt{0.3}$ (mod. $2\pi$), and that $r_\ell = e^{j2\pi\ell/H}$, $\ell = 1,2,\ldots,H$, $|\nu_1+r_\ell\mu_1|$ is always less than 1. Consequently G is a stable matrix and the components $\hat{x}_1$ of local states in $\mathcal{X}_{\mathcal{C}_i}$ eventually decay to 0 as $i\to+\infty$.

A similar argument holds for $\hat{x}_2$ components. This shows that any periodic global state $\mathcal{X}_{\mathcal{C}_0}$ determines a stable free evolution.

## 4. STABILITY CONDITIONS BY FOURIER ANALYSIS

By Corollary 1, $\Sigma = (A_1,A_2,-,-,-)$ is internally stable if and only if the Lyapunov equation

$$P(\omega) = I + (A_1 + e^{-j\omega}A_2)^T P(\omega)(A_1 + e^{j\omega}A_2) \tag{5}$$

admits a positive definite Hermitian solution $P(\omega)$ for every real $\omega$.

The entries of $P(\omega)$ are rational functions of $e^{j\omega}$, periodic in $\omega$ with period $2\pi$. The positive definite character of $P(\omega)$ can be checked by applying Sturm's test to the principal minors of $P(\omega)$.

The Fourier coefficients $P_k$ of the expansion $P(\omega)=\sum_{k=-\infty}^{+\infty} P_k e^{j\omega k}$ satisfy the following properties:

i) since the entries of $P(\omega)$ are in $L^2[-\pi,\pi]$, the sequence $\{P_k\}$ belongs to $\ell^2(\mathbb{R}^{n\times n})$

ii) for any integer k, $P_k = P_{-k}^T$, and the following set of equalities holds:

$$\begin{aligned} P_o &= I + A_1^T P_o A_1 + A_2^T P_o A_2 + A_1^T P_1^T A_2 + A_2^T P_1 A_1 \\ P_k &= A_1^T P_k A_1 + A_2^T P_k A_2 + A_1^T P_{k-1} A_2 + A_2^T P_{k+1} A_1, \quad k = 1,2,\ldots \end{aligned} \tag{6}$$

iii) the doubly infinite block Toeplitz matrix

$$\mathcal{P} = \begin{bmatrix} \ddots & \ddots & \ddots & \ddots & & \\ \ddots & P_{-1} & P_o & P_1 & \ddots & \\ & \ddots & P_{-1} & P_o & P_1 & \ddots \\ & & \ddots & P_{-1} & P_o & P_1 \\ & & & \ddots & \ddots & \ddots \end{bmatrix} \tag{7}$$

induces a positive definite scalar product in the space $\ell^2(\mathbb{C}^n)$.

In fact for every non zero vector function $v(\cdot):[-\pi,\pi]\to\mathbb{C}^n$, with elements in $L^2[-\pi,\pi]$ we have

$$\int_{-\pi}^{\pi} \check{v}^T(\omega)P(\omega)v(\omega)\,d\omega > 0$$

Then $\sum_{h,k} \check{v}_h^T P_{h-k} v_k > 0$ where $v_k$ are the Fourier coefficients of $v(\cdot)$: $v(\omega) = \sum_{k=-\infty}^{+\infty} v_k e^{j\omega k}$.

By Riesz-Fischer theorem, there is a bijection between $\ell^2(\mathbb{C}^n)$ and the space of n-tuples with elements in $L^2[-\pi,\pi]$. Therefore the relation

$$(u,w) = \sum_{h,k} \check{u}_h^T P_{h-k} w_k$$

defines a positive definite scalar product in $\ell^2(\mathbb{C}^n)$.

Conversely, assume that there exists a sequence $\{P_k\}$ of matrices in $\mathbb{R}^{n\times n}$ satisfying conditions i), ii), iii). Then the series $\sum_{k=-\infty}^{+\infty} P_k e^{j\omega k}$ defines a.e. a matrix function $P(\omega)$ with elements in $L^2[-\pi,\pi]$ which solves equation (5). In fact, using (6), we check directly that

$$I+A_1^T P(\omega)A_1 + A_2^T P(\omega)A_2 + A_1^T e^{j\omega} P(\omega)A_2 + A_2^T e^{-j\omega} P(\omega)A_1 = \sum_k P_k e^{j\omega k} = P(\omega)$$

Hence the entries of $P(\omega)$ are a.e. real rational functions of $e^{j\omega}$. Since

$$P_o = \int_{-\pi}^{\pi} P(\omega)d\omega$$

is finite, and $P(\omega)$ is a.e. positive definite, the analytic extension $Q(z)$ of $Q(e^{j\omega}) \triangleq P(\omega)$ has no poles on the unit circle.

Since by continuity $P(\omega)$ is at least positive semidefinite and satisfies equation (5) for all real $\omega$, then it is positive definite for all real $\omega$.

Assume $\Sigma = (A_1,A_2,-,-,-)$ to be internally stable and let $P(\omega) = \sum_{k=-\infty}^{+\infty} P_k e^{j\omega k}$ be the solution of equation (5). Then equations (6) can be used to investigate the structure of matrices $P_k$ in terms of the family of matrices (4).

First notice that the Toeplitz matrix $\mathcal{P}$ satisfies the equation

$$\mathcal{P} = \mathcal{J} + \mathcal{A}^* \mathcal{P} \mathcal{A} \tag{8}$$

where $\mathcal{J}$ is the (infinite) identity matrix and $\mathcal{A}$ and $\mathcal{A}^*$ are the doubly infinite block Toeplitz matrices:

$$\mathcal{A} = \begin{bmatrix} \ddots & 0 & A_1 & A_2 & 0 & & & \\ & & 0 & A_1 & A_2 & 0 & & \\ & & & 0 & A_1 & A_2 & 0 & \\ & & & & & & & \ddots \end{bmatrix} \qquad \mathcal{A}^* = \begin{bmatrix} \ddots & 0 & A_2^T & A_1^T & 0 & & & \\ & & 0 & A_2^T & A_1^T & 0 & & \\ & & & 0 & A_2^T & A_1^T & 0 & \\ & & & & & & & \ddots \end{bmatrix}$$

Whe therefore have the following Theorem.

Theorem 6. *Let $\mathcal{P}$ as in (7). Then $\mathcal{P}$ is the sum of the series*

$$\mathcal{J} + \sum_{i=1}^{\infty} \mathcal{A}^{*i} \mathcal{A}^i$$

*of Toeplitz matrices and the blocks $P_k$ are expressed as*

$$P_k = \sum_{r,s=0}^{\infty} (A_1^{r+k} \sqcup\!\sqcup\, {}^s A_2)^T (A_1^r \sqcup\!\sqcup\, {}^{s+k} A_2) \qquad k = 0,1,2,\ldots \tag{9}$$

Proof. Since the P's are the Fourier coefficients of the continuous bounded function $P(\omega)$, the operator $\mathcal{P}: \ell^2(\mathbb{C}^n) \to \ell^2(\mathbb{C}^n)$ is continuous [12]. Hence there exists a positive integer M such that $\check{v}^T \mathcal{P} v < M \check{v}^T v$ for every v in $\ell^2(\mathbb{C}^n)$. Then by (8) we have

$$\check{v}^T v + \check{v}^T \mathcal{A}^* \mathcal{A} v + \ldots + \check{v}^T \mathcal{A}^{*M} \mathcal{A}^M v < M \check{v}^T v$$

$$\sigma \overset{\Delta}{=} \sup_{\|v\|=1} \min\{\check{v}^T \mathcal{A}^* \mathcal{A} v, .., \check{v}^T \mathcal{A}^{*M} \mathcal{A}^M v\} < 1$$

Let t be a positive integer such that $\sigma^{\frac{t}{M} - 1} < \frac{1}{2M}$.

For any $\nu > t$ and v in $\ell^2(\mathbb{C}^n)$, $\|v\| = 1$, there exists a partition of $\nu = \nu_1 + \nu_2 + \ldots + \nu_{r+1}$, $\nu_i \leqslant M$, $i = 1,2 \ldots r+1$ such that

$$\check{v}^T \mathcal{A}^{*\nu_1} \mathcal{A}^{\nu_1} v \leq \sigma$$

$$\check{v}^T \mathcal{A}^{*\nu_1} \mathcal{A}^{*\nu_2} \mathcal{A}^{\nu_1} \mathcal{A}^{\nu_2} v \leq \sigma^2$$

$$\ldots\ldots\ldots\ldots\ldots\ldots\ldots$$

$$\check{v}^T \mathcal{A}^{*\nu_1} \ldots \mathcal{A}^{*\nu_r} \mathcal{A}^{\nu_r} \ldots \mathcal{A}^{\nu_1} v \leq \sigma^r$$

and

$$\check{v}^T \mathcal{A}^{*\nu} \mathcal{A}^{\nu} v \leq \sigma^r M \leq \sigma^{\frac{\nu}{M} - 1} M \leq \sigma^{\frac{t}{M} - 1} M < \frac{1}{2}$$

Thus the sequence of operators $\mathcal{P}_0 = \mathcal{I}, \mathcal{P}_i = \mathcal{P}_{i-1} + \mathcal{A}^{*i} \mathcal{A}^i$, $i > 0$, is a Cauchy sequence, as the following inequalities

$$\|(\mathcal{P}_n - \mathcal{P}_m) v\| \leq \sum_{i=m+1}^{n} \|\mathcal{A}^{*i} \mathcal{A}^i v\| < \sum_{i=m+1}^{n} \frac{2\, M^2}{2^{i/t}}$$

hold for $n > m > t$ and $\|v\| = 1$. Since $\ell^2(\mathbb{C}^n)$ is complete, the operator $\sum_{i=1}^{\infty} \mathcal{A}^{*i} \mathcal{A}^i + \mathcal{I}$ is well defined and solves equation (8).

Finally $\mathcal{P}$ coincides with the series $\mathcal{I} + \sum_{i=1}^{\infty} \mathcal{A}^{*i} \mathcal{A}^i$.

In fact $\Delta = \mathcal{P} - (\mathcal{I} + \Sigma \mathcal{A}^{*i} \mathcal{A}^i)$ satisfies the equations chain

$$\Delta = \mathcal{A}^* \Delta \mathcal{A} = \ldots = \mathcal{A}^{*i} \Delta \mathcal{A}^i \ldots$$

and $\Delta = 0$, as $\lim_{i \to \infty} \mathcal{A}^i = 0$.

<u>Example</u> (scalar case). If the local state space is one dimensional, the matrices $A_1$ and $A_2$ become scalars $a_1$ and $a_2$ respectively, and a closed form computation of the Fourier coefficients $P_k$ is possible when the 2-D system is internally stable.

The solution of equation (5), given by

$$P(\omega) = (1 - a_1^2 - a_2^2 - 2a_1 a_2 \cos\omega)^{-1}$$

is positive for every real $\omega$ if and only if

$$1 - a_1^2 - a_2^2 > |2a_1 a_2| \tag{10}$$

Incidentally note that condition (10) can be restated in the following equivalent forms

i) $|a_1| + |a_2| < 1$

ii) $q(z_1,z_2) = 1 - a_1z_1 - a_2z_2 \neq 0$ in $\mathcal{P}_1$

iii) $\begin{bmatrix} 1-a_1^2-a_2^2 & 2a_1a_2 \\ 2a_1a_2 & 1-a_1^2-a_2^2 \end{bmatrix} > 0$

When condition (10) is satisfied, we obtain

$$P_o = \frac{1}{2\pi}\int_{-\pi}^{\pi}(1-a_1^2-a_2^2-2a_1a_2\cos\omega)^{-1}d\omega = \left[(1-a_1^2-a_2^2)^2-4a_1^2a_2^2\right]^{1/2}$$

When once $P_o$ is computed, the coefficients $P_k$, k=1,2... are obtained recursively from eqns. (6).

Observe that in general the sum of the series (9) is not a rational function of the matrices $A_1$ and $A_2$.

REFERENCES

[1] J.L.Shanks, S.Treitel, J.H.Justice: IEEE Trans. Audio Elec., AU-20, 1972, pp. 115-128.

[2] T.S.Huang: IEEE Trans. Audio Elec., AU-20, 1972, pp.158-163.

[3] S.Attasi: IRIA Rapport LABORIA, n. 31, 1973

[4] "Multidimensional Systems" Special Issue of IEEE Proc., 65, n. 6, 1977.

[5] D. Goodman: IEEE Trans. Circuit and Syst., CAS 24, n.4, 1977 pp. 201-208.

[6] E.Fornasini, G.Marchesini: Mathematical Systems Theory, 12, n. 1, 1978.

[7] E.Fornasini, G.Marchesini: Lect.Notes in Econ. and Math.Sys. 162, Springer Verlag 1978, pp. 123-129.

[8] R.Eising: IEEE Trans. Autom. Control, AC-23, n. 5, 1978, pp. 793-99.

[9] E.Fornasini: Journées sur l'Analyse des Systèmes, Bordeaux, Sept. 11-15,1978; to appear in "Astérisque".

[10] E.Fornasini, G.Marchesini: IEEE Trans. Autom. Control, AC-24, n. 1, 1979, pp. 129-30, and AC-24, n. 4, 1979, pg. 678.

[11] R.E.Bellman "Introduction to Matrix Analysis" McGraw-Hill, New York, 1970.

[12] U. Grenander, G.Szegö "Toeplitz forms and their applications" Berkeley, Univ. of California Press, 1958.

# ON THE STABILITY ANALYSIS OF COMPLEX DISCRETE DYNAMICAL SYSTEMS

Georges BITSORIS

Systems and Measurements Laboratory
University of Patras
Patras - Greece

ABSTRACT: A new comparison principle for the stability analysis of complex discrete dynamical systems is developed. To establish the comparison system, vector valued functions with positive definite maximal component instead of vector positive definite functions are used. Then an augmented comparison system composed of the studied system and a low order subsystem with particular properties is established. The stability properties of the studied system are derived from the corresponding partial stability properties of the comparison system with respect to the new state variables of its second subsystem. Finally sufficient conditions for partial stability in the sense of Lyapunov , uniform asymptotic stability, exponential stability and boundedness of the comparison system are developed.

INTRODUCTION: The extended use of digital computers for the analysis of dynamical systems has increased the interest of the development of mathematical methods for the analysis of descrete dynamical systems. As a first step to this direction, the existing mathematical approaches to the analysis of continuous time dynamical systems was extended to discretetime systems. Thus in the stability analysis, Lyapunov's second method and the comparison principle approach initially established for continuous time systems has now been extended to discrete time systems [1-3], [5-9]. This paper is concerned with the development of a new comparison principle for the stability analysis of complex discrete systems analogous to that developed for continuous time systems [4].

In the proposed method, to establish the comparison system,vector valued functions with positive definite functions are used. The comparison system is composed of two subsystems: the studied system itself and a low order subsystem with a non decreasing



*D. G. Lainiotis and N. S. Tzannes (eds.), Advances in Control, 227–235.*

transition function. Then it is proved that the partial stability properties of the comparison system with respect to the state variables of its second subsystem imply the corresponding stability properties of the studied system. Finally sufficient conditions for partial stability in the sense of Lyapunov, uniform asymptotic stability, exponential stability and boundedness of the comparison are developed.

## THE COMPARISON PRINCIPLE

Let $R^n$ denote the n-dimensional Euclidean space, and $\|\cdot\|$ be the norm $\|X\| = |x_1| + |x_2| + \ldots + |x_n|$ where $X = [x_1, x_2, \ldots, x_n]^T$, $X \in R^n$. In $R^n$ we introduce the partial order relations $X \leq Y : x_i \leq y_i$ $i=1,2,\ldots,n$ and $X < Y : x_i < y_i$ $i=1,2,\ldots,n$. Finally $R^n_+$ denotes the set $R^n_+ = \{X \in R^n : X \geq 0\}$.
The discrete dynamical systems considered in this paper are described by a vector difference equation of the form

$$\Delta X(k+1) = F[k, X(k)] \qquad (S)$$

where

$$\Delta X(k) = X(k+1) - X(k)$$

In the above equation $X(k)$ represents the state of the system at the instant $k \in T$, $T = \{0,1,2,\ldots\}$. It is assumed that the function $F(k,X)$, $F : TxR^n \longrightarrow R^n$ satisfies the relation $F(k,0) = 0$ for all $K \in T$, so that the origin $X=0$ to be an equilibrium state of the system S.

<u>Definition 1.</u> The vector valued function $V(k,X), V: TxR^n \longrightarrow R^s$ is said to be <u>of class $\mathcal{V}$</u> if the function $\max_i [v_i(k,X)]$ is positive definite.

A subclass of the above functions is the class of vector positive definite functions possessing the additional property that $v_i(k,X) \geq 0$ $i=1,2,\ldots,s$ for all $(k,X) \in TxR^n$, [2].

<u>Definition 2.</u> A function $V(k,X)$, $V: TxR^n \longrightarrow R^s$ is said to be <u>decrescent</u> if there exists a continuous positive definite function <u>$v^*(X)$</u>, $v^*: R^n \longrightarrow R_+$ such that

$$\max_i ( v_i(k,X) ) \leq v^*(X)$$

for all $(k,X) \in TxR^n$.

<u>Definition 3.</u> A function $V(t,X), V: TxR^n \longrightarrow R^s$ is said to be <u>radially unbounded</u> if for any real positive number a there exists a positive number $\beta$ such that $\|X\| \geq \beta$ implies $\max_i |v_i(t,X)| \geq a$ $\forall t \in T$.

Let us now assume that for a function $V(k,X)$, $V: TxR^n \longrightarrow R^s$ of class $\mathcal{V}$, there exists a function $H(k,X,W)$, $H: TxR^n xR^s \longrightarrow R^s$ such that

1. The function $H(k,X,W)$ posseses the property $\mathcal{H}^*$, namely all the components $h_i(k,X,w_1,\ldots,w_s)+w_i$ $i=1,2,\ldots,s$ of the function $H(k,X,W)+W$ are non decreasing with respect to $w_j$ $j=1,2,\ldots,s$
2. The inequality

$$\Delta V(k,X)_{(S)} \leq H[k,X,V(k,X)] \tag{1}$$

where

$$\Delta V[k,X]_{(S)} \leq V[k,F(k,X)+X]-V(k,X)$$

is satisfied for all $(k,X)\in TxR^n$.

Under the above hypothesis we can easily prove by induction that if

$$V(k_o,X_o) \leq W_o \tag{2}$$

then

$$V[k,X(k,k_o,X_o)] \leq W(k,k_o,X_o,W_o) \tag{3}$$

for all $k \geq k_o$, $W(k,k_o,X_o,W_o)$ being the solution of the augmented comparison system

$$\Delta X(k) = F[k,X(k)] \tag{4a}$$
$$\Delta W(k) = H[k,X(k),W(k)] \tag{4b}$$

From (3) it follows that for any $(k_o,X_o)\in TxR^n$ the inequality

$$0 \leq \max_i[v_i(k,k_o,X_o)] \leq \max_i[w_i(k,k_o,X_o,W_o)].$$

is satisfied for all $k \geq k_o$. Thus, it can be easily seen that, under some conditions on $V(k,X)$, the stability properties of the origin $X=0$ of S may be derived from the analogous W-stability properties of the origin $X=0$, $W=0$ of the augmented comparison system (4). This is established in the following theorem.

<u>Theorem 1</u>. Assume that there exists a function $V(k,X), V:TxR^n \longrightarrow R^s$ of class $\mathcal{V}$ continuous on the origin $X=0$ with respect to X and a function $H(k,X,W)$, $H : TxR^n xR^s \rightarrow R^s$ possessing the property $\mathcal{H}^*$ which satisfies inequality (1) for all $(k,X)$ $TxR^n$.

Assume also that $H(k,X,0) \equiv 0$.

If the equilibrium state $X=0$, $W=0$ of (4) is

1. W-stable in the sense of Lyapunov in $R^n xR^s_+$, then the equilibrium $X=0$ of S is stable.
2. Uniformly W-stable in $R^n xR^s_+$ and V is decrescent, then the equilibrium $X=0$ of S is uniformly stable.
3. W-attractive in $R^n xR^s_+$, then the equilibrium $X=0$ of S is attractive.
4. Uniformly W-attractive in $R^n xR^s_+$, then the equilibrium $X=0$ of S is uniformly attractive.

5. Asymptotically W-stable in $R^n x R^s_+$, then the equilibrium $X = 0$ of S is asymptotically W-stable.
6. Uniformly assymptotically W-stable in $R^n x R^s_+$, then the equilibrium $X = 0$ of S is uniformly asymptotically stable.
7. Exponentially W-stable in $R^n x R^s_+$ and V satisfies for all $(k,X) \in T x R^n$ the inequalities

$$\max_i [v_i(k,X)] \geq \eta_1 \|X\| \quad (5a)$$
$$V(k,X) \leq \eta_2 \|X\| \quad (5b)$$

$\eta_1$, $\eta_2$ being positive real numbers, then the equilibrium $X = 0$ of S is exponentially stable.

We give below the proof of the parts 1 and 2 of the theorem concerning the stability in the sense of Lyapunov and the uniform stability. The proofs of the other part of the theorem are established in [2].

Proof

1. Since the equilibrium $X = 0$, $W = 0$ of (4) has been assumed to be W-stable in the sense of Lyapunov, in $R^n x R^s_+$, for any $\varepsilon' > 0$ and $k_o \in T$ there exists a $\delta'(k_o, \varepsilon') > 0$ such that

$$\|X_o\| + \|W_o\| < \delta' \Rightarrow \max_i [w_i(k,k_o,X_o,W_o)] < \varepsilon' \quad \forall k \geq k_o \quad (7)$$

whith $(X_o, W_o) \in R^n x R^s_+$. Then from (2) and (3) follows that

$$\|X_o\| + \|V(k_o,X_o)\| < \delta' \Rightarrow \max_i [v_i(k, X(k,k_o,X_o)] < \varepsilon' \quad (8)$$

for all $k \geq k_o$. On the other hand, from the continuity of V(k,X) which is also of class $\mathcal{V}$, it follows that ginen a $\varepsilon > 0$ there exists a $\varepsilon' > 0$ such that for any $k \in T$

$$\max [v_i(k,X)] < \varepsilon' \Rightarrow \|X\| < \varepsilon \quad (9)$$

Finally, given a $\delta' > 0$ and $k_o \in T$ there exists a $\delta(k_o, \delta') > 0$ such that

$$\|X_o\| < \delta \Rightarrow \|X_o\| + \|V(k_o,X_o)\| < \delta' \quad (10)$$

because V(k,X) is continuous and $V(k_o,0) \equiv 0$.
Now, by combining (8)-(10) we can state that geven an $\varepsilon > 0$ and $k_o \in T$ there exists a $\delta(k_o, \varepsilon) > 0$ such that

$$\|X_o\| < \delta \Rightarrow \|X(k,k_o,X_o)\| < \varepsilon \quad \forall k \geq k_o$$

that is the equilibrium $X = 0$ of S is stable in the sense of Lyapunov.

2. If, in addi tion, the equilibrium state $X = 0$, $W = 0$ is uniformly W-stable and the function V(k,X) is decrescent, then the po-

sitive real numbers $\delta'$ and $\delta$ do not depend on $k_o$, thus the stability of $X = 0$ of S is uniform.

STABILITY CRITERIA

Although the comparison system established in the previous section is more complex than the studied system itself, the stability analysis with respect to its W-components is a rather simple problem because of the particular properties of the function H(k,X,W).

An important property of the solutions of the comparison system is stated in the following Lemma which can be easily proven by induction.

Lemma

If

$$W_o \leq W_o^*$$

$W_o, W_o \in R_+^s$, then

$$0 \leq W(k,k_o,X_o,W_o) \leq W(k,k_o,X_o,W_o^*)$$

for all $k_o \in T$, $X_o \in R^n$ and $k \geq k_o$.

Based on this result we can establish a condition of W-boundedness of the solutions of the comparison system in $R^n x R_+^s$.

Theorem 2. If there exists a vector $W_o \in R_+^s$ with positive components such that

$$H(k,X,W_o^*) \leq 0$$

for all $(k,X) \in T x R^n$, then

$$0 \leq W(k,k_o,X_o,W_o) \leq W_o^*$$

for all $k_o \in T, X_o \in R^n$, $k \geq k_o$ and $W_o \in R(W)$ where

$$R(W) = \{W_o \in R_+^s : W_o \leq W_o^*\}$$

Proof

From the result of the Lemma, follows that it is sufficient to prove that

$$W(k,k_o,X_o,W_o^*) \leq W_o^*$$

for all $k_o \in T$, $X_o \in R^n$ and $k \geq k_o$.
This inequality is verified for $k = k_o$. Let us assume that it is also verified for a $k' \geq k_o$, that is

$$W(k',k_o,X_o,W_o^*) \leq W_o^*$$

Then

$$W(k'+1,k_o,X_o,W_o^*) = H[k', X(k',k_o,X_o,W_o^*),W(k',k_o,X_o,W_o^*)] +$$
$$+W(k',k_o,X_o,W_o^*) \leq H[k', X(k',k_o,X_o,W_o),W_o^*] +W_o \leq W_o^*.$$

because the function H(k,X,W)+W has been assumed to be non decreasing with respect to all the components of W and $H(k,X,W_o^*) \leq 0$. This completes the proof of the theorem.

It is to be noted that the condition of W-boundedness of the comparison system established in the above theorem involves only the properties of the function H thus overcoming the problems due to the assumed complexity of system S whose behaviour is described by the function F. This is also the case in the following theorem relative to the uniform asymptotic W-stability of the comparison system.

Theorem 3. If there exists a vector $W_o^* \in R^s$ with positive components and a vector valued function $\Phi(k,X,r)$, $\Phi^+ : TxR^n x[0,\bar{r}] \to R_+^s$ with positive definite components with respect to r, $\bar{r}$ being a positive real number, such that

$$H(k,X,rW_o^*) \leq -\Phi(k,X,r) \qquad \forall (k,X,r) \in TxR^n x[0,\bar{r}] \quad (11)$$

then

$$R(X,W) = \{(X,W) \in R^n x R_+^s : W \leq \bar{r} W_o^*\}$$

is a region of uniform asymptotic W-stability of the equilibrium X = 0, W = 0 of (4).

Proof

Let us choose the function

$$v^*(X,W) = \max_i \left(\frac{w_i}{w^*_{oi}}\right) \quad (12)$$

which is W-positive definite and decrescent, as a candidate Lyapunov function. Then

$$\Delta v^*(X,W)_{(4)} = \max_i \left[\frac{h_i(k,X,W)+w_i}{w^*_{oi}}\right] - \max_j \left[\frac{w_j}{w^*_{oj}}\right]$$
$$\leq \max_i \left[\frac{h_i(k,X,v^*(X,W)W_o^*)+v^*(X,W)w^*_{oi}}{w^*_{oi}}\right] - v^*(X,W)$$
$$\leq \max_i \left[\frac{h_i(k,X,v^*(X,W)W_o^*)}{w^*_{oi}}\right]$$

because

$$W \leq v^*(X,W)W_o^*$$

and the function H possesses the property $\mathcal{H}^*$. Now, from (11) it follows that

$$\Delta v^*(X,W)_{(4)} \leq \max_i \left[\frac{-\Phi_i\left[k,X,v^*(X,W)\right]}{w_{oi}^*}\right]$$

provided that $v^*(X,W) \leq \bar{r}$ or, equivalently, that $(X,W) \in R(X,W)$. Thus $R(X,W)$ is a region of uniform asymptotic W-stability of the equilibrium $X = 0$, $W = 0$ of (4) because all the components $\Phi_i$ of $\Phi$ have been assumed to be positive definite.

If, in the above theorem, the components $\Phi_i(k,X,r)$ of $\Phi(k.W,r)$ were semi-positive definite then $v^*(X,W)_{(4)}$ would be semi-negative definte. Hence, we can establish the following theorem:

Theorem 4

If there exists a vector $W_o^* \in R_+^s$ with positive components and a vector valued function $\Phi(k.X,\bar{r})$, $\Phi : TxR^n x[0,\bar{r}] \longrightarrow R_+^s$ with semi-positive definite components with respect to r, $\bar{r}$ being a positive real number, such that

$$H(k,X,rW_o^*) \leq -\Phi(k,X,r) \qquad \forall (k,X,r) \in TxR^n x[0,\bar{r}]$$

then the equilibrium $X = 0$, $W = 0$ of (4) is uniformly W-stable.

Finally we establish a condition of exponential W-stability for the augmented comparison system.

Theorem 5

If there exists a vector $W_o^* \in R_+^s$ with positive components and a real number $\beta$, $0 < \beta < 1$ such that

$$H(k,X,rW_o^*) \leq -r\beta W_o^* \qquad \forall (k,X,r) \in TxR^n x[0,\bar{r}] \tag{13}$$

$\bar{r}$ being a positive real number, then

$$R(X,W) = \left\{(X,W) \in R^n x R_+^s \; : \; W \leq \bar{r} W_o^*\right\}$$

is a region of exponential W-stability of the equilibrium $X = 0$, $W = 0$ of (4).

Proof

By a procedure analogous to that adopted in the proof of Theorem 3 we obtain the inequality

$$\Delta v^*(X,W)_{(4)} \leq \max_i \left[\frac{h_i(k,X,v^*(X,W)W_o^*)}{w_{oi}^*}\right]$$

and taking into account (13) we get

$$\Delta v^*(X,W)_{(4)} \leq -\beta v^*(X,W)$$

provided that $v^*(X,W) \leq \bar{r}$, or equivalently that $(X,W) \in R(X,W)$. Thus

$$v^*\left[X(k,k_o,X_o), W(k,k_o,X_o,W_o)\right] \leq (1-\beta)^{k-k_o} v^*(X_o,W_o)$$

or

$$v^*\left[X(k,k_o,X_o) W(k,k_o,X_o,W_o)\right] \leq v^*(X_o,W_o) \exp\left[-\beta^*(k-k_o)\right] \quad (14)$$

where

$$\beta^* = -\ln(1-\beta)$$

On the other hand, the function $v^*(X,W)$ satisfies the inequalities

$$\frac{\|W\|}{\|W_o^*\|} \leq v^*(X,W) \leq \frac{\|W\|}{\min_i w_{oi}^*}.$$

From these inequalities and (14) follows that

$$\|W(k,k_o,X_o,W_o)\| \leq a \cdot \|W_o\| \cdot \exp\left[-\beta^*(k-k_o)\right]$$

for all $(X_o,W_o) \in R(X,W)$, $k_o \in T$ and $k \geq k_o$, with

$$a^* = \frac{\|W_o^*\|}{\min_i w_{oi}^*}.$$

## CONCLUSION

A comparison principle for the stability analysis of large-scale or complex nonlinear systems has been established. In the proposed method it is not necessary the vector Lyapunov functions to have non-negative components. On the other hand the state vector of comparison system's second sybsystem can appear explicitly in the right side member of inequality (1), thus increasing the flexibility of construction of the comparison system.

In the last part of the paper, partial stability coditions of the comparison system has been established. It is to be noted that all these conditions involve only the properties of comparison system's second subsystem.

## REFERENCES

(1) ARAKI,M., ANDO,K., KONDO,B., "Stability analysis of sampled - data composite systems with many non-linearities", 1971,IEEE Trans. AC, 1, p.p. 22-27

(2) BITSORIS, G., "Principe de comparaison et stabilité des systèmes complexes". 1978, Thèse d'Etat, Univ. de Toulouse.

(3) BITSORIS, G., "Stability analysis of complex discrete systems" (to be published).

(4) BITSORIS, G., "A new comparison principle for the stability analysis of complex systems", 1979, Proc. 2nd IFAC Symp. Large Scale Syst. Appl.
(5) GRUJIC,Lj.T., "Uniform asymptotic stability of discrete large-scale systems", 1973, IEEE Trans. SMC, 6, p.p. 636-643
(6) GRUJIC,Lj.T., SILJAK,D.D., "Exponential stability of large - scale discrete systems", 1974, Int.J.Control,, 19, p.p. 481-491.
(7) HAHN,W., "On the application of the method of Lyapunov to difference equations"(In German), 1958, Mathematische Annalen, 136, p.p. 403-441.
(8) KALMAN, R.E., BERTRAM, J.E., "Control system analysis and design via the second Method of Lyapunov" II, 1962, Trans. ASME J. Basic Engin. 82, p.p. 371-393.
(9) MICHEL, A.N., PORTER, D.W., "Stability analysis of composite systems", 1972, IEEE Trans AC, 2, p.p. 222-226.

# USING INNERS TO DETERMINE MULTIPLE ROOTS OF POLYNOMIALS *

T. Millett, School of Electrical Engineering

University of New South Wales, Sydney, Australia

and

I. Tang, Department of Mathematical Studies

Hong Kong Polytechnic, Kowloon, Hong Kong

An example is given of the application of the theory of the inners to the detection and determination of multiple roots of polynomials.

## 1. INTRODUCTION

The theory and applications of inners have been discussed extensively by Jury (1), (2). This article introduces yet another example of application – that of using inners to determine multiple roots of a polynomial.

The theory on which the discussion in this article is based follows the work of the authors on multiple root detection (3). However, consideration is now being given to approaching the problem by using inners.

---

* To appear also in Bull. SE Asian Math. Soc.



*D. G. Lainiotis and N. S. Tzannes (eds.), Advances in Control, 236–246.*

## 2. DEFINITION

The term <u>inners</u> is given to certain square submatrices obtained in a square matrix. If $D_N$ is an NxN matrix, then the (N-2)x(N-2) matrix $D_{N-2}$, formed by deleting the first and last rows and the first and last columns of $D_N$, is an inner. Repeat this process on $D_{N-2}$ and an inner in the form of an (N-4)x(N-4) matrix $D_{N-4}$ results. Continuing in the manner, we get for the matrix $D_N$ a set of inners $D_j$, $j = N-2, N-4, \ldots, 2$, if N is even or $j = N-2, N-4, \ldots, 1$ if N is odd.

The subscript j is the order of the inner, and the inner of the lowest order is called the innermost inner. Thus, for N=7, as shown in Eq. 1, we have the three inners $D_5$, $D_3$ and $D_1$, with $D_1$ being the innermost inner.

$$D_7 = \begin{bmatrix} a_{11} & a_{12} & a_{13} & a_{14} & a_{15} & a_{16} & a_{17} \\ a_{21} & a_{22} & a_{23} & a_{24} & a_{25} & a_{26} & a_{27} \\ a_{31} & a_{32} & a_{33} & a_{34} & a_{35} & a_{36} & a_{37} \\ a_{41} & a_{42} & a_{43} & a_{44} & a_{45} & a_{46} & a_{47} \\ a_{51} & a_{52} & a_{53} & a_{54} & a_{55} & a_{56} & a_{57} \\ a_{61} & a_{62} & a_{63} & a_{64} & a_{65} & a_{66} & a_{67} \\ a_{71} & a_{72} & a_{73} & a_{74} & a_{75} & a_{76} & a_{77} \end{bmatrix} \qquad (1)$$

Consider a set of inners $D_1, D_3, D_5, \ldots, D_{N-2}$, or $D_2, D_4, D_6, \ldots, D_{N-2}$ of a matrix $D_N$, depending on whether N is odd or even. If none (all) of the determinants $\Delta_j$ of the inners $D_j$, $j < N$ and of the matrix $D_N$ itself are zero, then the matrix is termed <u>non-null (null) innerwise</u>. If only some of the determinants are zero, we designate the matrix as being <u>semi-innerwise</u>. Furthermore, in the case of a Sylvester-type matrix referred to in what is to follow, if J is the smallest integer for which $\Delta_J$ is zero, then it can be shown that all higher orders of $\Delta_j$, $j > J$ are zero.

Let two polynomials

$$B(x) = b_m x^m + b_{m-1} x^{m-1} + \ldots + b_1 x + b_0 \ ,$$

and

$$A(x) = a_n x^n + a_{n-1} x^{n-1} + \ldots + a_1 x + a_0 \ , \qquad (2)$$

be given, where $a_n$, $b_m \neq 0$ and $n \geq m$.

Take a specific case where n=5 and m=4 and form what is known as a Sylvester matrix of dimension (m+n)x(m+n) which is characteriz by a symmetrical triangular array of zeros as shown:

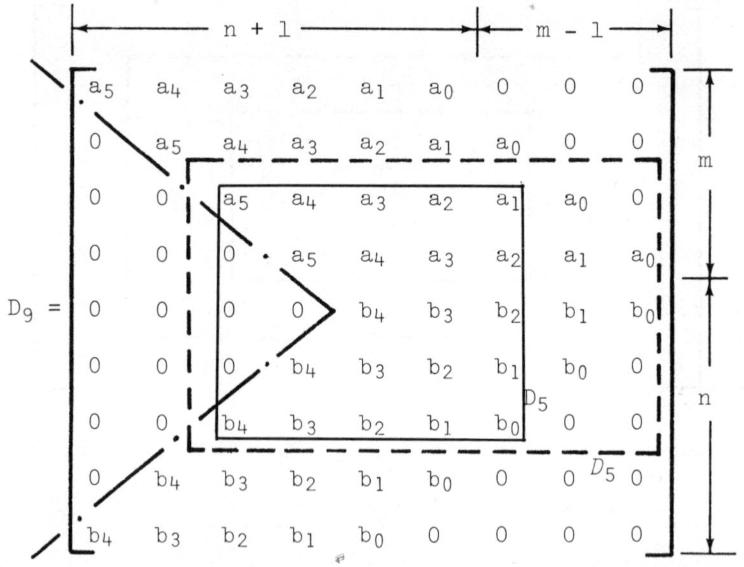

$$D_9 = \begin{bmatrix} a_5 & a_4 & a_3 & a_2 & a_1 & a_0 & 0 & 0 & 0 \\ 0 & a_5 & a_4 & a_3 & a_2 & a_1 & a_0 & 0 & 0 \\ 0 & 0 & a_5 & a_4 & a_3 & a_2 & a_1 & a_0 & 0 \\ 0 & 0 & 0 & a_5 & a_4 & a_3 & a_2 & a_1 & a_0 \\ 0 & 0 & 0 & 0 & b_4 & b_3 & b_2 & b_1 & b_0 \\ 0 & 0 & 0 & b_4 & b_3 & b_2 & b_1 & b_0 & 0 \\ 0 & 0 & b_4 & b_3 & b_2 & b_1 & b_0 & 0 & 0 \\ 0 & b_4 & b_3 & b_2 & b_1 & b_0 & 0 & 0 & 0 \\ b_4 & b_3 & b_2 & b_1 & b_0 & 0 & 0 & 0 & 0 \end{bmatrix} \qquad (3)$$

(Column widths marked: $n+1$, $m-1$; row heights marked: $m$, $n$; inner boxes labelled $D_5$.)

When the two polynomials do not have a common factor, according to Trudi (4) as cited in Householder (5), the Sylvester matrix as formed in Eq. 3 in non-null innerwise. However, if K of the determinants $\Delta_j$ are zero, then the two polynomials have a common factor of degree K. It is not difficult to see from Eq. 3 that if N is the dimension of the Sylvester matrix and $\Delta_L$ is the largest non-zero determinant, then K=(N-L)/2. By appropriate manipulations of the columns of the matrix, the common factor can furthermore be determined.

## 3. APPLICATION

Given a polynomial of degree m which is monic:

$$g_m(x) = x^m + a_{m-1}x^{m-1} + \ldots + a_1x + a_0 \ . \qquad (4)$$

In order to determine whether multiple roots exist, we generate the sequence

$$g_i(x) = g'_{i+1}(x)/(i + 1) = \sum_{j=0}^{i} \frac{i!(m - j)!}{m!(i - j)!} a_{m-j}x^{i-j}, \qquad (5)$$

for i = m - 1, m - 2, ..., 1.

It is well-known that if $g_m(x) = 0$ has a root of multiplicity r, then $g_m$, $g_{m-1}$, ..., $g_{m-r+1}$ have a common factor. Thus we start with $g_m$ and $g_{m-1}$, generated from Eq. 5, and consider the Sylvester matrix formed using the coefficients. When the matrix is not semi-innerwise in the sense of Trudi, $g_m$ and $g_{m-1}$ are relatively prime, and $g_m$ has no multiple factor. However, if the matrix is semi-innerwise, a multiple factor exists, since there is a common divisor between $g_m$ and $g_{m-1}$. We shall call the common divisor $q_{m-1}$, which is found following the procedure described subsequently. Then, we inspect the Sylvester matrix of $q_{m-1}$ and $g_{m-2}$, and obtain the common divisor $q_{m-2}$, if one exists. In this manner, we find, for i = 1, 2, 3, ...,

$$q_{m-i} = \text{h.c.f. of } q_{m-i+1} \text{ and } g_{m-i} \quad , \tag{6}$$

where h.c.f. stand for highest common factor. Then when at last a value r is reached such that $q_{m-r}$ can no longer be found other than a constant, we terminate the process and conclude that $q_{m-(r-1)}$ is an r-tuple factor of $g_m$. Other factors of $g_m$ with multiplicity $1<k<r$ can now be found from the sequence generated from Eq. 6.

Once the existence of a common factor is established, it may be found according to Trudi's consideration as described in Householder. The procedure can best be illustrated, without loss of generality, by an example. Given two polynomials

$$g_4(x) = b_4x^4 + b_3x^3 + b_2x^2 + b_1x + b_0 \quad ,$$

and $$g_5(x) = a_5x^5 + a_4x^4 + a_3x^3 + a_2x^2 + a_1x + a_0 \quad . \tag{7}$$

Suppose we have established from the Sylvester matrix $D_9$ as given in Eq. 3 that $\Delta_1$, $\Delta_3$ and $\Delta_5$ are non-zero, but that $\Delta_7$, and hence $\Delta_9$, are zero. Then the degree of the common factor is two, and the largest non-zero inner is

$$\left[\begin{array}{ccc|cc} a_5 & a_4 & a_3 & a_2 & a_1 \\ 0 & a_5 & a_4 & a_3 & a_2 \\ 0 & 0 & b_4 & b_3 & b_2 \\ 0 & b_4 & b_3 & b_2 & b_1 \\ b_4 & b_3 & b_2 & b_1 & b_0 \end{array}\right] \begin{array}{l} \left.\vphantom{\begin{array}{c}a\\a\end{array}}\right\} i = 2 \\ \left.\vphantom{\begin{array}{c}a\\a\\a\end{array}}\right\} j = 3 \end{array} \tag{8}$$

which is bracketed as $D_5$ in Eq. 3.

In this case, according to Trudi, if the five elements in the last column of the largest non-zero inner were replaced respectively by the five polynomials $xg_5(x)$, $g_5(x)$, $g_4(x)$, $xg_4(x)$ and $x^2g_4(x)$, the following matrix would be obtained after several simple column operations:

$$D_{5*} = \left[\begin{array}{cccc|c} a_5 & a_4 & a_3 & a_2 & a_1x^2 + a_0x + 0 \\ 0 & a_5 & a_4 & a_3 & a_2x^2 + a_1x + a_0 \\ 0 & 0 & b_4 & b_3 & b_2x^2 + b_1x + b_0 \\ 0 & b_4 & b_3 & b_2 & b_1x^2 + b_0x + 0 \\ b_4 & b_3 & b_2 & b_1 & b_0x^2 + 0 \quad + 0 \end{array}\right] \tag{9}$$

The first four elements in Column 5 of Eq. 9 may be reduced to zero by further column manipulations. The last element in the column will, however, result in the form $p_2(x) = \alpha x^2 + \beta x + \gamma$, where the coefficients $\alpha$, $\beta$ and $\gamma$ are constants depending on $a_0$, ..., $a_5$; $b_0$, ..., $b_4$. The determinant value of $D_{5*}$ is hence

$$\Delta_{5*} = \text{constant} \cdot (\alpha x^2 + \beta x + \gamma) \tag{10}$$

and Trudi has shown that the common factor between $g_4$ and $g_5$ is given by the polynomial in $\Delta_{5*}$ of Eq. 10 with a constant multiplier.

The procedure in reducing the four elements in Column 5 of Eq. 9 to zero is accomplished in this instance by first inspecting the elements of the augmented (5x7) matrix $D_5$ shown in Eq. 3. We note that the last three elements in each row of $D_5$ are the coefficients of the polynomial of the last element of the corresponding row in Column 5 of Eq. 9. The complete reduction of the first four elements in each of the last three columns of $D_5$ would therefore be equivalent to reducing to zero the first four elements in Column 5. Yet, at the same time, the last three elements resulted in the bottom row give the coefficients α, β and γ respectively. It can be shown that these values are simply constant multiples of the coefficients that would result from the Euclid algorithm (6)

## 4. EXAMPLE

Consider

$$f(x) = x^5 + 2x^4 - 2x^3 - 4x^2 + x + 2 \ . \tag{11}$$

Generate the sequence as in Eq. 5:

$$\begin{aligned} g_5(x) &= x^5 + 2x^4 - 2x^3 - 4x^2 + x + 2 \ , \\ g_4(x) &= x^4 + 1.6x^3 - 1.2x^2 - 1.6x + .2 \ , \\ g_3(x) &= x^3 + 1.2x^2 - .6x - .4 \ , \\ g_2(x) &= x^2 + .8x - .2 \ , \\ \text{and} \qquad g_1(x) &= x + .4 \ . \end{aligned} \tag{12}$$

We first test the Sylvester matrix of $g_5(x)$ and $g_4(x)$ for semi-innerwise. A convenient way of carrying out the test is by the double triangularization procedure outlined in the Appendix. It was found that $g_5(x)$ and $g_4(x)$ are not relatively prime, the common divisor $q_4(x)$ being $0.8(x^2 - 1)$. We next proceed to test the matrix of $q_4(x)$ and $g_3(x)$, but the matrix was found to be not semi-innerwise. Thus the multiplicity of the common factor can be determined to be 2.

The Sylvester matrix in this case is shown in Eq. 13.

$$\begin{bmatrix} 1 & 2 & -2 & -4 & 1 & 2 & 0 & 0 & 0 \\ 0 & 1 & 2 & -2 & -4 & 1 & 2 & 0 & 0 \\ 0 & 0 & 1 & 2 & -2 & -4 & 1 & 2 & 0 \\ 0 & 0 & 0 & 1 & 2 & -2 & -4 & 1 & 2 \\ 0 & 0 & 0 & 0 & 1 & 1.6 & -1.2 & -1.6 & .2 \\ 0 & 0 & 0 & 1 & 1.6 & -1.2 & -1.6 & .2 & 0 \\ 0 & 0 & 1 & 1.6 & -1.2 & -1.6 & .2 & 0 & 0 \\ 0 & 1 & 1.6 & -1.2 & -1.6 & .2 & 0 & 0 & 0 \\ 1 & 1.6 & -1.2 & -1.6 & .2 & 0 & 0 & 0 & 0 \end{bmatrix} \qquad (13)$$

The situation after double triangularisation is shown in Eq. 14. We see from the two elements underlined that $D_7$, and hence $D_9$, are zero. Thus, K = 2.

$$\begin{bmatrix} 1 & 2 & -2 & -4 & 1 & -20.4 & 0 & 0 & 0 \\ 0 & 1 & 2 & -2 & -4 & -3 & 20 & 0 & 0 \\ 0 & 0 & 1 & 2 & -2 & 9.6 & 0 & 0 & 0 \\ 0 & 0 & 0 & 1 & 2 & 0 & 0 & 0 & 0 \\ 0 & 0 & 0 & 0 & 1 & 0 & 0 & 0 & 0 \\ 0 & 0 & 0 & 1 & 1.6 & 1.4 & 0 & 0 & 0 \\ 0 & 0 & 1 & 1.6 & -1.2 & 8.6 & .8 & 0 & 0 \\ 0 & 1 & 1.6 & -1.2 & -1.6 & -3.5 & 18.4 & \underline{0} & 0 \\ 1 & 1.6 & -1.2 & -1.6 & .2 & -8.6 & -4 & 0 & \underline{0} \end{bmatrix} \qquad (14)$$

Retracing the double triangularisation process, we find that as we reach Eq. 15, there is an augmented matrix, from the bottom row of which the multiple root may be determined to a constant multiple. In this example,

$$\begin{bmatrix} 1 & 2 & -2 & -4 & 1 & -20.4 & 0 & -2 & -1 \\ 0 & 1 & 2 & -2 & -4 & -3 & 20 & -1 & -20 \\ 0 & 0 & 1 & 2 & -2 & 9.6 & 0 & 0 & 0 \\ 0 & 0 & 0 & 1 & 2 & 0 & 0 & 0 & 0 \\ 0 & 0 & 0 & 0 & 1 & 0 & 0 & 0 & 0 \\ 0 & 0 & 0 & 1 & 1.6 & 1.4 & 0 & 0 & 0 \\ 0 & 0 & 1 & 1.6 & -1.2 & 8.6 & \underline{.8} & \underline{0} & \underline{-.8} \\ 0 & 1 & 1.6 & -1.2 & -1.6 & -3.5 & 18.4 & -1 & -18.4 \\ 1 & 1.6 & -1.2 & -1.6 & .2 & -8.6 & -4 & -1.6 & 3 \end{bmatrix} \quad (15)$$

we determine from the elements underlined in Eq. 15 that the multiple root is of order 2 and is proportional to $0.8x^2 + 0x - 0.8 = 0.8(x^2 - 1)$. Since $f(x)$ is monic, and the multiplicity is 2, the factor is therefore $(x^2 - 1)^2$.

## 5. CONCLUSION

The procedure described yields the factor of highest multiplicity, where it exists. Having found such a factor and extracted same, the procedure can be re-applied to find further factors of lesser multiplicity, once again if they exist.

The procedure is particularly suited for computer implementat and finds application in the detection of multiple roots where the extraction of such roots successively might be difficult, time consuming or open to doubt as to their existence.

## 6. APPENDIX

### Double Triangularisation Procedure

Given the Sylvester matrix below:

$$\begin{bmatrix} a_{11} & a_{12} & a_{13} & a_{14} & a_{15} & a_{16} & a_{17} & a_{18} & a_{19} \\ 0 & a_{22} & a_{23} & a_{24} & a_{25} & a_{26} & a_{27} & a_{28} & a_{29} \\ 0 & 0 & a_{33} & a_{34} & a_{35} & a_{36} & a_{37} & a_{38} & a_{39} \\ 0 & 0 & 0 & a_{44} & a_{45} & a_{46} & a_{47} & a_{48} & a_{49} \\ 0 & 0 & 0 & 0 & a_{55} & a_{56} & a_{57} & a_{58} & a_{59} \\ 0 & 0 & 0 & a_{64} & a_{65} & a_{66} & a_{67} & a_{68} & a_{69} \\ 0 & 0 & a_{73} & a_{74} & a_{75} & a_{76} & a_{77} & a_{78} & a_{79} \\ 0 & a_{82} & a_{83} & a_{84} & a_{85} & a_{86} & a_{87} & a_{88} & a_{89} \\ a_{91} & a_{92} & a_{93} & a_{94} & a_{95} & a_{96} & a_{97} & a_{98} & a_{99} \end{bmatrix}$$

First Pivot -

Pivot on $a_{55}$ and by column operations reduce successively $a_{56}$, $a_{57}$, $a_{58}$, $a_{59}$ to zero. All the elements $a_{i6}$, $a_{i7}$, $a_{i8}$ and $a_{i9}$ become $\bar{a}_{i6}$, $\bar{a}_{i7}$, $\bar{a}_{i8}$ and $\bar{a}_{i9}$, respectively.

Second Pivot -

Pivot on $a_{44}$ and reduce $\bar{a}_{46}$, $\bar{a}_{47}$, $\bar{a}_{48}$, $\bar{a}_{49}$ to zero. Elements $\bar{a}_{i6}$, $\bar{a}_{i7}$, $\bar{a}_{i8}$ and $\bar{a}_{i9}$ become $\bar{\bar{a}}_{i6}$, $\bar{\bar{a}}_{i7}$, $\bar{\bar{a}}_{i8}$ and $\bar{\bar{a}}_{i9}$.

Third Pivot -

Pivot on $\bar{\bar{a}}_{66}$ and reduce $\bar{\bar{a}}_{67}$, $\bar{\bar{a}}_{68}$, $\bar{\bar{a}}_{69}$ to zero.

Fourth Pivot -

Pivot on $a_{33}$, etc.

until we get the matrix shown:

$$\begin{bmatrix} a_{11} & a_{12} & a_{13} & a_{14} & a_{15} & a^*_{16} & a^*_{17} & a^*_{18} & 0 \\ 0 & a_{22} & a_{23} & a_{24} & a_{25} & a^*_{26} & a^*_{27} & 0 & 0 \\ 0 & 0 & a_{33} & a_{34} & a_{35} & a^*_{36} & 0 & 0 & 0 \\ 0 & 0 & 0 & a_{44} & a_{45} & 0 & 0 & 0 & 0 \\ 0 & 0 & 0 & 0 & a_{55} & 0 & 0 & 0 & 0 \\ 0 & 0 & 0 & a_{64} & a_{65} & a^*_{66} & 0 & 0 & 0 \\ 0 & 0 & a_{73} & a_{74} & a_{75} & a^*_{76} & a^*_{77} & 0 & 0 \\ 0 & a_{82} & a_{83} & a_{84} & a_{85} & a^*_{86} & a^*_{87} & a^*_{88} & 0 \\ a_{91} & a_{92} & a_{93} & a_{94} & a_{95} & a^*_{96} & a^*_{97} & a^*_{98} & a^*_{99} \end{bmatrix}$$

It can thus be seen that

$$\Delta_1 = a_{55} ,$$

$$\Delta_3 = a^*_{66}\Delta_1 a_{44} ,$$

$$\Delta_5 = a^*_{77}\Delta_3 a_{33} ,$$

.
.
.

etc., where $\Delta_j = \det D_j$ and $D_j$ is the jth order inner.

REFERENCES

1. E.I. Jury, "The Theory and Applications of Inners," Proc. IEEE, Vol. 63, No. 7, July, 1975.

2. E.I. Jury, Inners and Stability of Dynamic Systems, Wiley Interscience, 1974.

3. I. Tang, "Detecting Multiple Roots of Polynomials," Z. Angew. Math. Mech., Vol. 55, Nos. 7/8, 1975.

4. N. Trudi, "Teori dė determinanli e loro applicazioni," Libreria Scientifica e Industriale de B. Pellerano, Napoli,18

5. A.S. Householder, "Bygradients and the Problem of Routh and Hurwitz," SIAM Review, Vol. 10, No. 1, January, 1968

6. C.C. MacDuffee, Theory of Equations,John Wiley & Sons, Inc., 19

# PROLONGATION STRUCTURE AND STABILITY OF CERTAIN BILINEAR SYSTEMS

N. Kalouptsidis

Division of Electronics
University of Athens

ABSTRACT

This paper is concerned with the stability analysis of certain classes of bilinear systems. The approach to be followed is based on some previous work of the author on stability theory of general nonlinear control systems. It was shown that various stability phenomena are related to questions of topological nature about certain sets naturally associated with the system. These sets are the so-called controllability orbits and their prolongations. The present paper specializes the general stability theory to bilinear systems whose the Lie algebra associated with the exact time controllability orbits is either compact or one-dimensional. The structure of the prolongations is analyzed and subsequently is used to explore the stability behaviour of the corresponding systems.

## 1. INTRODUCTION

A smooth, finite dimensional autonomous control system is modelled by the equation

$$\dot{x}= f(x,u)$$

The evolution of the system takes place on the state space N which is a smooth manifold. The class of admissible inputs U consists of functions defined on various intervals of the real line and with values in the control space $\Omega$, a subset of $R^m$. Without loss of generali-



*D. G. Lainiotis and N. S. Tzannes (eds.), Advances in Control, 247–255.*

ty we may assume that U consists of piecewise constant functions. The dynamics f is assumed to behave well with respect to Ω, so that existence and uniqueness of solutions of the system equation is guaranteed for any admissible input.

We denote by $X^u$ the vector field $f(\cdot u)$ that corresponds to the constant input u. Moreover, $t \longrightarrow X_t(x)$ is used to denote the integral curve of $X^u$ which starts from x at t=0. Finally we shall assume that the set of vector fields $D = \{X^u : u \in \Omega\}$ is complete.

For any finite collection $\{u_1, u_2, \ldots, u_k\}$ of Ω and any real numbers $t_1, t_2, \ldots, t_k$ the mapping:

$$x \longrightarrow X_{t_2}^{u_1}(X_{t_2}^{u_2}(\ldots(X_{t_k}^{u_k}(x)\ldots)$$

is a diffeomorphism on N. The set of all diffeomorphisms obtained this way defines a group which we denote by G. For each t in R, let $G^t$ be the subset of the group G consisting of elements of the form:

$$X_{t_1}^{u_1} \ldots X_{t_k}^{u_k}, \quad t_1, t_2, \ldots, t_k \text{ in } R, \ \sum t_i = t$$

For t=0, it is easily seen that $G^0$ is a normal subgroup of G. Moreover each $G^t$ is a coset of $G^0$ in G.The grou groups of diffeomorphisms $G, G^0$ act on N in a natural way. The orbits induced by these actions provide the controllability orbits:

$$G(x) = \{gx : g \in G\}, \quad G^o(x) = \{gx : g \in G^o\}$$

For any x in N and t in R, the set

$$G^t(x) = \{gx : g \in G^t\}$$

is the <u>t-orbit of x.</u>

Given any t,s in R the corresponding orbits $G^t(x)$, $G^s(x)$, are diffeomorphic. In particular the zero time orbit $G^o(x)$ is sufficient to determine any other orbit of x, since for any u in Ω $G^t(x) = X_t^u G^o(x)$.
Notice also that

$$G(x) = \cup\{G^t(x) : t \in R\}$$

Finally the set $G^+(x) = \cup\{G^t(x) : t \geq 0\}$ is called the <u>positive orbit of x</u>.

The state space is assumed to be a paracompact manifold and hence it is metrizable. Let d be a metric defining the topology of N. We shall use the following notation. For any x in N, M subset of N and $\varepsilon>0$:

$$d(x,M)=\inf\{d(x,y):y\in M\}$$

$$S(M,\varepsilon)=\{x\in N:d(x,M)\leq\varepsilon\}$$

Next we recall some definitions of certain basic types of stability. See [1].[2]

Definitions. Let M be a closed subset of N.

1. M is called stable if for any $\varepsilon>0$ and each x in M there is $\delta>0$ possibly depending on x and $\varepsilon$ such that for any y in $S(x,\delta)$, the positive orbit $G^+(y)$ is contained in $S(M,\varepsilon)$.

2. M is uniformly stable if for any $\varepsilon>0$ there is $\delta>0$ such that for any x in $S(M,\delta)$, $G^+(x)$ is contained in $S(M,\varepsilon)$.

3. The set

$$A(M)=\{x\in N:\sup\{d(y,M):y\in G^t(x)\}\to 0,\text{as } t\to+\infty\}$$

is called the region of attraction of M.

4. The set $A_u(M)=\{x\in N;$ for any $\varepsilon>0$ there is $\delta>0$ and $T>0$ such that for any $y\in S(x,\delta)$, $G^t(y)$ is contained in $S(M,\varepsilon)$ for all $t\geq T\}$ is called the region of uniform attraction of M.

5. M is called an attractor if A(M) contains a neighborhood $S(M,\varepsilon)$.

6. M is a uniform attractor if $A_u(M)$ contains $S(M,\varepsilon)$ for some $\varepsilon>0$.

7. M is asymptotically stable if it is stable and an attractor.

8. M is uniformly asymptotically stable, if it is uniformly stable and a uniform attractor.

A decisive role to the understanding of the previous stability patterns is played by the concept of $G^o$-prolongations. $G^o$ prolongations are naturally introduced in the attempt of analyzing the way that the zero time orbits are glued together. They are defined in the following way:

Definition. For any x in N, the set

$P(x)=\{y\in N:$ there are sequences $\{x_n\}$ in N, $\{g_n\}$ in $G^o$, with $x_n\to x$, $g_nx_n\to y\}$ is called the $G^o$ prolongation of x, (of order one).

Another family of sets useful in analyzing stability, consists of $G^+$ prologations.

<u>Definition</u>. For any x in N, the set

$D^+(x) = \{y \in N$: there are sequences $\{x_n\}$ in N, $t_n \geq 0$, $g_n \in G^{t_n}$, with $x_n \longrightarrow x$, $g_n x_n \longrightarrow y\}$ is called the $G^{\pm}$ <u>prolongation of x</u>, and the set

$J^+(x) = \{y \in N$: there are sequences $\{x_n\}$in N, $\{t_n\}$ in R, $g_n \in G^{t_n}$, with $t_n \longrightarrow +\infty$, $x_n \longrightarrow x$, $g_n x_n \rightarrow y\}$ is called the <u>$G^+$ prolongational limit set</u> of x.

It can be proved that a compact set M is stable if and only if $D^+(M)=M$. Moreover the region of uniform attraction of M coincides with the set $\{x \in N: J^+(x)$ is nonempty subset of M$\}$. See [1].

Besides the concepts of prolongations introduced above higher prolongations can also be defined and used to describe more uniform types of stability. [3] [4] [5] Higher prolongations are defined via the operators $\mathcal{D}$, $\mathcal{L}$. If $2^N$ denotes the set of all subsets of N and $\Gamma$ is an operator from N into $2^N$, then $\mathcal{D}\Gamma$ is defined by $\mathcal{D}\Gamma(x)=\{y \in N$: there exist sequences $\{x_n\}$,$\{y_n\}$in N with $x_n \longrightarrow x$, $y_n \longrightarrow y$ and $y_n \in \Gamma(x_n)\}$ and the transitizing operator $\mathcal{L}$ by

$$\mathcal{L}\Gamma(x) = \cup\{\Gamma^n(x): n=1,2,\dots\}$$

where $\Gamma^1(x)=\Gamma^n(x)=\Gamma(\Gamma^{n-1}(x)) \quad n=2,3,\dots$

If we view $\Gamma$ as a relation in N, $\Gamma \subseteq N \times N$ and $\Gamma(x)=\{y \in N: (y,x) \in \Gamma\}$, then clearly $\mathcal{D}\Gamma(x)$ corresponds to the closure of $\Gamma$ in NxN, $\mathcal{D}\Gamma(x)=\{y \in N: (y,x) \in \bar{\Gamma}\}$ while $\mathcal{L}\Gamma(x)$ corresponds to the smallest relation that contains $\Gamma$ and is transitive.

It is evident that the $G^{\circ}$prolongations of order one are given by $P_1=P=\mathcal{D}G^{\circ}$. Using induction we define the prolongation of order n by

$$P_n=\mathcal{D}\mathcal{L}P_{n-1}$$

Thus successive application of the operators $\mathcal{D}$, $\mathcal{L}$ leads to higher prolongations. For the general case this procedure will terminate after $\gamma$ steps, where $\gamma$ is the first uncountable ordinal number. The significance of higher prolongations was demonstrated in [5], where it is shown that existence of Lyapunov functions correspond to the type of asympotic stability that is described in terms of prolongations of order $\gamma$.

Next we restrict ourselves to bilinear systems on Eucledean spaces. A bilinear system on $R^n$ is described by the law:

$$x(t) = (A + \sum_{i=1}^{m} u_i B_i) x(t), \qquad (1)$$

where $A, B_1, B_2, \ldots, B_m$ are matrices of dimensions $n \times n$ and $u_i$ are scalar controls. The group of diffeomorphisms G consists of elements of the form

$$e^{(A+\sum u^1{}_i B_i)t_1} e^{(A+\sum u^2{}_i B_i)t_2} \ldots e^{(A+\sum u^k{}_i B_i)t_k}$$

This is a linear Lie group whose corresponding Lie algebra L(G) is generated by the matrices $A, B_1, B_2, \ldots\ldots, B_m$. Similarly the group $G^\circ$ is a Lie subgroup of G. Its Lie algebra is generated by the matrices $B_1, B_2, \ldots, B_m$ and the derived algebra of L(G). (Recall that the derived algebra of L(G) is spanned by matrices of the form

$$[X,Y]=XY-YX, X, Y \text{ in } L(G).$$

Thus $L(G^\circ)$ is the smallest ideal in L(G), containing $B_1, B_2, \ldots, B_m$. In particular each element of $G^\circ$ can be expressed as

$$e^{H_1 t_1} e^{H_2 t_2} \ldots e^{H_n t_n}, \quad H_i \in L(G^\circ), t_i \in R.$$

To gain familiarity with the above concepts the reader is referred to [6] [7].

## 2. BILINEAR SYSTEMS WITH COMPACT $G^\circ$

We consider the bilinear system (1) and we assume that the Lie group $G^\circ$ is compact. Under the compactness assumption, the prolongations can be readily identified as the next theorem indicates.

Theorem 2.1. If for the bilinear system (1) the Lie group $G^\circ$ is compact, then the $G^\circ$ prolongations coincide with the $G^0$ orbits, namely

$$P(x)=G^\circ(x) \quad \text{for any } x \text{ in } N=R^n$$

Proof. Let y be in P(x). Then there exist sequences

$\{x_n\}$ in N and $\{g_n\}$ in $G^0$ such that $x_n \to x$, $x_n \to y$. Since $G^0$ is compact the sequence $g_n$ contains a convergent subsequence. Thus $g_{nk} \to g$, g in $G^0$. Therefore $x_{nk} \to x, g_{nk} \to g$ and by continuity, $g_{nk}x_{nk} \to gx$. We conclude that $y=gx$, or in other words, $P(x) \subset G^0(x)$. Hence $P(x)=G^0(x)$.

A consequence of the previous theorem is that all higher prolongations are identical to the $G^0$ orbits, or equivalently, the equivalence relation induced by $G^0$ is closed. Notice also that

$$P(0)=G^0(0)=\{0\}$$

Therefore $\{0\}$ is asymptotically stable if and only if it is absolutely asymptotically stable.

We denote by $J_A^+(x)$, $D_A^+(x)$, the prolongations of the dynamical system that correspond to the matrix A. The following lemma is easily proved.

Lemma 2.2 For any x in N.

$$J^+(x)=J_A^+(G^0(x)), \quad D^+(x)=D_A^+(G^0(x)).$$

Proof Let y in $J^+(x)$. Then there exist sequences $\{x_n\}$ in N, $\{t_n\}$ in **R**, $g_n \in G^{tn}$, with $x_n \to x$, $t_n \to +\infty$, $g_n x_n \to y$. The sequence $g_n$ can be written as $g_n = e^{Atn}g_n$, with $g_n$ in $G^0$, and by compactness we may assume without loss of generality $g_n \to g$, $g \in G^0$. Hence $g_n x_n \to gx$ and $g_n x_n = e^{Atn}g_n x_n \to y$. This implies that $y \in J_A^+(gx)$. Conversely let y in $J_A^+(G^0(x))$. Then y belongs to $J_A^+(gx)$, for some g in $G^0$. Furthermore there are sequences $\{x_n\}$in N, $\{t_n\}$in R, with $x_n \to gx$, $e^{Atn}x_n \to y$, $t_n \to +\infty$. Then clearly $g^{-1}x_n \to x$, $(e^{Atn}g)g^{-1}x_n \to y$ and $e^{Atn}g$ belongs to $G^{tn}$ for each n. We conclude that y is in $J^+(x)$. Similar reasoning proves the second equality.

The following corollary describes the stability behaviour of bilinear systems with compact $G^0$.

Corollary 2.3 The following statements are true

1) The origin is stable for the bilinear system (1), if and only if it is stable for A.

2) The origin is asymptotically stable if and only if it is asymptotically stable for A.

Proof: Necessity for both statements is immediate. Sufficiency is implied by lemma 2.2 and the characterizations of stability and asymptotic stability of compact sets in terms of $G^+$ prolongations, discussed in the introduction, since

$$J^{+}(0)=J_{A}^{+}(G^{0}(0))=J_{A}^{+}(0), \quad D^{+}(0)=D_{A}^{+}(0).$$

The previous analysis leads to the conclusion that if the matrix A is asymptotically stable, then, no matter what control is applied, the system will eventually approach the origin. Moreover asymptotic stability of the system is determined solely in terms of asymptotic stability of A.

## 3. BILINEAR SYSTEMS WITH $G^0$ ONE-DIMENSIONAL

This paragraph deals with bilinear systems of the form:

$$x(t)=(A+uB)x(t), \quad x\in R^{n} \tag{2}$$

whose $G^0$ orbits are simply the solution curves of B Clearly this will be the case if and only if for any x in $R^n$ holds, either $[A,B]x=0$ or $[A,B]x=Bx$. This in turn occurs if and only if the Lie algebra of $G^0$, $L(G^0)$ is one dimensional. It is evident that systems of this form possess the lowest nontrivial degree of controllability. Thus the simplicity of the zero time orbits allows for a complete description of the prolongations structure. It will be seen that if we abandon the compact case treated in the previous chapter, the prolongations can be considerably bigger than their corresponding $G^0$ orbits.

For the brevity of exposition we shall look only at the case when B has nonzero eigenvalues. Let $V_B^+, V_B^-$ denote the stable and unstable part of B respectively, namely,

$$V_B^{+}=\{x\in N:\ e^{Bt}x\to 0,\ t\to+\infty\},\quad V_B^{-}=\{x\in N:\ e^{Bt}x\to 0,\ t\to-\infty\}$$

<u>Theorem 3.1</u>: Suppose that B has nonzero eigenvalues. Then the prolongations are given as follows:

i) $P(0)=V_B^{+}\cup V_B^{-}$

ii) $P(x)=G^{0}(x)$ if x does not belong to $P(0)$.

iii) $\mathcal{L}P(x)=\begin{cases}P(0) & \text{if } x \text{ is in } P(0)\\ G^{0}(x) & \text{otherwise}\end{cases}$

iv) $\mathcal{DL}P(x)=\mathcal{L}P(x)$ for any x in N.

The above theorem clearly implies that the smallest closed equivalence relation that contains $G^0$ is simply

$$P_{\gamma}(x)=\begin{cases}V_B^{+}\cup V_B^{-} & \text{if } x\in V_B^{+}\cup V_B^{-}\\ G^{0}(x) & \text{otherwise}\end{cases}$$

We observe that the algorithm yielding the higher prolongations will terminate after at most two steps. The case of one step occurs when either $V_B^+$ or $V_B^-$ covers the whole space. Notice also that P(0) can be considerably bigger than $G^0(0)=\{0\}$. As a matter of fact, P(0) contains only the origin, when all eigenvalues of B have zero real part or equivalently, $G^0$ is compact. This is consistent with the results of the previous paragraph.

The detailed information obtained above for the $G^0$ prolongations can be utilized to explore the stability behaviour of the system and in particular its attraction to the origin. This is illustrated next for the case when B has nonzero real part eigenvalues, while similar in spirit results can be obtained for the general case of any matrix B. Since $G^0$ is assumed one dimensional and B possesses nonzero real part eigenvalues, A and B necesserily commute. We choose a basis which simultaneously diagonalizes the semisimple parts of A,B and let $\lambda_1,\lambda_2,\ldots,\lambda_n$, $p_1,p_2,\ldots,p_n$ the eigenvalues of A,B respectively, in the fixed order determined by the basis. Then the next theorem holds:

<u>Theorem 3.2</u> The following statements are equivalent:

1) P(0) is globally asymptotically stable (in case $V_B^+ \neq \{0\}$ or $V_B^- \neq \{0\}$, P(0) can never possess uniform stability).

2) P(0) is a global attractor

3) For any u in R, A+uB is asymptotically stable either on $V_B^+$ or on $V_B^-$.

4) The system is stabilizable by a constant control.

5) It holds:

$$\max_{\mathrm{Re}\,p_i<0}\left\{-\frac{\mathrm{Re}\lambda_i}{\mathrm{Re}\,p_i}\right\} \leq \min_{\mathrm{Re}\,p_j}\left\{-\frac{\mathrm{Re}\lambda_j}{\mathrm{Re}\,p_j}\right\}$$

REFERENCES

1) N. Kalouptsidis. "Accessibility and Stability Theory of Nonlinear Control Systems" $P_HD$ dissertation, Washington Univ. St. Louis.

2) N. Kalouptsidis and D. Elliot, "A New Approach to Stability Theory of Control Systems", to appear in Math. System Theory.

3) N.P. Bhatia and G.P. Szegö, Stability Theory of Dynamical Systems". Springer- Verlag, Berlin, 1970.

4) J. Auslander and P. Seibert, "Prolongations and Stability on Dynamical Systems", Ann. Inst. Fourier (Grenoble). Vol. 14, pp. 237-267, (1964).

5) N. Kalouptsidis, "Prolongations and Lyapunou functions in Control Systems", to appear in Math.System Theory.

6) R.N.Brockett, "System Theory on Group Manifolds and Coset Spaces", SIAM J. Control. Vol. 10, 1972, pp. 265-284.

7) V. Jurdjevic and H. Sussmann ,"Control Systems on Lie Groups", J. Diff. Eqs. Vol 16, 1972,pp.313-329.

# SOME NEW RESULTS ON THE GENERAL DECOUPLING PROBLEM OF LINEAR (A,B,C,D) - QUADRUPLES

J. DESCUSSE

Laboratoire d'Automatique de l'E.N.S.M.
44072 NANTES Cedex - FRANCE

The aim of this paper is to deal with the general decoupling problem of linear constant (A, B, C, D) - quadruples from two points of view. The first one is devoted to a characterization of the control laws which decouple quadruples by state feedback, which has been lacking up until now, in spite of the results previously obtained by Morse in [4]. Here, we give a complete characterization of solutions by state feedback by means of a necessary and sufficient condition of existence.

## 1. INTRODUCTION

Over the past few years, the theory of decoupling has been well developed for linear multivariable systems described by the classical (A,B,C) - triple representation [2]. This is no longer true when the system includes a direct feedtrough, in other words when it is described by an (A,B,C,D) - quadruple representation. For the latter, previous results have already been obtained by using either a direct algebraic approach [3] or a geometric one [4]. For the latter an incorrect result relation to the existence of state feedback decoupling, has been stated by Morse (Lemma 2 of [4]). The objective of this paper is to explain not only why it is incorrect but also to provide a necessary and sufficient condition concerning the existence of state feedback solutions for the decoupling problem.

The mathematical framework is purely geometric. The paper is organized as follows.



*D. G. Lainiotis and N. S. Tzannes (eds.), Advances in Control, 256–265.*

In Section 2, we specify the main notations and symbols used throughout the text, and we recall basic definitions from linear algebra which are necessary for understanding the paper.

In Section 3, we recall fundamental geometric concepts including controllability subspaces and output controllability subspaces.

In Section 4, the concept of output controllability subspace is used to formulate the restricted (static compensation) decoupling problem (R.D.P.). The main result of this section is provided by theorem 4.1 which gives a necessary and sufficient condition in order that a solution by state feedback may exist.

## 2. PRELIMINARIES

### 2.1. Notations and background algebra

Below, capital letters in italics A,B,C,..., denote matrices ; script letters $\mathcal{A},\mathcal{B},\mathcal{C}$..., denote finite dimensional vector spaces. The same symbol is used to denote both a matrix and its map. All matrices, maps and vector spaces are defined over the field of real numbers. Elements of a vector space are denoted by x,y,z,... The zero space is written as 0. The dimension of a space $\mathcal{X}$ is denoted by $d(\mathcal{X})$.

We recall basic definitions from linear algebra [7]. Two spaces are called independent it their intersection is zero. The sum of two spaces, denoted by +, is the space spanned by the union of their respective bases.

The sum is called a direct sum, denoted by $\oplus$, if the two spaces are independent. If two spaces $\mathcal{V}$ and $\mathcal{W}$ are isomorphic, we write $\mathcal{V} \approx \mathcal{W}$. If $\mathcal{V} \subset \mathcal{W} \subset \mathcal{X}$ any subspace $\hat{\mathcal{V}} \subset \mathcal{X}$ such that $\mathcal{W} = \mathcal{V} \oplus \hat{\mathcal{V}}$ is called a completion of $\mathcal{V}$. If $\mathcal{V} \subset \mathcal{X}$, $\mathcal{X}/\mathcal{V}$ denotes the factor space of $\mathcal{X}$ relative to $\mathcal{V}$.

The notation $M : \mathcal{X} \to \mathcal{U}$ means M is a map from $\mathcal{X}$ into $\mathcal{U}$. If $M : \mathcal{X} \to \mathcal{U}$ and $\mathcal{V} \subset \mathcal{X}$ the restriction of M to $\mathcal{V}$ is written as $M|\mathcal{V}$. If M is a map, {M} or $\mathcal{M}$ denote the Image of M and Ker M denotes its null space. $M^{-1}\mathcal{V}$ denotes the inverse image of $\mathcal{V}$ under M.

If $A : \mathcal{X} \to \mathcal{X}$, $\mathcal{B} \subset \mathcal{X}$ and $d(\mathcal{X}) = n$, $\{A|\mathcal{B}\} = \mathcal{B}+......+A^{n-1}\mathcal{B}$. If $\mathcal{V} \subset \mathcal{X}$ and $A\mathcal{V} \subset \mathcal{V}$, then $\mathcal{V}$ is called an A-invariant subspace. If $A\mathcal{V} \subset \mathcal{V} + \mathcal{B}$, $\mathcal{V}$ is called an (A,B) - invariant subspace.

If k is a positive integer $\underline{k} = \{1,2,...,k\}$. A family of subspaces $\mathcal{R}_i \subset \mathcal{X}$, $i \in \underline{k}$, is denoted by $\{\mathcal{R}_i\}_k$. For this family $\tilde{\mathcal{R}}_i = \sum_{j \neq i} \mathcal{R}_j$, $i, j \in \underline{k}$, and $\tilde{\mathcal{R}} = \bigcap_i \tilde{\mathcal{R}}_i$, $i \in \underline{k}$, is the radical of $\{\mathcal{R}_i\}_k$.

2.2: System

The system of interest is described by the differential equation

$$\dot{x} = A\,x + B\,u \tag{2.1}$$

and the output relation

$$y = C\,x + D\,u \tag{2.2}$$

Here $u \in \mathcal{U}$ is an r-vector, $x \in \mathcal{X}$ is an n-vector and $y \in \mathcal{Y}$ is an m-vector ; $\mathcal{U},\mathcal{X},\mathcal{Y}$, denote the input, state, and output spaces, respectively, for the above system. In the sequel it is assumed that $\{A|\mathcal{B}\} = \mathcal{X}$, $C\{A|\mathcal{B}\} + \mathcal{D} = \mathcal{Y}$ and $\bigcap_i \operatorname{Ker} CA^{i-1} = 0 \; i \in \underline{n}$.

In the following sections, it will be convenient to regard $\mathcal{X}$ and $\mathcal{Y}$ as subspaces of $\mathcal{E} = \mathcal{X} \oplus \mathcal{Y}$. With $\mathcal{E}$ thus defined, P (respectively Q) denotes the projection from $\mathcal{E}$ onto $\mathcal{X}$ (respectively $\mathcal{Y}$) along $\mathcal{Y}$ (respectively $\mathcal{X}$). We shall use the extended maps

$$A_e : \mathcal{E} \to \mathcal{E}, \quad e \to APe$$

$$B_e : \mathcal{U} \to \mathcal{E}, \quad u \to Bu$$

$$C_e : \mathcal{E} \to \mathcal{E}, \quad e \to CPe$$

$$D_e : \mathcal{U} \to \mathcal{E}, \quad u \to Du$$

Note, in particular, that $A_e = A_eP = PA_e$, $C_e = QC_e = C_eP$, $B_e = PB_e$ and $D_e = QD_e$.

## 3. GEOMETRIC CONCEPTS

We recall certain basic geometric concepts used in Section 4. With regard to (2.1), consider the control law

$$u = F\,x + G\,v \tag{3.1}$$

where $v \in \mathcal{V}$ is an q-vector, $F : \mathcal{E} \to \mathcal{U}$ and $G : \mathcal{V} \to \mathcal{U}$. With (3.1) applied to (2.1) and (2.2), the largest subspaces $\mathcal{R} \subset \mathcal{X}$ and $\mathcal{S} \subset \mathcal{Y}$ which v can completely control are given by

$$\mathcal{R} = \{A + BF \mid \{BG\}\} \tag{3.2}$$

and

$$\mathcal{S} = (C + DF)\mathcal{R} + \{DG\} \tag{3.3}$$

$\mathcal{R}$ is called a controllability subspace of (A,B) while $\mathcal{S}$ is called an output controllability subspace of (A,B) or even a controllability subspace of (A,B,C,D). These subspaces will play a key role in the sequel. Existence conditions providing a characterization for them are given in [4], [5]. They are assumed to be known as are the specific properties of controllability and output controllability subspaces.

Note only the important property of the subspaces provided by the following theorem due to A.S. MORSE [4].

Theorem 3.1 : Let (A,B,C,D), $\mathcal{S} \subset \mathcal{Y}$ be fixed. Then $\mathcal{S}$ is a controllability subspace of (A,B,C,D) if there exists a controllability subspace $\mathcal{T} \subset \mathcal{E}$ of $(A_e + B_e, C_e + D_e)$ such that

$$\mathcal{S} = Q\,\mathcal{T} \tag{3.4}$$

If (3.4) holds, $\mathcal{R} = P\,\mathcal{T}$ is a generator of $\mathcal{S}$.

In the rest of the paper we shall be using this fact constantly.

## 4. THE PROBLEM OF DECOUPLING BY STATE FEEDBACK

Consider a partition of the output vector y into p nonempty subsets $y_i$, which are to be independently controlled. Algebraically, the structure of y leads us to assume that

$$\mathcal{Y} = \mathcal{Y}_1 \oplus \mathcal{Y}_2 \oplus \ldots . \oplus \mathcal{Y}_p$$

where $\mathcal{Y}_i$ is the output subspace associated with $y_i$. Decoupling the dynamical system (2.1), (2.2) by state feedback reverts to the search for a control

$$u = Fx + \sum_i G_i\, v_i \qquad\qquad i \in \underline{p}$$

so that $v_i$ can control $y_i$ without influencing $y_j$ for $j \neq i$. If $\mathcal{R}_i$ is the controllability subspace of (A,B) associated with $v_i$, these requirements imply that :

$$\mathcal{R}_i = \{A + BF \mid \{BG_i\}\} \qquad i \in \underline{p}$$

$$\mathcal{Y}_i = (C + DF)\mathcal{R}_i + \{DG_i\} \quad i \in \underline{p}$$

Expressed in this form, the problem of decoupling is somewhat complicated, but, if one uses theorem (3.1) to characterize the output controllability subspaces $\mathcal{Y}_i$ in the extended space $\mathcal{E}$, it may be easily treated.

In the extended space $\mathcal{E}$ the problem of decoupling may be formulated as follows [4], [7].

"Find conditions for the existence of a map $F : \mathcal{E} \to \mathcal{U}$ and a family $\{T_i\}_p$ such that :

$$T_i = \{A_e + C_e + (B_e + D_e)F \mid \{B_e + D_e\} \cap T_i\} \qquad i \in \underline{p} \qquad (4.1)$$

$$T_i + N_i = E \qquad i \in \underline{p} \qquad (4.2)$$

$$T_i \subset \bigcap_{j \neq i} N_j \qquad i,j \in \underline{p} \qquad (4.3)$$

where

$$N_i = X \oplus \sum_{j \neq i} Y_j \qquad i,j \in \underline{p}''$$

Moreover, if one searches for a solution by state feedback, F has to satisfy the constraint

$$F = F\,P \qquad (4.4)$$

The main advantage of this last formulation, defined by (4.1) - (4.3), lies in the possibility of using directly the results of [1], relative to the usual (A,B,C) - triples. However, it must be noted that, when a solution exists, the family $\{T_i\}_p$ is not necessarily compatible (i.e. there exists one F satisfying (4.1) for one F verifying (4.4), which is contradictory to lemma 2 of [4]. Indeed this lemma is not true in the usual case, what we have already pointed out using a counterexample in [6], and we justify this, from a theoretical viewpoint, a little further on. Here, we shall deal with the only solutions verifying (4.4) and more particularly we shall focus our attention on the determination of a necessary and sufficient condition for their existence. Such a condition has been lacking up until now. Therefore, we need some preliminary lemmas which follow.

Before going further, we shall allow that $\{T_i\}_p$ is a solution, in other words it satisfies (4.2), (4.3). The class of all the F which satisfy (4.1) will be denoted by $\mathcal{F}$. On the other hand, since we shall be working presently in the extended space $E$, we shall drop the index, and so, a map and its extension will be denoted by the same symbol.

Lemma 4.1 : Let $\{T_i\}_p$ be a compatible solution for R.D.P. There then exists at least one $F \in \mathcal{F}$ which satisfies (4.4) if

$$Y \cap \sum_i T_i^o = 0 \qquad (4.5)$$

where $T_i^o$ is any completion such that

$$T_i = T_i \cap Y \oplus T_i^o \qquad i \in \underline{p}$$

Proof : By (4.5) it is clear that $\sum_i T_i^o$ is isomorphic to its projection on $X$. There then exists, $F_o \neq F_o P \in F$, one $F_1$ such that

$$F_o \mid \sum_i T_i^o = F_1 \mid P \sum_i T_i^o$$

write $F = F_1 P$. Then for $i \in \underline{p}$

$$(A + C + (B + D)F)T_i = (A + C + (B + C)F_o)\, T_i^o \subset T_i$$

Consequently F belongs to $F$ and moreover it satisfies (4.4).

Remark : In [4], Morse claims that (4.5) has always been true. To prove this, he has written that

$$\begin{aligned} Y \cap \sum_i T_i^o &= Y \cap \sum_i ((Y_i + X) \cap T_i^o) \\ &= \sum_i (T_i^o \cap Y) \end{aligned}$$

but nothing implies that this last equality holds in every case, and consequently lemma 2 of [4] is not generally true. However it is interesting to search for conditions in which (4.5) is fulfilled.

Lemma 4.2 : Let $\tilde{R}$ be the radical of $\{PT_i\}_p$ and $\tilde{T}$ that of $\{T_i\}_p$. Then

$$\tilde{R} = \tilde{T} \tag{4.6}$$

if (4.5) is fulfilled.

Proof : First, note that, by construction, $\tilde{T} \subset \tilde{R}$. To prove equality it is enough to show that the reverse inclusion holds. In order to proceed start from (4.5) and note that $\tilde{T}^o$ is included in $\sum_i T_i^o$. Then

$$Y \cap ((T_i^o + \tilde{T}^o) + (\tilde{T}_i^o + \tilde{T}^o)) = 0 \qquad i \in \underline{p}$$

It follows that

$$P((T_i^o + \tilde{T}^o) \cap (\tilde{T}_i^o + \tilde{T}^o)) = P(T_i^o + \tilde{T}^o) \cap P(\tilde{T}_i^o + \tilde{T}^o)$$

Obviously we have, by definition of $\tilde{T}^o$,

$$(T_i^o + \tilde{T}^o) \cap (\tilde{T}_i^o + \tilde{T}^o) = \tilde{T}^o$$

On the other hand, we have also by construction

$$\tilde{T}^o \subset \tilde{T} \subset \tilde{R} \subset X$$

and so

$$\tilde{T}^o = (R_i + \tilde{T}^o) \cap (\tilde{R}_i + \tilde{T}^o)$$

which means that the quotient subspaces $\mathcal{R}_i + \tilde{\mathcal{T}}^o / \tilde{\mathcal{T}}^o$ are mutually independent. From lemma 10.1., iii of [5] it is clear that the smallest subspace of $E$ with the above property is precisely $\tilde{\mathcal{R}}$. Consequently $\tilde{\mathcal{R}} \subset \tilde{\mathcal{T}}^o \subset \tilde{\mathcal{T}}$ and then $\tilde{\mathcal{R}} = \tilde{\mathcal{T}}$ is established. Now starting from $\tilde{\mathcal{R}} = \tilde{\mathcal{T}} = \tilde{\mathcal{T}}^o$ it is easy to show (4.5) by means of the preceeding steps taken in the opposite direction.

Unfortunately we have failed in proving the necessity of (4.6) for the existence of one $F = F P \in \mathcal{F}$.

To go further in our study we first note the following result.

Lemma 4.3 : Let $\{\mathcal{T}_i\}_p$ be any solution for the decoupling problem. Then if $\tilde{\mathcal{T}}$ is (A + C, B + D) invariant, the family $\{\mathcal{W}_i\}_p$, where $\mathcal{W}_i$ denotes the maximal controllability subspace in $\mathcal{T}_i + \tilde{\mathcal{T}}$, is a compatible solution to the decoupling problem. Moreover there always exist maps $F = FP$ in $\bigcap_i \mathcal{F}(\mathcal{W}_i)$.

Proof : Since $\tilde{\mathcal{T}}$ is (A + C, B + D) invariant it is clear, by lemma 10.4 of [5], that the family $\tilde{\mathcal{T}}$, $\{\mathcal{T}_i + \tilde{\mathcal{T}}\}_p$ is compatible. Now define $\mathcal{W}_i$ as the maximal controllability subspace of (A + C, B + D) in $\mathcal{T}_i + \tilde{\mathcal{T}}$.

Then, for any $F \in \mathcal{F}(\tilde{\mathcal{T}}) \cap \bigcap_i \mathcal{F}(\mathcal{T}_i + \tilde{\mathcal{T}})$ we have, by theorem 5.5 of [5]

$$\mathcal{W}_i = \{A + C + (B+D) F \mid \{B+D\} \cap (\mathcal{T}_i + \tilde{\mathcal{T}})\} \qquad (4.6)$$

Clearly we have $\mathcal{T}_i \subset \mathcal{W}_i \subset \mathcal{T}_i + \tilde{\mathcal{T}}$, and so $\mathcal{W}_i$ satisfies (4.2) for $i \in \underline{p}$. On the other hand since decoupling is available with $\{\mathcal{T}_i\}_p$, it can be easily shown that $\tilde{\mathcal{T}} \subset \bigcap_i \mathcal{N}_i = X$. Consequently $\mathcal{T}_i + \tilde{\mathcal{T}}$ satisfies (4.3), which is also true for $\mathcal{W}_i$. Then $\{\mathcal{W}_i\}_p$ is a compatible solution to the decoupling problem.

Any $F \in \mathcal{F}(\tilde{\mathcal{T}}) \cap \bigcap_i \mathcal{F}(\mathcal{T}_i + \tilde{\mathcal{T}})$ may be written as

$$F = F_o + \overline{F}_1 \pi$$

where $F_o$ is any element of $\mathcal{F}(\tilde{\mathcal{T}})$, $\overline{F}_1 : E/\tilde{\mathcal{T}} \to \mathcal{U}$ is suitably chosen (See Lemma 10.4 of [5]) and $\pi : E \to E/\tilde{\mathcal{T}}$ denotes the canonical projection. It can be stated that $F_o = F_o'P + F_o''Q$, $\overline{F}_1\pi = F_1'P + F_1''Q$, and $F = (F_o' + F_1')P + (F_o'' + F_1'')Q$.

But since $\tilde{T} \subset X$ it is clear that $F''_o$ is arbitrary. In particular we may choose $F''_o = - F''_1$ and thus $F = FP$. The proof is now complete.

In fact, lemma 4.4. states that there always exists a solution by state feedback for the decoupling problem in so far as there exists a family $\{T_i\}_p$ satisfing (4.2), (4.3) and of which the radical $\tilde{T}$ is $(A + C,\ B + D)$ invariant and different from zero. Indeed if $\tilde{T} = 0$ we know that $\{T_i\}_p$ is compatible (the $T_i$'s are mutually independent), but nothing can be said about the existence of one $F = F\,P \in \bigcap_i F(T_i)$, since the map $F_o$ does not exist any longer. In order to take into account the most common situation we have established the following theorem.

Theorem 4.1 : There exists a solution by state feedback to the decoupling problem if and only if one can find a family $\{T_i\}_p$, satisfying (4.2) (4.3), such that

$$(A + C)\tilde{R} \subset \tilde{T} + \{B + D\} \tag{4.7}$$

Proof : The necessity of (4.7) is obvious. For sufficiency, let $F_o$ denote any map such that $(A + C + (B + D)\ F_o)\ \tilde{R} \subset \tilde{T}$. Since (4.7) is true, $F_o$ always exists.

Let $\Omega : E \to E/\tilde{R} = \overline{E}$ denote the canonical projection. Write $\overline{A}_o$ for the induced map by $A + C + (B + D)\ F_o$ in $\overline{E}$. It is evident that we have

$$\overline{A}_o\ \overline{T}_i \subset \overline{T}_i + \{\overline{B}_o\} \qquad i \in \underline{p}$$

with $\overline{T}_i = \Omega\ T_i$, $i \in \underline{p}$, and $\overline{B}_o = \Omega(B + D)$. Moreover, by proposition 1.2 of [5] the pair $(\overline{A}_o, \overline{B}_o)$ is completely controllable.

Now show that the subspaces $\overline{T}_i$ are mutually independent.

Lemma 4.3 : The factor spaces $T_i + \tilde{R}/\tilde{R}$, $i \in \underline{p}$, are mutually independent in $E/\tilde{R}$.

Proof : Since $\tilde{R}$ is the radical of $\{R_i\}_p$, the factor spaces $R_i + \tilde{R}/\tilde{R}$ are mutually independent in $E/\tilde{R}$ (Lemma 10.1 [5]). In other words

$$(R_i + \tilde{R}) \cap (\tilde{R}_i + \tilde{R}) = \tilde{R}$$

But $\tilde{R} \subset X$ and so $P\ \tilde{R} = \tilde{R}$. Then

$$P(T_i + \tilde{R}) \cap P(T_i + \tilde{R}) = \tilde{R}$$

which imples

$$P((\mathcal{T}_i + \tilde{\mathcal{R}}) \cap (\tilde{\mathcal{T}}_i + \tilde{\mathcal{R}})) \subset \tilde{\mathcal{R}}$$

or

$$P((\tilde{\mathcal{R}} + \mathcal{T}_i) \cap (\tilde{\mathcal{T}}_i + \tilde{\mathcal{R}})) \subset \tilde{\mathcal{R}}$$

So, this inclusion is necessarily an equality. Moreover it implies that $\mathcal{T}_i \cap (\tilde{\mathcal{T}}_i + \tilde{\mathcal{R}}) \subset \tilde{\mathcal{R}}$ which leads finally to

$$(\mathcal{T}_i + \tilde{\mathcal{R}}) \cap (\tilde{\mathcal{T}}_i + \tilde{\mathcal{R}}) = \tilde{\mathcal{R}} \qquad i \in \underline{p}$$

and the proof has been established.

Now, since the $\overline{\mathcal{T}}_i = \mathcal{T}_i + \tilde{\mathcal{R}}/\tilde{\mathcal{R}}$ are mutually independent there exists $\overline{F}_1 : \overline{\mathcal{E}}/\tilde{\mathcal{R}} \rightarrow \mathcal{U}$ such that

$$(\overline{A}_o + \overline{B}_o \overline{F}_1)\, \overline{\mathcal{T}}_i \subset \overline{\mathcal{T}}_i \qquad i \in \underline{p}$$

Let us write $F = F_o + \overline{F}_1\Omega$. Then it can be shown that (See Lemma 10.4 of [5]) $F \in \mathcal{F}(\tilde{\mathcal{R}}) \cap \bigcap_i \mathcal{F}(\mathcal{T}_i + \tilde{\mathcal{R}}) \neq \emptyset$. Now, by using the same procedure as the one developped in lemma 4.4, it can be easily proved that $\{\mathcal{W}_i'\}_p$, where $\mathcal{W}_i'$ denotes the maximal controllability subspace in $\mathcal{T}_i + \tilde{\mathcal{R}}$ is a compatible solution to the decoupling problem. Moreover $\bigcap_i \mathcal{F}(\mathcal{W}_i') \neq \emptyset$ always includes elements of the type $F = F P$, which completes the proof.

To prove theorem 4.1 we have assumed that $\tilde{\mathcal{R}}$ differs from 0. Indeed if $\tilde{\mathcal{R}} = 0$, the map $F_o$ does not exist and the proof cannot hold any longer. However when $\tilde{\mathcal{R}} = 0$, we have also $\tilde{\mathcal{T}} = 0$. Since $\tilde{\mathcal{R}} \supset \tilde{\mathcal{T}}$, by construction. There then exists one $F = FP \in \bigcap_i \mathcal{F}(\mathcal{W}_i')$ by lemma 4.1.

## 5. CONCLUSION

In this paper we have tackled the decoupling problem of linear constant (A, B, C, D) - quadruples by focusing our attention on the determination of a necessary and sufficient condition in order that solutions by state feedback may exist. Since in practice, mere existence is not sufficient and has to be consistent with stability, we should have also to investigate the stability problem under the assumption of state feedback decoupling . Other problems of interest have yet to be investigated. In particular, they include the problem of achieving state-feedback decoupling with stability with a suitable dynamical extension. However it has to be noted that, when state feedback decoupling is not possible, we can always use elements $F \neq FP \in \mathcal{F}$, when they exist. Indeed from an algebraic viewpoint they are still convenient for decoupling, but they deal with the dynamical system $\dot{e} = (A + C)e + (B + D)u$, where e is a vector of which the components in $\mathcal{X}$ and $\mathcal{Y}$ are x and $\int y\,dt$ respectively, the latter being the new output. For such a situation the stability problem may be investigated as in section 5 for the state feedback case.

## REFERENCES

[1] W.M. WONHAM, A.S. MORSE
"Decoupling and pole assigment in linear multivariable systems : a geometric approach"
SIAM J. Contr., vol. 8, pp. 1-18, Feb. 1970.

[2] A.S. MORSE, W.M. WONHAM
"Status of noninteracting control"
IEEE Trans. Automat. Contr., vol. AC 16, pp. 568-580, Dec. 1971.

[3] L.M. SILVERMAN, H.J. PAYNE
"Input-output structure of linear systems with application to the decoupling problem"
SIAM J. Contr., vol. 9, pp. 199-233, May 1971.

[4] A.S. MORSE
"Output controllability and system synthesis"
SIAM J. Contr., vol. 9, pp. 64-67, May 1971.

[5] W.M. WONHAM
"Linear multivariable control. A geometric approach"
Lecture notes in economics and mathematical systems, vol. 101. Springer Verlag, 1974.

[6] J. DESCUSSE
"Sur la classe des lois de commande qui découplent un système linéaire invariant (A,B,C,D)"
Revue Française d'Automatique, d'Informatique et de Recherche Opérationnelle, Systems Analysis and Control, vol. 13, n°3, Sep. 1979.

[7] N. JACOBSON
"Lectures in abstract algebra"
Vol. 2 : linear algebra. Springer Verlag, 1975.

[8] G. BASILE - G. MARRO
"A state space approach to non interacting controls"
Ricerche di Automatica, vol. 1, pp. 68-77, 1970.

TOOLS FOR DESIGN AND ANALYSIS OF MODEL REFERENCE ADAPTIVE SYSTEMS

I.D. LANDAU

Laboratoire d'Automatique (C.N.R.S.)
Institut Polytechnique de Grenoble
ENSIEG - BP 46
38402 Saint Martin d'Hères - France

The paper presents a theorem which gives explicit conditions for MRAS (model reference adaptive systems) with a general adaptation algorithm in order to assure global asymptotic stability of the adaptation error and which gives important information about the behaviour of various other variables involved (i.e : parameter errors observation vector entering in the adaptation algorithm). This theorem is directly applicable for the design and analysis of recursive identifiers and adaptive state observers and model reference adaptive control schemes. However for model reference adaptive control scheme an additional analysis in needed in order to show the boundness of the observation vector and this is developped in the last part of the paper.

## 1. INTRODUCTION

The results given in [1] concern the conditions assuring that the error used for adaptation in the case of M.R.A.S. (denoted by $\nu_k$) is globally asymptotically stable. Such results are enough for the design of recursive identifiers and adaptive state observers, but not enough for fully analyzing model reference adaptive control scheme designed from stability point of view. The reason is that for such schemes in addition to the stability of the adaptation error one should show that the input and output of the process which enter in the observation vector (denoted by $v_k$ in [1]) are bounded and that the various transient adaptation signals vanishes (which is in fact a consequence of the boundness of $v_k$). The results needed for making such an analysis can in fact be obtained by determining the behaviour of the variables entering in the time varying feedback path of the equivalent feedback representation of M.R.A.S. This aspect has not been covered in [1] and



*D. G. Lainiotis and N. S. Tzannes (eds.), Advances in Control, 266–275.*

in this paper we will state all the results which can be obtained for the system of equations formed by the adaptation algorithm and the equation generating the error used for adaptation when a stability point of view is taken. The notations are those of [1] and for the proofs reference will often be made to [1]. At the end of the paper, the results of this theorem in connection with a result from [2] [3] are used to show that $v_k$ is bounded in certain types of model reference adaptive contro schemes.

## II A THEOREM FOR A CLASS OF ADAPTIVE SYSTEMS.

Theorem 2.1.

Assume that the following adaptation algorithm for up-dating the parameter vector $\hat{p}(k)$ is used.

$$\hat{p}(k+1) = p(k) + F_k\, v_k\, \nu_k \tag{2.1}$$

where

$$F_{k+1}^{-1} = \lambda_1(k)\, F_k^{-1} + \lambda_2(k)\, v_k\, v_k^T \tag{2.2}$$

with

$$F_o > 0 \;;\; 0 < \lambda_1(k) \leq 1 \;;\; 0 \leq \lambda_2(k) < 2 \;;\; \forall\, k \tag{2.3}$$

Assume that the relation between $v_k$ and $\nu_k$ is given by :

$$\nu_k = H\,(q^{-1})\ [p - \hat{p}(k+1)]^T\ v_k \tag{2.4}$$

where $v_k$ is a bounded or unbounded vector sequence, $H(z^{-1})$ is a rational discrete transfer normalized under the form :

$$H\,(z^{-1}) = \frac{1 + h_1'\, z^{-1} + \ldots.\ h_\alpha'\, z^{-\alpha}}{1 + h_1\, z^{-1} + \ldots.\ h_\beta\, z^{-\beta}} \tag{2.5}$$

and p is a constant parameter vector. Then if the transfer function :

$$H'(z^{-1}) = H(z^{-1}) - \frac{\lambda}{2} \tag{2.6}$$

is strictly positive real where :

$$2 > \lambda > \max_{0 \leq k \leq \infty} [\lambda_2(k)] \tag{2.7}$$

one has for any $\nu_o$ and $\hat{p}(o)$ bounded :

1. $$\lim_{k\to\infty} \nu_k = 0 \tag{2.8}$$

2. $\lim_{k\to\infty} x_k = 0$ (2.9)

(where $x_k$ is the state vector of any minimal state realization of $H(z^{-1})$).

3. $\lim_{k\to\infty} v_k^T[\,(\hat{p}(k) - p) + F_k\, v_k\, \nu_k\,] = 0$ (2.10)

4. $\lim_{k\to\infty} [1 - \lambda_1(k)]\,||(\hat{p}(k) - p) + F_k\, v_k\, \nu_k||^2_{F_k^{-1}} = 0$ (2.11)

5. $\lim_{k\to\infty} ||v_k\, \nu_k||^2_{F_k} = \lim_{k\to\infty} \Delta\, \hat{p}(k)\; F_k^{-1}\; \Delta\, \hat{p}(k) = 0$ (2.12)

where $\Delta\, \hat{p}(k)=\hat{p}(k+1)-\hat{p}(k)$ (2.13)

6. $[\hat{p}(k) - p]^T\, F_k^{-1}\, [\hat{p}(k) - p] \leq M_1 < \infty \quad \forall\, k > 0$ (2.14)

7. $\lim_{k\to\infty} [\hat{p}(k) - p]^T\, F_k^{-1}\, [\hat{p}(k) - p] = \text{const.}$ (2.15)

Corrolary 2.1. If in addition to the hypotheses of Theorem 2.1. one considers

a. $F_k^{-1} > \varepsilon\, F_o^{-1}\ ;\ F_o > 0 \quad \varepsilon > 0\ ;\ \forall\, k > 0$ (2.16)

b. $F_k^{-1}$ is non decreasing for $k > k_o$ (2.17)

then in addition to 1 through 7 one has :

8. $\lim_{k\to\infty} \Delta\, \hat{p}(k) = \lim_{k\to\infty} F_k\, v_k\, \nu_k = 0$ (2.18)

9. $\lim_{k\to\infty} \hat{p}(k) = p^*$ (2.19)

10. $||\hat{p}(k)|| < M_2 < \infty \qquad \forall\, k > 0$ (2.20)

Remarks :

1) The adaptation algorithm considered corresponds to an integral adaptation with a general form of the time varying gain matrix $F_k$ [1]. The results can be extended to the case of proportional + integral adaptation.

2) The results (1) and (2) of Theorem 2.1. are of the type given in [1], however the positivity condition of Eqs (2.6) and (2.7) is weaker than in [1] for $\lambda_1(k) < 1$.

3) The results (1) (or (2)) of Theorem 2.1. is used for the design of all type of M.R.A.S., the other results of Theorem 2.1. and its corollary are useful for proving boundness of

input and output of the process in the case of model reference adaptive control schemes.

## III PROOF OF THEOREM 2.1. AND COROLLARY 2.1.

From Eqs (2.1) (2.2) and (2.4) one obtains the following equivalent feedback system.

$$x_{k+1} = A\, x_k + b\, u_k \tag{3.1}$$

$$\nu_k = c^T x_k + \delta\, u_k \tag{3.2}$$

$$r_{k+1} = r_k + F_k\, v_k\, \nu_k \tag{3.3}$$

$$\omega_k = -\, u_k = v_k^T r_k + v_k^T F_k\, v_k\, \nu_k \tag{3.4}$$

where Eqs (3.1) and (3.2) are a state space realization of H(z)

$$H(z) = \delta + c^T (zI - A)^{-1} b \tag{3.5}$$

$$r_{k+1} = \hat{p}(k+1) - p \tag{3.6}$$

and $F_k$ is given by Eq. (2.2).

In addition to the classes L(Λ) and N(Γ) defined in [1] and the Lemmas 1 and 2 given in [1] one will use for proving Theorem 2.1. the following theorem which is proven in Appendix.

Theorem 3.1.

Consider a discrete linear time invariant block

$$x_{k+1} = A\, x_k + B\, u_k \tag{3.7}$$

$$y_k = C\, x_k + D\, u_k \tag{3.8}$$

belonging to the class L(Λ) in feedback connection with a discrete linear time varying block.

$$\bar{x}_{k+1} = A(k)\, \bar{x}_k + B(k)\, \bar{u}_k \tag{3.9}$$

$$\bar{y}_k = C(k)\, \bar{x}_k + D(k)\, \bar{u}_k \tag{3.10}$$

belonging to the class N(Γ) and satisfying Lemma 2 of [1]. If :

$$\Lambda - \Gamma(k) \geq 0 \tag{3.11}$$

The following properties hold :

1. $\lim_{k\to\infty} x_k = 0$ ; $\lim_{k\to\infty} y_k = 0$ ; $\lim_{k\to\infty} u_k = -\lim_{k\to\infty} \bar{y}_k = 0$ (3.12)

with P(k) Q(k) S(k) R(k) bounded or unbounded for k > 0.

2. $$\lim [\bar{x}_k^T \ \bar{u}_k^T] \begin{bmatrix} Q(k) & S(k) \\ S^T(k) & R(k) \end{bmatrix} \begin{bmatrix} \bar{x}_k \\ \bar{u}_k \end{bmatrix} = 0 \quad (3.13)$$

with Q(k), S(k), R(k) bounded or unbounded for k > 0.

3. $\bar{x}_k^T P(k) \bar{x}_k \leq C_1 < \infty \qquad \forall k > 0$ (3.14)

with P(k) bounded or unbounded for k > 0.

4. $\lim_{k\to\infty} \bar{x}_k^T P(k) \bar{x}_k = \text{const.}$ (3.15)

with P(k) bounded or unbounded.

The following correspondance exists between Eqs. (3.1) through (3.4) and Eqs. (3.7) through (3.10) :

$$\begin{gathered} u_k = u_k \ ; \ y_k = \nu_k \ ; \ A = A \ ; \ B = b \ ; \ C = c^T \ ; \ D = \delta \\ \bar{y}_k = -u_k = \omega_k \ ; \ \bar{x}_k = r_k \ ; \ \bar{u}_k = \nu_k \ ; \ A(k) = I \ ; \\ B(k) = F_k v_k \ ; \ C(k) = v_k^T \ ; \ D(k) = v_k^T F_k v_k \end{gathered} \quad (3.16)$$

Therefore in order to apply theorem 3.1. one should verify if the block defined by Eqs (3.3) and (3.4) belongs to the class N(Γ) and then if the block of Eqs (3.1) (3.2) belongs to the class L(Λ) for Λ satisfying Eq (3.11).

Lemma 2 of [1] is satisfied by the system of Eqs. (3.3) (3.4) by choosing

$$\begin{gathered} P(k) = F_k^{-1} \ ; \ P_{(k+1)} = F_{k+1}^{-1} = \lambda_1(k) F_k^{-1} + \lambda_2(k) v_k v_k^T \\ \Gamma(k) = \lambda_2(k) \end{gathered} \quad (3.17)$$

which leads to

$$\begin{gathered} Q(k) = [1-\lambda_1(k)] F_k^{-1} \ ; \ S(k) = [1-\lambda_1(k)] v_k \\ R(k) = [2-\lambda_1(k)] v_k^T F_k v_k \end{gathered} \quad (3.18)$$

The condition of Eq. (2.19) of Lemma 2 from [1] can be alternatively expressed by :

$$R(k) - S^T(k) Q(k)^{-1} S(k) > 0 \quad (3.19)$$

which becomes in this case :

$$v_k^T F_k v_k > 0 \tag{3.20}$$

This condition is always satisfied because of Eqs (2.2) and (2.3) therefore the feedback block of Eqs (3.3) and (3.4) belong to the class $N(\Gamma)$ for $\Gamma(k) = \lambda_2(k)$.

The feedborward block of Eqs (3.1) and (3.2) will belong to the class $L(\Lambda)$ if it exists a $\lambda$ such that

$$H(z^{-1}) - \frac{\lambda}{2} \tag{3.21}$$

is strictly positive real. On the other hand in order to satisfy the condition of Eq. (3.11) from Theorem 3.1. one should have :

$$\lambda - \lambda_2(k) > 0 \qquad \forall\, k \tag{3.22}$$

which lead to the conditions of Eqs (2.6) and (2.7) from Theorem 2.1.

Using the correspondance established in Eq. (3.16) one obtains applying the results of Theorem 3.1. The results stated in Theorem 2.1. Corollary 2.1 is obtained from Eqs (2.12), (2.1) and (2.14) by taking in account the additional hypotheses.

## IV APPLICATION

In the case of model reference adaptive control the sequence $v_k$ is no more independent of the previous values of $\hat{p}(k)$ and $v_k$, and can be expressed in general by :

$$v_{k+d} = W(q^{-1})\ [\rho_k + (\hat{p}(k)-p)^T\ v_k] \tag{4.1}$$

where d is the delay of the process to be controlled, $\rho_k$ depends on the reference input and the zeros of the reference model and $W(z^{-1})$ is an asymptotically stable transfer vector. To draw conclusions upon the boundness of $v_k$ in addition to Theorem 2.1 and corollary 2.1 one need the following result [2] [3] ;

Lemma 4.1 Consider the system :

$$v_{k+d} = W(q^{-1})\ [\alpha(k) + \beta(k)\ v_k] \tag{4.2}$$

where $W(z^{-1})$ is an asymptotically stable transfer function and $||\beta_k|| < C_\beta < \infty$ $\forall\, k$ and $\lim_{k\to\infty} \beta_k = 0$ then :

$$||\ v_k\ || \leq C < \infty \quad \forall\, k \text{ if } |\ \alpha(k)\ | \leq C_\alpha < \infty \tag{4.3}$$

We will next consider the case $d = 1$ ; the extension for $d > 1$ being straight forward.

From Eqs (2.1) and (2.4) one has :

$$[\hat{p}(k) - p]^T v_k = -\frac{1}{H(q^{-1})} \nu_k + [F_k v_k \nu_k]^T v_k \tag{4.4}$$

and from Eqs(2.8) and (2.10) one has :

$$\lim_{k\to\infty} v_k^T[(\hat{p}(k) - p)^T + F_k v_k \nu_k] = -\lim_{k\to\infty} \frac{1}{H(q^{-1})} \nu_k = 0 \tag{4.5}$$

Taking $F_k$ such it satisfies conditions (2.16) and (2.17) one has from Eqs (2.18) and (2.20)

$$\lim_{k\to\infty} F_k v_k \nu_k = \lim_{k\to\infty} \Delta \hat{p}(k) = 0 \tag{4.6}$$

$$|| \Delta \hat{p}(k) || < M < \infty \qquad \forall k \tag{4.7}$$

Using Eq (4.4), Eq (4.1) becomes :

$$v_{k+1} = W(q^{-1}) [\rho_k - \frac{1}{H(q^{-1})} \nu_k + \Delta \hat{p}(k)^T v_k] \tag{4.8}$$

Taking in account Eqs (4.6) (4.7) and the fact $\rho_k$ is bounded one concludes applying Lemma 4.1 to the (4.8) that $v_k$ will be bounded.

APPENDIX :

Proof of Theorem 3.1.

In addition to the results given in [1] one will use the following two Lemmas :

Lemma A.1. The sum of input-output products for the system of the form of Eqs. (3.7) and (3.8) belonging to the class $L(\Lambda)$ can be expressed by :

$$\sum_{k=0}^{k_1} y_k^T u_k = \frac{1}{2} x_{k_1+1}^T P x_{k_1+1} - \frac{1}{2} x_o P x_o +$$

$$+ \frac{1}{2} \sum_{k=0}^{k_1} || L^T x_k + K u_k ||^2 + \frac{1}{2} \sum_{k=0}^{k_1} x_k^T M x_k + \frac{1}{2} \sum_{k=0}^{k_1} u_k^T \Lambda u_k \tag{A.1}$$

Lemma A.2. The sum of the input-output products for the time varying system of Eqs. (A.2.16) (A.2.17) belonging to the class $N(\Gamma)$ and satisfying Lemma A.2.2. can be expressed by

$$\sum_{k=0}^{k_1} \bar{y}_k^T \bar{u}_k = \frac{1}{2} \bar{x}_{k_1+1}^T P(k_1+1) \bar{x}_{k_1+1} - \frac{1}{2} \bar{x}_o^T P_o \bar{x}_o + \frac{1}{2} \sum_{k=0}^{k} [\bar{x}_k^T \; \bar{u}_k^T] \, M(k) \begin{bmatrix} \bar{x}_k \\ \bar{u}_k \end{bmatrix} - \frac{1}{2} \sum_{k=0}^{k} \bar{y}_k^T \Gamma(k) \bar{y}_k \tag{A.2}$$

The proofs of these two lemmas results immediately from Fig. 1 and 2 of [1] and the expression of input-output products for positive systems [1] [4].

The feedback connection of the block of Eqs (3.7) and (3.8) with the block of Eqs (3.9) and (3.10) implies that :

$$\sum_{k=0}^{k_1} u_k^T y_k = - \sum_{k=0}^{k_1} \bar{y}_k^T \bar{u}_k \tag{A.3}$$

1) Introducing Eqs (A.1) and (A.2) and rearranging the various terms one obtains :

$$x_{k_1+1}^T P x_{k_1+1} + \sum_{k=0}^{k_1} x_k^T M x_k = \bar{x}_o^T P_o \bar{x}_o - \bar{x}_{k_1+1}^T P_{k_1+1} \bar{x}_{k_1+1} - \sum_{k=0}^{k_1} [\bar{x}_k^T \; y_k^T] \, M(k) \begin{bmatrix} \bar{x}_k \\ y_k \end{bmatrix} - \sum_{k=0}^{k_1} \bar{y}_k^T (\Lambda - \Gamma(k)) \bar{y}_k - \sum_{k=0}^{k_1} || L^T x_k - K \bar{y}_k ||^2 + x_o P x_o \tag{A.4}$$

From this one obtains that

$$\lim \left( x_{k_{1+1}}^T P x_{k_1+1} + \sum_{k=0}^{k_1} x_k^T M x_k \right) \leq x_o^T P x_o + \bar{x}_o^T P_o \bar{x}_o \tag{A.5}$$

with $P(k)$, $Q(k)$, $S(k)$, $R(k)$ bounded or unbounded for $k > 0$.

Since $P > 0$, $M > 0$ this implies that

$$\lim_{k \to \infty} x_k = 0 \qquad \forall x_o , \forall \bar{x}_o \tag{A.6}$$

and since the system of Eqs. (3.7) and (3.8) is globally asymptotically stable one has also taking in account that $u_k = \bar{y}_k$.

$$\lim_{k\to\infty} y_k = 0 \; ; \; \lim_{k\to\infty} u_k = - \lim_{k\to\infty} \bar{y}_k = 0 \tag{A.7}$$

2+3) Rearranging again Eq. (A.4) one gets

$$\bar{x}^T_{k_1+1} P_{k_1+1} \bar{x}_{k_1+1} + \sum_{k=0}^{k_1} [\bar{x}^T_k \; y^T_k] M(k) \begin{bmatrix} \bar{x}_k \\ y_k \end{bmatrix} = \bar{x}^T_o P_o \bar{x}_o -$$

$$- \sum_{k=0}^{k_1} \bar{y}^T_k [\Lambda - \Gamma(k)] \bar{y}_k - \sum_{k=0}^{k_1} ||L^T x_k - K \bar{y}_k||^2 + x^T_o P x_o -$$

$$- x_{k_1+1} P x_{k_1+1} - \sum_{k=0}^{k_1} x^T_k M x_k \tag{A.8}$$

From Eq. (A.2.43) one obtains :

$$\lim (\bar{x}^T_{k_1+1} P_{k_1+1} \bar{x}_{k_1+1} + \sum_{k\to\infty}^{k_1} [\bar{x}^T_k \; \bar{u}_k] \begin{bmatrix} Q_k & S_k \\ S^T_k & R_k \end{bmatrix} \begin{bmatrix} \bar{x}_k \\ \bar{u}_k \end{bmatrix} \leq x^T_o P x_o +$$

$$+ \bar{x}^T_o P_o \bar{x}_o \tag{A.9}$$

with P(k), S(k), Q(k), R(k) bounded or unbounded for k > 0.

This implies that :

$$\lim [\bar{x}^T_k \; y^T_k] \begin{bmatrix} Q(k) & S(k) \\ S^T(k) & R(k) \end{bmatrix} \begin{bmatrix} \bar{x}_k \\ y_k \end{bmatrix} = 0 \tag{A.10}$$

and

$$\bar{x}^T_k P_k \bar{x}_k \leq C_1 < \infty \quad \forall k > 0 \tag{A.11}$$

4) Denotes

$$V(k) = x^T_k P x_k + \bar{x}^T_k P_k \bar{x}_k \tag{A.12}$$

Again from Eq. (A.2.39) one obtains :

$$V(k+1) - V(k) = - \bar{y}^T_k [\Lambda - \Gamma(k)] \bar{y}_k - x^T_k M_k x_k -$$

$$- ||L^T x_k - K \bar{y}_k||^2 - [\bar{x}^T_k \; y_k] M(k) \begin{bmatrix} \bar{x}_k \\ y_k \end{bmatrix} \leq 0 \tag{A.13}$$

Since $V(k) > 0 \; \forall \; x_k \neq 0 \; ; \; \forall \; \bar{x}_k \neq 0$ and $V(k+1) - V(k) < 0$ $V(k)$ will converge to a fixed value. One has proven already that $x_k \to 0$. One concludes that :

$$\lim_{k\to\infty} \bar{x}^T_k P_k \bar{x}_k = \text{const.} \tag{A.14}$$

## REFERENCES

1. I.D LANDAU, H.M. SILVEIRA, "A stability theorem with applications to adaptive control" I.E.E.E. Trans. on Aut. Contr., vol AC-24, n° 2, April 1979 pp. 305-312.

2. K.S. NARENDRA, Y.H. LIN, L.S. VALAVANI, "Stable adaptive controller design Part II : Proof of stability" Report 7904 Yale Univ., April 1979.

3. I.D. LANDAU, R. LOZANO, "On the design of discrete time explicit MRAC for tracking and regulation" 18th I.E.E.E. C.D.C. Dec. 1979, Fort Lauderdale.

4. I.D. LANDAU, "An addendum to unbiased recursive identification using model reference adaptive techniques" I.E.E.E. Trans. on Aut. Contr., Vol. AC-23, n° 1, pp. 97-99.

# A THEOREM FOR THE ANALYSIS OF THE UNCONDITIONAL OVERALL STABILITY OF MODEL REFERENCE ADAPTIVE SYSTEMS

Ph. de LARMINAT

Laboratoire d'Automatique de l'E.N.S.M.
44072 NANTES Cedex - FRANCE

Abstract : A model reference adaptive system is proposed. Its general form is quite similar to many other algorithms, that have been studied in the litterature. In this paper, a theorem is given, whose application allows one to prove the unconditional overall stability of these systems, so that they can be called "unconditional stabilizers".

## 1. INTRODUCTION

In this paper, an adaptive tracking process control is analysed. In a deterministic context, it can be viewed as a Model Reference Adaptive System (M.R.A.S.). In the stochastic case, the same controller is strictly identical to a Self-Tuning Regulator (STURE).

This algorithm is given as an application example : the main object of the paper is to provide a general convergence theorem for some commonly used prediction error adaptive systems : if $e_k$ is a prediction error, and $\phi_k$ an observation vector, $e_k$ doesn't necessarely converge to zero, if $\phi_k$ is not assumed bounded. Nevertheless, in the deterministic case, it is proved that $\mu_k$ and $\nu_k$ exist, with $e_k = \mu_k^T \phi_k + \nu_k$ and with $\mu_k$ and $\nu_k$ converging to zero.

The application of this theorem to the above M.R.A.S. permits to easily prove the global stability in a very general case, including *non decreasing gains* and simplified gains as well. It must be pointed out that the hyperstability theory, which seems to be fundamental tool for M.R.A.S. analysis is not used here.



*D. G. Lainiotis and N. S. Tzannes (eds.), Advances in Control, 276–281.*

No assumption is needed about the open-loop stability of the process, so that the tracking algorithm could be viewed as an "unconditional stabiliser".

## 2. PROCESS AND CONTROLLER EQUATIONS

The process to be controlled (input $u_k$, output $y_k$) is modelled by the linear $n^{th}$ order equation :

$$y_k = a_1 y_{k-1} + \dots + a_n y_{k-n} + b_1 u_{k-1} + \dots + b_n u_{k-n} \quad (1)$$

Using the shift operator $q^{-1}$, eqn. 1 could be rewritten as :

$$[1 - a_1 q^{-1} - \dots - a_n q^{-n}]\, y_k = [b_1 q^{-1} + \dots + b_n q^{-n}]\, u_k$$

or :

$$A(q^{-1})\, y_k = B(q^{-1})\, u_k .$$

It is assumed that the process is unit time lag ($b_1 \neq 0$), and minimum phase ($1/q\ B(q^{-1})$ is stable).

A set point sequence $d_k$ is given, with a reference model $M(q^{-1})$. The goal for the control is to eliminate the following deviation $x_k$ :

$$x_k \triangleq M(q^{-1})\, d_k - y_k \quad (2)$$

It is assumed that $m_o = 0$ in $M(q^{-1})$ :

$$M(q^{-1}) = [m_1 + m_2 q^{-1} + \dots]\, q^{-1} \triangleq N(q^{-1})\, q^{-1}$$

with the notation : $w_k \triangleq N(q^{-1})\, d_k$, eqn. 2 becomes :

$$x_k = q^{-1} w_k - y_k = w_{k-1} - y_k \quad (3)$$

In the sequel, $d_k$ will not be used in the control law, but only $w_k$, which is assumed bounded. The adaptive control law is defined by eqns. 4, 5, 6 :

$$r_k = r_{k-1} + l_{k-1}\,(y_k - w_{k-1}) \quad (= r_{k-1} - l_{k-1}\, x_k) \quad (4)$$

where
- $r_k$ is a (2n) - vector
- $z_{k-1}^T \triangleq [y_{k-1} \dots y_{k-n} \;\; u_{k-1} \dots u_{k-n}] \quad (5)$
- $l_k$ is a convenient gain vector sequence to be defined.

$u_k$ is then computed as the solution of :

$$r_k^T z_k - w_k = 0 \tag{6}$$

Remark : One can recognize in (4), (5), (6) the classical equations of a Self Tuning Regulator (STURE), if :

- there is no set point sequence : $d_k = w_k = 0$
- the $(n+1)^{th}$ component of $l_k$ and $r_k$ are respectively kept equal to 0 and 1.

In the stochastic case (Model (1) replaced by $Ay_k = Bu_k + C\varepsilon_k$, white sequence $\varepsilon_k$), the properties of STURE are well known (Minimum variance adaptive control).

In this paper, the stability will be proved, in the deterministic case, i.e. if $\varepsilon_k = 0$.

## 3. THEOREME

If : a vector $\theta_k$ and a scalar $e_k$ are given by :

$$\theta_{k+1} = \theta_k - l_k \phi_k^T \theta_k \tag{7}$$

$$e_k = \phi_k^T \theta_k \tag{8}$$

where $\phi_k$ is a given vector sequence, *not necessarely bounded*, and $l_k$ a gain vector, given by :

$$l_k = \Lambda_k \phi_k R_k^{-1} \tag{9}$$

$$\Lambda_k = P_k - P_k \phi_k (\phi_k^T P_k \phi_k + R_k)^{-1} \phi_k^T P_k \tag{10}$$

$$P_{k+1} = \Lambda_k + Q_k \tag{11}$$

with :

- matrix $P_o > 0$, and $\theta_o^T P_o^{-1} \theta_o$ bounded (12)
- scalar $R_k$ bounded : $0 \leqslant R_k \leqslant \sigma^2$ (13)
- matrix $Q_k \geqslant 0$, chosen such that : trace $(P_k) \leqslant \tau^2$ (14)

THEN : there exists a vector $\mu_k$ and a scalar $\nu_k$ such that

$$e_k = \mu_k^T h_k + \nu_k, \text{ with : } \lim_{k \to \infty} \begin{bmatrix} \mu_k \\ \nu_k \end{bmatrix} = 0 \tag{15}$$

: and $\lim(\theta_{k+1} - \theta_k) \to 0$

PROOF : Straightforward computations from eqns. 7 to 10 lead to :

$$\theta_{k+1}^T \Lambda_k^{-1} \theta_{k+1} = \theta_k^T P_k^{-1} \theta_k - e_k^2 [\phi_k^T P_k \phi_k + R_k]^{-1}$$

eqn. 11 yield :

$$\theta_{k+1}^T \Lambda_k^{-1} \theta_{k+1} \geqslant \theta_{k+1}^T P_{k+1}^{-1} \theta_{k+1}$$

and, from 13, 14 :

$$e_k^2 [\phi_k^T P_k \phi_k + R_k]^{-1} \geqslant e_k^2 [\tau^2 \phi_k^T \phi_k + \sigma^2]^{-1}$$

Therefore, defining $\alpha_k^2 \triangleq e_k^2 [\tau^2 \phi_k^T \phi_k + \sigma^2]^{-1}$ :

$$\alpha_k^2 \leqslant \theta_k^T P_k^{-1} \theta_k - \theta_{k+1}^T P_{k+1}^{-1} \theta_{k+1}$$

and, by sumation :

$$\sum_{k=1}^{\infty} \alpha_k^2 \leqslant \theta_o^T P_o^{-1} \theta_o - \theta_\infty^T P_\infty^{-1} \theta_\infty \leqslant \theta_o^T P_o^{-1} \theta_o \text{ (bounded)}$$

therefore : $\lim_{k \to \infty} \alpha_k^2 = 0$

Define now :

$$\mu_k \triangleq e_k [\tau^2 \phi_k^T \phi_k + \sigma^2]^{-1} \tau^2 \phi_k, \text{ and } \nu_k \triangleq e_k [\tau^2 \phi_k^T \phi_k + \sigma^2]^{-1} \sigma^2$$

Then : $e_k = \mu_k^T \phi_k + \nu_k$

and : $\tau^{-2} \mu_k^T \mu_k + \sigma^{-2} \nu_k^2 = \alpha_k^2 \to 0$. Therefore $\mu_k$ and $\nu_k \to 0$.

From eqns. 7 to 10, $\alpha_k \to 0$ yields immediatly that $(\theta_{k+1} - \theta_k) \to 0$, and the theorem is proved.

## 4. APPLICATION

Let us define :

$$\theta_k^T \triangleq r_{k-1}^T - [a_1 \dots a_n \; b_1 \dots b_n], \quad \phi_k \triangleq z_{k-1}, \quad x_k \triangleq e_k$$

Then eqns. 7, 8 are satisfied. If $1_k$ satisfies eqns. 9 to 14, eqn. 15 can be written, with $\mu_k$ and $\nu_k \to 0$ :

$$x_k = \mu_k^T z_{k-1} + \nu_k \tag{16}$$

From eqns. 3, 16 :

$$y_k = - \mu_k z_{k-1} - \nu_k + w_{k-1} \tag{17}$$

From eqns. 1, 5 :

$$y_{k+1} = a_1 y_k + b_1 u_k + [a_2 \dots a_n \; 0 \; b_2 \dots b_n \; 0] \, z_{k-1} \tag{18}$$

By successive substitution of $y_{k+1}$ and $y_k$ by eqn. 17 in eqn. 18 :

$$\begin{bmatrix} b_1 + \mu_{n+1} \\ \quad k+1 \end{bmatrix} u_k = - \begin{bmatrix} a_2 \dots a_n \; 0 \; b_2 \dots b_n \\ + [\mu_2 \dots \mu_n \; 0 \; \mu_{n+2} \dots \mu_{2n} \; 0]_{k+1} \\ - (a_1 + \mu_{k+1}) \, \mu_k^T \end{bmatrix} z_{k-1} \tag{19}$$

$$- \nu_{k+1} + w_k + [a_1 + \mu_{k+1}] (\nu_k - w_{k-1})$$

eqn. 19 can be written as :

$$u_k = \lambda_k^T z_{k-1} + \rho_k + w_k - \frac{a_1}{b_1} w_{k-1} \tag{20}$$

where : $\lim_{k \to \infty} \lambda_k^T = - [\frac{a_2}{b_1} \dots \frac{a_n}{b_1} \; 0 \; \frac{b_2}{b_1} \dots \frac{b_n}{b_1} \; 0]$

$$\lim_{k \to \infty} \rho_k = 0$$

Finally, from eqns. 5, 17 and 20 :

$$z_k = F_k \, z_{k-1} + g_k + v_k$$

where :

$$F_k = \left[\begin{array}{c|c} \multicolumn{2}{c}{-\mu_k^T} \\ \hline \begin{matrix} 1 & & \\ & \ddots & \\ & & 1 \; 0 \end{matrix} & 0 \\ \hline \multicolumn{2}{c}{\lambda_k^T} \\ \hline 0 & \begin{matrix} 1 & & \\ & \ddots & \\ & & 1 \; 0 \end{matrix} \end{array}\right], \quad g_k = \begin{bmatrix} -\nu_k \\ \hline 0 \\ \hline \rho_k \\ \hline 0 \end{bmatrix}, \quad v_k = \begin{bmatrix} w_{k-1} \\ \hline 0 \\ \hline w_k - \frac{a_1}{b_1} w_{k-1} \\ \hline 0 \end{bmatrix}$$

and with :

$$\lim_{k\to\infty} F_k = F = \left[\begin{array}{ccc|ccc} 0 & \cdots & 0 & & & \\ 1 & \ddots & & & 0 & \\ & & 1 \ 0 & & & \\ \hline \multicolumn{3}{c|}{\frac{-1}{b_1}[a_2 \ldots a_n 0} & \multicolumn{3}{c}{b_2 \ldots b_n 0]} \\ & & & 1 & & \\ & 0 & & & \ddots & \\ & & & & & 1 \ 0 \end{array}\right], \quad \lim_{k\to\infty} g_k = 0$$

Due to the minimum phase hypothesis, the system $\{z_k = F z_{k-1} + v_k\}$ is stable, with $v_k$ bounded, and therefore, $z_k$ is bounded.

Only, now, eqn. 16 yields that $x_k \to 0$.

## 5. CONCLUSION

The above result is quite strong. It shows that under the main condition of minimum phase, the deviation $x_k$ reachs 0, for any parameters $a_i$, $b_i$ of the process (unconditional stability), and for any initial coefficient $r_o$ and initial condition $z_o$ (overall stability). The theorem could also be used to analyse other features of adaptive control [2], [3], [4].

For instance, one can see that, if there is no set-point, the controller equations are exactly the same as that for the well known self tuning regulators [1]. So, our theorem is also convenient for the analysis of this kind of adaptive control.

## REFERENCES

[1] ÅSTRÖM K.J., BORISSON U., LJUNG L. and WITTENMARK
Theory and application of Self Tuning Regulators.
Automatica, 13, 1977. Page 457.

[2] MONOPOLI R.V., and TROIANI M.
Discrete model reference adaptive systems with measurement noise. IFAC World Congress - HELSINKI, 1978. Paper 46 A.2.

[3] de LARMINAT Ph., JEANNEAU J.L.
Metodo de regulacion adaptativa para sistemas de fase no minima. 3e Congreso nacional informatica y automatica. MADRID, Oct. 1975.

[4] LANDAU I.D., and BETHOUX G.
Algorithms for discrete time model reference adaptive systems.
IFAC World Congress - BOSTON 1975. Paper 58.4.

# PARAMETER IDENTIFICATION FOR AUTOREGRESSIVE MOVING AVERAGE MODELS

G.ALENGRIN

Université de Nice - Laboratoire de Signaux et Systèmes -
Equipe de Recherche Associée au CNRS n°835 -
41 Bd. Napoléon III - 06041 NICE Cedex - FRANCE.

## ABSTRACT

We consider the problem of estimating the parameters of ARMA stationary processes from covariance sequences. Using a projection method we first derive an identification method for the autoregressive parameters.
This procedure is based upon an increasing order construction so that we can obtain the order of the process together with the estimation of the AR parameters.

Then, using the particular form obtained, we derive a very simple canonical state space model for the residuals. A stochastic realization algorithm is given in order to estimate the statistical parameters.

Finally we introduce a fast algorithm for estimating the statistical part based upon a fast projection method.

## I - INTRODUCTION

We consider the problem of parameter estimation of a stationary, scalar AutoRegressive Moving Average (ARMA) from covariance sequences.
The general model may be written

$$y(k) + a_{n-1}y(k-1)+\ldots+a_{o}\, y(k-n) = e(k) + b_{n-1}\, e(k-1) +\ldots+ b_{o}\, e(k-n) \quad (1)$$

where $\{e(k)\}$ is a white sequence with zero mean and variance $\tilde{\gamma}_o$ .
$\{y(k)\}$ is the observed signal, with covariance coefficients:

$$E[y(k+i)\ y(k)] = \gamma_i$$



*D. G. Lainiotis and N. S. Tzannes (eds.), Advances in Control, 282–291.*

Equation (1) may also be written as :

$$A(Z^{-1})\ y(k) = B(Z^{-1})\ e(k) \tag{2}$$

where $A(Z^{-1}) = 1 + a_{n-1} Z^{-1} + \ldots + a_o Z^{-n}$

and $B(Z^{-1}) = 1 + b_{n-1} Z^{-1} + \ldots + b_o Z^{-n}$

are polynomials in the backward operator $Z^{-1}$ .

We assume that $A(Z^{-1})$ is stable (all its roots are located outside the unit circle), and that $A(Z^{-1})$ and $B(Z^{-1})$ are relatively prime.

Equation (1) may be considered as the input-output description of an input-state-output system.
Among the variety of state space models, let us examine first a filter form :

$$\begin{cases} x(k+1) = \Phi\, x(k) + K\ e(k) \\ y(k) = H\ x(k) + e(k) \end{cases}$$

given the $\{Y_i\}$ sequence, Mehra [11] has developed a procedure for estimating the coefficients of the characteristic polynomial of $\Phi$ which are also the $\{a_i\}$ parameters of equation (1).

Using a state variable canonical form G.Salut [14] , G.Favier [5] has proposed a canonical filter form in which the parameters $\{a_i\}$ and $\{b_i\}$ of the ARMA model (1) appear explicitely:

$$\begin{cases} x(k+1) = J\,x(k) + A\,y(k) + B\,e(k) \\ y(k) = H\,x(k) + e(k) \end{cases} \tag{3}$$

with

$$J = \begin{pmatrix} o & \cdots & & o \\ 1 & \ddots & & o \\ & \ddots & \ddots & \\ o & & 1 & o \end{pmatrix} \quad A = \begin{pmatrix} -a_o \\ \vdots \\ -a_{n-1} \end{pmatrix} \quad B = \begin{pmatrix} b_o \\ \vdots \\ b_{n-1} \end{pmatrix}$$

$H = (o\ o \ldots o\ 1)$

the innovation process $\{e(k)\}$ being a white noise sequence of covariance $\tilde{Y}_o$.
Using this formulation, G.Favier has given several procedures for the estimation of the statistical parameters $\{b_i\}, \tilde{Y}_o$ of the model (3). Then G.Favier-G.Alengrin [6] have developed a fast algorithm for this estimation.

In this paper we consider the global problem of estimating both the dynamical parameters $\{a_i\}$ and the statistical parameters $\{b_i\}$, $\tilde{Y}_o$ from the sequence of covariances $\{Y_i\}$.

First we present a recursive-in-order algorithm for identifying the AutoRegressive part (AR) of equation (1), that is, the dynamical parameters.

Then we define a residual process by

$$\epsilon(k) = e(k) + b_{n-1}\, e(k-1) + \ldots + b_o\, e(k-n)$$

It is a Moving Average process (M.A.) and, using the above results, we can evaluate the order of this process, together with its covariances.
We propose a canonical state model for the residuals, and, using this special form, we develop a fast algorithm for the estimation of the statistical parameters $\{b_i\}, \widetilde{Y}_o$ .

## II - IDENTIFICATION OF THE DYNAMICAL PARAMETERS

Consider the residual process

$$\epsilon(k) = e(k) + b_{n-1}\, e(k-1) + \ldots + b_o\, e(k-n) \tag{4}$$

If we assume $\epsilon(k)$ to be a M.A. process of maximal order n, then by sufficient shifting, we could obtain independent processes,not only for the residual processes but also between residual and observed process y(k).
If the shift is at least equal to n+1 we obtain a residual $\epsilon(k+n+1)$ independent of y(k), y(k-1),...
This property has been used to develop the well known equations for estimating the parameters $\{a_i\}$ :

$$\epsilon(k+n+1) = y(k+n+1) + a_{n-1}\, y(k+n) + \ldots + a_o\, y(k+1) .$$

The orthogonality property leads to

$$\begin{pmatrix} Y_{n+1} & Y_n & \ldots & Y_1 \\ Y_{n+2} & Y_{n+1} & \ldots & Y_2 \\ \vdots & & & \\ Y_{2n} & \ldots & \ldots & Y_n \end{pmatrix} \begin{pmatrix} 1 \\ a_{n-1} \\ \vdots \\ a_o \end{pmatrix} = \begin{pmatrix} o \\ o \\ o \\ o \\ o \end{pmatrix} \tag{5}$$

Or in a more familiar form:

$$\begin{pmatrix} Y_1 & \ldots & Y_n \\ Y_2 & \ldots & Y_{n+1} \\ \vdots & & \\ Y_n & \ldots & Y_{2n-1} \end{pmatrix} \begin{pmatrix} a_o \\ \vdots \\ a_{n-1} \end{pmatrix} = - \begin{pmatrix} Y_{n+1} \\ \vdots \\ Y_{2n} \end{pmatrix} \tag{6}$$

where we notice that the $\{a_i\}$ parameters can be obtained by inversion of a Hankel matrix [11].
For a recursive solution of these equations, consider the Hankel matrix as a covariance matrix between the past and the future [2].

$$E = \left[ \begin{pmatrix} y(k+1) \\ y(k+2) \\ \vdots \\ y(k+n) \\ y(k+n+1) \end{pmatrix} \left( y(k)\ y(k-1) \ldots y(k-n) \right) \right] = \begin{bmatrix} Y_1 & Y_2 \ldots Y_{n+1} \\ Y_2 & Y_3 \ldots Y_{n+2} \\ \vdots & \vdots \\ Y_{n+1} & \ldots Y_{2n} \end{bmatrix} \quad (7)$$

The problem is to find a linear combination of $y(k+n+1), \ldots, y(k+1)$ orthogonal to $y(k)$, $y(k-1) \ldots y(k-n)$.
This problem is solved recursively by the following procedure:

Define $\epsilon^{(1)}(k+2) = y(k+2) + \alpha\, y(k+1)$ (8)

so that $\epsilon^{(1)}(k+2)$ is orthogonal to $y(k)$ and test if it is orthogonal to $y(k-1)$ otherwise define :

$$\eta^{(1)}(k+3) = \epsilon^{(1)}(k+3) + \beta\, y(k+1) \quad (9)$$

such that it is orthogonal to $y(k)$, then define:

$$\epsilon^{(2)}(k+3) = \eta^{(1)}(k+3) + \gamma\, \epsilon^{(1)}(k+2) \quad (10)$$

$\gamma$ being such that $\epsilon^{(2)}(k+3)$ is orthogonal to $y(k-1)$, and test if $\epsilon^{(2)}(k+3)$ is orthogonal to $y(k-2)$.
Otherwise continue.
This orthogonalization procedure is very similar to that of Akaike[1], the only difference is that in our case we consider the covariance between the past $\{y(k), y(k-1), \ldots\}$ and the future: $\{y(k+1), y(k+2), \ldots\}$.
We may also notice that the same procedure can be applied to an AutoRegressive (AR) model.

## III- IDENTIFICATION OF STATISTICAL PARAMETERS

Once the $\{a_i\}$ parameters are identified and assuming that we have obtained the exact values, the problem remains to identify the MA process:

$$\epsilon(k) = (1\ a_{n-1} \ldots a_0) \begin{pmatrix} y(k) \\ y(k-1) \\ \vdots \\ y(k-n) \end{pmatrix} \quad (11)$$

First let us define the autocovariance function of this process:

$$\varphi_{\epsilon\epsilon}(i) \triangleq E\left[\epsilon(k+i)\ \epsilon(k)\right] \quad (12)$$

### III-1 State space model for the M.A. process

For a state space description of the Moving Average process we use a very simple canonical filter form.

$$\begin{cases} x(k+1) = J\, x(k) + B\, e(k) \\ \epsilon(k) = H\, x(k) + e(k) \end{cases} \quad (13)$$

where x(k) is a p-vector $(p \leq n)$, e(k) is the innovation process with covariance $\tilde{Y}_o$.

$$J = \begin{pmatrix} o & \dots & o \\ 1 & & \\ & \ddots & \\ & 1 & o \end{pmatrix} \qquad B = \begin{pmatrix} b_o \\ \vdots \\ b_{p-1} \end{pmatrix}$$

$$H = (\, o \; o \; \dots o \; 1)$$

By hypothesis, $\epsilon(k)$ is a stationary process, and it is easy to verify that (13) is equivalent to :

$$\epsilon(k) = e(k) + b_{p-1}\, e(k-1) + \dots + b_o\, e(k-p)$$

Notice that the state vector for the model (13) is not the same as in equation (3).
A stochastic realization algorithm, using the very simple state space model (13) has been given by Favier [7].

## III-3 Fast algorithm for the statistical parameters estimation

Consider the state variable model (13), we can alternatively define a model with time varying parameters (similarly to a Kalman filter)

$$\begin{cases} x(k+1) = J\, x(k) + B(k+1)\, v(k) \\ \epsilon(k) = H\, x(k) + v(k) \end{cases} \tag{14}$$

where v(k) is the innovation process, with covariance $\tilde{Y}_o(k)$, depending on time, and $\epsilon(k)$ a stationary process.

Let $\hat{x}(k+1 \,|\, [o,k])$ be the optimal estimation (in the sense of conditional expectation) of x(k+1) given the sequence $\{\epsilon(o), \dots \epsilon(k)\}$.
We thus have :

$$\begin{cases} \hat{x}(k+1 \,|\, [o,k]) = J\, \hat{x}(k \,|\, [o,k-1]) + B(k+1)\, v(k) \\ \epsilon(k) = H\, \hat{x}(k \,|\, [o,k-1]) + v(k) \end{cases} \tag{15}$$

with

$$v(k) = \epsilon(k) - \hat{\epsilon}(k \,|\, [o,k-1]) \tag{16}$$

being the innovation process.
In the same way, define the following process [Kailath, Morf, Sidhu] [Rissanen].

$$c(k-1) \triangleq \epsilon(k) - \hat{\epsilon}(k \,|\, [1,k-1]) \tag{17}$$

$\{c(k-1)\}$ is a sequence of independent random variables and, by the property of stationary of the $\epsilon(k)$ process, c(k) and v(k) have the same variance.
The optimal estimator of x(k+1) given the observations $\{\epsilon(1), \dots \epsilon(k)\}$ is thus:

$$\begin{cases} \hat{x}(k+1 \,|\, [1,k]) = J\, \hat{x}(k \,|\, [1,k-1]) + B(k)\, c(k-1) \\ \epsilon(k) = H\, \hat{x}(k \,|\, [1,k-1]) + c(k-1) \end{cases} \tag{18}$$

One can verify easily that the gain of the filter equation (18) must be equal to B(k), in order to obtain compatibility between the models (15) and (18).
Let us finally define the following process:

$$r(k) \triangleq \epsilon(o) - \hat{\epsilon}(o \,|\, [1,k]) \tag{19}$$

It follows from the above definition that the space Y(o,k) generated by the observations $\{\epsilon(o), \epsilon(1), \ldots, \epsilon(k)\}$ can be decomposed as: $Y(o,k) = Y(1,k) \oplus r(k)$ and the projection of x(k+1) on the space Y(o,k) may be written as:

$$\hat{x}(k+1 \,|\, [o,k]) = \hat{x}(k+1 \,|\, [1,k]) + L(k+1)\, r(k) \tag{20}$$

where L(k+1) is a smoothing gain to be determined.
Finally, we have:

$$\hat{x}(k \,|\, [o,k-1]) = \hat{x}(k \,|\, [1,k-1]) + L(k)\, r(k-1) \tag{21}$$

Combining equations (15),(18),(20) and (21) we obtain the following relationship between the filters gains and the smoothing gains:

$$\boxed{J\, L(k)\, r(k-1) + B(k+1)\, v(k) = B(k)\, c(k-1) + L(k+1)\, r(k)} \tag{22}$$

Furthermore the combination of (15),(18) and (21) gives

$$c(k-1) = H\, L(k)\, r(k-1) + v(k) \tag{23}$$

Equations (22) and (23) are the basic relationships for determining recursive equations for the gains B(•) and L(•) .

First we examine orthogonality (or independence) properties of the processes v(k), c(k), r(k).

$$v(k) \text{ is orthogonal to } r(k-1) \tag{24}$$

(By definition (19),,$r(k-1) \in Y(o,k-1)$)

$$r(k) \text{ is orthogonal to } c(k-1) \tag{25}$$

(By definition (17), $c(k-1) \in Y(1,k)$)

These orthogonality properties are then applied to equation (22) to successively obtain the recursive relationships for the B(•) and then for the L(•).

$$B(k+1)\, R_v(k) = B(k)\, R_{cv}(k-1,k) + L(k+1)\, R_{rv}(k,k) \tag{26}$$

where $R_v(k) \triangleq E[v(k)\, v(k)]$ (which is scalar in the problem considered) and

$$Rcv(k-1,k) \triangleq E[c(k-1)\, v(k)]$$
$$Rrv(k,k) \triangleq E[r(k)\, v(k)]$$

Notice that we could obtain another equation for B(•) using orthogonality property (25) and multiplying (22) by c(k-1).

• Relationships between the various covariances

From equation (23) we get :

- After multiplication by v(k), and using orthogonality property (24): $$R_v(k) = R_{cv}(k-1,k) \tag{27}$$

- After squaring (23):

$$R_v(k) = R_c(k-1) - 2H\,L(k)\,R_{rc}(k-1,k-1) + [H\,L(k)]^2\,R_r(k-1) \quad (28)$$

- Multiplying by r(k-1) and using again property (24):

$$R_{cr}(k-1,k-1) = H\,L(k)\,R_r(k-1) \quad (29)$$

- Multiplying by r(k) and using property (25):

$$R_{rv}(k,k = -H\,L(k)\,R_{rr}(k-1,k) \quad (30)$$

Recalling that, by the stationarity property of $\epsilon(k)$, we have:

$$R_v(k) = R_c(k)$$

and using (28) and (29) we can write:

$$\boxed{R_v(k) = R_v(k-1) - R^2_{cr}(k-1,k-1)\,R_r^{-1}(k-1)} \quad (31)$$

Furthermore $Y(1,k) = Y(1,k-1) \oplus c(k-1)$

so that

$$r(k) = r(k-1) + M(k)\,c(k-1) \quad (32)$$

with M(k) to be determined.

From the above equation we get:

- After multiplication by c(k-1), and using property (25):

$$o = R_{rc}(k-1,k-1) + M(k)\,R_c(k-1) \quad (33)$$

- Multiplying by r(k) and using again property (25):

$$R_r(k) = R_{rr}(k-1,k) \quad (34)$$

- Squaring (32):

$$R_r(k) = R_r(k-1) + 2M(k)\,R_{cr}(k-1,k-1) + M^2(k)\,R_c(k-1) \quad (35)$$

Combination of (33) and (35) leads to

$$R_r(k) = R_r(k-1) - R^2_{cr}(k-1,k-1)\,R_c^{-1}(k-1)$$

OR

$$\boxed{R_r(k) = R_r(k-1) - R^2_{cr}(k-1,k-1)\,R_v^{-1}(k-1)} \quad (36)$$

Notice that equations (31) and (36) have the same structure and the same initial condition:

$$R_v(o) = R_r(o) = \varphi_{\epsilon\epsilon}(o) \quad \text{thus}$$

$$R_v(k) = R_r(k) \quad (37)$$

and, by (33) and (29):

$$M(k) = -H\,L(k)\ . \quad (38)$$

• Recursive equation for the gain B(•)

By (27),(30),(34),(37), equation (26) can be written :

$$\boxed{B(k+1) = B(k) - L(k+1)\,[H\,L(k)]} \quad (39)$$

• Recursive equation for the gain L(•)

Consider equation (22). If we multiply both sides of (22) by r(k) and using orthogonality condition (25), we obtain:

$$J\,L(k)\,R_{rr}(k-1,k) + B(k+1)\,R_{vr}(k,k) = L(k+1)\,R_r(k) \quad (40)$$

then, by (34) and (30), we can write :

$$\boxed{L(k+1) = J\,L(k) - B(k+1)\,[H\,L(k)]} \tag{41}$$

Equations (39) and (41) are recursive coupled equations for the determination of the gains $B(\bullet)$ and $L(\bullet)$.
For computational purposes we arrange (39) and (41) to obtain

$$\boxed{L(k+1) = \left[1 - [H\,L(k)]^2\right]^{-1}\left[J\,L(k) - B(k)\,[H\,L(k)]\right]} \tag{42}$$

These equations are of the same form as equations given in [Lindquist] for a KALMAN-filter gain estimation.

• **Recursive equation for the variance of the innovation process**
The variance of the innovation process has been defined as $R_v(k) \triangleq E[v(k)\,v(k)]$.
By (31),(29) and (37) we can write the recursive equation

$$\boxed{R_v(k) = \left[1 - [H\,L(k)]^2\right] R_v(k-1)} \tag{43}$$

Equations (39),(42) and (43) form a complete set of equations for the estimation of the filter gain $B(k)$, the smoothing gain $L(k)$ and the variance $R_v(k)$ of the innovation process $v(k)$.
Let us notice, as it has been previously pointed out [6], that these equations are very easily implemented owing to the particular structure of state-space model (15).

• Initial conditions for the algorithm
We have defined the processes $v(k)$, $c(k)$ and $r(k)$ with the following initial values:

$$v(o) = r(o) = \epsilon(o) \quad \text{and} \quad c(o) = \epsilon(1) \quad .$$

Consider now the relationship :

$$\epsilon(k) = v(k) + H\,B(k)\,v(k-1) + H\,J\,B(k-1)\,v(k-2) + \dots \tag{44}$$

which is deduced from model (15).
Let $k = 1$, we have : $\epsilon(1) = v(1) + H\,B(1)\,v(o)$
or $\epsilon(1) = v(1) + H\,B(1)\,\epsilon(o)$ .
Using the fact that $v(1)$ is orthogonal to $\epsilon(o)$ we get

$$\varphi_{\epsilon\epsilon}(1) = H\,B(1)\,\varphi_{\epsilon\epsilon}(o) \quad .$$

Using the same procedure for

$$\epsilon(2) = v(2) + H\,B(2)\,v(1) + H\,J\,B(1)\,\epsilon(o)$$

and noting that $v(2)$ and $v(1)$ are orthogonal to $\epsilon(o)$ we obtain:

$$\varphi_{\epsilon\epsilon}(2) = H\,J\,B(1)\,\varphi_{\epsilon\epsilon}(o)$$

then we obtain:

$$\begin{pmatrix} \varphi_{\epsilon\epsilon}(p) \\ \varphi_{\epsilon\epsilon}(p-1) \\ \\ \varphi_{\epsilon\epsilon}(1) \end{pmatrix} = \underbrace{\begin{pmatrix} H\,J^{p-1} \\ \vdots \\ H \end{pmatrix}}_{\text{identity matrix}} B(1)\,\varphi_{\epsilon\epsilon}(o)$$

thus

$$B(1) = \begin{pmatrix} \varphi_{\epsilon\epsilon}(p) \\ \varphi_{\epsilon\epsilon}(1) \end{pmatrix} \varphi_{\epsilon\epsilon}^{-1}(o) \tag{45}$$

*Remark* : The same result can be obtained by using the filter form (18) with the innovation process c(k).

- Initial condition for L(•) :
combining (21) and (18) we see that

$$\hat{x}(1/o) = L(1)\, r(o)$$

Furthermore, model (15) with $v(o) = \epsilon(o)$ gives the first step predictor

$$\hat{x}(1/o) = B(1)\, v(o)$$

then, initial condition for the smoothing gain, is given by

$$L(1) = B(1) \tag{46}$$

- Initial condition for the variance equation:
for k=1 the variance equation (43) is written:

$$R_V(1) = \left[1 - [H\, L(1)]^2\right] R_V(o)$$

with

$$R_V(o) = \varphi_{\epsilon\epsilon}(o)$$

(39),(42),(43) together with initial conditions (45),(46) are the equations of the fast algorithm for the ARMA model estimation of the statistical parameters.
When $k \to \infty$, B(k) tends to the constant gain B of the filter model (13), and $R_V(k)$ tends to $\tilde{Y}_o$ , the covariance of the innovation process e(k).
It should be pointed out that the filter form (13) asymptotically stable and consequently the fast algorithm gives always the parameters of a stable inverse ARMA model, (that is the polynomial $B(Z^{-1}$ in (2) has all its roots l cated outside the unit circle).

## IV - CONCLUSION

We have studied the estimation of parameters for an ARMA model in the case of an exact knowledge of the covariance coefficients of the process.
We have derived from previous results a procedure for a recursive-in-order estimation of the dynamical parameters.
Then we have presented a two filters approach of the statistical parameters estimation, using a canonical filter form and leading to very simple equations for the fast algorithm implementation.

## REFERENCES

(1) H.AKAIKE-1974 "Markovian representation of stochastic processes and its application to the analysis of autoregressive moving average processes". Annals of the Institute of Statistical Mathematics,tome 26,pp.363-387,(1974).

(2) H.AKAIKE-1975 "Markovian representation of stochastic processes by canonical variables". SIAM J.Control,vol.13,n°1, pp.162-173, January 1975.

(3) G.E.P.BOX, G.M.JENKINS-1970 "Time series analysis: Forecasting and Control". Holden-Day Revised Edition,(1976).

(4) P.FAURRE-1973 "Réalisations markoviennes de processus stationnaires". Rapport LABORIA,n°13, mars 1973.

(5) G.FAVIER-1977 "Identification d'une représentation gaussienne markovienne - Algorithmes de réalisation stochastique". Thèse de Docteur-Ingénieur,Nice, Juin 1977.

(6) G.FAVIER, G.ALENGRIN-1979 "Algorithmes de filtrage - Application à l'identification des paramètres statistiques d'un modèle ARMA". 7e Colloque GRETSI sur le Traitement du Signal et ses applications,Nice, juin 1979.

(7) G.FAVIER-1979 "A comparison of stochastic realization algorithms for identification of ARMA models". 2nd International Conference on Information Sciences and Systems,Patras,July 1979.

(8) T.KAILATH, M.MORF, G.S.SIDHU-1973 "Some new algorithms for recursive estimation in constant,linear,discrete-time systems". Proc.7th Princeton Symposium on Information and System Science, pp.344-352,(1973).

(9) R.E.KALMAN, R.S.BUCY-1961 "New results in linear filtering and prediction theory". Trans.of ASME,Journal of Basic Engineering,vol.83D,pp.95-108, March 1961.

(10) A.LINDQUIST-1974 "A new algorithm for optimal filtering of discrete-time stationary processes". SIAM J.Control,vol.12, n°4, pp.736-746, November 1974.

(11) R.K.MEHRA-1970 "On-line identification of linear dynamic systems with applications to Kalman filtering". Proc.Joint Automatic Control Conference,pp.373-382,(1970)

(12) J.RISSANEN-1975 "Canonical markovian representations and linear prediction". Proc. 6th IFAC World Congress, Boston (1975).

(13) G.RUCKEBUSCH-1975 "Représentations markoviennes de processus gaussiens stationnaires". Thèse de 3e cycle,Université de Paris VI,(1975).

(14) G.SALUT-1976 "Identification optimale des systèmes linéaires stochastiques". Thèse de Doctorat d'Etat,Université Paul Sabatier,Toulouse,(1976).

# AN INNOVATIONS APPROACH TO FAST IDENTIFICATION ALGORITHMS OF LEVINSON TYPE FOR ARMA PROCESSES

Albert BENVENISTE Christian CHAURE

IRISA, Université de RENNES
Avenue du Général Leclerc
35042 RENNES CEDEX (FRANCE)

Fast identification procedures for both AR and ARMA processes (e.g. Chandrasekhar, Square Root algorithms,...) are available since few years; those algorithms realize the desired transfer functions in the classical polynomial or rational form. On the other hand, the synthesis of polynomial and rational transfer function in "lattice and ladder" form has becomed great interest in network theory; this type of synthesis is related to the theory of orthogonal polynomials on the unit circle; identification procedures with realization in lattice form were available for AR processes. We give here a simple approach for obtaining those algorithms, obtaining in the same way ARMA identification procedures with the same features (new algorithms); the connection with classical AR and ARMA identification procedures is indicated. The so obtained algorithms are fast and well-conditioned ones.

## 1. INTRODUCTION : AN ABSTRACT ORTHOGONALIZATION LEMMA OF LEVINSON TYPE

The purpose of this section is to give efficient formulae for obtaining an orthogonal basis of the linear space involved in any estimation problem.

Let $\mathsf{H}$ be a real Hilbert space, and $\mathsf{G}$ a closed subspace; let $\mathbb{E}(X/\mathsf{G})$ denote the orthogonal projection of $X \in \mathsf{H}$ onto $\mathsf{G}$; we write also $\mathbb{E}(X/Y_1,\ldots Y_N)$ when $\mathsf{G}$ is spaned by the elements $Y_1,\ldots Y_N$ of $\mathsf{H}^d$ $(d \geq 1)$.

Let $Y_0,\ldots,Y_{-N}$ belong to $\mathsf{H}^d$, we define the *forward and backward innovations*



*D. G. Lainiotis and N. S. Tzannes (eds.), Advances in Control, 292–299.*

$$\begin{aligned} E_n^+(k) &= Y_k - \mathbb{E}(Y_k/Y_{k-1},\ldots,Y_{k-n}) \\ E_n^-(k) &= Y_{k-n} - \mathbb{E}(Y_{k-n}/Y_k,\ldots,Y_{k-n+1}) \end{aligned} \tag{1.1}$$

for $0 \leqq n \leqq N$ and $0 \geqq k \geqq n-N$. We have then the following formulae of Levinson type:

$$\begin{aligned} E_{n+1}^+(k) &= E_n^+(k) - \mathbb{E}(E_n^+(k)/E_n^-(k-1)) \\ E_{n+1}^-(k) &= E_n^-(k-1) - \mathbb{E}(E_n^-(k-1)/E_n^+(k)) \end{aligned} \qquad E_o^{\pm}(k) = Y_k \tag{1.2}$$

for $0 \leqq n \leqq N$ and $0 \geqq k \geqq n-N$. Those formulae can be obtained by standard Hilbert space manipulations; in the same way, we get the further properties:

$$E_0^-(k),\ldots,E_{N+k}^-(k) \text{ is an orthogonal basis of } Sp(Y_k,\ldots,Y_{-N}) \tag{1.3}$$

$$\mathbb{E}(Y_0/Y_{-1},\ldots,Y_{-N}) = \sum_1^N \mathbb{E}(E_{n-1}^+(0)/E_{n-1}^-(-1)). \tag{1.4}$$

Formulae (1.2) are nothing but abstract recursion formulae of Levinson type; on the other hand, (1.3) indicates how the formulae (1.2) provide us with an orthogonal basis of the given linear space, whereas (1.4) is a general form of a one step ahead predictor. We shall now use (1.2),(1.3),(1.4) for obtaining various identification procedures.

## 2. REALIZATIONS OF AR AND ARMA PROCESSES IN LATTICE AND LADDER FORM

Let $(y_t)$ be a wide-sense stationary zero mean random sequence, the spectrum of which is assumed (roughly speaking) to have no poles nor zeros on the unit circle. This process can be realized as the output of an all-poles (AR realization) or poles-zeros (ARMA realization) filter, the input of which is as white as possible. Possible techniques for achieving this are the following:

### 2.1. AR realization

$$e_t = P(z^{-1})y_t = y_t - \sum_1^N a_i\, y_{t-i}\ ; \tag{2.1}$$

the AR realization is achieved via the Least Squares method, by minimizing

$$J(a_1,\ldots,a_N) = \mathbb{E}(y_t - \sum_1^N a_i\, y_{t-i})^2 \tag{2.2}$$

2.2. ARMA realization

$$e_t = \frac{P(z^{-1})}{Q(z^{-1})}\, y_t = y_t - \sum_1^N a_i y_{t-i} - \sum_1^P b_j e_{t-j}; \tag{2.3}$$

the ARMA relaization can be achieved via the two following methods:
-- Maximum Likelihood method (ML), by minimizing

$$\begin{aligned} &J(a_1,\ldots,a_N;b_1,\ldots,b_P) \\ &= \mathbb{E}\,(y_t - \sum_1^N a_i y_{t-i} - \sum_1^P b_j e_{t-j})^2 \end{aligned} \tag{2.4}$$

-- Extended Least Squares method (ELS), by satisfying the orthogonality condition:

$$e_t \perp y_{t-1},\ldots,y_{t-N};e_{t-1},\ldots,e_{t-P}. \tag{2.5}$$

Not e that both approach give the same result when $(y_t)$ is exactly an ARMA process of orders N,P, whereas the results of both methods are approximately the same when N,P are sufficiently large; anyway, the corresponding recursive procedures (RML,ELS) are different. (1).

We shall give realizations of $(y_t)$ corresponding to LS and ELS techniques, based upon the abstract Levinson formulae (1.2), (1.3),(1.4).

2.3. AR realization in lattice and ladder form (2)

Introducing the forward and backward innovations

$$\begin{aligned} e_n^+(t) &= y_t - \mathbb{E}\,(y_t/y_{t-1},\ldots,y_{t-n}) \\ e_n^-(t) &= y_{t-n} - \mathbb{E}\,(y_{t-n}/y_t,\ldots,y_{t-n+1}) \end{aligned} \tag{2.6}$$

we can translate the results of the introduction, obtaining the classical Levinson formulae:

$$\begin{aligned} e_{n+1}^+(t) &= e_n^+(t) - k_{n+1} e_n^-(t-1) \\ e_{n+1}^-(t) &= e_n^-(t-1) - k_{n+1}\, e_n^+(t) \\ e_0^\pm(t) &= y_t,\quad k_{n+1} = \frac{\mathbb{E}\, e_n^+(t)\; e_n^-(t-1)}{s_n^2} \end{aligned} \tag{2.7}$$

where $s_n^2 = \mathbb{E}\, e_n^+(t)^2 = \mathbb{E}\, e_n^-(t-1)^2$, because we are in the stationary scalar case;

$$e_0^-(t-1),\dots,e_{N-1}^-(t-1) \text{ is an orthogonal basis of } Sp(y_{t-1},\dots,y_{t-N}); \tag{2.8}$$

$$e(y_t/y_{t-1},\dots,y_{t-N}) = \sum_0^{N-1} k_{n+1}\, e_n^-(t-1) \tag{2.9}$$

Note that $e_N^+(t) = e_t$ given by (2.1),(2.2). Writing $e_n^{\pm}(t)=A_n^{\pm}(z^{-1})y_t$, the recursion (2.7) can be translated in the frequency domain for obtaining the classical recursion formulae of the theory of orthogonal polynomials on the unit circle:

$$\left.\begin{aligned} A_{n+1}^+(z^{-1}) &= A_n^+(z^{-1}) - k_{n+1} z^{-1} A^-(z^{-1}) \\ A_{n+1}^-(z^{-1}) &= z^{-1} A_n^-(z^{-1}) - k_{n+1} A_n^+(z^{-1}) \end{aligned}\right\},\quad A_0(z^{-1}) = 1, \tag{2.10}$$

where the polynomials $A_n^+$ and $A_n^-$ are reciprocals $(A_n^-(z^{-1})=z^{-n}A_n^+(z))$. It is a very pleasant property of the AR realization of $(y_t)$ based upon formulae (2.7),(2.8),(2.9) that the following simple criterion is available for ensuring $P(z^{-1}) = A_N^+(z^{-1})$ is minimum phase (this criterion can be obtained via a simple analytic function argument):

$$P(z^{-1}) \text{ is minimum phase if and only if } |k_n|<1 \text{ for } n=1,\dots,N \tag{2.11}$$

Moreover, contrary to the realization procedures in classical form, it is clear that for obtaining an AR realization of order N+1, only the new *reflection coefficient* $k_{N+1}$ has to be evaluated, the previous ones, namely $k_1,\dots,k_N$ being still valid. Finally, we have obtained a realization of $(y_t)$ with the following features:

-- an orthogonal basis of the linear information space $Sp(y_{t-1},\dots,y_{t-N})$ is used;

-- the realization method is recursive in the order of the model;

-- the whitening filter $P(z^{-1})$ is minimum phase if and only if the reflection coefficients are of modulus smaller than one.

The reason for the name "in lattice and ladder form" of this type of realization is due to the particular form of the block-diagrams realizing the whitening filter $P(z^{-1})$ according to the recursion formulae (2.10) (in "lattice form") and the formula (2.9) for the one step ahead predictor (in "ladder form").

## 2.4. ARMA realization in lattice and ladder form

Following the ELS method, we shall try to satisfy the orthogonality condition (2.5); for simplicity, we restrict ourselves to the case N=P, referring the reader to (3,4) for the general case, as well as for the MA realizations.

We shall use the following method:

-- orthogonalize the information space $Sp(y_{t-1},\dots,y_{t-N};\ e_{t-1},\dots,e_{t-N})$ using the abstract Levinson formulae (1.2);

-- use (1.3),(1.4) for obtaining the one step ahead predictor of $y_t$.

For that purpose, let us introduce the forward and backward vector innovations :

$$E_n^+(t) = \begin{pmatrix} y_t \\ e_t \end{pmatrix} - \mathbb{E}\left(\begin{pmatrix} y_t \\ e_t \end{pmatrix} \middle| y_{t-1},\dots,y_{t-n};e_{t-1},\dots,e_{t-n}\right)$$
$$E_n^-(t) = \begin{pmatrix} y_{t-n} \\ e_{t-n} \end{pmatrix} - \mathbb{E}\left(\begin{pmatrix} y_{t-n} \\ e_{t-n} \end{pmatrix} \middle| y_t,\dots,y_{t-n+1};e_t,\dots,e_{t-n+1}\right) \tag{2.12}$$

The abstract Levinson formulae of the introduction can be translated in our situation

$$\begin{aligned} E_{n+1}^+(t) &= E_n^+(t) - K_{n+1}^+ E_n^-(t-1) \\ E_{n+1}^-(t) &= E_n^-(t-1) - K_{n+1}^- E_n^+(t) \end{aligned} \quad , \ E_0(t) = \begin{pmatrix} y_t \\ e_t \end{pmatrix} \tag{2.13}$$

whereas the *reflection matrix coefficients* $K_{n+1}^{\pm}$ are here different for forward and backward innovations, since those innovations have no more the same covariance matrix, contrary to the scalar case.

The one step ahead predictor is obtained, using the orthogonal basis of our linear information space, that are given in (1.3), by translating the abstract formula (1.4):

$$\begin{aligned} &\mathbb{E}(y_t/y_{t-1},\dots,y_{t-N};e_{t-1},\dots,e_{t-N}) \\ &= \text{first coordinate of } \textstyle\sum_0^{N-1} K_{n+1}^+ E_n^-(-1). \end{aligned} \tag{2.14}$$

Formulae (2.13),(2.14) give the desired realization, which turns out to be equivalent to the estimation of the reflection matrix coefficients $K_n^{\pm}$. Note that, contrary to the AR case, the relationships (2.3),(2.4) characterizing the desired realization are *implicit* ones, as customary in ARMA realization (in fact, both minimization of (2.4) and orthogonality condition (2.5) are implicit conditions).

With the realization algorithm we have indicated, we loss the stability criterion, as well as the recursiveness in the order of the model, keeping the use of an orthogonal basis of our information space.

The corresponding block-diagrams, for which we refer the reader to (3,4) are of the same form as in AR case, namely:

-- a 2-port lattice part, corresponding to the recursion formulae (2.13);
-- a ladder part, corresponding to formula (2.14).

## 3. AR AND ARMA (RECURSIVE) IDENTIFICATION PROCEDURES IN LATTICE AND LADDER FORM

It can be shown (3,4) that the abstract Levinson formulae applied to the empirical situation, when

-- a finite sample of the process $(y_t)$ is given, requiring a best AR or ARMA fitting of this sample (non recursive identification problems),
-- an infinite sample of $(y_t)$ is available, requiring a recursive adjustment of the AR or ARMA model (recursive identification problems),

provide us with (recursive) identification procedures in lattice and ladder form, that are equivalent to the (R)LS procedure (AR identification), or R(ELS) procedure (ARMA identification), apart from asymptotically neglectible modifications.

The technique is simply: *replace in the appropriate realization scheme the reflection (matrix) coefficients by there estimates given by the law of large numbers, or recursive implementations of there.*

Let us indicate an example of those estimators, that can be used, in the recursive case.

### 3.1. RLS algorithm in lattice and ladder form

$$\begin{aligned} e^+_{n+1}(t) &= e^+_n(t) - k_{n+1}(t)\, e^-_n(t-1) \\ &\qquad\qquad , \quad e_0(t) = y_t, \quad e^+_N(t) = e_t \qquad (3.1) \\ e^-_{n+1}(t) &= e^-_n(t-1) - k_{n+1}(t)\, e^+_n(t) \end{aligned}$$

where the reflection coefficients are recursively updated as follows:

$$\begin{aligned} a_{n+1}(t) &= a_{n+1}(t-1) + e^+_n(t-1)\, e^-_n(t-2) \\ b_{n+1}(t) &= b_{n+1}(t-1) + (e^+_n(t-1)^2 + e^-_n(t-2)^2) \qquad (3.2) \\ k_{n+1}(t) &= 2\,\frac{a_{n+1}(t)}{b_{n+1}(t)}. \end{aligned}$$

Note that, because of Schwarz inequality, we have always $|k_{n+1}(t)|<1$, so that, during the identification, the whitening filter is ensured to remain minimum phase.

Finally, we obtain a *fast recursive identification procedure* (6N multiplications per step) which is also *numerically well-*

*conditionned* (robust in presence of roundoff noise), *recursive in the order of the model*, and which provides us with a simple *stability guaranty*.

### 3.2. RELS algorithm in lattice and ladder form

$$E^{+}_{n+1}(t) = E^{+}_{n}(t) - K^{+}_{n+1}(t)\, E^{-}_{n}(t-1)$$

$$E^{-}_{n+1}(t) = E^{-}_{n}(t-1) - K^{-}_{n+1}(t)\, E^{+}_{n}(t) \quad E^{\pm}_{0}(t) = \begin{pmatrix} y_t \\ e_t \end{pmatrix}, \tag{3.3}$$

$$E^{\pm}_{0}(t) = \begin{pmatrix} y_t \\ e_t \end{pmatrix}$$

$$e_t = y_t - \text{first coordinate of } \sum_{0}^{N-1} K^{+}_{n+1}(t)\, E^{-}_{n}(t-1), \tag{3.4}$$

whereas the reflection matrix coefficients $K^{\pm}_{n+1}(t)$ are updated using only information available at time t-1, so that *the recursion (3.3),(3.4) is feasible and no more implicit* (the same situation is encountered when the classical ELS, or ML, procedure is replaced by its recursive implementation).

Finally, we obtain a *fast recursive identification procedure for ARMA processes (34N multiplications per step)*, which is also *numerically well-conditionned*, with unfortunately the lack of both stability criterion and recursiveness in the order of the model.

Furthermore, let us indicate that we obtain also *an ELS-type identification procedure in lattice and ladder form for MA processes*, which is numerically more efficient than the corresponding one for ARMA processes, involving only 1-port lattice structure and scalar innovations.

## 4. NUMERICAL RESULTS

We refer the reader to (3,4) where the so obtained algorithms are compared to the corresponding classicale ones; the results are comparable, and sometimes better, especially when the process has zeros or poles near to the unit circle.

## REFERENCES

(1) Soderström, T., Ljung, L., Kailath, T. : *A theoretical Analysis of Recursive Identification Methods;* Automatica 14, pp. 231-244, 1978.

(2) Makhoul, J. : *A Class of All-zero Lattice Digital Filters: Properties and Applications;* IEEE-ASSP 26 n° 4, pp. 304-314, 1978.

(3) Benveniste, A., Chauré, C.: *Une méthode rapide pour arriver à des algorithmes rapides d'estimation pour des fonctions de transfert AR et ARMA synthétisées en treillis;* Rapport IRISA n° 116, 1979.

(4) Benveniste, A., Chauré, C.: *AR and ARMA Identification Algorithms of Levinson Type: an Innovations Approach";* presented for publication to IEEE-AC, 1979.

A NONLINEAR IDENTIFICATION METHOD USING A POLYNOMIAL GENERALIZATION OF REGULAR (OR BILINEAR) SYSTEMS

D. Cyrot-Normand*, H. Dang Van Mien**

* Laboratoire des Signaux et Systèmes, CNRS-ESE, Plateau du Moulon, 91190 Gif-sur-Yvette, France

** Electricité de France, Direction des Etudes et Recherches, 1, avenue du Général de Gaulle, 92141 Clamart, France

Abstract :

A nonlinear identification technique for polynomial generalization of regular (or bilinear) systems using the least squares method is developped. The interest of this representation is to keep the usually wellknown canonical Companion or Jordan forms in the regular system. Several applications concerning electrical power plants are given in the single input-single output case. Simulation results are obtained and illustrate the validity of our approach for large variations of the input and of the working point.

The proposed identification method uses a polynomial generalization of regular (or bilinear) systems (Sontag [7]). A strong motivation to this approach is provided by the approximation theorem established by Fliess [4] and Sussmann [8] which states that all nonlinear systems in continuous time can be approximated by finite-dimensional regular systems. A regular system is described in discrete time by the following equations :

$$(1) \quad \begin{cases} q(t+1) = (A_o + \sum_i u_i(t)\, A_i)\, q(t) \\ y(t) = \lambda\, q(t) \end{cases}$$

where

- The state q belongs to $\mathbb{R}^N$

- $(u_1, \ldots, u_m) : \{0, 1, \ldots\} \to \mathbb{R}$ are the inputs



D. G. Lainiotis and N. S. Tzannes (eds.), Advances in Control, 300–307.

- $A_o, \ldots, A_m$ and $\lambda$ are matrices

of appropriate dimensions.

Some applications were already given for distillation columns (España and Landau [3]) or biological models (Mohler [6]). This is one of the first, if not the first regular identification brought to a successful conclusion without resorting to physical equations.

The present identification algorithm applies to nonlinear systems which can be represented for several definite working points by linear models. These nonlinear systems are known either by transfer functions with varying coefficients depending on parameters or by step responses to input signals. In both cases we can associate to the studied system, for each definite working point, a linear state space representation with varying matrices. Several linear identification algorithms exist but we have prefered the Kalman-Ho algorithm which was improved by Dang Van Mien [2] and Irving [5].

A least squares method to arrange all the linear models in a nonlinear model of the following form.

$$\begin{cases} q(t+1) = A_o + \sum_{\nu=o}^{K} \sum_{i_\nu,\ldots,i_o=1}^{M} A_{i_\nu \ldots i_o} u_{i_\nu}(t) \ldots u_{i_o}(o) \\ y(t) = \lambda\, q(t) \end{cases}$$

where :

- the matrices $A_{i_\nu,\ldots i_o}$ are independent of u and t.

- M = number of parameters plus number of controllable inputs.

It can be seen that systems of type (2) are linear in the state and polynomial in the inputs as those introduced by Sontag [7] and which he called state-affine. It should be noticed that state-affine systems are equivalent to regular systems modulo an elementary change of variables.

The originality of this work is not to distinguish parameters or variables from controllable inputs. This property allows a non-linear system to be obtained capable of representing variations of the controllable inputs and of the parameters.

A disadvantage of this method could be the state space linear representations for several working points. But this objection is greatly compensated by the fact that these structures in the identified nonlinear model enable practical considerations such as gain and time constants or poles and zeros of companion or Jordan canonical forms to be kept.

This method is successfully applied to many practical single input - single output examples concerning electrical, thermic or hydraulic power plants. In all cases, limited order (three or four) state-affine systems are obtained which remain valid over a large scale of working.

The identified models are tested in several ways. Step responses to input signals are given for various values of parameters. A pseudo-random binary signal on the input simultaneous with a linear variation of parameters shows the validity of the models for transcient working points.

We have, for example studied the temperature evolution in super-heaters in thermal power plants. These superheaters are second or third order systems with the time constant depending on a parameter which represents the power of the plant. In that case, the controllable input is the spray water flow.

Fig. 1 shows the step responses of the linear and state-affine systems for values of the power stepped between 60 MW and 108 MW. In that case, the static gain is constant, but we remark that the time constant evolution is produced by the state-affine system perfectly.

Fig. 2 illustrates the validity of the model for a PRBS input signal and a linear variation of the power. Nonlinear and linear simulation graphs are almost confused.

In Fig. 3 we can observe that a static gain evolution can be perfectly represented by a state affine model. We have studied a second order system, the gain and the two time constants of which were dependent on a parameter.

Others applications [1] have given good results, such as an oscillating system as wave propagation in an hydraulic power plant and more generally as a part of a heat exchanger in a nuclear power plant.

In the nuclear case, this method has been extended to decoupled multi inputs - multi outputs systems.

The recurrent character of regular systems suggests an interesting development such as the implement on microprocesseurs.

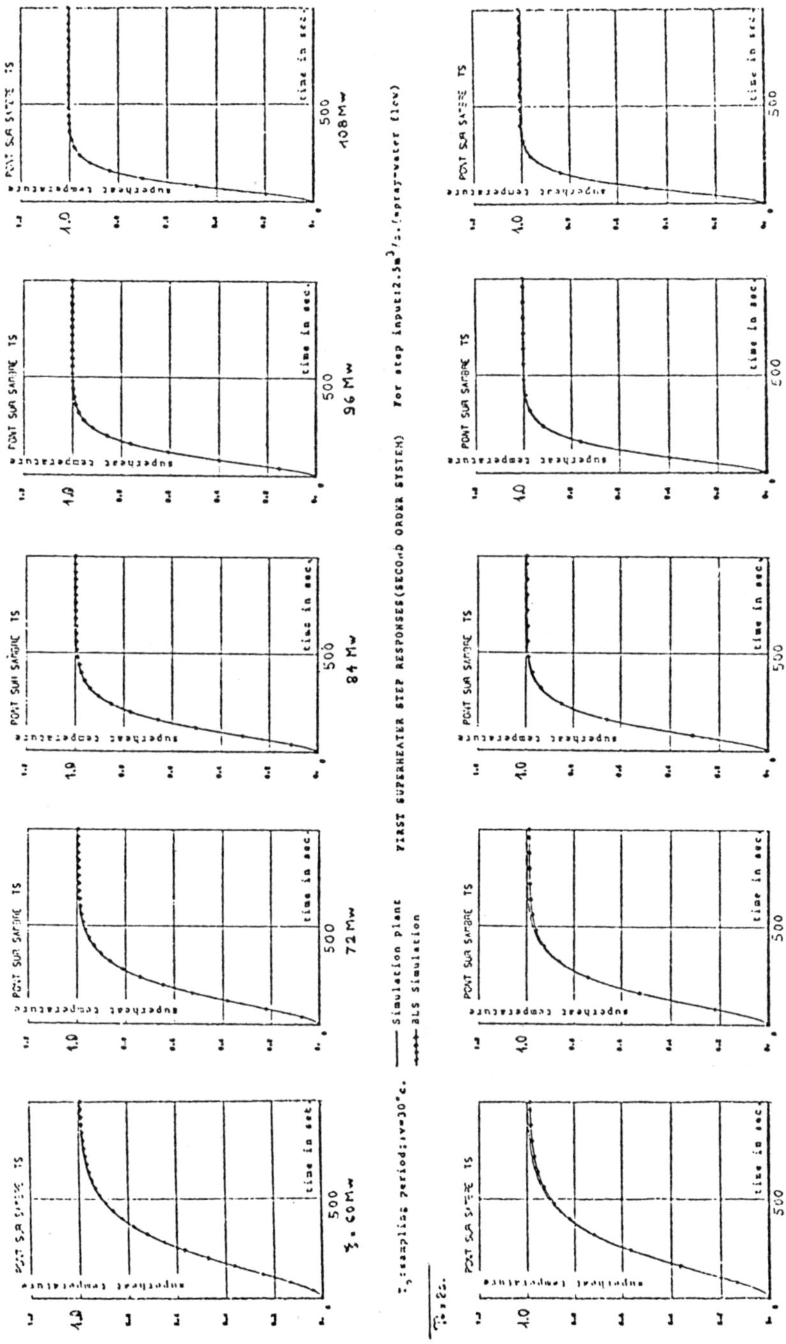
FIRST SUPERHEATER STEP RESPONSES(SECOND ORDER SYSTEM)
Simulation plant
RLS Simulation
superheat temperature
time in sec.
500
108Mw
96 Mw
84 Mw
72 Mw
1.0

fig:1

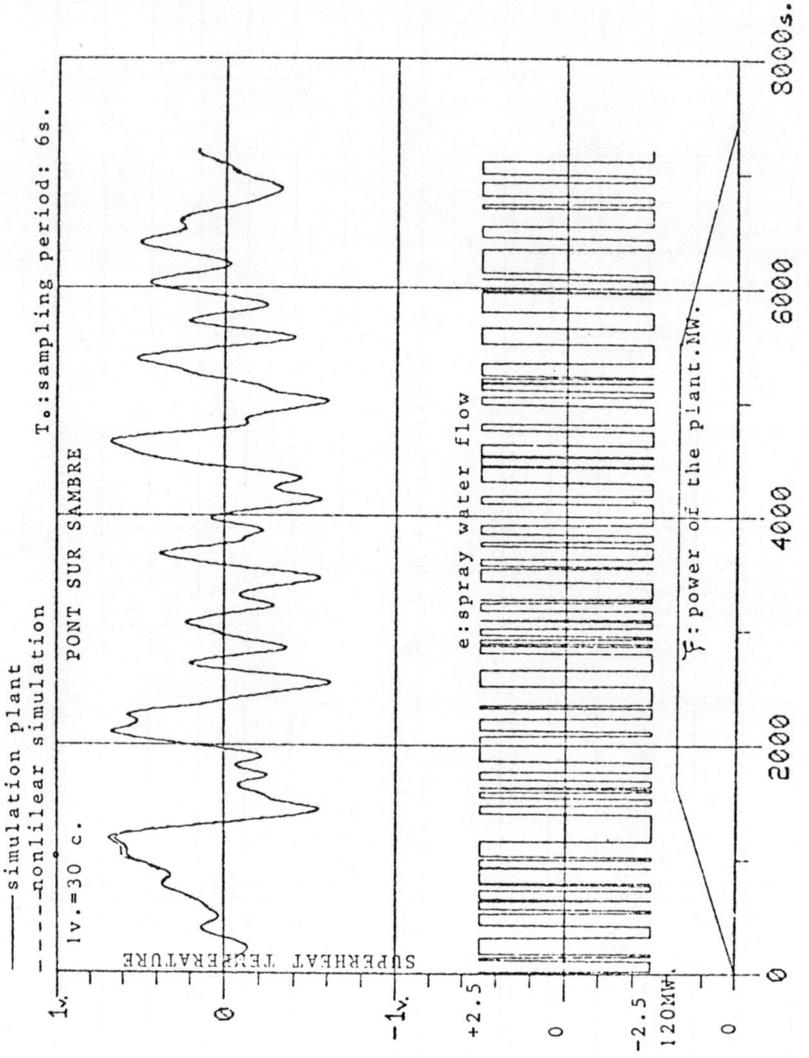
simulation plant
nonlilear simulation
1v.=30 c.
PONT SUR SAMBRE
T₀:sampling period: 6s.
SUPERHEAT TEMPERATURE
e:spray water flow
F: power of the plant.MW.
1v.
0
-1v.
+2.5
0
-2.5
120MW.
0
0
2000
4000
6000
8000s.

fig:2

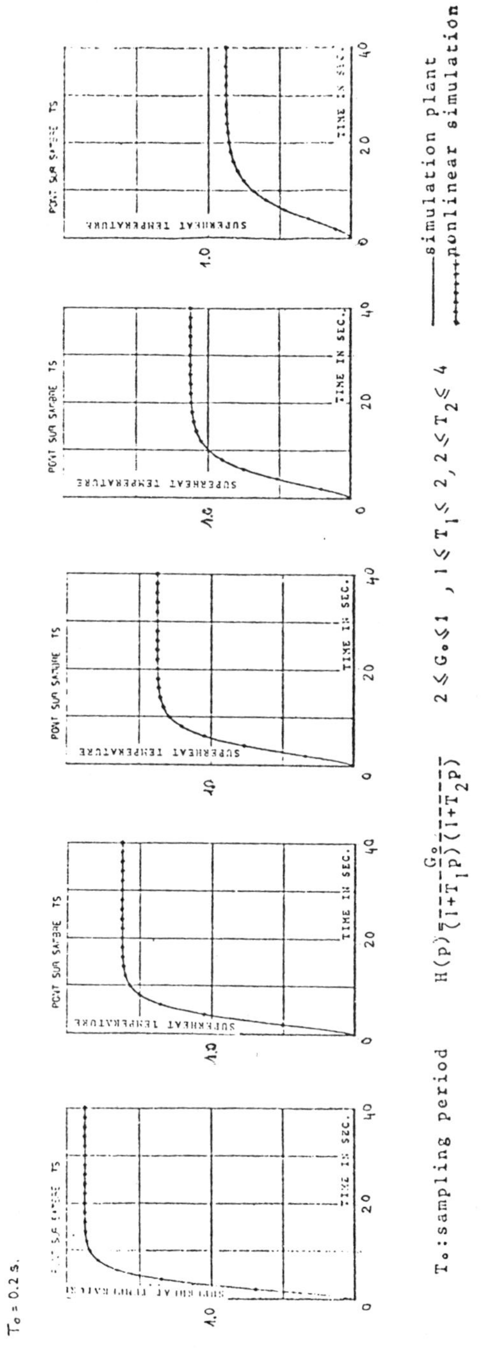

$T_o = 0.2$ s.
$T_o = 0.4$ s.
SUPERHEAT TEMPERATURE
TIME IN SEC.
1.0
0
20
40
$T_o$: sampling period
$H(p) = \frac{G_o}{(1+T_1 p)(1+T_2 p)}$
$2 \leqslant G_o \leqslant 1$ , $1 \leqslant T_1 \leqslant 2$, $2 \leqslant T_2 \leqslant 4$
simulation plant
nonlinear simulation


fig:3

REFERENCES:

[1] D. CYROT-NORMAND : Utilisation de certaines familles algébriques de systèmes non linéaires à quelques problèmes de filtrage et d'identification.
Thèse de Doctorat de 3ème cycle, Université Paris VII, 1978.

D. CYROT-NORMAND and H. DANG VAN MIEN : Nonlinear state-affine identification methods; applications to electrical power plants. Symposium IFAC Power generation Distribution and Protection,

D. CYROT-NORMAND and H. DANG VAN MIEN, to be published.

[2] H. DANG VAN MIEN : Régulation numérique de la température de surchauffe d'un générateur de vapeur, Centrale de Pont-sur-Sambre. Revue automatisme n° 12, 1970, pp.617-626.

Etude de la commande des systèmes multidimensionnels. Rapport EDF, Etudes et Recherches, AMP 117, Déc. 1972.

Utilisation pratique de l'algorithme de Kalman-Ho, Application à la réduction des systèmes multidimensionnels. Amélioration par la méthode des moindres carrés. EDF, Etudes et Recherches, Série C, Maths. Info, n° 2, 1973, pp. 23-46.

[3] M. ESPANA and I.D. LANDAU : Reduced Order Bilinear Models for Distillation Columns. Automatica , Vol. 14, 1978, pp. 345-355.

[4] M. FLIESS : Un outil algébrique : les séries formelles non commutatives, in Mathematical Systems Theory (G. Marchisini and S.K. Mitter, eds.) Lect. Notes Econom. Math. Syst. 131, pp.122-148, Springer-Verlag, Berlin, 1976.

Fonctionnelles causales non linéaires et indéterminées non commutatives (to appear).

[5] E. IRVING : Identification des systèmes.EDF, Bulletin des Etudes et Recherches, Série C, n° 1-2, 1969, n° 1, 1971.

[6] P. A. FRICK and R. R. MOHLER : Case Studies in Bilinear System Identification. Proc. 1977. Joint Automatic Control Conference, San Francisco.

R. R. MOHLER, C. F BARTON and V. R. KARAMAN : BLS Identification by Orthogonal Functions with Application to Immulogy. Proc. I.E.E.E. Conf. Decision and Control. Clear water (FL), Dec. 1976.

[7] E. SONTAG : Realization Theory of Discrete-Time Nonlinear Systems Part I : The Bounded Case. I.E.E.E. Trans. on Circuits and Systems. CAS-26, 1979, pp. 342-356.

[8] H.J. SUSSMANN : Semi-group representations, Bilinear Approximation of Input-Output Maps and Generalized inputs. In Mathematical Systems Theory (G. Marchisini and S.K. Mitter, eds.) Lect. Notes Econom. Math. Syst. 131, pp. 172-191, Springer Verlag, Berlin 1976.

# A COMPARISON OF STOCHASTIC REALIZATION ALGORITHMS FOR IDENTIFICATION OF ARMA MODELS

G. FAVIER

Université de Nice- Laboratoire de Signaux et Systèmes - Equipe de Recherche Associée au CNRS n°835 - 41 Bd. Napoléon III - 06041 NICE Cedex - FRANCE.

## ABSTRACT

In this paper we first present an algorithm to estimate the dynamical parameters of AutoRegressive Moving Average models, and then we give new algorithms to solve the problem of the statistical parameters estimation by use of stochastic realization algorithms. These algorithms are based upon a special canonical state-space representation, directly and simply related to ARMA models. The two first stochastic realization algorithms presented here need the solution of an $n \times n$ matrix Riccati-type equation, while the third one needs the solution of another set of only 2n difference equations, called Chandrasekhar-type equation. These algorithms are compared on simulated examples (operations count and precision).

## I - INTRODUCTION

Signal processing applications and control system design need to identify a mathematical model first. The linear least-squares estimation or linear filtering theory are the tools commonly employed to solve this problem (see [1] for a survey).
However there exists different degrees of complexity in the linear discrete-time models.

The simplest structure is the least-square model, also called AutoRegressive (AR) model or all-pole model, that corresponds to the case without dynamical noise in the state-space representation. This structure was extensively studied by many authors: the recursive least-squares method is very well suited to identify such a model, but, when the data is given in terms of covariance functions, Levinson [2] first formulated a "recursive in order" method for the parameters estimation. The same procedure was used by



*D. G. Lainiotis and N. S. Tzannes (eds.), Advances in Control, 308–325.*

Durbin[3] for recursive fitting of autoregressive models to scalar time-series data. Whittle[4] generalized this method for the solution of multidimensional Yule-Walker equations. Wiggins and Robinson[5] applied and extended this procedure to the multichannel filtering problem, in geophysical applications.
In the scalar case, Anderson[6], Carayannis[7], Markel and Gray[8] gave other formulations for solving the Yule-Walker equations.The two first papers are based on matrix inversion by partitioning and on the basic triangularization theorems of symmetric matrices,respectively,while the last one is based on a scalar product definition. In the multidimensional case, Akaike[9] gave an iterative procedure for the inversion of a general block Toeplitz matrix,and this result includes the Whittle's algorithm as a special case.

A more general linear model is the AutoRegressive Moving-Average (ARMA) model. Such a model is characterized by two kinds of parameters: the dynamical (AR part) and the statistical (MA part) parameters corresponding to the spectral denominator and the spectral numerator respectively.

Many authors[10]-[14] presented and compared the most common on-line identification methods for such models.

But in the case of stochastic realization algorithms,i.e. with covariance information, Gersch[15] derived an algorithm for identifying the AR part by an Hankel matrix inversion. Mehra[16] and Favier[17] obtained the same algorithm by use of special canonical state-space representations. Alengrin[18] proposed a recursive in order procedure to solve this problem.

However while the AR part can be estimated by solving a set of linear equations, the MA part estimation needs to solve a set of non linear equations[19]. Durbin[20] avoided this non linear problem by fitting a high order autoregressive scheme and estimating the MA parameters from the fitted AR coefficients, by solution of a set of linear algebraic equations. Graupe and Cline[21] generalized this work by identifying the orders of the ARMA models too, from pure AR models.

As mentioned before, Favier[17] used a special canonical state-space representation to show that this set of non linear equations is equivalent to a Riccati equation provided by the related filtering problem.

However in the state-space formulation[22],[23] and for computing the matrix gain, new algorithms were indepently obtained by Kailath[24] and Lindquist[25], in continuous and discrete time respectively. Kailath's algorithm was obtained by Lindquist[26] in a different way using a backward innovation process, while Morf, Sidhu and Kailath[27] derived an other lightly different form of the discrete time version.

The main feature of these new algorithms is to replace the Riccati-type equation by another set of non linear equations called of Chandrasekhar-type, which have numerical advantages because they need less operations than the standard Riccati equation of the Kalman filter, so that they are called "fast algorithms".

Lindquist [28] gave the connections between these fast algorithms and Levinson's and Whittle's work.
When the noise covariances are unknown, Ruckebusch [29] derived a fast version of the stochastic realization algorithm of Faurre [30] to compute the Kalman gain. Favier and Alengrin [31] developed and compared several stochastic reslization algorithms using Chandrasekhar-type equations to identify the Kalman gain and the MA part of ARMA models.

In this paper we first present an algorithm to estimate the AR parameters and then we give new algorithms to solve the problem of MA parameters estimation by use of stochastic realization algorithms.
These algorithms are based upon special canonical state-space representations for the residual sequence, directly and simply related to the MA part of the ARMA models.

The two first algorithms are derived from [17] and [32] while the last one is obtained by a factorization of increments of a covariance matrix, in the same way than in [27].

## II - CANONICAL STATE-SPACE REPRESENTATION FOR INPUT-OUTPUT RELATIONS

We shall assume that the process under study can be described mathematically by an input-output relation with the following general structure :

$$\sum_{j=o}^{n} a_{n-j}\, y(k-j) = \sum_{j=1}^{n} d_{n-j}\, u(k-j) + \sum_{j=o}^{n} b_{n-j}\, e(k-j) \qquad (2.1)$$

Equation (2.1) may be written more concisely in operator notation as

$$\mathscr{A}(z^{-1})y(k) = \mathscr{D}(z^{-1})u(k) + \mathscr{B}(z^{-1})\, e(k) \qquad (2.2)$$

where:

1) $y(k)$ and $u(k)$ are observed stationary zero-mean scalar processes corresponding to the output and input signals, respectively

2) $e(k)$ is an unobserved stationary zero-mean scalar white noise, whose variance is $\sigma^2$, and assumed uncorrelated with $\{u(j),\ j \leqslant k\}$.

3) $\mathscr{A}(z^{-1})$, $\mathscr{D}(z^{-1})$ and $\mathscr{B}_1(z^{-1})$ are polynomials in the unit backward shift operator $z^{-1}$ [i.e. $z^{-1}y(k) = y(k-1)$] given by :

$$\begin{cases} \mathscr{A}(z^{-1}) = a_n + a_{n-1} z^{-1} + \ldots + a_o z^{-n} & (2.3\,a) \\ \mathscr{D}(z^{-1}) = d_{n-1} z^{-1} + \ldots\ldots. + d_o z^{-n} & (2.3\,b) \\ \mathscr{B}(z^{-1}) = b_n + b_{n-1} z^{-1} + \ldots + b_o z^{-n} & (2.3\,c) \end{cases}$$

Without loss of generality, we take :

$$a_n = b_n = 1$$

Moreover we shall assume that the order n of the model (2.1) is known a priori.

Such a model (2.1) is characterized by :

- 2n dynamical parameters: $a_o, a_1, \ldots, a_{n-1}, d_o, d_1, \ldots, d_{n-1}$
- (n+1) statistical parameters: $b_o, b_1, \ldots, b_{n-1}, \sigma^2$

We shall call the model (2.1) ARMA model.

Using a special canonical form of the state equations[33], we introduce a Canonical Innovation Representation (CIR) directly and simply related to the input-output relation (2.1) :

$$\begin{cases} x(k+1) = J\,x(k) + A\,y(k) + D\,u(k) + B\,e(k) & (2.4\,a) \\ y(k) = H\,x(k) + e(k) & (2.4\,b) \end{cases}$$

where:

$x(k)$ = n-dimensional state vector
$e(k)$ = innovation process whose variance is $\sigma^2$

and

$$\begin{cases} J = \begin{bmatrix} o & \cdots & \cdots & o \\ 1 & \ddots & & \vdots \\ \vdots & \ddots & \ddots & \vdots \\ o & \cdots & 1 & o \end{bmatrix} \quad A = -\begin{bmatrix} a_o \\ \vdots \\ a_{n-1} \end{bmatrix} \quad D = \begin{bmatrix} d_o \\ \vdots \\ d_{n-1} \end{bmatrix} \quad B = \begin{bmatrix} b_o \\ \vdots \\ b_{n-1} \end{bmatrix} \\ H = \begin{bmatrix} o & \cdots & o & 1 \end{bmatrix} \end{cases} \qquad (2.5)$$

This CIR is characterized by the same parameters $(A, D, B, \sigma^2)$ than the model (2.1).

Below we give two properties of the J matrix:

- First Property:

The right (left) multiplication by J of a M matrix corresponds to a shifting of columns (rows) of M to the left (to the bottom), with zeros on the last column (first row).
So that we have

$$J^i = o \qquad \forall\, i \geq n \qquad (2.6)$$

- Second Property:

For $i < n$

$$H\,J^i = \begin{bmatrix} o & \cdots\cdots & 1 & \cdots\cdots & o \\ \text{column} & & \text{column} & & \text{column} \end{bmatrix} \qquad (2.7)$$
$$\qquad\quad 1 \qquad\qquad\qquad (n-i) \qquad\qquad\quad n$$

then:

$$\begin{bmatrix} H\,J^{n-1} \\ \vdots \\ H\,J \\ H \end{bmatrix} = I \text{ (Identity matrix of order n)} \qquad (2.8)$$

## III - DYNAMICAL PARAMETERS ESTIMATION ALGORITHM

### III-1 Expression of the covariance function of the process y(k)

By definition:

$$\varphi_{yy}(i) = E\left[y(k+i)\ y^T(k)\right] \qquad (3.1\ a)$$

$$\varphi_{uy}(i) = E\left[u(k+i)\ y^T(k)\right] \qquad (3.1\ b)$$

The notation $E[\bullet]$ indicates the expectation operation and a superscript "T" specifies the transposition operation.

▪ *Case* $i > o$ : From equations (2.4), we obtain:

$$\varphi_{yy}(i) = E\left\{\left[H\ x(k+i) + e(k+i)\right]\ y^T(k)\right\} \qquad (3.2)$$

with

$$x(k+i) = J^i x(k) + \sum_{j=o}^{i-1} J^{i-1-j}\left[A\,y(k+j) + D\,u(k+j) + B\,e(k+j)\right] \qquad (3.3)$$

Using the orthogonality property of e(k):

$$E\left[e(j)\ y^T(k)\right] = o \qquad \forall\ j > k \qquad (3.4)$$

and defining:

$$X = E\left[x(k)\ x^T(k)\right] \qquad (3.5)$$

we get:

$$\varphi_{yy}(i) = H\left\{J^i X\ H^T + \sum_{j=o}^{i-1} J^{i-1-j}\left[A\varphi_{yy}(j) + D\varphi_{uy}(j)\right] + J^{i-1} B\,\sigma^2\right\} \qquad (3.6)$$

or:

$$\varphi_{yy}(i) = H\left\{J^{i-1}(J\,X\,H^T + B\,\sigma^2) + \sum_{j=o}^{i-1} J^{i-1-j}\left[A\varphi_{yy}(j) + D\varphi_{uy}(j)\right]\right\} \qquad (3.7)$$

▪ *Case* $i = o$ :

$$\varphi_{yy}(o) = H\ X\ H^T + \sigma^2 \qquad (3.8)$$

### III-2 Dynamical parameters estimation algorithm

Let us consider M coefficients $\{\varphi_{yy}(i),\ i = n+m,\ m \in [1,M],\ M \geqslant 2n\}$.

By use of the first property of the J matrix, we have:

$$\varphi_{yy}(n+m) = \sum_{j=m}^{n+m-1} H\, J^{n+m-1-j}[A\,\varphi_{yy}(j) + D\varphi_{uy}(j)] \quad (3.9)$$

and, from the second property, the expression (3.9) becomes:

$$\varphi_{yy}(n+m) = [\varphi_{yy}(m)\ldots\varphi_{yy}(n+m-1]\,A + [\varphi_{uy}(m)\ldots\varphi_{uy}(n+m-1]D \quad (3.10)$$

Consequently the dynamical parameters (A,D) verifies a system of linear equations and thus we obtain the following dynamical parameters estimation formula:

$$\begin{bmatrix} \hat{A} \\ \\ \hat{D} \end{bmatrix} = \mathcal{H}^{\#} \begin{bmatrix} \varphi_{yy}(n+1) \\ \vdots \\ \varphi_{yy}(n+M) \end{bmatrix} \quad (3.11)$$

where # denotes the inversion (if $M = 2n$) or pseudo-inversion (if $M > 2n$) operations.
And:

$$\mathcal{H} = \begin{bmatrix} \varphi_{yy}(1) \ldots\ldots \varphi_{yy}(n) & \varphi_{uy}(1) \ldots\ldots \varphi_{uy}(n) \\ \vdots & \vdots \\ \varphi_{yy}(M) \ldots \varphi_{yy}(n+M-1) & \varphi_{uy}(M) \ldots\ldots \varphi_{uy}(n+M-1) \end{bmatrix} \quad (3.12)$$

The dynamical parameters estimation algorithm can be summarized in the following steps:

1) Given $\{y(k),u(k),\ k\in[1,N]\}$ estimate the covariance functions $\{\varphi_{yy}(i)\ ,\ \varphi_{uy}(j),\ i\in[1,n+M],\ j\in[1,n+M-1],\ M\geqslant 2n\}$ by use of the ergodic estimator:

$$\hat{\varphi}_{yy}(i) = \frac{1}{N}\sum_{k=o}^{N} y(k+i)\ y(k) \quad (3.13\,a)$$

$$\hat{\varphi}_{uy}(i) = \frac{1}{N}\sum_{k=o}^{N} u(k+i)\ y(k) \quad (3.13\,b)$$

2) Compute $(\hat{A},\hat{D})$ from equations (3.11) and (3.12) with the estimated values $\{\hat{\varphi}_{yy}(i),\hat{\varphi}_{uy}(i)\}$

## IV - STATISTICAL PARAMETERS ESTIMATION ALGORITHM

### IV-1 The residual sequence

The dynamical parameters (A,D) having been estimated in the previous paragraph, we can filter the output signal y(k) through the deterministic part of the model (2.1), generating a new signal $\epsilon(k)$ as follows:

$$\epsilon(k) = \sum_{j=o}^{n} a_{n-j}\, y(k-j) - \sum_{j=1}^{n} d_{n-j}\, u(k-j) \qquad (4.1)$$

$\epsilon(k)$ is called the residual sequence.

Comparing the model (4.1) to the model (2.1) gives:

$$\epsilon(k) = \sum_{j=o}^{n} b_{n-j}\, e(k-j) \qquad (4.2)$$

Consequently $\epsilon(k)$ corresponds to the MA part of the model (2.1).

Let:

$$\varphi_{\epsilon\epsilon}(i) = E\,[\epsilon(k+i)\, \epsilon^T(k)] \qquad (4.3)$$

From equation (4.2), we get :

$$\varphi_{\epsilon\epsilon}(i) = \begin{cases} \sum_{\ell=o}^{n-|i|} b_\ell\, b_{\ell+|i|} & \text{if } o \leqslant |i| \leqslant n \qquad (4.4\,a) \\ 0 & \text{if } |i| > n \qquad (4.4\,b) \end{cases}$$

Or using (4.1) and (3.10) we also have (for: $o \leqslant |i| \leqslant n$):

$$\varphi_{\epsilon\epsilon}(i) = \sum_{\ell=o}^{n-i} a_{n-\ell}\left[\sum_{j=o}^{n} a_{n-j}\, \varphi_{yy}(i+\ell-j) - \sum_{j=1}^{n} d_{n-j}\, \varphi_{uy}(i+\ell-j)\right]$$
$$+\sum_{\ell=1}^{n} d_{n-\ell}\left[\sum_{j=1}^{n} d_{n-j}\, \varphi_{uu}(i+\ell-j) - \sum_{j=o}^{n} a_{n-j}\, \varphi_{yu}(i+\ell-j)\right] \qquad (4.5)$$

To estimate the statistical parameters $(B,\sigma^2)$, we must solve the system of non linear equations (4.4 a), with $\varphi_{\epsilon\epsilon}(i)$ calculated from (4.5), or estimated by use of the ergodic estimator:

$$\hat{\varphi}_{\epsilon\epsilon}(i) = \frac{1}{N} \sum_{k=1}^{N-i} \epsilon(k+i)\, \epsilon(k) \qquad (4.6)$$

Using a special canonical state-space representation for the residual sequence $\epsilon(k)$, we derive new techniques to solve these non linear equations.

## IV-2 RICCATI - type algorithm

By comparing equations (2.1) and (4.1), the model (2.4) becomes

$$\begin{cases} \chi(k+1) = J\chi(k) + B\, e(k) & (4.7\,a) \\ \epsilon(k) = H\chi(k) + e(k) & (4.7\,b) \end{cases}$$

Defining :

$$P = E[\chi(k)\ \chi^T(k)] \qquad (4.8)$$

the expression (3.6) with (A = D = o) becomes :

$$\varphi_{\epsilon\epsilon}(i) = H\{J^i P H^T + J^{i-1} B \sigma^2\} \qquad i > o \tag{4.9}$$

and we also have

$$\varphi_{\epsilon\epsilon}(o) = H P H^T + \sigma^2 \tag{4.10}$$

where P is solution of the matrix equation :

$$P = J P J^T + B \sigma^2 B^T \tag{4.11}$$

Let:

$$G = J P H^T + B \sigma^2 \tag{4.12}$$

By use of the second property of the J matrix, (4.9) gives :

$$G = \begin{bmatrix} \varphi_{\epsilon\epsilon}(n) \\ \vdots \\ \varphi_{\epsilon\epsilon}(1) \end{bmatrix} \tag{4.13}$$

so that we obtain the following expressions for the statistical parameters

$$\sigma^2 = \varphi_{\epsilon\epsilon}(o) - H P H^T \tag{4.14}$$

$$B = [G - J P H^T][\varphi_{\epsilon\epsilon}(o) - H P H^T]^{-1} \tag{4.15}$$

and P obeys the RICCATI-type equation :

$$P = J P J^T + [G - J P H^T][\varphi_{\epsilon\epsilon}(o) - H P H^T]^{-1}[G - J P H^T]^T \tag{4.16}$$

Using the same procedure than in [30] and [34] we can solve this equation as follows:

$$\begin{cases} \sigma^2(m) = \varphi_{\epsilon\epsilon}(o) - H P(m) H^T & (4.17\text{ a}) \\ B(m) = [G - J P(m) H^T][\sigma^2(m)]^{-1} & (4.17\text{ b}) \\ P(m+1) = J P(m) J^T + B(m) \sigma^2(m) B^T(m) & (4.17\text{ c}) \end{cases}$$

with the initial condition: $P(o) = o$.

Then the estimation algorithm for the statistical parameters can be summarized in the following steps:

1) Compute the (n+1) first covariance coefficients $\varphi_{\epsilon\epsilon}(i)$ either from the expression (4.5) or by use of the ergodic estimator (4.6). In this last case, it is necessary to generate the residual sequence previously using equation (4.1).

2) Compute the G vector by use of equation (4.13).

3) Solve the RICCATI equation (4.16) by finding the stationary solution $P(\infty)$ of the recursive equations set (4.17).

4) Estimated values for the statistical parameters are :

$$\hat{\sigma}^2 = \varphi_{\epsilon\epsilon}(o) - H P(\infty) H^T \tag{4.18}$$

$$\hat{B} = [G - J P(\infty) H^T][\varphi_{\epsilon\epsilon}(o) - H P(\infty) H^T]^{-1} \tag{4.19}$$

*Remarks*: - The above calculations are very easy to implement, owing to the special structure of the J and H matrices.

- This algorithm is closed to our previous one [17], but the use of the residual model instead of the process model simplifies the RICCATI equation and the computation of the G vector.

## IV-3 Other form of the RICCATI-type algorithm

Following new results recently obtained to show the link between the spectral factorization problem and the solution of two uncoupled RICCATI equations [32], we can introduce an artificial state-space model to represent the residual sequence:

$$\begin{cases} \xi(k+1) = J\,\xi(k) + w(k) & (4.20\,\mathrm{a}) \\ \epsilon(k) = H\,\xi(k) + v(k) & (4.20\,\mathrm{b}) \end{cases}$$

with

$$E[w(k)] = E[v(k)] = o \qquad (4.21)$$

and :

$$E = \left\{ \begin{bmatrix} w(k) \\ v(k) \end{bmatrix} [w^T(i) \;\; v^T(i)] \right\} = \begin{bmatrix} Q & S \\ S^T & R \end{bmatrix} \delta(k-i) \qquad (4.22)$$

The Innovation Representation related to the model (4.20) is given by equations (4.7), and the corresponding parameters $(B,\sigma^2)$ now constitute the KALMAN gain vector and the innovation process variance for the KALMAN filter built on the model (4.20), so that we have:

$$\begin{cases} B = (J\tilde{P}H^T + S)(H\tilde{P}H^T + R)^{-1} & (4.23\,\mathrm{a}) \\ \sigma^2 = H\tilde{P}H^T + R & (4.23\,\mathrm{b}) \end{cases}$$

Where $\tilde{P}$ is defined as:

$$\tilde{P} = E\,\{[\xi - \chi][\xi - \chi]^T\} \qquad (4.24)$$

and $\tilde{P}$ obeys the following RICCATI equation :

$$\tilde{P} = J\tilde{P}J^T - (J\tilde{P}H^T + S)(H\tilde{P}H^T + R)^{-1}(H\tilde{P}J^T + S^T) + Q \qquad (4.25)$$

In order to obtain the values of the covariances (Q,R,S), we consider the generalized noise vector:

$$w_G(k) = \begin{bmatrix} w(k) \\ v(k) \end{bmatrix} \qquad (4.26)$$

whose covariance matrix is:

$$Q_G = \begin{bmatrix} Q & S \\ S^T & R \end{bmatrix} \qquad (4.27)$$

and we choose this covariance matrix with a special structure which is the TOEPLITZ form:

$$Q_G = \begin{bmatrix} q_o & q_1 & \cdots & q_n \\ q_1 & \ddots & \ddots & \vdots \\ \vdots & \ddots & \ddots & q_1 \\ q_n & \cdots & q_1 & q_o \end{bmatrix} \qquad (4.28)$$

Applying the Z-transform to equations (4.20), we get:

$$E(z) = H(zI-J)^{-1}\ W(z) + V(z) \qquad (4.29)$$

where E(z), W(z) and V(z) are the z-transforms of $\epsilon(k)$, $w(k)$ and $v(k)$, respectively.

Noticing that (zI-J) has the JORDAN form, we can deduce that:

$$H(zI-J)^{-1}\ [z^{-n}\ z^{-n+1} \dots z^{-2}\ z^{-1}] \qquad (4.30)$$

then the spectrum $\Phi_{\epsilon\epsilon}(z)$ is obtained as follows:

$$\Phi_{\epsilon\epsilon}(z) = [z^{-n} \dots z^{-1}\ 1]\ Q_G [z^{n} \dots z\ 1]^T \qquad (4.31)$$

and using the expression (4.28) for $Q_G$, we get:

$$\Phi_{\epsilon\epsilon}(z) = (n+1)q_o + \sum_{i=1}^{n} (n+1-i)\ q_i (z^{i} + z^{-i}) \qquad (4.32)$$

Moreover by definition:

$$\Phi_{\epsilon\epsilon}(z) = \sum_{i=-\infty}^{+\infty} \varphi_{\epsilon\epsilon}(i)\ z^{-i} \qquad (4.33)$$

and using (4.4) it is easily seen that:

$$q_i = \frac{\varphi_{\epsilon\epsilon}(i)}{n+1-i} \qquad (4.34)$$

The resulting algorithm is:

1) Compute the (n+1) first covariance coefficients $\varphi_{\epsilon\epsilon}(i)$ either from the expression (4.5) or by use of the ergodic estimator (4.6).

2) Compute the $q_i$ coefficients by use of expression (4.34).

3) Solve the RICCATI equation (4.25) by finding the stationary solution $\tilde{P}(\infty)$ of the recursive equations set:

$$\begin{cases} \sigma^2(m) = H\ \tilde{P}(m)\ H^T + R & (4.35\text{ a}) \\ B(m) = [J\tilde{P}(m)\ H^T + S][\sigma^2(m)]^{-1} & (4.35\text{ b}) \\ \tilde{P}(m+1) = J\ \tilde{P}(m)\ J^T - B(m)\ \sigma^2(m)\ B^T(m) + Q & (4.35\text{ c}) \end{cases}$$

with the initial condition: $\tilde{P}(o) = P_o$, where $P_o$ is solution of the LYAPOUNOV equation:

$$P_o = J\ P_o\ J^T + Q \qquad (4.36)$$

4) Estimated values for the statistical parameters are:

$$\begin{cases} \hat{\sigma}^2 = H\ \tilde{P}(\infty)\ H^T + R & (4.37\text{ a}) \\ \hat{B} = [J\tilde{P}(\infty) H^T + S][H\tilde{P}(\infty)\ H^T + R]^{-1} & (4.37\text{ b}) \end{cases}$$

*Remark:* Using the special structure of the J and Q matrices, we have:

$$P_o(i,j) = i\ q_{j-i} \qquad \forall\ i \leqslant j \qquad (4.38)$$

with:

$$P_o(j,i) = P_o(i,j) \qquad (4.39)$$

IV-4 Fast-type algorithm

In this paragraph, we shall use the technique of [27] consisting in factorizing the increments of the RICCATI equation variable (4.17).

Let :
$$\delta P(m) = P(m+1) - P(m) \quad (4.40)$$
with the initial value:
$$\delta P(o) = P(1) - P(o) = B(o)\,\sigma^2(o)\,B^T(o) \quad (4.41)$$
Thus, we can see that $\delta P(o)$ has the following form:
$$\delta P(o) = L(o)\,\sigma^2(o)\,L^T(o) \quad (4.42)$$
with
$$\begin{cases} L(o) = B(o) \\ \dim L(o) = (n \times 1) \end{cases} \quad (4.43)$$
Proceeding recursively and assuming that:
$$\begin{cases} \delta P(m-1) = L(m-1)\,\sigma^2(m-1)\,L^T(m-1) \\ \dim L(m-1) = (n \times 1) \end{cases} \quad (4.44)$$
we can show that:
$$\begin{cases} \delta P(m) = L(m)\,\sigma^2(m)\,L^T(m) \\ \dim L(m) = (n \times 1) \qquad \forall\, m \geqslant o \end{cases} \quad (4.45)$$
Indeed, from equations (4.17), we have
$$\sigma^2(m) = \sigma^2(m-1) - H\,\delta P(m-1)\,H^T \quad (4.46)$$
$$B(m) = [B(m-1\ \sigma^2(m-1) - J\,\delta P(m-1)\,H^T][\sigma^2(m)]^{-1} \quad (4.47)$$
$$\delta P(m) = J\,\delta P(m-1)\,J^T + B(m)\,\sigma^2(m)\,B^T(m) - B(m-1)\,\sigma^2(m-1)\,B^T(m-1) \quad (4.48)$$
From the hypothesis (4.44) and noticing that:
$$H\,L(m-1) = L^T(m-1)\,H^T \qquad \text{(scalar)} \quad (4.49)$$
we have:
$$\sigma^2(m) = \{1 - [H\,L(m-1)]^2\}\,\sigma^2(m-1) \quad (4.50)$$
and the expressions (4.47) and (4.48) become:
$$B(m) = \{1 - [H\,L(m-1)]^2\}^{-1}\{B(m-1) - [H\,L(m-1)]\,J\,L(m-1)\} \quad (4.51)$$
$$\delta P(m) = \{1 - [H\,L(m-1)]^2\}^{-1}\{J\,L(m-1) - [H\,L(m-1)]\,B(m-1)\}$$
$$\sigma^2(m)\,\{J\,L(m-1) - [H\,L(m-1)]\,B(m-1)\}^T\{1 - [H\,L(m-1)]^2\}^{-1} \quad (4.52)$$
thus we can conclude that (4.45) is verified with:
$$\begin{cases} L(m) = \{1 - [H\,L(m-1)]^2\}^{-1}\{J\,L(m-1) - [H\,L(m-1)]\,B(m-1)\} \\ \dim L(m) = (n \times 1) \end{cases} \quad (4.53)$$
Therefore the estimation of the statistical parameters can be made by solving the following non linear equations set:
$$\begin{cases} \sigma^2(m) = \{1 - [H\,L(m-1)]^2\}\,\sigma^2(m-1) & (4.54\text{ a}) \\ B(m) = \{1 - [H\,L(m-1)]^2\}^{-1}\{B(m-1) - [H\,L(m-1)]\,J\,L(m-1)\} & (4.54\text{ b}) \\ L(m) = \{1 - [H\,L(m-1)]^2\}^{-1}\{J\,L(m-1) - [H\,L(m-1)]\,B(m-1)\} & (4.54\text{ c}) \end{cases}$$
with the initial conditions:
$$L(o) = B(o) = G\,[\varphi_{\epsilon\epsilon}(o)]^{-1} \quad (4.55)$$
It is interesting to notice that other forms of these equations can be given.

From (4.53) we have:

$$\{1-[H\,L(m-1)]^2\}^{-1} J\,L(m-1) = L(m)+\{1-H\,L(m-1)]^2\}^{-1}[H\,L(m-1)]\,B(m-1) \quad (4.56)$$

Replacing this expression in equation (4.51), we obtain a first other form of the equations (4.54):

$$\begin{cases} \sigma^2(m) = \{1-[H\,L(m-1)]^2\}\,\sigma^2(m-1) & (4.57\text{ a}) \\ L(m) = \{1-[H\,L(m-1)]^2\}^{-1}\{J\,L(m-1)-[H\,L(m-1)]\,B(m-1)\} & (4.57\text{ b}) \\ B(m) = B(m-1)-[H\,L(m-1)]\,L(m) & (4.57\text{ c}) \end{cases}$$

with the same initial conditions (4.55).

In the same way, from (4.51), we have :

$$\{1-[H\,L(m-1)]^2\}^{-1} B(m-1) = B(m)+\{1-[H\,L(m-1)]^2\}^{-1}[H\,L(m-1)]\,J\,L(m-1) \quad (4.58)$$

and by replacing this expression in equation (4.53), we obtain a second other form of the equations (4.54) :

$$\begin{cases} \sigma^2(m) = \{1-[H\,L(m-1)]^2\}\,\sigma^2(m-1) & (4.59\text{ a}) \\ B(m) = \{1-[H\,L(m-1)]^2\}^{-1}\{B(m-1)-[H\,L(m-1)]\,J\,L(m-1)\} & (4.59\text{ b}) \\ L(m) = J\,L(m-1)-[H\,L(m-1)]\,B(m) & (4.59\text{ c}) \end{cases}$$

with the same initial conditions (4.55).

However we must notice that if equations (4.54) are appropriate for parallel computation, that is no more the case with equations (4.57) and (4.59).

Moreover, owing to the special structure of the J and H matrices, we can explicitly develop the equations (4.54).

Let us define $L_i(m)$ and $B_i(m)$ the i-th component of the vectors L(m) and B(m);

we have :

$$H\,L(m-1) = L_n(m-1) \quad (4.60)$$

and

$$J\,L(m-1) = \begin{bmatrix} o \\ L_1(m-1) \\ \\ L_{n-1}(m-1) \end{bmatrix} \quad (4.61)$$

we can deduce the scalar form of equations (4.54) :

$\forall\ i \in [1,n]$

$$\begin{cases} \sigma^2(m) = \{1-[L_n(m-1)]^2\}\,\sigma^2(m-1) & (4.62\text{ a}) \\ B_i(m) = \dfrac{B_i(m-1)-[1-\delta(i-1)]L_n(m-1)\,L_{i-1}(m-1)}{1-[L_n(m-1)]^2} & (4.62\text{ b}) \\ L_i(m) = \dfrac{[1-\delta(i-1)]\,L_{i-1}(m-1)-L_n(m-1)\,B_i(m-1)}{1-[L_n(m-1)]^2} & (4.62\text{ c}) \end{cases}$$

where

$$\delta(j) = \begin{cases} 1 & \text{if } j=o \\ o & \text{if } j\neq o \end{cases} \quad (4.63)$$

*Remark:* As it is easily seen from equations (4.62), when the algorithm converges we have :

$$L \longrightarrow o$$

That can be used as a convergence test for the algorithm.
A block diagram of the resulting identification scheme is shown in figure 1.

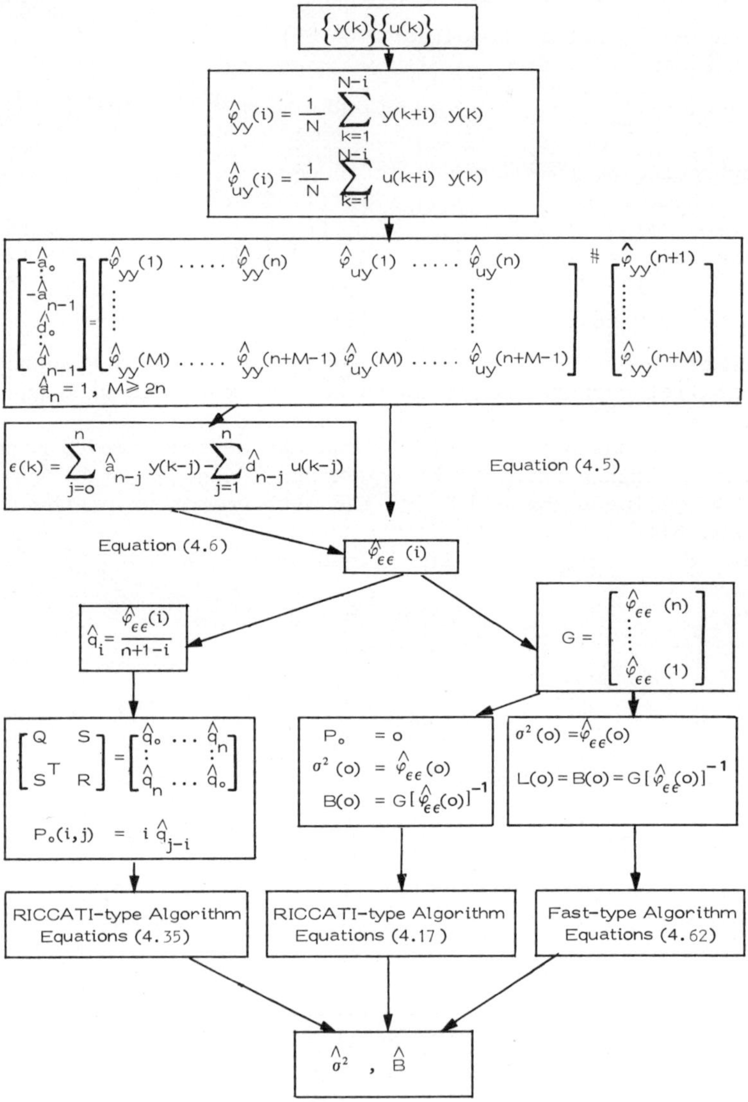


Figure 1

## V - PRACTICAL ASPECTS

First, we must notice that all the algorithms presented in this paper give the stable factor of the spectral numerator, so that only minimum phase systems can be identified by these methods, i.e. systems for which the polynomial $\mathcal{B}(z^{-1})$ has all its zeros outside the unit circle.

A comparison of arithmetic operations count for the MA parameters estimation is summarized in Table 1, with the following notation: MD ≡ Multiplications or Divisions • AS ≡ Additions or Subtractions

Table 1

| Algorithms / Steps | 1st RICCATI-type algorithm. Equations (4.17) | | 2nd RICCATI-type alg.Equat. (4.35) | | Fast-type algorithm.Equations (4.62) | |
|---|---|---|---|---|---|---|
| Initial Values Computations from $\varphi_{\epsilon\epsilon}(i)$ | n<br>o | MD<br>AS | $\frac{(n^2+n+2)}{2}$<br>o | MD<br>AS | n<br>o | MD<br>AS |
| Computations in One Iteration | $n(n+2)$<br>$\frac{n(n+1)}{2}$ | MD<br>AS | $n(n+2)$<br>$n(n+1)$ | MD<br>AS | $(4n+1)$<br>$(2n+1)$ | MD<br>AS |

As it was pointed out in the introduction, for the fast-type algorithm the number of non linear equations to be solved in each iteration is of order n rather than $n^2$ as for the RICCATI-type algorithms.

## VI - SIMULATION RESULTS

A fourth order linear system was simulated on an CII-IRIS 50 digital computer. We present numerical results obtained with each algorithm previously defined for the identification of MA parameters. The exact model was a pure MA model and 5000 points y(k) were generated from:

$$y(k) = e(k) - o.6\ e(k-1) + o.38\,e(k-2) - o.43\ e(k-3) + o.1325\ e(k-4)$$

with $\sigma^2 = 1$.

In the operator notation this model can be written:

$$y(k) = (1-o.6\ z^{-1} + o.38\ z^{-2} - o.43\ z^{-3} + o.1325\ z^{-4})\ e(k)$$

or in a factorizing form:

$$y(k) = (1-o.5z^{-1})^2\ (1+o.4\,z^{-1} + o.53\ z^{-2})\ e(k)$$

the corresponding statistical parameters vector is :

$$B = \begin{bmatrix} o.1325 \\ -o.43 \\ o.38 \\ -o.6 \end{bmatrix}$$

In table 2 we give the estimated initial values $\left\{\begin{matrix} B(o) \\ \sigma^2(o) \end{matrix}\right\}$ and the estimated values of parameters $\begin{Bmatrix} B(IT) \\ \sigma^2\ (IT) \end{Bmatrix}$ after IT iterations, for each algorithm using the true covariance function $\varphi_{\epsilon\epsilon}(i)$.

In table 3, the same results are given when the estimated covariance function $\hat{\varphi}_{\epsilon\epsilon}(i)$ is used.

Table 2

| IT<br>algorithms | Initial values | IT=1 | IT=6 | IT=10 | IT=20 |
|---|---|---|---|---|---|
| 1st RICCATI-type algorithm | o,0776<br>-o,2985<br>o,4033<br>-o,6142<br>1,707 | o,0997<br>-o,3473<br>o,3798<br>-o,6022<br>1,330 | o,1284<br>-o,4204<br>o,3801<br>-o,5987<br>1,032 | o,1317<br>-o,4282<br>o,3801<br>-o,5998<br>1,006 | o,1325<br>-o,43<br>o,38<br>-o,6<br>1 |
| 2nd RICCATI-type algorithm | o,3881<br>-o,7463<br>o,6721<br>-o,7678<br>1,707 | 1,421<br>-2,267<br>-1,223<br>-o,0873<br>o,0933 | o,0852<br>-o,3205<br>o,4151<br>-o,6253<br>1,556 | o,1237<br>-o,4106<br>o,3845<br>-o,5976<br>1,071 | o,1324<br>-o,4297<br>o,38<br>-o,6<br>1,001 |
| Fast-type algorithm | o,0776<br>-o,2985<br>o,4033<br>-o,6142<br>1,707 | o,1247<br>-o,4028<br>o,3532<br>-o,5885<br>1,063 | o,1319<br>-o,4278<br>o,3778<br>-o,5994<br>1,005 | o,1325<br>-o,4298<br>o,3799<br>-o,6001<br>1 | o,1325<br>-o,43<br>o,38<br>-o,6<br>1 |

Table 3

| IT<br>algorithms | Initial values | IT=1 | IT=6 | IT=10 | IT=20 |
|---|---|---|---|---|---|
| 1st RICCATI-type algorithm | o,0755<br>-o,2996<br>o,3958<br>-o,6016<br>1,650 | o,0926<br>-o,3339<br>o,3517<br>-o,5611<br>1,345 | o,1108<br>-o,3857<br>o,3581<br>-o,5518<br>1,125 | o,1123<br>-o,3898<br>o,3589<br>-o,5516<br>1,110 | o,1125<br>-o,3905<br>o,3589<br>-o,5515<br>1,107 |
| 2nd RICCATI-type algorithm | o,3774<br>-o,7491<br>o,6597<br>-o,7520<br>1,650 | 1,317<br>-2,227<br>-1,352<br>-o,0028<br>o,0946 | o,0825<br>-o,3225<br>o,4129<br>-o,6178<br>1,51 | o,1199<br>-o,4106<br>o,3701<br>-o,5795<br>1,046 | o,1272<br>-o,4294<br>o,3647<br>-o,5818<br>o,9792 |
| Fast-type algorithm | o,0755<br>-o,2996<br>o,3958<br>-o,6016<br>1,650 | o,1183<br>-o,3984<br>o,3379<br>-o,5696<br>1,053 | o,1255<br>-o,4250<br>o,3645<br>-o,5808<br>o,9925 | o,1261<br>-o,4272<br>o,3667<br>-o,5816<br>o,9878 | o,1262<br>-o,4274<br>o,3667<br>-o,5815<br>o,9874 |

## VII - CONCLUSION

Several new stochastic realization algorithms have been presented for identifying the parameters of ARMA models.
Firstly we derived an algorithm to estimate the dynamical parameters by a simple inversion of a matrix built with the autocorrelation function of the process and the cross correlation function between the output and the control signals.
For the statistical parameters estimation, we introduced a CIR for the residual sequence and we first obtained two algorithms based upon the solution of a RICCATI-type equation, as in the standard KALMAN filter. Then we deduced a fast version for one of these algorithms, by use of a factorization of covariance matrix increments. In this last case, the number of elementary arithmetic operations is proportional to $n$ instead of $n^2$ as in RICCATI-type equation. All these algorithms are very simple to implement because they can be written with scalar equations very easily.
Numerical results showed the good convergence of the MA estimation algorithms but it will be necessary to simulate other examples to conclude about the convergence properties of such algorithms.

## REFERENCES

(1) T.KAILATH-1974 "A view of three decades of linear filtering theory". I.E.E.E. Tr.,vol.IT-20,n°2,pp.146-180,March 1974.

(2) N.LEVINSON-1947 "The Wiener RMS (Root Mean Square) error criterion in filter design and prediction". Jnl.Math.Phys.,vol.25, n°4,pp.261-278,(1947).

(3) J.DURBIN-1960 "The fitting of time-series models". Revue Inst. Int. de Stat.,vol.28,pp.233-243,(1960).

(4) P.WHITTLE-1963 "On the fitting of multivariate autoregressions, and the approximate canonical factorization of a spectral density matrix". Biometrika,vol.50,pp.129-134,(1963).

(5) R.A.WIGGINS, E.A.ROBINSON-1965 "Recursive solution to the multichannel filtering problem". Journal of Geophysical Research, vol.70,n°8,pp.1885-1891,April 1965.

(6) T.W.ANDERSON-1971 "The Statistical Analysis of Time Series". John Wiley & Sons, Inc.,(1971).

(7) G.CARAYANNIS-1977 "An alternative formulation for the recursive solution of the covariance and autocorrelation equations". I.E.E.E.,Tr.,vol.ASSP-25,n°6,pp.574-577,December 1977.

(8) J.D.MARKEL, A.H.GRAY-1973 "On autocorrelation equations as applied to speech analysis". I.E.E.E.,Tr.,vol.AU-21,n°2,pp. 69-79,April 1973.

(9) H.AKAIKE-1973 "Block Toeplitz matrix inversion". SIAM J.Appl. Math.,vol.24,n°2,pp.234-241,March 1973.

(10) K.J.ÅSTRÖM, P.EYKHOFF-1971 "System identification: A survey". Automatica,vol.7,pp.123-162,(1971).

(11) R.ISERMAN, U.BAUR, W.BAMBERGER, P.KNEPPO, H.SIEBERT-1973 "Comparison and evaluation of six on-line identification and parameter estimation methods with three simulated processes". Proc. 3rd.Symposium on Identification,pp.1081-1100, The Hague, (1973).

(12) G.N.SARIDIS-1974 "Comparison of six on-line identification algorithms". Automatica,vol.10,pp.69-79,(1974).

(13) T.SÖDERSTRÖM, L.LJUNG, I.GUSTAVSSON-1974 "A comparative study of recursive identification methods". Report 7427, Lund Institute of Technology, Department of Automatic Control,Lund, Sweden,December 1974.

(14) T.SÖDERSTRÖM, L.LJUNG, I.GUSTAVSSON-1978 "A theoretical analysis of recursive identification methods". Automatica,vol.14, pp.231-244,(1978).

(15) W.GERSCH-1970 "Estimation of the autoregressive parameters of a mixed autoregressive moving average time series". I.E.E.E., Tr.on Automatic Control,pp.583-588,October 1970.

(16) R.K.MEHRA-1971 "On line identification of linear dynamic systems with applications to Kalman filtering". I.E.E.E.,Tr.,vol. AC-16,n°1,pp.12-21,February 1971.

(17) G.FAVIER-1977 "Identification d'une représentation gaussienne markovienne - Algorithmes de réalisation stochastique". Thèse de Docteur-Ingénieur,Nice,juin 1977.

(18) G.ALENGRIN-1979 "Parameters identification for autoregressive moving-average models". 2nd.International Conference on Information Sciences and Systems, Patras Greece, July 1979.

(19) G.E.BOX, G.M.JENKINS-1976 "Time-series analysis: Forecasting and Control". Holden-Day revised edition,(1976).

(20) J.DURBIN-1959 "Efficient estimation of parameters in moving-average models". Biometrika,vol.46,pp.306-316,(1959).

(21) D.GRAUPE, W.K.CLINE-1975 "Derivation of ARMA parameters and orders from pure AR models". Int.J.Systems Sci.,vol.6,n°2, pp.101-106,(1975).

(22) R.E.KALMAN-1960 "A new approach to linear filtering and prediction problem". Trans.of ASME,Jnl. of Basic Engineering,vol. 82D,pp.35-44, March 1960.

(23) R.E.KALMAN, R.S.BUCY-1961 "New results in linear filtering and prediction theory'. Trans.of ASME,Jnl. of Basic Engineering,vol.83D,pp.95-108, March 1961.

(24) T.KAILATH-1973 "Some new algorithms for recursive estimation in constant linear systems". I.E.E.E. Tr.,vol.IT-19,n°6,pp. 750-760, November 1973.

(25) A.LINDQUIST-1974 "A new algorithm for optimal filtering of discrete-time stationary processes". SIAM J.Control,vol.12, n°4,pp.736-746, November 1974.

(26) A.LINDQUIST-1974 "Optimal filtering of continuous-time stationary processes by means of the backward innovation process". SIAM J.Control,vol.12,n°4,pp.747-754, November 1974.

(27) M.MORF, G.S.SIDHU, T.KAILATH-1974 "Some new algorithms for recursive estimation in constant linear, discrete-time systems". I.E.E.E.,Tr., vol AC-19,n°4,pp.315-323, August 1974.

(28) A.LINDQUIST-1975 "On Fredholm integral equations, Toeplitz equations and Kalman-Bucy filtering".Applied Math.& Opt.,vol. 1,n°4,pp.355-373,(1975).

(29) G.RUCKEBUSCH-1975 "Représentations markoviennes de processus gaussiens stationnaires". Thèse de 3e cycle, Université de Paris VI,(1975).

(30) P.FAURRE-1973 "Réalisations markoviennes de processus stationnaires". Rapport Laboria,n°13,IRIA, 78 Rocquencourt,mars 1973.

(31) G.FAVIER, G.ALENGRIN-1979 "Algorithmes de filtrage rapide - Application à l'identification des paramètres statistiques d'un modèle ARMA". 7ème Colloque sur le Traitement du Signal et ses applications, Nice,(1979).

(32) G.FAVIER, G.SALUT-1979 "Factorisation spectrale et équation de Riccati". 7ème Colloque sur le Traitement du Signal et ses applications, Nice,(1979).

(33) G.SALUT-1976 "Identification optimale des systèmes linéaires stochastiques". Thèse de Doctorat d'Etat, Université Paul Sabatier, Toulouse, juin 1976.

(34) E.TSE, H.L.WEINERT-1973 "Structure determination and parameter identification for multivariable stochastic linear systems". Proc.of J.A.C.C. Conference,(1973).

# REALIZATION OF INTELLIGENT CONTROLS

G. N. Saridis

Purdue University
W. Lafayette, IN 47907 U.S.A.

## 1. INTRODUCTION

The hierarchically intelligent control approach is proposed to handle the problems of control of complex nonlinear large scale systems, logically decomposable into smaller subsystems, functioning under various modes of operation and accomodating a number of performance criteria in real-time [12]. It is based on the principle of "Increasing Intelligence with Decreasing Precision" which is actually evident in all hierarchical management systems. According to this principle the lowest level in the hierarchy must execute a task with precision by satisfying certain local performance standards, thus requiring a rather sophisticated and precise model. The next level of the hierarchy is that of a coordinator where the individual subtasks are put to work together by appropriately selecting their performance evaluation requiring less precise models and modes of operations while the need of higher level decision making capabilities is apparent. In addition to coordinating the lower levels, the coordinator should be endowed with learning capabilities to improve its performance under reappearing control situations. An automaton capable of executing fuzzy commands or the equivalent formal language, directly implementable by digital microcomputers is appropriate for such a function. Finally a master computer may serve as a linguistic organizer or general manager at the top of the hierarchy where the highest level decision making takes place in order to organize the tasks, identify the control situations, and assign the proper coordination pattern without much knowledge of the execution details of the actual process. Fig. 1 gives a block diagram of the hierarchically intelligent control algorithm. A more technical description of the method is given in the sequel.



*D. G. Lainiotis and N. S. Tzannes (eds.), Advances in Control, 326–333.*

## 2. DYNAMIC SYSTEM DECOMPOSITION AND CONTROL

Many large scale systems are aggregates of simpler subsystems with varying degrees of interaction due to dynamic coupling. Decomposition, into subsystems may be performed by clustering either the variables of a high amount of interaction or those pertaining to the same physical subsystem.

The dynamics of the ith subsystem may be described by a set of nonlinear differential equations.

$$\frac{dx_i}{dt} = A_i^{mi} x_i + f_i^{mi}(x) + B_i^{mi} u_i + C_i^{mi} w_i$$

$$Z_i = g_i^{mi}(x_i) + v_i \qquad \begin{aligned} m_i &= 1,\ldots s_i \\ i &= 1,\ldots p \end{aligned} \tag{1}$$

where p is the number of subsystems, and $s_i$ is the number of possible patterns associated with the particular situation, $x_i$ is the state vector of the subsystem, $u_i$ the control input, $Z_i$ is the measurement vector and $w_i$ are the respective samples from the two statistically independent random processes. The parameter matrices $A_i^{mi}$, $B_i^{mi}$, $C_i^{mi}$ and the nonlinear functions $f_i^{mi}$ and $g_i^{mi}$ correspond to the situation $m_i$ of the section and need not be known a priori.

In an analogous way, the equations (1) may be given in a discrete-time form:

$$x_i(k+1) = F_i^{mi} x_i(k) + f_i^{mi}(x_i) + G_i^{mi} u_i + D_i^{mi} w_i$$

$$Z_i = g_i^{mi}(x_i) + v_i \qquad \begin{aligned} m_i &= 1,\ldots s_i \\ i &= 1,\ldots p \end{aligned} \tag{2}$$

where they symbols used have an analogous meaning with the symbols of equation (1).

After the decomposition is performed, each subsystem may be optimized with respect to an intrasubsystem, or local criterion that neglects all coupling effects. Each subprocess is optimized with respect to a quadratic criterion with variable parameters that ensure smooth trajectories. Assume a feedback controller structure

$$u_i = L_i(z_i - z_i^d) + N_i\, f_i(z) = u_{Li} + u_{Ni} \qquad i = 1,\ldots,p \tag{3}$$

An expanding subinterval performance adaptive algorithm is used to asymptotically minimize the ith local criterion through an on-line search for $L_i$ and $N_i$.

The learning process sequentially improves a goal

$$J(\alpha,k) = \sum_{i=1}^{p} \mu_i J_i(\alpha_i,k) \tag{4}$$

where $\mu_i$ are weighting coefficients and $J_i(\alpha_i,k)$ are the individual local performance criterion of the ith subsystem. This criterion $J_i(\alpha_i,k)$ and the coefficient $\mu_i$ are selected in each occasion by the linguistic organizer to satisfy an appropriate condition imposed by control situations at the time of the operation. For example a minimum time could be selected when maximum flow is desired or minimum flow criterion may be selected when the occasion arises. The additional parameter $\alpha_i$ serves for a finer adjustment of the performance. Each performance criterion $J_i(\alpha_i k)$ may be modified to include penalty functions to minimize or divert flow though areas of the system.

## 3. LINGUISTIC COORDINATION

The variable parameters, or coordination inputs, in the performance criteria of the controllers are selected, among a finite set of values, by a learning automaton in the middle level of the functional hierarchy (Figure 2). The automaton coordinates the lower level controllers by minimizing an intersubsystem, or overall criterion of the form

$$K(\alpha) = \sum_{i=1}^{p} \mu_i K_i(\alpha_i)$$

where $K_i(\alpha_i)$ is a function representing the cost of operation of the ith subsystem. The analytical solution of this coordination problem is very difficult to solve, if not impossible. Statistical pattern recognition methods have been applied to control problems with a finite number of inputs, or "control actions". The major difficulty encountered in this approach lies in the selection, both structural and parametric, of proper discriminant functions. A linear function is desirable, but does not provide sufficient accuracy for most applications. A nonlinear discriminant polynomial results in a better partition of the feature space, but is objectionable when used in conjunction with a learning algorithm, since the addition of a nonlinear term to the series requires the readjustment of the coefficients of all the existing terms.

A syntactic approach was used to bypass these difficulties in the solution of the coordination problem [11]. A feature space is selected as a subset of $S^2$, where S is the state space, and partitioned into several classes, each grouping all the "control situations" corresponding to the same coordination input. Each class is described by a context-free grammar in Chomsky normal form. At the onset of motion, the control situation is coded into a string which is parsed by an appropriate parsing algorithm, and associated with one of the coordination classes. The CYK parsing algorithm may be used for its simplicity and adaptability to learning, since the parsed strings are of short length. The grammatical production rules for the coordination classes are listed in [11,12].

## 4. THE ORGANIZATION LEVEL

In the proposed method, the higher level of the hierarchy is a man-machine interface. Its task oriented activity consists of the translation of the operator's linguistic commands into a "machine-language" information pattern that determines the operation of the lower two levels by direct intervention in their performance indices. The input to the first level is a linguistic string of symbols representing primitives in a task space. In general, the operator's commands can be expressed in strings of spoken words, or typed characters. The command string is parsed by a bottom-up procedure. If it is recognized as syntactically valid, the parser output is mapped into a sequence of goals (in state space) that determines the performance of the lower levels (Figure 3).

The parse interpretation consists in the mapping of the command string left parse into the set of command variables, thus determining their numerical values. This operation can be realized by a modified pushdown automaton $\Pi$, i.e., an 8 tuple $\Pi = (q,\Sigma,\Gamma,\delta,q_0,\gamma_0,Z,F)$ where

1. Q is the set of possible states.
2. $\Sigma$ is the input alphabet consisting of the set of production numbers from the parser output.
3. $\Gamma$ is the alphabet pushdown symbols consisting of the command variables.
4. Z is the output alphabet consisting of the numerical of the values of the command variables.
5. $\delta$ is a mapping from Q X $\Gamma$ X $\Sigma$ to the finite subsets of Q X $\Gamma^*$ X Z.
6. $q_0 \in Q$ is the initial state.
7. $\gamma_0 \in \Gamma$ is the symbol that appears initially on the pushdown list.
8. F is the final state.

The possible moves of $\Pi$ are listed [11]. The pushdown symbol in any configuration represents the command variable under consideration. The output symbol in the new configuration resulting from a move of the automaton represents the numerical value of this command variable.

## 5. REDUCTION OF UNCERTAINTIES BY LEARNING

Two types of reducable uncertainties are to be distinguished: those pertaining to the identification of the plan dynamics, and those pertaining to the effective control of the process. The first category includes uncertainties due to factors such as task dependent parameters, and incompletely known plant dynamics due to simplifying approximations. The second category includes the unknown gains in the assumed controller structure, which is chosen heuristically in the absence of a closed form solution. The selection of the controller structure was motivated by the speculative biological principle of minimum interaction [11], and by the accompanying simplicity of stability and controllability analysis. The above uncertainties are reduced by the self-organizing algorithm in the third level of the hierarchy. This on-line learning process can be accelerated by a judicious choice of the order in which different tasks are performed. Learning is also required to determine the production rules for the characteristic grammar of each coordination class, as little a priori information is available to the designer.

The reader is advised to check Ref. [11] and [12] for performance evaluation of the method as applied to the control of a prosthetic arm.

## 6. APPLICATION OF THE METHOD TO FREEWAY AND URBAN TRAFFIC CONTROL AND PROSTHESES

The application of the hierarchically intelligent control method for the control and management of vehicular urban traffic systems and their interconnecting freeways may be formulated. The discrete time model (2) may be directly used to describe the traffic flow at an arterial link with appropriate local performance criteria selected to accommodate various control situations. Modern high speed, low cost microcomputers may be used to coordinate the flow of traffic within neighborhoods of similar traffic patterns. A second level of coordination may be added if necessary. the continuous time model (1) or even a discretized version of it [6] may be used to describe the traffic flow or a section of the limited access freeway lay between two access points. Appropriate performance criteria may be chosen to accomodate various control situations depending on the weather, the part of the city they serve, the time of the day and week, congestion conditions, accidents etc. Coordination among the various sections and with the arteries of the city can

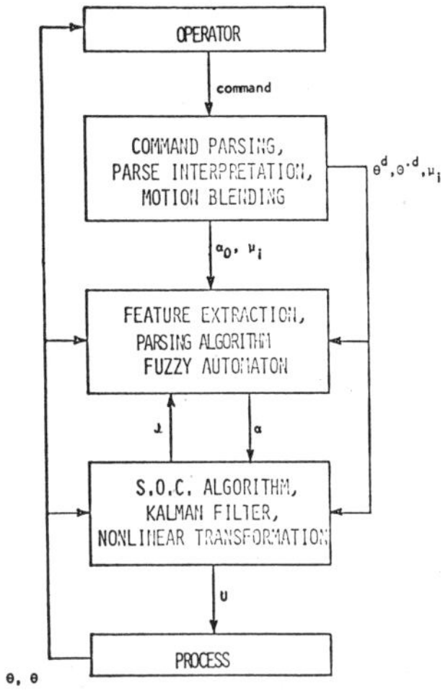


Figure 1. General block diagram of the hierarchical control system

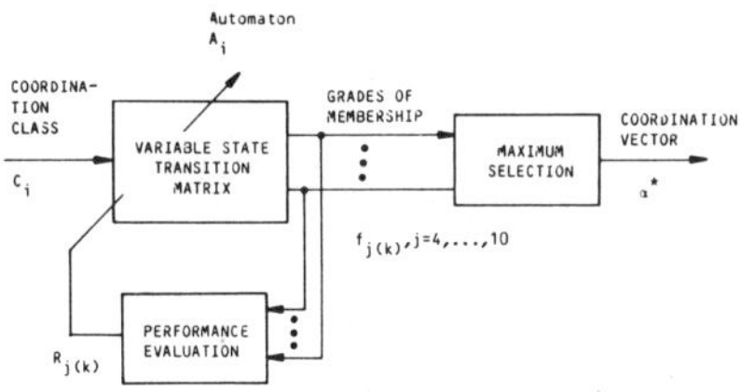


Figure 2. Structure of the Learning Automaton

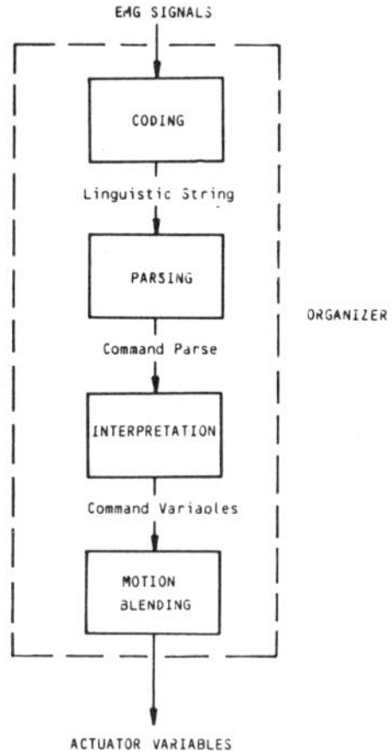


Figure 3. Block Diagram of the Organization Level

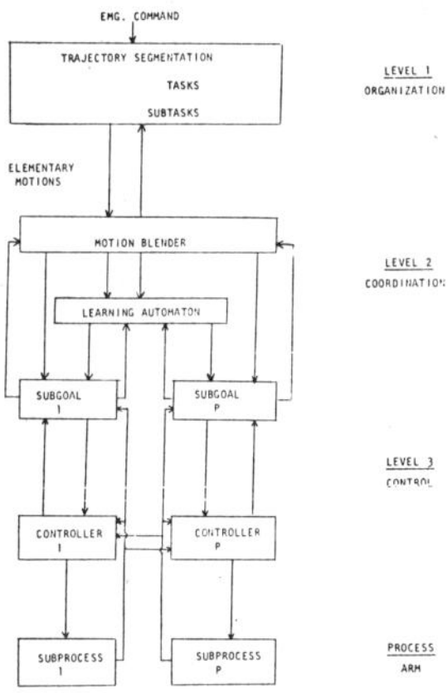


Figure 4. Hierarchically Intelligent Control of a Prosthetic Arm

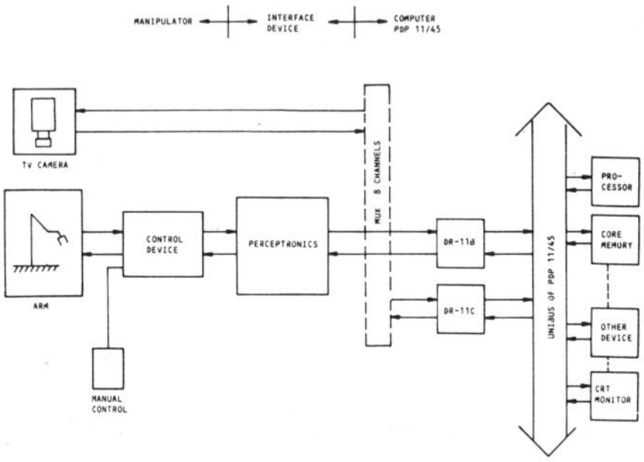


Figure 5. Purdue's Manipulator System Configuration

be obtained through computers composed of low cost microprocessors locally installed.

Other applications of the above theory of intelligent control was found in human Prostheses (Figure 4) and General Purpose Manipulators (Figure 5), [11,12].

## 7. CONCLUSION

The proposed algorithm is based on a hierarchical decomposition of the decision-making process into three levels associated with organization, coordination, and direct control, respectively. It is directly applicable to manage and control vehicular traffic flows in urban environments and their interconnecting freeways.

The top level is a man-machine interface. Its task oriented activity consists in the translation of the operator's linguistic commands into a "machine-language" information pattern that determines the operation of the lower two levels by direct intervention in their performance indices. The processing of detailed, precise information for the execution of the task is accomplished by the third level, which applies direct control inputs to the process actuators. The control process is decomposed into several subprocesses, and each is optimized with respect to a local criterion. The second level adjusts the variable parameters in the local criteria so that an overall criterion is minimized. The suggested hierarchy is characterized by an increasing order of intelligence and decreasing order of precision as one goes from the lower to the upper level.

Current research in the Advanced Automation Research Laboratory at Purdue is directed towards the realization of appropriate algorithms especially adapted to the traffic management and control problem and assess their applicability to real life situations.

ACKNOWLEDGEMENT

This research was supported by a Grant of the Department of Transportation RC92001.

## REFERENCES

[1] Athans M. Levine W., "Optimal Vehicle Control for the Merging Problem", MIT Report ESL-R-327, Cambridge, Mass., Nov. 1967.

[2] Baras J., Levine W., "Estimation of Traffic Flow Parameters in Urban Traffic Networks", Proceeding of 1977 JACC, San Francisco, CA, June 1977, pp. 428-435.

[3] Chong C. Y., "On the Decentralized Control of Large Scale Systems", MIT Report ESL-R-503, Cambridge, Mass., June 1973.

[4] Gazis D. C., Ed., Traffic Science, John Wiley and Sons, New York, 1974.

[5] Ho Y. C., Chu K. C., "Team Decision Theory and Information Structures in Optimal Control Problem - Part I", IEEE Transaction on AC., Vol. AC-17, No. 1, Feb. 1972.

[6] Houpt P. K., "Decentralized Stochastic Control of Finite-State Systems with Application to Vehicular Traffic Control", MIT Report ESL-R-579, Cambridge, Mass., Nov. 1974.

[7] IEEE Transactions on Automatic Control, "Special Issue on Large Scale Systems and Decentralized Control", Vol AC-23, No. 2, April 1978.

[8] Mesarovic M. D., Macko, Takahara Y., Theory of Hierarchical Multilevel Systems, Academic Pren, New York, 1970.

[9] Proceedings of 1977 Joint Automatic Control Conference, "Theme on Automatic Control in Transportation Vehicles", San Francisco, CA, June 1977, pp. 332-500.

[10] Proceedings of 1978 Conference on Decision and Control, "Sessions WA1, TA3, TP3, TP4, FA1, FP1, San Diego, CA, Jan. 1979.

[11] Saridis, G. N., Stephanou, H. E., "A Hierarchical Approach to the Control of a Prosthetic Arm", IEEE Transactions on SMC, Vol. SMC-7, No. 6, June 1977, pp. 407-420.

[12] Stephanou, H. E., Saridis, G. N., "A Hierarchically Intelligent Method for the Control of Complex Systems", Journal of Cybernetics, Vol. 6, 1976, pp. 249-261.

[13] Transportation Research Board, "Summary of Progress through 1977 of Highway Cooperative Highway Research Program", National Research Council 1977.

[14] U. S. Department of Transportation, FHA, "1977 Federally Coordinated Program of Highway Research and Development", DOT, FHA, 1977.

[15] U. S. Department of Transportation FHA, "Traffic Control Systems Handbook", DOT, FHA, Feb. 1976.

[16] U. S. Department of Transportation, FHA and UMTA, "The Urban Traffic Control System", DOT, FHA, UMTA, Sept. 1974.

[17] Weintranl, A. F., "The Assignment and the Shortest Route Problems", Tech. Report ORC 72-20 Operation Research Center, Univ. of CA, Berkeley, Berkeley, CA, Aug. 1972.

# MUSMAR: MULTIVARIABLE ADAPTIVE REGULATORS BASED ON MULTISTEP COST FUNCTIONALS (1)

Giuseppe Menga* and Edoardo Mosca**

* CENS and Istituto di Elettronica Generale, Politecnico di Torino, Corso Duca degli Abruzzi, 24 10129 TORINO, Italy

** SAIS, Istituto di Elettronica, Facoltà di Ingegneria, Università di Firenze, Via S. Marta, 3 50139 FIRENZE, Italy

Abstract. This paper introduces an extension of self-tuning regulators (STURE) to unknown but time-invariant plants with an arbitrary number of inputs and outputs, to be regulated according to a quadratic-cost functional of input-output samples over a control horizon of arbitrary length. The resulting multistep-multivariable adaptive regulator (MUSMAR) is singled out, by enforcing the certainty equivalence (CE) principle, as an approximation to the classic linear-quadratic-gaussian (LQG) regulator by exploiting a reinterpretation of the LQG regulation problem.

## 1. A REINTERPRETATION OF THE LQG PROBLEM

Let us consider a plant with inputs $u(t)\varepsilon R^m$ and outputs $y(t)\varepsilon R^p$ whose dynamic behavior can be described by a discrete-time Gauss-Markov stochastic system in its innovations-representation form (Åström, 1970)

$$\begin{aligned} x(k+1) &= Ax(k) + Bu(k) + Ke(k) \\ y(k) &= Cx(k) + e(k) \end{aligned} \tag{1}$$

where: $x(k)\varepsilon R^n$; $e(k)$ is the innovations process of $y(k)$; A,B,C,K are known matrices of consistent dimensions. The LQG problem considered hereafter is to find the input sequence $u(t), u(t+1),\dots, u(t+T)$ to the system (1) minimizing the quadratic-cost functional

$$E[J(t,t+T)] = E\{\sum_{k=t}^{t+T}[\, \| y(k+1)\|^2_{Q_y} + \| u(k) \|^2_{Q_u} ]\} \tag{2}$$



D. G. Lainiotis and N. S. Tzannes (eds.), Advances in Control, 334–341.

where: E denotes expectation; $Q_y$ and $Q_u$ are symmetric non-negative definite matrices; $\| y \|^2_{Q_y} = y'Q_y y$; $\| u \|^2_{Q_u} = u'Q_u u$. According to the admissible control strategy, u(k) must be a non-anticipative linear function of input-output pairs, viz., $u(k)\varepsilon[I^k]$, where $I^k=\{y(k), u(k-1), y(k-1), u(k-2),\ldots\}$, and $[I^k]$ denotes the linear subspace of random vectors generated by $I^k$. It is known that the above control strategy amounts to requiring that u(k) be given by a linear state-feedback law

$$u(k) = F_k s(k) \tag{3}$$

where

$$\begin{aligned} s(k) &= E[x(k+1)\mid I^k, u(k)=0] \\ &= Ax(k) + Ke(k). \end{aligned} \tag{4}$$

We want to show that the choice of the first vector u(t) in the control sequence solution of the LQG problem (1)-(4) can equivalently be made by minimizing the quadratic functional (14) of cross-correlations between the future of the input-output joint process and the present state s(t). To this end, let us introduce the following conditional means

$$\begin{aligned} \hat{x}(k) &= E[x(k)\mid s(t)] & \tilde{x}(k) &= x(k) - \hat{x}(k) \\ \hat{u}(k) &= E[u(k)\mid s(t)] & \tilde{u}(k) &= u(k) - \hat{u}(k) \\ \hat{y}(k) &= E[y(k)\mid s(t)] & \tilde{y}(k) &= y(k) - \hat{y}(k) \end{aligned} \tag{5}$$

Since $E[e(k)s'(t)]=0$ for all $k\geq t+1$, and by (3) $\hat{u}(t)=u(t)$, we get

$$\begin{aligned} \hat{x}(k+1) &= \begin{cases} s(t) + Bu(t), & k=t \\ A\hat{x}(k) + B\hat{u}(k), & k\geq t+1 \end{cases} \\ \hat{y}(k+1) &= C\hat{x}(k+1), \quad k\geq t. \end{aligned} \tag{6}$$

Further,

$$\begin{aligned} \tilde{x}(k+1) &= \begin{cases} 0 & , k=t \\ A\tilde{x}(k) + B\tilde{u}(k) + Ke(k), & k\geq t+1 \end{cases} \\ \tilde{y}(k+1) &= C\tilde{x}(k+1) + e(k+1), \quad k\geq t \end{aligned} \tag{7}$$

Moreover, since $E[\hat{y}(k+1)\tilde{y}'(k+1)]=0$ and $E[\hat{u}(k)\tilde{u}'(k)]=0$, the cost functional (2) can be rewritten as follows

$$E[J(t,t+T)] = E[\hat{J}(t,t+T)] + E[\tilde{J}(t,t+T)] \tag{8}$$

where:

$$E[\hat{J}(t,t+T)] = E\{\sum_{k=t}^{t+T} [\ \|\hat{y}(k+1)\|^2_{Q_y} + \|\ \hat{u}(k)\ \|^2_{Q_u}]\} \ ; \tag{9}$$

$$E[\widetilde{J}(t,t+1)] = E\{\sum_{k=t}^{t+T} [\ \|\ \widetilde{y}(k+1)\|^2_{Q_y} + \|\ \widetilde{u}(k)\ \|^2_{Q_u}]\} \ . \tag{10}$$

Eqs. (6)-(10) show that the LQG problem (1)-(4) splits into two separate problems: problem (6), (9), which is purely deterministic once s(t) is given; and problem (7), (10), of purely stochastic nature. In particular, the latter is not affected by the first vector u(t) in the control sequence solution of the LQG problem (1)-(4). Consequently, in order to find the optimal u(t) it suffice to limit our consideration to problem (6)-(9).
Let us define the following crosscorrelations

$$\begin{aligned} R_{\hat{u}s}(k,t) &= R_{us}(k,t) = E[u(k)s'(t)] \\ R_{\hat{x}s}(k,t) &= R_{xs}(k,t) = E[x(k)s'(t)] \\ R_{\hat{y}s}(k,t) &= R_{ys}(k,t) = E[y(k)s'(t)] \end{aligned} \tag{11}$$

If S(t) denotes the covariance matrix of s(t) that hereafter is assumed to be non-singular, we have

$$\begin{aligned} \hat{u}(k) &= R_{us}(k,t)S^{-1}(t)s(t) \\ \hat{x}(k) &= R_{xs}(k,t)S^{-1}(t)s(t) \\ \hat{y}(k) &= R_{ys}(k,t)S^{-1}(t)s(t) \end{aligned} \tag{12}$$

Further, postmultiplying each vector in (6) by s'(t) and taking expectation, we get

$$\begin{aligned} R_{xs}(k+1,t) &= \begin{cases} S(t) + BR_{us}(t,t), & k=t \\ AR_{xs}(k,t) + BR_{us}(k,t), & k\geq t+1 \end{cases} \\ R_{ys}(k+1,t) &= CR_{xs}(k+1,t),\ k\geq t. \end{aligned} \tag{13}$$

Consider now each term in the cost functional (9). We find

$$\begin{aligned} E[\|\hat{y}(k+1)\ \|^2_{Q_y}] &= E[\|R_{ys}(k+1,t)S^{-1}(t)s(t)\ \|^2_{Q_y}] \\ &= \|\ R_{ys}(k+1,t)P(t)\ \|^2_{Q_y} \end{aligned}$$

where $S^{-1}(t)=P(t)\,P'(t)$, and $\| L \|_Q^2$ denotes the Frobenious norm (Lawson and Hanson, 1974) of L, for any matrix L and $Q=Q'\geq 0$ of consistent dimensions

$$\| L \|_Q^2 = \mathrm{Tr}(L'QL).$$

Similarly,

$$E[\| \hat{u}(k) \|]_{Q_u}^2 = \| R_{us}(k,t)P(t) \|_{Q_u}^2 .$$

Therefore, (9) can be written as follows

$$E[\hat{J}(t,t+T)] = \sum_{k=t}^{t+T} [\| R_{ys}(k+1,t)P(t) \|_{Q_y}^2 + \| R_{us}(k,t)P(t) \|_{Q_u}^2]. \tag{14}$$

The above results are summed up in the following

Theorem 1 - The first vector u(t) in the optimal control sequence u(t),u(t+1),...,u(t+T) solution of the LQG control problem (1)-(4) is given by

$$u(t) = R_{us}(t,t)S^{-1}(t)s(t) \tag{15}$$

where s(t) is the system state (4) at time t, $S(t)=E[s(t)s'(t)]$, and $R_{us}(t,t)$ minimizes (14) and satisfies the constraints (13).

In other terms, the LQG optimal control action must be chosen so as to minimize the effects due to the present state s(t) over the future of the input-output joint process as measured by the quadratic functional (14) of output-state and input-state cross-correlations.

## 2. A RECURSIVE ALGORITHM TO COMPUTE THE LQG OPTIMAL FEEDBACK-GAIN MATRIX

It is assumed hereafter that the system (1) together with the experimental conditions be such that

$$R_{ys}(t+i+1,t) = R_{ys}(i+1)$$
$$R_{us}(t+i,t) = R_{us}(i) \tag{16}$$

E.g., this assumption holds true if: the matrices A, B, C, K in (1) are constant; the innovations process e(k) is stationary; and

$$u(k) = Fs(k) \tag{17}$$

with F a constant feedback-gain matrix. Under assumptions(16), (14) becomes

$$E[\hat{J}(t,t+T)] = \sum_{i=0}^{T} [\| R_{ys}(i+1)P \|^2_{Q_y} + \| R_{us}(i)P \|^2_{Q_u}] \tag{18}$$

where we set P(t)=P.
Our next goal is to find a convenient estimate of (18). This estimate must be in a form that lends itself to some suitable minimization with respect to the feedback-gain matrix F. To this end, let us first assume that the system is fed by a sequence of inputs $u(k)=Fs(k)$, $k=\ldots,t,t+1,\ldots,t+T$. Consider the corresponding sequence of the input-output joint process, and let the two matrices $\theta(F)$ and $\psi(F)$ in (19) be computed according to the model (6) and (17)

$$[y'(t+T+1)u'(t+T)\ldots y'(t+2)u'(t+1) \mid y'(t+1)u'(t)] = [u'(t) \mid s'(t)] \begin{bmatrix} \bar{\theta}(F) & \theta \\ \bar{\psi}(F) & \psi \end{bmatrix} + v(t) \tag{19}$$

where

$$\theta(F) = [\bar{\theta}(F) \mid \theta] \qquad \psi(F) = [\bar{\psi}(F) \mid \psi]$$

and $E[v(t)s'(t)]=0$. Notice that $\theta(F)$ and $\psi(F)$ are made up of blocks depending on F, viz. $\bar{\theta}(F)$ and $\bar{\psi}(F)$, and blocks not depending on F, viz. $\theta$ and $\psi$. Denoting the L.H.S. of (19) by $z'(t,t+T)$, premultiplying $z'(t,t+T)$ by $s(t)$ and taking expectation, we get

$$E[s(t)z'(t,t+T)] = [R'_{ys}(T+1)\ R'_{us}(T)\ \ldots\ R'_{ys}(1)\ R'_{us}(0)]\,.$$

Therefore, taking into account (18),

$$E[\hat{J}(t,t+T)] = \| E[z(t,t+T)s'(t)P] \|^2_{Q_{T+1}} \tag{$\overline{19}$}$$

where

$$Q_1 = Q = \begin{bmatrix} Q_y & 0 \\ 0 & Q_u \end{bmatrix}$$

and

$$Q_k = \text{Block-diag}\{Q,Q,\ldots,Q\} \text{ (k-times)}.$$

In order to underline the dependence of ($\overline{19}$) by the feedback-gain matrix F, we write
$E[\hat{J}(t,t+T;F)]$ instead of simply $E[\hat{J}(t,t+T)]$.
Thus,

$$E[\hat{J}(t,t+T;F)] = E[\hat{J}(t+1,t+T;F)] + E[\hat{J}(t,t)] \tag{20}$$

where

$$E[\hat{J}(t+1,t+T;F)] = \| E[z(t+1,t+T)s'(t)]P \|^2_{Q_T}$$

$$= \| \{\bar{\theta}'(F)E[u(t)s'(t)] + \bar{\psi}'(F)S\}P \|^2_{Q_T}$$

where $S = E[s(t)s'(t)]$ and

$$E[\hat{J}(t,t)] = \|\{\theta' E[u(t)s'(t)] + \psi' S\}P\|_Q^2 \quad .$$

In order to minimize (20) with respect to u(t), we take into account that, according to the admissible control strategy, u(t) = Ls(t). Therefore, we get

$$\min_L \{E[\hat{J}(t+1,\ t+T;F)] + E[\hat{J}(t,t)]\} =$$
$$= \min_L \{\|[\theta'(F)L + \psi'(F)]SP\|^2_{Q_{T+1}}\}.$$

It can be shown that, as long as SP is non-singular, the matrix L minimizing the above expression coincides with the matrix solution of the following simplified problem

$$\min_L \|\theta'(F)L + \psi'(F)\|^2_{Q_{T+1}}. \tag{21}$$

If, in view of an iterative minimization procedure based on (21), F is replaced by $F_k$, the minimum is attained at $L = F_{k+1}$, where

$$F_{k+1} = -[\theta(F_k)Q_{T+1}\theta'(F_k)]^{-1}[\theta(F_k)Q_{T+1}\psi'(F_k)] \ . \tag{22}$$

By iterating (2), a sequence of feedback-gain matrices $\{F_k\}$ is obtained. A natural question at this point is:
Assuming that the matrix sequences $\{\theta(F_k)\}$ and $\{\psi(F_k)\}$ are known, what are the properties of $\{F_k\}$ relatively to the solution of the initial LQG problem (1)-(4)? An answer is given by the following:
Theorem 2 - Let the matrix sequences $\{\theta(F_k)\}$ and $\{\psi(F_k)\}$ in (19) be given. Thus, the feedback-gain matrix sequence $\{F_k\}$ iteratively generated by (22) approximates for $T\to\infty$ the feedback-gain matrix sequence $H_k$ obtained by Kleinmann's algorithm (Kleinmann, 1968; Kwakernaak and Sivan, 1972) for the LQG problem (1)-(4) over an infinite horizon ($T=\infty$).
Further indication on the relationship between the LS problem (21) and the initial LQG problem (1)-(4) is given for the opposite situation (shortest control interval) to that of Theorem 1 (infinite control horizon), by the following:
Theorem 3 - Assume that T=0. Thus, the matrix $L=\hat{F}$ minimizing $E[\hat{J}(t,t)]$ for u(t)=Ls(t) is the solution of the LS problem

$$\min_L \|\theta' L + \psi'\|_Q^2 \qquad \text{i.e.,}\ \hat{F} = -(\theta Q\theta')^{-1}\theta Q\psi' \ .$$

3. MUSMAR

It is assumed hereafter that the dynamic stochastic model (1) of the plant to be regulated is unknown. Thus, in order to be able to use the results of Sect. 3, a suitable estimate of the matrices $\theta(F)$ and $\psi(F)$ of (19) must be obtained from the plant operating records. To begin with, we replace the unknown plant state s(t) of (4) by the following vector of past input-output samples

$$s(t) = [y'(t)u'(t-1)\dots y'(t-\hat{n}+1)u'(t-\hat{n})] \quad (23)$$

Further, let

$$Z(t,t+T) = [z(t,t+T)z(t-1,t+T-1)\dots z(t-N+1,t+T-N+1)]$$

where

$$z(k,k+T) = [y'(k+T+1)u'(k+T)y'(k+T-1)\dots y'(k+1)u'(k)]'$$

Newt define

$$\Sigma(t) = [s(t)s(t-1)\dots s(t-N+1)]$$
$$U(t) = [u(t)u(t-1) \quad u(t-N+1)] .$$

Consider the LS estimates $\hat{\theta}(F)$ and $\hat{\psi}(F)$ of $\theta(F)$ and $\psi(F)$ specified by

$$Z'(t,t+T) \sim [U'(t) \mid \Sigma'(t)] \begin{bmatrix} \hat{\theta}(F) \\ \hline \hat{\psi}(F) \end{bmatrix} \quad (24)$$

where $Y \sim A\hat{X}$ is a shorthand notation to indicate that $\hat{X}$ minimizes $\| Y-A\hat{X} \|$ among all admissible X. Eq. (24) corresponds to N successive equations similar to (19). The input-output samples contained in Z(t,t+T), U(t), and $\Sigma(t)$ are obtained by feeding the plant input by u(k)=Fs(k)+n(k), k=t-N+1-$\hat{n}$,..., t+T, where n(k) a low-intensity zero-mean white noise independent of the plant, superimposed to the feedback-control component so as to make $[U'(t) \mid \Sigma'(t)]$ a full column-rank matrix. $\hat{\theta}(F)$ and $\hat{\psi}(F)$ can be obtained in principle as follows:

$$\begin{bmatrix} \hat{\theta}(F) \\ \hat{\psi}(F) \end{bmatrix} = - \begin{bmatrix} UU' & U\Sigma' \\ \Sigma U' & \Sigma\Sigma' \end{bmatrix}^{-1} \begin{bmatrix} UZ' \\ \Sigma Z' \end{bmatrix} \quad (25)$$

where, by the sake of simplicity, the arguments of U, $\Sigma$ and Z were dropped.

The MUSMAR based on the results of Sect. 2 operates according to the following procedure (or one of its many possible variants):

1) Feed the plant by the inputs $u(t)=\hat{F}_k s(t)+n(t)$ over T+N+$\hat{n}$ successive steps;

2) Compute the LS estimates $\hat{\theta}(\hat{F}_k)$ and $\hat{\psi}(\hat{F}_k)$ according to (24);

3) Compute the solution $\hat{F}_{k+1}$ of the LS problem

$$\min_{L} \| \hat{\theta}'(\hat{F}_k)L + \hat{\varphi}'(\hat{F}_k) \|^2_{Q_{T+1}} \quad ;$$

4) Feed the plant by the inputs $u(t) = \hat{F}_{k+1}s(t)+n(t)$ over the next $T+N+\hat{n}$ successive steps, and repeat the cycle.

(1) This work was supported in part by the Italian CNR under contracts 78.02422.07 and 78.01638.07.

REFERENCES

Åström, K.J. (1970). "Introduction to Stochastic Control Theory". Academic Press, New York.

Åström, K.J., U. Borisson, L. Ljung, and B. Wittenmark(1977). "Theory and Applications of Self-Tuning Regulators". Automatica, 13, pp. 457-476.

Åström, K.J., and B. Wittenmark(1973). "On Self-Tuning Regulators". Automatica, 9, pp. 185-199.

Bierman, G.J.(1977). "Factorization Methods for Discrete Sequential Estimation". Academic Press, New York.

Borisson, U.( 1976). "Self-Tuning Regulators for a Class of Multivariable Systems". Preprints, 4th IFAC Symp. Identification and System Parameter Estimation, Tbilisi, pp. 420-429.

Clarke, D.W., and P.J. Gawthrop(1975). "Self-Tuning Controller". Proc. IEE, 122, pp. 929-934.

Gawthrop, P.J.(1977). "Some Interpretation of Self-Tuning Controller". Proc. IEE, 124, pp. 889-894.

Keviczky, L., J. Hetthéssy, M. Hilger, and J. Kolostori(1978). "Self-Tuning Adaptive Control of Cement Raw Material Blending". Automatica, 14, pp. 525-532.

Kleinmann, D.L.( 1968). "On an Iterative Technique for Riccati Equation Computation". IEEE Trans. Autom. Control, 13, pp. 114-115.

Kwakernaak, H., and R. Sivan (1972). "Linear Optimal Control Systems". Wiley, New York.

Lawson, C.L., and R.J. Hanson (1974). "Solving Least Squares Problems". Prentice-Hall, Englewood Cliffs, N.J.

Peterka, V., and K.J. Åström ( 1973). "Control of Multivariable Systems with Unknown but Constant Parameters". Preprints, 3rd IFAC Symp. Identification and System Parameter Estimation, The Hague.

Wittenmark, B. (1975). "Stochastic Adaptive Control Methods : A Survey". Int. J. Control, 21, pp. 705-729.

ON TRAJECTORY SENSITIVITY MINIMIZATION WITH AN ADAPTIVE CONTROLLER

M. Eslami and R. S. Marleau

In this paper an algorithm is presented for establishing the optimal controller gain $K = K(p)$, which minimizes a combined-quadratic cost and is adaptively sensitive to system parameters by feeding back only the system state. This algorithm subsumes all the existing procedures for deriving the optimal controller gain in similar problems.

## 1. INTRODUCTION

The inherent uncertainity which exists in any system identification leads to discrepancies between the nominal and the actual values of the system physical parameters, hence the system trajectory. Additional discrepancies may be caused by further parameter changes stimulated by environmental changes, aging, etc. In spite of all these parameter changes modern technology demands system designs with more reliable and self-correcting operations than ever before. To respond to such demands, sensitivity analysis and synthesis of systems has been advanced and is commonly employed. A variety of measures of system sensitivity are in common use toda This paper details and expands on the one measure termed the "system - trajectory sensitivity function," and its minimization.

Perhaps the main shortcomings of some of the earlier works in this area are: (i) the erroneous assumption of $\partial^2 x/\partial p_j^2=0$ (where $p_j$ $j=1,2,\ldots,r$, are the system physical parameters and $x(t)$ is the system state vector), [15]; (ii) the direct attempt of applying th results of linear optimal control theory (regulator problem) to minimization of an integral of a quadratic sensitivity function, [13]; (iii) the lack of establishment of the requirements for the convergence of the proposed algorithms, [8] and [9]; (iv) the computational burden of solving the corresponding Riccati equation (v) the nonexactness of the entire computational process due to th inclusion of various above mentioned approximations; and (vi) inclusion of all the above approximations and resultant computationa inaccuracies leading to additional design difficulties in the already expensive process of the reconstructions of the system traje tory sensitivity function for the proposed closed-loop control pol icy. Effort has been extended to derive an optimal controller gai which depends only on the system state vector.

## 2. PROBLEM STATEMENT AND PREVIEW OF RESULTS

Consider the completely reconstructible linear system describ by $\qquad \dot{x}(t;p) = A(p)x(t;p) + B(p)u(t;p), x(0;p) = x_o \neq 0, \qquad (1)$



D. G. Lainiotis and N. S. Tzannes (eds.), Advances in Control, 342–350.

where $x(t;p) \in R^n$ is the system state vector; $u(t;p) \in R^m$ is the system input vector; $p \in R^r$ is the system-physical parameter vector; and A and B are the real coefficient matrices of appropriate dimensions. The plant-quadratic cost is denoted as follows.

$$J_p = \int_0^\infty \{x^T(t)W_0x(t) + u^T(t)Wu(t)\}dt; \tag{2}$$

Where $W_0=W_0^T \geq 0$, and $W=W^T>0$. Let the $J_{sr}^a$ be the adaptive system sensitivity cost as follows.

$$J_{sr}^a = \int_0^\infty \sum_{j=1}^r (\frac{\partial x}{\partial p_j})^T W_j (\frac{\partial x}{\partial p_j})dt, \tag{3}$$

where each $W_j=W_j^T>0$, $j=1,2,\ldots,r$ (or $W_j=0$, for some j), and the $\partial x/\partial pj$, $j=1,2,\ldots,r$, are the system-trajectory sensitivity functions. Then the objective is to minimize the following combined-quadratic cost $J_r'^a = J_p + \alpha J_{sr}^a$, (4)
where $\alpha > 0$ is a constant, useful in computational algorithm. The main result of this paper is to establish a numerical algorithm to determine this optimal controller gain. Before advancing a solution to this optimization problem, however, it is appropriate to discuss further this problem and the corresponding extensions to a plant with disturbances, substituting $u(t;p)= -K(p)\,x(t;p)$ into (1), results in $\dot{x}(t;p) = (A(p) - B(p)K(p))x(t;p)$. (5)
Equation (5) is used to generate all the system-trajectory sensitivity functions; i.e., $\frac{\partial x}{\partial p_j}$, $j=1,2,\ldots,r$. Wherein

$$\frac{\partial \dot{x}(t;p)}{\partial p_j} = (A(p) - B(p)K(p))\frac{\partial x(t;p)}{\partial p_j} + \frac{\partial (A(p)-B(p)K(p))}{\partial p_j}x(t;p). \tag{6}$$

Augmenting these differential equations results in the following "system model", $\dot{m}_r^a(t;p) = A_r^a(p)m_r^a(t;p)$. (7)

Where $m_r^a(t;p) \triangleq [x^T(t;p), \frac{\partial x^T}{\partial p_1}, \ldots, \frac{\partial x^T}{\partial p_r}]^T$, with $m_r^a(0;p) \triangleq [x_o^T, 0^T, \ldots, 0^T]^T$, and

$$A_r^a(p) \triangleq \begin{pmatrix} A-BK & 0 & \cdots & \cdots & 0 \\ \frac{\partial}{\partial p_1}(A-BK) & A-BK & \ddots & & \vdots \\ \vdots & 0 & \ddots & \ddots & 0 \\ \vdots & \vdots & \ddots & \ddots & \\ \frac{\partial}{\partial p_r}(A-BK) & 0 & \cdots & 0 & A-BK \end{pmatrix}. \tag{8}$$

Here

$$\frac{\partial}{\partial p_j}(A-BK) = \frac{\partial A}{\partial p_j} - \frac{\partial B}{\partial p_j}K - B\frac{\partial K}{\partial p_j}, \quad j=1,2,\ldots,r. \tag{9}$$

Considering the system model (7), $J_r'^a$, given in (4), becomes

$$J_r'^a = \min \int_0^\infty \{x^T(t)(W_0 + K^TWK)x(t) + \alpha \sum_{j=1}^r (\frac{\partial x}{\partial p_j})^T W_j (\frac{\partial x}{\partial p_j})\}dt,$$

$$\triangleq \min Tr\{m_r^a(0;p)^T P_r^a m_r^a(0;p)\}. \tag{10}$$

Here $P_r^a = P_r^{aT} > 0$ is the unique solution of following Lapunov equation (provided that $A_r^a$ is a stability matrix and indeed this is required hereafter),

$$A_r^{aT}P_r^a + P_r^aA_r^a + \tilde{W}_r = 0, \tag{11}$$

$$\tilde{W}_r \triangleq \mathrm{diag}(W_0 + K^TWK, \alpha W_1, \ldots, \alpha W_r). \tag{12}$$

In order that the performance index does not depend on the initial condition, which generally is not available, it is a common practice to consider a statistical average of this performance index [7]; [11]. Thus the new performance index, with $M_r \triangleq E(m_r^a(0;p)m_r^a(0;p)^T)$, is

$$J_r^a = \min \mathrm{Tr}(P_r^a M_r), \tag{13}$$

There exists an analogy between the above optimization problem and the results given by [1], and [12]. This analogy is cited in the following, but first their results are reviewed.

Suppose the system given by (1) is corrupted by white noise $\omega_1(t)$, i.e., $\dot{x}(t;p) = A(p)x(t;p) + B(p)u(t;p) + \omega_1(t)$, (14)
and the system observation $y(t;p) \in R^o$ is available as

$$y(t;p) = Cx(t;p) + \omega_2(t), \tag{15}$$

wherein $\omega_2(t)$ is an observation noise and

$E\{\omega_1(t)\omega_1^T(\tau)\} = V_1\delta(t-\tau)$; (16): $E\{\omega_2(t)\omega_2^T(\tau)\} = V_2\delta(t-\tau)$; (17) and $E\{\omega_1(t)\omega_2^T(\tau)\} = 0$. (18)

They assume there exists a linear dynamic compensator $z(t) \in R^c$ such that $\dot{z}(t;p) = Pz(t;p) + Ny(t;p)$, $z(0) \equiv 0$, (19)
then the optimal input

$$u(t;p) = Hz(t;p) + Gy(t;p) \tag{20}$$

is chosen such that the expectation of a combined-quadratic cost similar to that of (4) is minimized. Here they also assumed that $\dim(z) < \dim(x)$. Thus the main application of the optimization problem given in [1] and [12], is to design a limited state variable feedback controller for linear time-invariant stochastic systems in which the dynamic compensator is of dimension much less than the dimension of system state.

In the trajectory sensitivity minimization of this paper the differential equation parallel to that of the dynamic compensator (19) is the trajectory sensitivity functions given by (6). The resultant augmented set of trajectory sensitivity differential equations is of dimension never less than that of the system state vector. Additionally, the direct feedback of these trajectory sensitivity functions is always prohibited due to their high-dimension and impracticality or high cost of on-line reconstruction. In fact, a major result of the present paper is the avoidance of this direct feedback. Thus the main difference between the objective of the optimization problem chosen in this paper and that chosen in [1] and [12], is the choice of the optimal input, the former of which does not contain trajectory sensitivity function (i.e., parallel of the dynamic compensator), while the latter has a component of the state of the dynamic compensator.

Our results can be extended to stochastic system by use of the relevant material of [10].

## 3. DERIVATION OF THE OPTIMAL ADAPTIVE CONTROLLER GAIN

The optimization problem set forth herein is minimization of (13) subject to (7). This minimization problem as it stands can be solved by several standard techniques, one of which Lagrange multipliers. In this paper it is first assumed that the solution exists, then necessarily this solution has to satisfy $\partial J_r^a/\partial K=0$.

This solution may be, of course, a local minimum. Thus the uniqueness of the solution, if it exists, is an open question. Also by inspection of (13) and (10) it is clear that $K = \infty$ is not a solution since as $K \to \infty$, $J \to 0$, while by construction of this J it is clear that $J > 0$, i.e., contradiction. Therefore, as the first step for derivation of the optimal adaptive controller gain, an attempt has to be made in the following to derive $\partial J_r^a/\partial K$ directly.

To derive $\partial J_r^a/\partial K$, the following approach is advanced. If $K \to K + \delta K$, with $\delta K$ being "small", then $\mathcal{A}_r^a \to \mathcal{A}_r^a + \delta \mathcal{A}_r^a$, where $\delta\mathcal{A}_r^a$ is "small" and $\delta\mathcal{A}_r^a \to 0$ as $\delta K \to 0$, similarly $\tilde{W}_r \to \tilde{W}_r + \delta\tilde{W}_r$, $\mathcal{P}_r^a \to \mathcal{P}_r^a + \delta\mathcal{P}_r^a$ such that $\mathcal{P}_r^a + \delta\mathcal{P}_r^a$ remains positive-definite and

$$\delta P_r^a = \int_0^\infty \exp(A_r^a t)^T (\delta A_r^{aT} P_r^a + P_r^a \delta A_r^a + \delta\tilde{W}_r)\exp(A_r^a t)\,dt. \quad (21)$$

Similarly, $J_r^a \to J_r^a + \delta J_r^a$, where $\delta J_r^a = \mathrm{Tr}(\delta\mathcal{P}_r^a M_r)$, or

$$\delta J_r^a = \mathrm{Tr}(\Pi_r^a E_r^a), \quad (22)$$

where

$$\Pi_r^a \triangleq \delta A_r^{aT} P_r^a + P_r^a \delta A_r^a + \delta\tilde{W}_r \in R^{n(1+r)\times n(1+r)}, \quad (23)$$

and

$$E_r^a \triangleq \int_0^\infty \exp(A_r^a t) M_r \exp(A_r^a t)^T dt \in R^{n(1+r)\times n(1+r)}. \quad (24)$$

From (8) and (12), $\delta A_r^a$ and $\delta\tilde{W}_r$ are:

$$\delta A_r^a \triangleq \begin{pmatrix} -B\delta K & 0 & \cdots & \cdots & 0 \\ -\frac{\partial}{\partial p_1}(B\delta K) & -B\delta K & \ddots & & \vdots \\ \vdots & 0 & \ddots & \ddots & 0 \\ -\frac{\partial}{\partial p_r}(B\delta K) & 0 & \cdots & 0 & -B\delta K \end{pmatrix}, \quad (25)$$

and

$$\delta\tilde{W}_r \triangleq \left(\begin{array}{c|c} \delta K^T W K + K^T W \delta K & O \\ \hline O & O \end{array}\right), \quad (26)$$

since $\partial(\delta K)/\partial p_j = 0$, from the assumtpion of $\delta K$ being "small" in the above. Substituting (25) and (26) into (23), and partitioning $\mathcal{P}_r^a$ accordingly, yields

$$\pi_{ij}^a \triangleq \delta K^T \phi_{ij}^{aT} + \psi_{ij}^a \delta K; \quad (27)$$

where $\pi_{ij}^a = \pi_{ji}^{aT}$ are the nxn block matrix partitions of $\Pi_r^a$; and $\phi_{ij}^a$, $\psi_{ij}^a$ are explicit functions of $K, B$, $\partial B/\partial p_j$ $(j=1,\ldots r)$, W and $\mathcal{P}_{ij}^a (i,j=1,\ldots,(1+r))$, and $\phi_{ij}^a = \psi_{ji}^a$. In particular, it can be shown that

$$\psi_{11}^{aT} = \phi_{11}^{aT} = WK - B^T P_{11}^a - \sum_{j=1}^{r} \frac{\partial B^T}{\partial p_j} P_{1(1+j)}^{aT}. \quad (28)$$

If $\mathcal{E}_r^a$ is also partitioned similarly, then it can be shown that (22) is as follows.

$$\delta J_r^a = \mathrm{Tr} \sum_{i,j=1}^{1+r} \pi_{ij}^a E_{ji}^a, \quad (29)$$

with $E^a_{ij} = E^{aT}_{ji}$. Substituting (27) into (29) and using certain trace operations, results in $\delta J^a_r = \mathrm{Tr} \sum_{i,j=1}^{1+r} (\delta K^T \phi^{aT}_{ij} E^{aT}_{ij} + E^a_{ji} \psi^a_{ij} \delta K)$. (30)

From $\frac{\partial}{\partial A} \mathrm{Tr}(AB) = B^T$ and $\frac{\partial}{\partial A} \mathrm{Tr}(A^T B) = B$, it can be shown that (30) yields

$$\frac{\partial J^a_r}{\partial K} = \lim_{\delta K \to 0} \frac{\delta J^a_r}{\delta K} = \sum_{i,j=1}^{1+r} (\phi^{aT}_{ij} E^{aT}_{ij} + \psi^{aT}_{ij} E^a_{ij}), \tag{31}$$

The optimality of K requires that $\partial J^a_r / \partial K = 0$. Thus

$$(\phi^{aT}_{11} + \psi^{aT}_{11}) E^a_{11} + \sum_{\substack{i,j=1 \\ (i,j) \neq (1,1)}}^{1+r} (\phi^{aT}_{ij} E^{aT}_{ij} + \psi^{aT}_{ij} E^a_{ij}) = 0. \tag{32}$$

Substituting for $\phi^{aT}_{11}$ and $\psi^{aT}_{11}$, from (28), into (32) results in

$$2(WK - B^T P^a_{11} - \sum_{j=1}^{r} \frac{\partial B^T}{\partial p_j} P^{aT}_{1(1+j)}) E^a_{11} + \sum_{\substack{i,j=1 \\ (i,j) \neq (1,1)}}^{1+r} (\phi^{aT}_{ij} E^{aT}_{ij} + \psi^{aT}_{ij} E^a_{ij}) = 0. \tag{33}$$

Therefore, recalling $W = W^T > 0$, the optimal gain satisfies

$$K^\circ E^a_{11} = W^{-1} \{ [B^T P^a_{11} + \sum_{j=1}^{r} \frac{\partial B^T}{\partial p_j} P^{aT}_{1(1+j)} ] E^a_{11} - \frac{1}{2} \sum_{\substack{i,j=1 \\ (i,j) \neq (1,1)}}^{1+r} (\phi^{aT}_{ij} E^{aT}_{ij} + \psi^{aT}_{ij} E^a_{ij}) \}. \tag{34}$$

To compute this optimal adaptive controller gain K numerically, it is to be recalled that the right hand side of (34) is a function of K, B, $\partial B / \partial p_j$ $(j=1,\dots,r)$; as well as $\phi^a_{ij}$, $E^a_{ij}$, $\psi^a_{ij}$, and $P^a_{ij}$ $(i,j=1,\dots,(1+r))$. More importantly, it is to be recalled that $P^a_{ij}$ and $E^a_{ij}$ are functions of $\partial K / \partial p_j$, $j=1,\dots r$. The natural question which arises is: how to compute $\partial K / \partial p_j$, $j=1,\dots r$? This question is studied subsequently.

## 4. OPTIMAL ADAPTIVE CONTROLLER GAIN FOR A SINGLE PARAMETER SYSTEM

Without any loss of generality, a single parameter system is considered in the following. For this case an answer to the question just raised in Section 3 is given. For the case of multi-parameter system a similar algorithm to that of the next section, with appropriate modification, can be established. Suppose r=1 (with $p_1 \triangleq p$, and dropping all the corresponding remaining subscripts r=1), then: from (8), (11), (12), (22), (23), and (24)

$$A^a(p) \triangleq \begin{pmatrix} A^a_{11}(p) & A^a_{12}(p) \\ A^a_{21}(p) & A^a_{22}(p) \end{pmatrix} \triangleq \begin{pmatrix} S & 0 \\ \frac{\partial S}{\partial p} & S \end{pmatrix},$$

$$S \triangleq A - BK, \quad \frac{\partial S}{\partial p} = \frac{\partial A}{\partial p} - \frac{\partial B}{\partial p} K - B \frac{\partial K}{\partial p},$$

$$A^{aT} P^a + P^a A^a + \tilde{W} = 0, \quad \tilde{W} \triangleq \begin{pmatrix} W_0 + K^T W K & 0 \\ 0 & \alpha W_1 \end{pmatrix},$$

$$\delta J^a = \mathrm{Tr}(\pi^a_{11} E^a_{11} + \pi^a_{12} E^{aT}_{12} + \pi^{aT}_{12} E^a_{12} + \pi^a_{22} E^a_{22}),$$

$$\pi^a_{11} \triangleq \delta K^T \phi^{aT}_{11} + \psi^a_{11} \delta K \triangleq \delta K^T (WK - B^T P^a_{11} - \frac{\partial B^T}{\partial p} P^{aT}_{12}) + (K^T W - P^a_{11} B - P^a_{12} \frac{\partial B}{\partial p}) \delta K,$$

$$\pi_{21}^{aT} = \pi_{12}^{a} \triangleq \delta K^T(-B^T P_{12}^a - \frac{\partial B^T}{\partial p} P_{22}^a) - P_{12}^a B \delta K,$$

$$\pi_{22}^{a} \triangleq -\delta K^T B^T P_{22}^a - P_{22}^a B \delta K ,$$

$$E^a \triangleq \begin{pmatrix} E_{11}^a & E_{12}^a \\ E_{12}^{aT} & E_{22}^a \end{pmatrix} \triangleq \int_0^\infty \exp(A^a t) M \exp(A^a t)^T dt, \tag{35}$$

where without any loss of generality it is assumed that

$$M = \begin{pmatrix} X & 0 \\ 0 & 0 \end{pmatrix} , \tag{36}$$

here $X = E(x(0)x^T(0))$. Substituting for $\pi_{ij}^a$ and $E_{ij}^a$, i,j=1,2,into $\delta J^a$, results in

$$\frac{\partial J^a}{\partial K} = 2[(WK-B^T P_{11}^a - \frac{\partial B^T}{\partial p} P_{12}^{aT})E_{11}^a - (B^T P_{12}^a + \frac{\partial B^T}{\partial p} P_{22}^a)E_{12}^{aT} - B^T P_{12}^{aT} E_{12}^a - B^T P_{22}^a E_{22}^a]. \tag{37}$$

Recalling $W = W^T > 0$, the optimal adaptive controller gain then satisfies

$$K^{\circ} E_{11}^a = W^{-1}[B^T(P_{11}^a E_{11}^a + P_{12}^a E_{12}^{aT} + P_{12}^{aT} E_{12}^a + P_{22}^a E_{22}^a) + \frac{\partial B^T}{\partial p}(P_{12}^{aT} E_{11}^a + P_{22}^a E_{12}^{aT})], \tag{38}$$

where $\mathbf{P}_{ij}^a$, i,j=1,2, are the partitions of $\mathcal{P}^a$ corresponding to $A_{ij}^a$. To compute $K^{\circ}$, from (38), $E_{ij}^a$ and $P_{ij}^a$ (i,j=1,2,), must be determined first (recall that A,B, $\partial A/\partial p$, $\partial B/\partial p$, X, $\alpha$, $W_0$, W and $W_1$ are known), Here $P_{ij}^a$, i,j=1,2, are the solutions of the following three sets of nxn matrix linear equations.

$$(A-BK)^T P_{22}^a + P_{22}^a (A-BK) + \alpha W_1 = 0, \tag{39}$$

$$(A-BK)^T P_{12}^a + P_{12}^a (A-BK) + (\frac{\partial}{\partial p}(A-BK)^T) P_{22}^a = 0, \tag{40}$$

$$(A-BK)^T P_{11}^a + P_{11}^a (A-BK) + W_0 + K^T W K + (\frac{\partial}{\partial p}(A-BK)^T) P_{12}^{aT} + P_{12}^a \frac{\partial}{\partial p}(A-BK) = 0. \tag{41}$$

Similarly $E_{ij}^a$, i,j=1,2, are the solutions of the following:

$$(A-BK)E_{11}^a + E_{11}^a (A-BK)^T + X = 0, \tag{42}$$

$$(A-BK)E_{12}^a + E_{12}^a (A-BK)^T + E_{11}^a \frac{\partial}{\partial p}(A-BK)^T = 0, \tag{43}$$

$$(A-BK)E_{22}^a + E_{22}^a (A-BK)^T + (\frac{\partial}{\partial p}(A-BK))E_{12}^a + E_{12}^{aT} \frac{\partial}{\partial p}(A-BK)^T = 0. \tag{44}$$

One problem remains, the expression $\partial K/\partial p$, in (40),(41),(43), and (44), to be determined. Addressing this problem the following algorithm is proposed.

## 5. ALGORITHM FOR THE NUMERICAL SOLUTION OF THE OPTIMAL ADAPTIVE CONTROLLER GAIN

The flow-chart given in Fig. 1, illustrates an iterative scheme which has been proposed for numerical solution of $K^{\circ}$ given in (38).

The following equations must be also considered in completion of this algorithm. The initial $\partial K/\partial p$ is derived as follows. Set $\alpha=0$, $J^a=J_p$, then from (39), $P_{22}^a = 0$, from (40), $P_{12}^a = 0$, and from (41),

$$(A-BK)^T P_{11}^a + P_{11}^a (A-BK) + W_0 + K^T W K = 0. \tag{45}$$

Correspondingly, the initial K becomes $K = W^{-1}B^TP_{11}^a$, and the initial $\frac{\partial K}{\partial p}$ is

$$\frac{\partial K}{\partial p} = W^{-1}\left(\frac{\partial B^T}{\partial p} P_{11}^a + B^T \frac{\partial P_{11}^a}{\partial p}\right). \tag{46}$$

Here $P_{11}^a$ and $\frac{\partial P_{11}^a}{\partial p}$, with $K = W^{-1}B^TP_{11}^a$, are given by

$$A^TP_{11}^a + P_{11}^aA + W_0 - P_{11}^aBW^{-1}B^TP_{11}^a = 0, \tag{47}$$

and

$$(A-BK)^T \frac{\partial P_{11}^a}{\partial p} + \frac{\partial P_{11}^a}{\partial p}(A-BK) + \frac{\partial A^T}{\partial p} P_{11}^a + P_{11}^a \frac{\partial A}{\partial p} - P_{11}^a \frac{\partial}{\partial p}(BW^{-1}B^T)P_{11}^a = 0. \tag{48}$$

The following equations are to be considered in completion of the algorithm (with $\partial^2K/\partial p^2 = 0$) as well:

$$\frac{\partial^2}{\partial p^2}(A-BK) = \frac{\partial^2 A}{\partial p^2} - \frac{\partial^2 B}{\partial p^2}K - 2\frac{\partial B}{\partial p}\frac{\partial K}{\partial p}, \tag{49}$$

$$(A-BK)^T \frac{\partial P_{22}^a}{\partial p} + \frac{\partial P_{22}^a}{\partial p}(A-BK) + \left(\frac{\partial}{\partial p}(A-BK)^T\right)P_{22}^a + P_{22}^a \frac{\partial}{\partial p}(A-BK) = 0, \tag{50}$$

$$(A-BK)^T \frac{\partial P_{12}^a}{\partial p} + \frac{\partial P_{12}^a}{\partial p}(A-BK) + \left(\frac{\partial}{\partial p}(A-BK)^T\right)\left(P_{12}^a + \frac{\partial P_{22}^a}{\partial p}\right) + P_{12}^a \frac{\partial}{\partial p}(A-BK) + \left(\frac{\partial^2}{\partial p^2}(A-BK)^T\right)P_{22}^a = 0, \tag{51}$$

$$(A-BK)^T \frac{\partial P_{11}^a}{\partial p} + \frac{\partial P_{11}^a}{\partial p}(A-BK) + \left(\frac{\partial}{\partial p}(A-BK)^T\right)\left(P_{11}^a + \frac{\partial P_{12}^{aT}}{\partial p}\right) + \left(P_{11}^a + \frac{\partial P_{12}^a}{\partial p}\right)\frac{\partial}{\partial p}(A-BK) + \frac{\partial}{\partial p}(K^TWK) + \left(\frac{\partial^2}{\partial p^2}(A-BK)^T\right)P_{12}^{aT} + P_{12}^a \frac{\partial^2}{\partial p^2}(A-BK) = 0, \tag{52}$$

$$(A-BK)\frac{\partial E_{11}^a}{\partial p} + \frac{\partial E_{11}^a}{\partial p}(A-BK)^T + \left(\frac{\partial}{\partial p}(A-BK)\right)E_{11}^a + E_{11}^a \frac{\partial}{\partial p}(A-BK)^T = 0, \tag{53}$$

$$(A-BK)\frac{\partial E_{12}^a}{\partial p} + \frac{\partial E_{12}^a}{\partial p}(A-BK)^T + \left(\frac{\partial}{\partial p}(A-BK)\right)E_{12}^a + \left(E_{12}^a + \frac{\partial E_{11}^a}{\partial p}\right)\frac{\partial}{\partial p}(A-BK)^T + E_{11}^a \frac{\partial^2}{\partial p^2}(A-BK)^T = 0, \tag{54}$$

$$(A-BK)\frac{\partial E_{22}^a}{\partial p} + \frac{\partial E_{22}^a}{\partial p}.(A-BK)^T + \left(\frac{\partial}{\partial p}(A-BK)\right)\left(E_{22}^a + \frac{\partial E_{12}^a}{\partial p}\right) + \left(E_{22}^a + \frac{\partial E_{12}^{aT}}{\partial p}\right)\frac{\partial}{\partial p}(A-BK)^T + \left(\frac{\partial^2}{\partial p^2}(A-BK)\right)E_{12}^a + E_{12}^{aT}\frac{\partial^2}{\partial p^2}(A-BK)^T = 0, \tag{55}$$

and finally

$$\frac{\partial K^\circ}{\partial p} = W^{-1}\left\{\frac{\partial B^T}{\partial p}P_{11}^a + B^T\frac{\partial P_{11}^a}{\partial p} + \frac{\partial^2 B^T}{\partial p^2}P_{12}^{aT} + \frac{\partial B^T}{\partial p}\frac{\partial P_{12}^{aT}}{\partial p} + \frac{\partial}{\partial p}\{[(B^TP_{12}^a + \frac{\partial B^T}{\partial p}P_{22}^a)E_{12}^{aT} + B^TP_{12}^{aT}E_{12}^a + B^TP_{22}^aE_{22}^a]E_{11}^{aI}\}\right\}. \tag{56}$$

Where $E_{11}^{aI}$ in (56) is the generalized inverse matris of $E_{11}^a$, recall that $E_{11}^a = E_{11}^{aT}$.

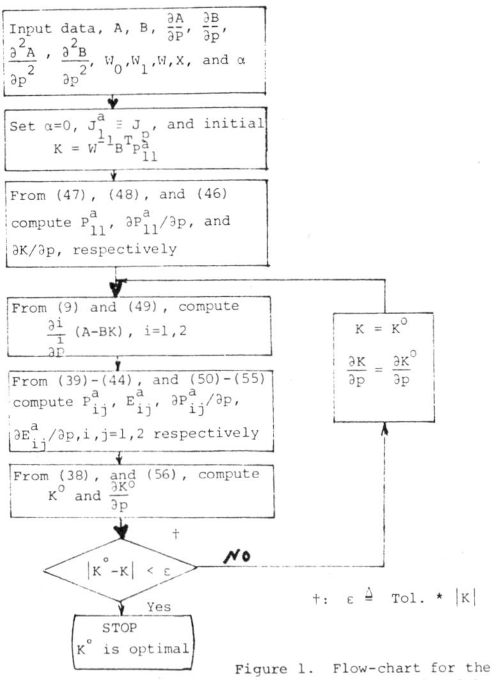

Figure 1. Flow-chart for the Proposed Algorithm.

When this algorithm reaches the end of its first iteration, for computation of an update $\partial K/\partial p$, it needs values such as $\partial P_{ij}^a/\partial p$ and $\partial E_{ij}^a/\partial p$ (i,j=1,2). These in turn require knoweldge of $\partial^2 K/\partial p^2$. This is the "heart" of the problem. This means that if it is desired that the optimal adaptive controller gain is to be computed as accurately as possible, then the computational cost will increase exponentially for computing the higher-order $\partial^i K/\partial p^i$, $i \geq 2$. However, in many problems $\partial^i A/\partial p^i$ and/or $\partial^i B/\partial p^i$, $i \geq 2$, are identically equal to zero. Thus the assumption $\partial^2 K/\partial p^2 = 0$, in the present algorithm, is not unrealistic. In fact in a practical problem, if the parameter variation is not "small" then the "small" sensitivity analysis, as carried out in this paper, does not serve usefully. Indeed, if the "small" sensitivity analysis can not be applied to a problem, then the entire computation carried out herein must be replaced by a corresponding analysis in terms of sensitivity in the "large".

In regard to the convergence of the proposed algorithm the following remarks can be stated. It is well known that the initial $K = W^{-1}B^T P_{11}$ (standard regulator problem), will stabilize the closed loop matrix (A-BK). Correspondingly, this stable matrix (A-BK) provides results that all the linear matrix equations (39)-(44) and (50) - (55) can be solved easily. Equations (39) -(44), then yield $P_{22}^a \geq 0$, $P_{11}^a \geq 0$, $E_{11}^a \geq 0$, and $E_{22}^a \gtrless 0$; and both $P_{12}^a$ and $E_{12}^a$ have the same signs, i.e., either $P_{12}^a \geq 0$ and $E_{12}^a \geq 0$, or $P_{12}^a \leq 0$ and $E_{12}^a \leq 0$. Substituting $P_{ij}^a$ and $E_{ij}^a$, i,j=1,2, into (38), yields, in general, a new $K^i$ such that $BK^i \geq 0$, where i is the ith-iteration. Correspondingly, $A-BK^i$ remains a stability matrix and similarly equations (39) - (44) and (50) - (55) can be easily solved, by using well known routines as those given in [14]. These observations eliminate some of the concerns regarding the solution of (39) - (44) and (50) - (55).

## 6. CONCLUSION

A new approach for deriving the optimal adaptive controller gain which results in a system, with minimum sensitivity cost, has been advanced. A supporting algorithm for numerical solution of the set of quadratic matrix equations, the result of which enters into the numerical evaluation of the optimal adaptive controller gain, has been proposed. Analysis of this algorithm reveals that the optimal controller gain can be derived only to a limited degree of accuracy.

Extensions of this result to the "large parameter variations," and to the "optimal closed-loop constrained policies for system identification versus the system sensitivity reduction with plant-operator," are discussed in Eslami [4].

## References

[1]Basuthakur,S.,and Knapp, C.H., 1975 IEEE Trans. Autom. Control, 20, pp. 664-666.

[2] Brandt, A.J., 1968, IEEE Trans. Autom. Control, 13, PP.110-111.

[3] Byrne, P.C., and Burke, M., 1976, IEEE Trans. Autom. Control, 21, pp. 282-283.

[4] Eslami, M., 1978, Ph.D. Dissertation, The University of Wisconsin-Madison.

[5] Fleming, P.J., and Newmann, M.M., 1977a, IEEE Trans. Autom. Control, 22, p. 151; 1977b, Int. J. Control, 25, PP. 965-978.

[6] Hendricks, T.C., and D'Angelo, H., 1967, Proc. 5th Ann. All. Conf. on Circuit and System Theory, University of Illinois, pp 142-151.

[7] Johnson, T.L., and Athans, M., 1970, IEEE Trans. Autom. Control, 15, pp. 658-660.

[8] Kleinman, D.L., and Rao, K.P., 1977, Proc. 16th Decision and Control Conf. of IEEE.

[9] Kreindler, E., 1968, Int. J. Control, 8, pp. 89-96; 1969, IEEE Trans. Autom. Control, 14, pp. 206-207.

[10] Kwakernaak, H., and Sivan, R., 1972, Linear Optimal Control Systems (Wiley-Interscience).

[11] Levine, W.S., and Athans, M., 1970, IEEE Trans. Autom. Control, 15, pp. 44-48.

[12] Mendel, J.M., and Feather, J., 1975, IEEE Trans. Autom. Control, 20, pp. 653-657.

[13] Newmann, M.M., 1970, Int. J. Control, 11, pp. 1079-1084.

[14] Sandell, N.R., and Athans, M., 1974 Computer Manual (Center for Advanced Engineering Study, M.I.T.).

[15] Sannuti, P., and Cruz, J.B., Jr., 1968, IEEE Trans. Autom. Control, 13, pp. 111-113.

[16] Stavroulakis, P., and Sarachik, P.E., 1974, Int. J. Control, 19, pp. 15-31.

[17] Subbayyan, R., Sarma, V.V.S., and Vaithilingam, M.C., 1977, IEEE Trans. Autom. Control, 22, pp. 657-659.

# WEAKER CONDITIONS FOR THE CONVERGENCE OF RECURSIVE STOCHASTIC ALGORITHMS AND THE SELF-TUNING PROPERTY OF (LS+MV) REGULATORS

V. Barbarani

SAIS, Facoltà di Ingegneria
via S. Marta 3, Florence

ABSTRACT The important problem of specifying sufficient conditions for the convergence of recursive stochastic algorithms is revisited. As is known, this problem is relevant to many system identification and adaptive control schemes.

Using Ljung's approach to the problem, it is shown that one of the known sufficient conditions for convergence is inadequate relatively to the applications of the theory to establish the self-tuning (ST) property of the most popular adaptive regulator, made up of a least squares (LS) parameter estimator plus a minimum-variance regulator ((LS+MV) regulator). The new condition is simpler than the one used up to now in the literature, and directly related to the stability of the closed loop regulated system.

## Introduction

The analysis of convergence of a recursive stochastic algorithm may be reduced to the stability analysis of an Ordinary Differential Equation (ODE), associated with the algorithm. These methods are presented in detail by Ljung [1] - [2] with some applications. Sect. I of this paper contains a brief resume of Ljung's results on sufficient conditions for convergence, and comments on some difficulties arising in applying the theory to adaptive regulators.

In Sect. II it is shown how these difficulties can be removed by modifying two regularity conditions on



*D. G. Lainiotis and N. S. Tzannes (eds.), Advances in Control, 351–360.*

the algorithm. and by introducing a stability assumption on the"system of observations".

In Sect. III it is shown that these new assumptions are satisfied in the LS+MV regulation algorithm, also known as self-tuning regulator, applied to an ARMAX process. Thus, as will be discussed, theconvergence analysis of the LS+MV regulator can be developed by removing some pitfalls present in the previous formulation, and used so as to get a weaker convergence condition. A further benefit of the slightly different formulation of this paper is that some theoretical conditions are given in a form closer to the most likely technical use of the theory.

## I . CONVERGENCE OF RECURSIVE STOCHASTIC SCHEMES

A general expression for recursive dicrete time stochastic algorithms is the following [1]:

$$x(t)= x(t-1)+\mu(t)Q(t;x(t-1),\varphi(t)) \tag{1}$$

$$\varphi(t)= A(x(t-1))\varphi(t-1)+B(x(t-1))e(t) \tag{2}$$

where $t=1,2,3,\ldots$ ; $x(.)\in\mathbb{R}^n$ is the estimate sequence; $\varphi(.)\in\mathbb{R}^m$ is the vector of "observations" containing input and output signals of the process whose parameters are estimated; $e(.)\in\mathbb{R}^r$ is a sequence of possibly stochastic vectors; $A(.)$ and $B(.)$ are matrices of appropriate dimensions; $\mu(.)\in\mathbb{R}$ is a sequence of positive scalars; $Q(.;.;.):\mathbb{N}\times\mathbb{R}^n\times\mathbb{R}^m\rightarrow\mathbb{R}^n$ is a deterministic function. Eq. (2) represents the "system of observations" [1]. Notice that in (1) and (2) the current estimate $x(t)$ depends on all previous estimates. Thus this formulation allows one to represent stochastic adaptive control schemes, where control inputs are calculated on the basis of current estimates. According to [1] , to which we refer the reader for full details and explanations, we adopt the following definitions and regularity conditions.

The set of stability $D_s$ is defined as

$$D_s = \{x\in\mathbb{R}^n / A(x) \text{ has all eigenvalues strictly inside the unit circle}\}.$$

Hence for $\bar{x}\in D_s$ the system defining the random variable $\varphi(t,\bar{x})$:

$$\varphi(t,\bar{x})= A(\bar{x})\varphi(t-1,\bar{x})+B(\bar{x})e(t),\ \varphi(0,\bar{x})=0 \tag{3}$$

is asymptotically stable. Similarly assuming that

$0 \leq \lambda < 1$, $0 \leq c < \infty$, the following system

$$v(t,\lambda,c) = \lambda v(t-1,\lambda,c) + c\|e(t)\|, \quad v(0,\lambda,c) = 0 \quad (4)$$

is asymptotically stable.
Consider the set: $D_R \subset D_S$ open connected.
By $S(\bar{x},\rho)$ we denote: $S(\bar{x},\rho) = \{x \in \mathbb{R}^n / \|x-\bar{x}\| < \rho\}$.
The following regularity conditions on the algorithm are introduced in [1] as the C-set of hypotheses:

C.1) The function $Q(t,x,\varphi)$ is locally Lipschitz continous in x and $\varphi$, viz. for all $x \in D_R$ there exists $\rho(x) > 0$ such that: $\|Q(t,x_1,\varphi_1) - Q(t,x_2,\varphi_2)\| < K_1(x,\varphi,\rho,v) \cdot \{\|x_1 - x_2\| + \|\varphi_1 - \varphi_2\|\}$ for all $x_i \in S(x,\rho)$, $\varphi_i \in S(\varphi,v)$, $v \geq 0$.

C.2) The matrix functions $A(\cdot)$ and $B(\cdot)$ are locally Lipshitz continous in $D_R$.

C.3) Defining

$$z(t,x) = z(t-1,x) + \mu(t)[Q(t,x,\varphi(t,x)) - z(t-1,x)] \quad (5)$$
$$z(0,x) = 0$$

$$\lim_{t\to\infty} z(t,x) = f(x).$$

Notice that if $\mu(t) = 1/t$, which is the most commonly used scalar sequence, $z(t,x)$ becomes:

$$z(t,x) = \frac{1}{t} \sum_{k=1}^{t} Q(k,x,\varphi(k,x)).$$

C.4) If the quantity $K_v$ is defined by

$$K_v(t,x,\lambda,c) = K_v(t-1,x,\lambda,c) + \mu(t)[K_1(x,\varphi(t,x),\rho,$$
$$v(t,\lambda,c))(1+v(t,\lambda,c)) - K\ (t-1,x,\lambda,c)] \quad (6)$$

$$K\ (0,x,\lambda,c) = 0$$

with $K_1$ and v defined respectively in C.1 and (4), then for all $x \in d_R$, $0 \leq \lambda < 1$, $0 \leq c < \infty$ there exists the limit:

$$\lim_{t\to\infty} K_v(t,x,\lambda,c) = \bar{K}_v.$$

C.5) $\sum_{s=1}^{\infty} \mu(s) = \infty$

C.6) $\lim_{t\to\infty} \mu(t) = 0$.

Further if $\{e(t)\}$ is a random sequence, then it is such that statements C.3 and C.4 follow w.p.1.

Under C.1-C.6, the ODE

$$\frac{dx^D(\tau)}{d\tau} = f(x^D(\tau)) \quad (7)$$

with $f(x)$ defined in C.3, can be associated to the algorithm (1)-(2) and the following result about convergence of estimates is proved in Theorem 1 of [1].

If:

a) there exists w.p.1 a random variable C and a subsequence $\{x(t_K)\}$ of the sequence of the estimates such that:

$x(t_K) \in \bar{D}$ , $\|\varphi(t)\| \leq C$

with $\bar{D} \subset D_R$ compact. By the sake of coinciseness this condition can be equivalently written as

$$x(t) \in \bar{D}, \|\varphi(t)\| \leq C \text{ i.o. (infinitely often) w.p.1,}$$

and hereafter will be referred to as the "boundedness condition".

b) $\bar{D}$ is such that the trajectories of (7) that start in it remain a closed subset $\bar{D}_R$ of $D_R$;

c) the differential equation (7) has a statinary point $x^*$, i.e. $f(x^*)=0$, globally asymptotically stable (or alternatively with domain of attraction $D_A \supset \bar{D}$);

then, with reference to (1) and (2), one has:

$$\lim_{t \to \infty} x(t) = x^* \quad \text{w.p.1.}$$

When the above result is applied in order to prove convergence of a certain recursive scheme, some difficulties arise in that sometimes it is not possible to assure that sufficient conditions a) and b) are a priori verified.

In particular condition a) is difficult to be verified in the analysis of stochastic control algorithms, where the "true" process is unknown and,consequently, the set $\bar{D}$ is also unknown. In such a case the projection procedure discussed in [1] cannot be used in order to assure the validity of a).

Let us illustrate this difficulty in connection with the LS+MV algorithm made up of a recursive least--squares parameter estimator plus a minimum-variance regulator [3,4] . Assume that the process to be regulated is described by a scalar ARMAX model

$$y(t)+a_1 y(t-1)+\ldots+a_n y(t-n) = b_1 u(t-1)+\ldots+b_n u(t-n) +e(t)+c_1 e(t-1)+\ldots+c_n e(t-n) \qquad (8)$$

where u is the input, y the output, and $\{e(t)\}$ a sequence of uncorrelated random variables with zero mean and variance $\sigma^2$. The LS+MV regulator is defined in the following way. A predictive model of the process is chosen

$$\hat{y}(t) = \hat{\theta}^T\psi(t) \tag{9}$$

with

$$\hat{\theta} = [\hat{a}_1 \ldots \hat{a}_m \hat{b}_1 \ldots \hat{b}_m]^T$$

$$\psi(t) = [-y(t-1) \ldots -y(t-n) \; u(t-1) \ldots u(t-n)]^T$$

where for the sake of simplicity the order of the predictive model is set equal to that of (8). At time t the LS estimate $\hat{\theta}(t)$ of the parameters in (9) is obtained by minimizing the functional

$$V_t(\hat{\theta}) = \sum_{1}^{t} {}_s[y(s) - \hat{\theta}^T\psi(s)]^2 \tag{10}$$

whereas MV control u(t) is calculated as follows

$$\hat{y}(t+1) = \hat{\theta}^T(t)\psi(t+1) = 0 \;. \tag{11}$$

The corresponding LS estimates of $\hat{\theta}$ are recursively given by:

$$\hat{\theta}(t) = \hat{\theta}(t-1) + \frac{1}{t} K(t)y(t) \tag{12.a}$$

$$K(t) = \frac{R^{-1}(t-1)\psi(t)}{1 + \frac{1}{t}[\psi^T(t)R^{-1}(t-1)\psi(t) - 1]} \tag{12.b}$$

$$R(t) = R(t-1) + \frac{1}{t}[\psi(t)\psi^T(t) - R(t-1)] \;. \tag{12.c}$$

This is an LS estimation algorithm in which the MV control (11) was taken into account. Let us introduce the unit-delay operator $q^{-1}$, $q^{-1}u(t) = u(t-1)$, and the polynomials:

$$\alpha(q^{-1}) = \hat{a}_1 + \hat{a}_2 q^{-1} + \ldots + \hat{a}_m q^{-m+1}$$

$$\beta(q^{-1}) = \hat{b}_1 + \hat{b}_2 q^{-1} + \ldots + \hat{b}_m q^{-m+1} \;.$$

Thus the control variable is given as the output of the following time-varying rational filter:

$$u(t) = \frac{\alpha(q^{-1}, \hat{\theta}(t))}{\beta(q^{-1}, \hat{\theta}(t))} y(t) \;.$$

Using this equation in (8) that can be written, with obvious definition of polynomials $\mathcal{A}$, $\mathcal{B}$, $\mathcal{C}$,

$$\mathcal{A}(q^{-1})y(t) = \mathcal{B}(q^{-1})u(t-1) + \mathcal{C}(q^{-1})e(t)$$

we get the following time-varying nonlinear stochastic equation for the closed-loop system:

$$[\mathcal{A}(q^{-1})\beta(q^{-1},\hat{\theta}(t-1))-q^{-1}\mathcal{B}(q^{-1})\alpha(q^{-1},\hat{\theta}(t-1))]y(t)=$$
$$=\beta(q^{-1},\hat{\theta}(t-1))\mathcal{C}(q^{-1})e(t) \ . \qquad (13.a)$$

The LS+MV algorithm can be represented by the general form (1)-(2) (see [1] ex.3) where the matrix A only depends on $\hat{\theta}$. Further the eigenvalues of $A(\hat{\theta})$ are the same as the poles of the system (13.a), that, in turn, are the zeros of its characteristic equation:

$$z^{2m-1}[\mathcal{A}(z^{-1})\beta(z^{-1},\hat{\theta})-z^{-1}\mathcal{B}(z^{-1})\alpha(z^{-1},\hat{\theta})]=0 \qquad (13.b)$$

plus a certain number of zero eigenvalues.
This is true of course at every time t. Hence for the LS+MV algorithm the requirement $x(t)\in\bar{D}$, which is part of the boundedness condition, amounts to assuming that the closed-loop system (13) has all its poles strictly inside the unit circle i.o. w.p.1. In the previous paper on the topic [1,2,5] this requirement was considered to be fulfilled in that implied by the stability of the regulated closed-loop system, discussed in [6]. This implication however does not appear to be valid in general unless the order of the process n is equal to one. In fact, for n greater than one, it is possible to find counterexamples of time-varying asymptotically stable systems having at every time at least one eigenvalue outside the unit circle. In [9] a more accurate proof of the boundedness condition is gven for LS+MV regulator but it seems to make use of not completely correct implications. As counterexample consider the system:

$$\begin{aligned} &x(t+1)=A(t)x(t) \\ &x(2)=\hat{x} \ , \ t=2,3,\ldots \end{aligned} \qquad (14)$$

with the time-varying matrix A(t) defined as follows:

$$A(2t)=\begin{bmatrix}1+\delta^{2t} & 0\\ 0 & 1-\delta^{2t}\end{bmatrix}, \ A(2t+1)=\begin{bmatrix}1-\delta^{2t} & 0\\ 0 & 1+\delta^{2t}\end{bmatrix}$$

$$t=1,2,\ldots \qquad 0<\delta<1.$$

One has $x(t)=\Phi(t,2)\hat{x}$ where it is not difficult to check that the transition matrix $\Phi(t,2)$ has limited norm. Hence the system (14) is stable even if at every time t it has an eigenvalue outside the unit circle and $\sup \| A(t)-A(t-1)\| \to 0$ as $t\to\infty$.

The other sufficient condition b) is difficult to be checked a priori too. Often in applications it must be assumed as a hypothesis, see for example the discussion of self-tuning regulator in [2].

## II. MODIFICATION OF SUFFICIENT CONDITIONS FOR CONVERGENCE

Hereafter it will be shown how the difficulties pointed out in the previous section can be removed by slightly modifying some of the initial hypotheses on the algorithm. A closer analysis of Ljung's work [1] shows that the boundedness condition a) and the condition b) must be introduced in order to guarantee:

1) the existence of a cluster point of the sequence of estimates;

2) the validity of lemma 1 of [1] which locally and asymptotically assures that the estimates $x(t)$ given by (1) follow the trajectories $x^D(\tau)$ of the associated differential equation (7). This property is assured only locally for $x(t)\in D_S$. In order to give sufficient conditions for convergence easier to be checked, the preliminary lemma discussed above is modified so as to avoid the requirement that $x(t)\in D_S$. this goal is achieved by introducing the following new version of hypotheses C1-2.

C'.1) For all $x\in D_R$ and $\rho$, $0\leq\rho<\infty$, there exists a constant $K_1(x,\varphi,\rho,v)$ such that:

$$\|Q(t,x_1,\varphi_1)-Q(t,x_2,\varphi_2)\|\leq K_1(x,\varphi,\rho,v)\{\|x_1-x_2\|+\|\varphi_1-\varphi_2\|\}$$

for $x_i\in S(x,\rho)$, $\varphi_i\in S(\varphi,v)$, $v\geq 0$, $i=1,2$.

C'.2) The matrix functions $A(.)$ and $B(.)$ are Lipschitz continous in $\mathbb{R}^n$.

Further suppose that the system of observation (2) is stable, in the sense that:

$$\|\varphi(t)\|\leq C<\infty \text{ for all } t, \text{ w.p.1 }. \qquad (15)$$

Then the preliminary lemma can be reformulated as stated above, and the following theorem about sufficient conditions for convergence can be proved.

Theorem 1'

Consider the algorithm (1-2) under the assumptions C'.1-2, C.3-6 and the hypothesis of stability (15). Further let us assume that

i) $x(t)\in \bar{W}$ i.o., w.p.1 (16)

where $\bar{W}\subset\mathbb{R}^n$ is a generic compact subset;

ii) the differential equation (7) has a stationary point $x^*$, globally asymptotically stable (or alternatively with domain of attraction $D_A\supset\bar{W}$).

Then, with reference to (1) and (2), one has:

$$\lim_{t \to \infty} x(t) = x^* \quad \text{w.p.1.}$$

The proof of the modified preliminary lemma and of theorem 1',will not be given here to keep the paper on a expositive level.

In order to get the above result,the stability of system (2) is assumed. This is a natural assumption in connection with stochastic adaptive controlschemes. In fact notice that when the LS+MV regulator is applied to an ARMAX process, the stability of system (2) is equivalent to that of the regulated closed-loop system. The latter property is discussed in [6] and it can be also extended [7] to the adaptive strategy based on a Maximum Likelihood parameter estimator working in conjunction with a Minimun Variance regulator. Notice that a stability hypothesis similar to (15) is assumed also by the boundedness condition a), though not for all times t. But, in this connection, condition (15) does not seem to be much more severe in practical applications. From a theoreticalpoint of view condition (15) implies that the noise is not gaussian.

The boundedness condition a) is now replaced by (16) which is less restrictive. Condition (16) amounts to the existence w.p.1 of a bounded subsequence of estimates, with no reference to the set $\bar{D}$ which is unknown in many practical applications. It is also realistic to think that once the stability of system (2) is assumed, then condition (16) is fullfilled too.

Theorem 1' shows the direct role played by the stability of system (2) as far as the convergence of estimates is concerned. It allows one to reduce the study of convergence of recursive schemes, to the study of two main stability problems: first the stability of the "system of observation", second the stability of the ODE associated to the algorithm. No specific assumption is required on the behaviour of its trajectories. The price to be paid for this semplification is represented by the modified conditions C'.1-2 stronger than C.1-2. Once C'.1-2 are satisfied, the theory can be applied in a remarkably direct way. As will be shown in Sect. III, this is the case for the LS+MV regulator.

## III. APPLICATION TO LS+MV REGULATOR

Let us apply the theory developed in Sect. II to the LS+MV regulator defined in (8-12). In order to avoid numerical problems and difficulties in the analysis, the parameter $\hat{b}_1$ in (9) is set equal to $b_1$. The ARMAX process under control is assumed to be minimum phase and subjected to the consistency condition, i.e. with polynomials $z^{m-1}\mathcal{B}(z^{-1})$ and $z^n \mathcal{C}(z^{-1})$ stable. As was already mentioned in Sect. I, the LS+MV algorithm can be reduced to the general form (1-2) and the deterministic function $Q(t,x,\varphi)$ in this case is:

$$Q(t,x,\varphi) = \begin{bmatrix} \dfrac{R^{-1}\psi}{1+\frac{1}{t}[\psi^T R^{-1}\psi - 1]} y \\ \mathrm{col}(\psi\psi^T - R) \end{bmatrix} \cong \begin{bmatrix} R^{-1}\psi y \\ \mathrm{col}(\psi\psi^T - R) \end{bmatrix}$$

where, remembering that, because of the stability of the regulated closed-loop system, $\psi^T R^{-1}\psi$ is finite, the approximation follows since our interest is in the asymptotic behaviour of the algorithm. Notice that Q does not depend directly on $\hat{\theta}$. Hence it is obviously Lipschitz continous relatively to the vector of estimates for all $\rho$, as requested in C'.1. More generally by using the simplification above, it is not difficult to show that:

$$\| Q(t,x_1,\varphi_1) - Q(t,x_2,\varphi_2) \| \leqslant [1+2(\|\varphi\|+v)(\| R^{-1}\|^2(\|\varphi\|+v)+1)] \cdot \{ \| R_1 - R_2 \| + \| \varphi_1 - \varphi_2 \| \}$$

where col $R_i \in S(\mathrm{col}\ R, \ )$, for all $\rho$, and $\varphi_i \in S(\varphi, v)$, $v \geqslant 0$ $i=1,2$. The previous inequality is valid for $R_i \in \mathcal{P}$ $i=1,2$, where $\mathcal{P}$ is the set of all positive definite matrices of appropriate dimensions. The proof of C'.2 is straightforward by observing [7] that, once $\hat{b}_1$ is fixed, the matrix $A(\hat{\theta})$ and $B(\hat{\theta})$ depend linearly on $\hat{\theta}$. The other conditions C.3-6 are proved in [1]. The set of stability is defined as: $D_s = \{ \hat{\theta} \in \mathbb{R}^{2m-1} / (13.b)$ has allits zeros inside the unit circle$\}$. For any $\hat{\theta} \in D_s$, consider the processes $\psi(t,\hat{\theta})$, $y(t,\hat{\theta})$, obtained by the system (13.a) when $\hat{\theta}(t) \equiv \hat{\theta}$, and the process $\tilde{\varphi}(t,\hat{\theta}) = (1/\mathcal{C}(q^{-1}))\varphi(t,\hat{\theta})$. Since they are provided by a stable system and a stable filter, they will become stationary asymptotically. Hence if one defines:

$$G(\hat{\theta}) = \lim_{t\to\infty} E[\psi(t,\hat{\theta})\psi^T(t,\hat{\theta})] \tag{17}$$

$$\tilde{G}(\hat{\theta}) = \lim_{t\to\infty} E[\psi(t,\hat{\theta})\tilde{\psi}^T(t,\hat{\theta})] \tag{18}$$

the ODE associated to the LS+MV algorithm can be written [1,2] as follows

$$d\hat{\theta}/d\tau = -R^{-1}\tilde{G}(\hat{\theta})(\hat{\theta} - \theta_{MV}) \tag{19}$$

$$dR/d\tau = G(\hat{\theta})-R \quad (20)$$

Here $\theta_{MV}$ is the point that gives the optimal MV control $\theta_{MV}=(a_1-c_1,\ldots,a_n-c_n,b_2\ldots b_m)^T$. The regulation algorithm has the self-tuning property if $\hat{\theta}(t)$ provided by (12.a) tends to $\theta_{MV}$ w.p.1 as t tends to infinity. Notice that $(\theta_{MV},G(\theta_{MV}))$ is always a stationary point of the ODE (19)-(20), and it is also asymptotically globally stable if [2] the ARMAX process is such that $[1/\mathcal{C}(q^{-1})-1/2]$ is strictly positive real, i.e. if

$$\mathrm{Re}\left[\frac{1}{(e^{i\omega})}-\frac{1}{2}\right]>0 \qquad -\pi<\omega\leq\pi \quad . \quad (21)$$

By applying the theorydeveloped in Sect.II it follows:

1) if there exists w.p.1 a bounded subsequence of the estimates, 2) if the process under control is minimum phase and (21) is valid, than the estimate $\hat{\theta}(t)\rightarrow\theta_{MV}$ w.p.1 as $t\rightarrow\infty$, i.e. LS+MV has the self-tuning property.

This statement follows from Theorem 1' which, as remarked above, does not require any assumption on the instantaneous eigenvalues of the closed-loop regulated system.

## REFERENCES

1) L.Ljung: Analysis of recursive stochastic algorithms. IEEE Trans. Aut. Control, AC-22 (1977)

2) L.Ljung: On positive real transfer function and the convergence of some recursive stochastic schemes. IEEE Trans. Aut. Control, AC-22 (1977)

3) K.J.Astrom, B.Wittenmark: On self-tuning regulators. Automatica 9, pp. 185-189 (1973)

4) K.J.Astrom, U.Borisson,L.Ljung,B.Wittenmark: Theory and applications of self-tuning regulators. Automatica 13, pp. 457-476 (1977).

5) L.Ljung,B.Wittenmark: Analysis of a class of adaptive regulators. IFAC Symposium, Budapest (1974)

6) L.Ljung,B.Wittenmark: On a stabilizing property of adaptive regulators. IFAC Symposium, Tbilisi (1976)

7) V.Barbarani: Ottimalità di tecniche di identificazione e controllo adattivo di processi. Thesis, Fac. Elec. Eng., Univ. Florence (1979)

8) W.Hahn: Theory and applications of Liapunov's direct method. Prentice-Hall, New York (1963)

9) L.Ljung, B.Wittenmark: Asymptotic properties of self-tuning regulators. Report 7404, Dept of Automatic Control, Lund Inst. of Techn. (1974)

# ADAPTIVE FILTER FOR INERTIAL NAVIGATION SYSTEMS

Leonard Chin

U. S. Naval Air Development Center
Warminster, Pennsylvania 18974

## ABSTRACT

A theoretical approach to reduce errors in an inertial navigation system (INS) is presented. Verification of the developed theory (through computer simulations) is reported. Implementation requirements and effectiveness in estimating INS reset parameters are given.

## I. INTRODUCTION

Covert operation is one of the major objectives for both attack and strategic submarines. The degree to which covert operation can be achieved directly influences the overall effectiveness of a submarine mission. To achieve covert operation the submarine's navigation system must operate in a self-contained manner (i.e., without requiring a near-surface operation) and still be able to calculate the boat's position accurately. A system that potentially meets this requirement and which is currently used is the INS (Inertial Navigation System). However, INS position errors grow with time and corrections or resets (removing accumulated errors and calibrating the predominant errors source - gyro drift rate) are required. The performance of a reset requires external reference information which can be acquired only when the boat is near the surface. Thus, INS is not completely self-contained and it is desirable to improve the system such that enhancement of covert operation can be realized. One way to approach this problem is to improve the reset technique because the rate at which INS errors grow is directly influenced by reset accuracy (i.e., improved reset accuracy yields a lower position error growth, thus lengthening the submerged period of the boat).



*D. G. Lainiotis and N. S. Tzannes (eds.), Advances in Control, 361–377.*

In current submarine operational INS, resets are performed using two different methods: Deterministic and Stochastic. The deterministic method was implemented earlier. An example in this category is the "3-fix" reset, in which gyro drift rates are assumed to be constant and external reference data are assumed to be perfectly accurate. The stochastic method was implemented later, with the intention of improving the reset accuracy. An example in this category is the Kalman filtering technique, in which random components of the gyro drift rate and random components of the error associated with external reference data are accounted for.

Consideration of random components in the Kalman filter technique theoretically improves the reset accuracy. However, the reset accuracy is improved if and only if mathematical descriptions of the random components are precise and the precise description, unfortunately, is never available. Experience with operational systems indicated that direct application of the Kalman filtering algorithm with imprecise error models always yields non-optimal results. In fact, reset accuracy can be degraded rather than enhanced. If the Kalman filtering algorithm could adapt itself to the prevailing conditions in actual operational use, this problem would be overcome. However, implementation of the adaptive Kalman filtering algorithm requires extra computer resources. The effectiveness of the adaptive filter is not independent of the boats' operation. These and other factors undoubtedly will have a direct impact on the design of an inertial navigation system. The objective of this paper is to seek appropriate conditions under which adaptive Kalman filtering techniques can be applied to long-term submarine inertial navigation systems such that covert operation may be improved through the improvement of the reset accuracy. Specifically, the design of an adaptive filter to identify the unknown first and second order statistics (Gaussian white noise) is discussed.

## II. BACKGROUND

An adaptive filter may be defined as a Kalman-Bucy filter that can successfully achieve its optimal performance even if the environment under which the filter operates is changing. If the filter can adapt to the environment, it must be able to perceive the current state and change of the state of the environment and then make the necessary decision to modify its behavior based on the latest observations of the environment.

In general, the adaptive filter may have the following two types of adaptations:

- Structure adaptation — in which the filter is capable of learning the *order* of the linear differential equations that describe the process and noise involved.

- Parameter adaptation — in which the filter is capable of learning the *coefficients* of the fixed order linear differential equations that describe the process and noise involved.

The problems concerning the combined adaptation of structure and parameter as well as adaptation of structure only were attacked by Lainiotis[1,2].

The problems concerning adaptation of parameters only have been studied by many investigators. For example, Krause and Graupe[3] solved the problem of simultaneously identifying both transition matrix parameters and covariances of the state error and of the measurement noise for the case where no a priori information on any of these parameters existed. Maybeck[4] and Weiss[5] derived methods which simultaneously estimated the states and parameters of a linear dynamic system model, based upon noise measurements of the actual system outputs.

The main interest of this section, however, is to review techniques that were developed for the adaptation of the system noise and measurement noise statistics. It is believed that this type of adaptation has the highest application potentials in long-term inertial navigation systems.

During the past decade, a great number of techniques for noise covariance adaptation have been developed. In general, these techniques can be classified into four categories:

(1) The Bayesian Approach

(2) The Maximum Likelihood Approach

(3) The Covariance Matching Method

(4) The Correlation Method

*(1) The Bayesian Approach*

The Bayesian Approach was first introduced by Magill[6], who studied the optimal estimation of a sampled, Gauss-Markov stochastic process in which certain parameters are initially unknown. The general problem was formulated as a finite number of processes generated by different noise covariances and model parameters. The solution to the problem of finding the best (in the sense of minimizing a quadratic error function) estimate of states x, is known to be the conditional mean[7]:

$$\hat{x} = \int_X x\, p(x/y_k)\, dx \tag{2.1}$$

where $\hat{x}$ is the best estimate of x, X is the space of all states x, $p(x/y_k)$ is the conditional probability of x given measurement data $y_k$. Note that $y_k$ is a set consisting of k elements denoted by $y_k = \{y^1, y^2, \ldots y^k\}$. In the case of parameter adaptation, it is required to obtain recursive equations for the a posteriori probability density of x, the unknown parameter $\alpha$, and the conditional probability function p $(x, \alpha/y_k)$, which can be written as:

$$p(x,\alpha/y_k) = p(x/\alpha, y_k)\, p(\alpha/y_k) \tag{2.2}$$

where $p(x/\alpha, y_k)$ is Gaussian with mean $x(\alpha)$ and covariance $p(\alpha)$. Reference 6 has shown that for a continuous time system:

$$\hat{x} = \int_A \hat{x}(\alpha)p(\alpha/y_k)\,d\alpha \tag{2.3}$$

in which all possible values of the unknown parameters, $\alpha$, form an infinite set of stochastic processes with a known a priori probability of each process occuring $p(\alpha)$, where A is the space of all $\alpha$. In a discrete time system:

$$\hat{x} = \sum_{i=1}^{N} \hat{x}(\alpha_i)p(\alpha_i/y_k) \tag{2.4}$$

in which it is assumed that all possible values of the unknown parameters form a finite set of possible stochastic processes ($\{\alpha_i\}$, $i = 1,2,...,N$) with a known a priori probability of each process occuring $p(\alpha_i)$. The main difficulty in using equation 2.3 is that it and $p(\alpha/y_k)$ require integration over a large dimension space. In practice, when $\alpha$ is derived from a finite set, equation 2.4 is used.

*(2) The Maximum Likelihood Approach*

Depending on what probability density function is to be maximized (i.e. what quantity is to be estimated), this approach may take different forms. If the marginal density function $p(x, \alpha/y_k)$ is maximized, it is called the Marginal Maximum Likelihood estimate, in which $\alpha$ is being estimated. On the other hand, if both x and $\alpha$ are to be estimated simultaneously, then the joint density function $p(x,\alpha/y_k)$ is maximized, and this is called the Joint Maximum Likelihood estimate. These and other forms (e.g., Conditional Mode estimate, in which the conditional density function $p(x/y_k)$ is maximized) were discussed in papers by Kashyap[8] and Mehra[9].

*(3) The Covariance Matching Method*

The basic concept of the Covariance Matching Method is to make the residuals consistent with their theoretical covariances. For example, consider the innovation sequence of $\nu$, which has a theoretical covariance P. If it is noticed that the actual covariance of $\nu$ is much larger than P that was obtained from the Kalman filter, then the process noise should be increased, which in turn brings the actual covariance closer to the theoretical covariance. However, the covergence of the Covariance Matching Method is doubtful, as it was discussed in reference 9.

*(4) The Correlation Method*

The basic concept of this method is to correlate the output or a function related to the output of the system. From this operation, a set of equations is generated relating the autocorrelation function to the unknown parameters, which can be solved algebraically. The Correlation Method has two variations (stochastic approximations are considered as special cases of the Correlation

Method, not variations) depending on the manner in which the autocorrelation function is formed. If it is generated directly from the output, the method is referred to as Output Correlation, in which the nth lag autocorrelation is formed by taking the expected value of the output $y_k$, i.e.

$C_n = E\left\{ y_k, y^T_{k-n} \right\}$. If it is generated from the innovation sequence,

the method is referred to as Innovation Correlation, in which the nth lag autocorrelation is formed by taking the expected value of the innovation sequence $\nu_k$, i.e. $C_n = E\left\{ \nu_k, \nu^T_{k-n} \right\}$.

Reference 9 indicated that the first method (Output Correlation) is applicable only if $y_k$ is stationary or if the transition matrix ($\Phi$) is stable, whereas the second method (Innovation Correlation) is applicable to cases in which $\Phi$ is not stable. However, both methods require the system to have constant coefficients and it must be completely controllable and completely observable. In general, the estimates obtained using the second method are more efficient than those obtained from the first method (innovation sequence is less correlated than the output sequence). For this reason, the method of Innovation Correlation will be reviewed in a greater detail.

The innovation sequence $\nu_i$ is the new information brought by the measurement $y_i$; their relationship is defined as:

$$\nu_i = y_i - H\hat{x}_i$$

Kailath [10] showed that $\nu_i$ and $y_i$ contained the same statistical information and are equivalent as far as linear operation is concerned. Mehra[11] showed that for an optimal filter, $\nu_i$ is a zero-mean Gaussian white noise sequence; for a suboptimal filter, $\nu_i$ is correlated and the autocorrelation function $C_k$ has been derived for a stationary system:

$$C_k = \begin{cases} HMH^T + R & k = 0 \\ H[\Phi(I - KH)]^{k-1}\,\Phi\,[MH^T - KC_o] & k > 0 \end{cases}$$

where M is the covariance of the state error

R is the covariance of the measurement noise

K is the filter gain

H and $\Phi$ as defined previously

It is clear that the necessary and sufficient condition for the optimality of a Kalman filter is that the innovation sequence be white. In other words, adaptive action is required when $\nu_i$ is not white. The basic idea is to find the optimal gain, K, so that $C_k$ vanish for all k except k = 0. This is known as the innovation property of the optimal filter. Mehra[11] has solved the problem in which the true value of the non-correlated second order noise statistics are unknown.

Carew and Belanger[12] solved the same problem using auto-correlation functions of the innovation sequence of the suboptimal filter to determine the optimum filter steady state gain directly, without the intermediate determination of the unknown covariances for the system and measurement noises. Their approach is to identify an output equivalent representation of the original system which does not involve the unknown covariances directly. The difference between this solution and Mehra's is that the solution obtained by Carew and Belanger requires only matrix multiplication whereas the solution obtained by Mehra requires solving the Lyapunov-type matrix equation. Additionally, the stability of the algorithm presented by Carew-Belanger appears to be less dependent on the assumed values of the unknown noise covariance.

The major disadvantage of both the Mehra[11] and Crew-Belanger[12] approaches to the adaptation problem is their implicit assumption that the sample correlation function can be considered as equivalent to the true correlation function for the purpose of developing the estimation algorithm. This assumption simplifies the algebraic manipulations, but at the same time rejects the probabilistic interpretation of the problem.

In order to account for the errors in the measured correlation function, Belanger[13] has shown that if the covariance matrices are linear in a set of parameters, then the correlation function of the innovation sequence is also a linear function of this set of parameters. This fact was used to perform a weighted least-squares fit to a sequence of measured correlation products, and an algorithm was developed to provide recursive estimates of noise covariance parameters. These estimates were shown to be unbiased and consistent.

Neethling and Young[14] performed simulations for various innovation correlation techniques. Their results indicated that the weighted least-squares approach yields considerable improvements (smaller error variances) over the approaches discussed in references 11 and 12. However, advantages and disadvantages of various techniques for practical applications are not well known at this point.

All adaptive schemes discussed thus far were concerned with the identification of the *second* order statistics of the white Gaussian noise that influenced the dynamics and measurements of a linear stochastic system. In other words, adaptive problems were formulated with a priori knowledge that the first order statistics are known exactly. The problem concerning the Kalman filtering with no a priori knowledge about the noise has been studied recently. For example, Friedland[15] has considered the unknown noise means as an additional state variable and shown how the augmented state vector can be estimated by decoupling the bias estimate from the state estimate. Lin and Sage[16] have applied the invariant imbedding technique for solving the same problem and obtained the same results. Godbole[17] has applied the innovation sequence concept to show that the identification of the noise covariance can be achieved without the knowledge of noise means. Thus, the estimation of Gaussian noise statistics can be treated separately in two stages: First the covariance, then the means. This approach seemed to gain significant savings in computation.

In conclusion, it should be pointed out that there are no general techniques currently available that can be applied with absolute confidence in designing adaptive schemes for noise statistics identification. Nonetheless, physical intuition and off-line simulations are considered to be effective tools used to obtain good designs.

## III. PROBLEM DEFINITION AND SOLUTION

### Problem Definition

Consider an augmented inertial navigation system which employs the Kalman filter to estimate and reset system errors, using external measurements. Assume that the error dynamics of the system can be represented by a set of *linear* difference equations which are suitable for state space formulations. That is, the *state* vector is composed of elements including errors in latitude, longitude, gyro drift rates, etc., and the *measurement* vector is viewed as the difference between the navigator's measurement and the measurement indicated by the reference source. Assume also that only *additive* random errors appear as driving forces in both the state and measurement vector equations. The driving forces are series of random sequences which can be generated from appropriate shaping filters with *white* noise inputs. However, *the first and second order statistics of the white noises are unknown.* In short, the *system* to be considered is represented by:

$$x_{i+1} = \Phi_i x_i + \Gamma_i u_i \tag{3.1}$$

$$y_i = H_i x_i + v_i \tag{3.2}$$

where the dimension of x is n, dimension of y is m, dimension of u is r, and other matrices have appropriate dimensions. The sequences $\{\Phi_i\}$, $\{\Gamma_i\}$, and $\{H_i\}$ are bounded and *time varying* in general. The white noise sequences, $\{u_i\}$ *and* $\{v_i\}$ are *correlated, stationary* and *Gaussian.* The first and second order statistics are defined:

$$E\{u_i\} = \mu_u \; ; \; E\{v_i\} = \mu_v$$

$$E\{(u_i - \mu_u)(u_j - \mu_u)^T\} = Q\,\delta_{ij}$$

$$E\{(v_i - \mu_v)(v_j - \mu_v)^T\} = R\,\delta_{ij}$$

$$E\left\{(u_i - \mu_u)(v_j - \mu_v)^T\right\} = S\,\delta_{ij}$$

in which $\mu_u$, $\mu_v$, Q, R, and S are unknown except for the fact that Q and R are *positive definite* and S is consistent with Q and R.

The filter to be considered is:

$$\hat{x}^*_{i+1/i} = \phi_i \hat{x}^*_{i/i} + \Gamma_i \mu^*_u + K^*_i \left[ y_i - H_i \hat{x}^*_{i/i} - \mu^*_v \right] \tag{3.3}$$

$$\hat{x}^*_{i/i} = \hat{x}^*_{i/i-1} + B^*_i \nu^*_i \tag{3.4}$$

$$\nu^*_i = y_i - H_i \hat{x}^*_{i/i-1} - \mu^*_v \tag{3.5}$$

$$K^*_i = \Gamma_i S^*_i (R^*_i)^{-1} \tag{3.6}$$

in which the starred notations indicate that the uncertain descriptions for the noise statistics are being used to operate the Kalman filter. That is, $\hat{x}^*_{i+1/i}$ denotes the optimal estimate of $x_{i+1}$ using measurements up to and including the $i^{th}$ interval, based on the assumed values $\mu_u$*, $\mu_v$*, Q*, R*, and S* for $\mu_u$, $\mu_v$, Q, R, and S. $B_i$* is the usual Kalman filter gain and $\nu_i$* is the innovation sequence. The problem may be stated as follows:

Given a linear time-varying system described by equations 3.1 and 3.2, and data from a steady state Kalman filter with uncertain descriptions for system and measurement noises, it is desired to improve the filter optimality by adapting the second order (Q, R, and S) and first order ($\mu_u$ and $\mu_v$) noise statistics using external measurements.

## Solution

The solution will be given in two parts. The algorithm for adapting the second order noise statistics is discussed in part one and the algorithm for adapting the first order noise statistics is discussed in part two. The reasons for this approach is based on the work of Godbole[17], who has shown that the estimation of the second order noise statistics does not require the explicit knowledge of the first order noise statistics.

### *Part One — Adaptation of Second Order Noise Statistics*

Substituting equation 3.4 into 3.3:

$$\hat{x}^*_{i+1/i} = \phi_i \hat{x}^*_{i/i-1} + \Gamma_i \mu^*_u + G^*_i \nu^*_i \tag{3.7}$$

where

$$G_i^* = \phi_i B_i^* + K_i^* - K_i^* H_i B_i^* \tag{3.8}$$

Define

$$e_{i+1}^* = x_{i+1} - \hat{x}_{i+1/i}^* \tag{3.9}$$

Using equations 3.1, 3.5, and 3.7 to obtain a recursive expression for $e_i$*:

$$e_{i+1}^* = (\phi_i - G_i^* H_i) e_i^* + \Gamma_i (u_i - \mu_u^*) - G_i^* (v_i - \mu_v^*). \tag{3.10}$$

Using the definition for $e^*_{i+1}$, equation 3.5 becomes

$$\nu_i^* = H_i e_i^* + v_i - \mu_v^* \tag{3.11}$$

Equations 3.10 and 3.11 contain all necessary information for generating an adaptive algorithm. Prior to developing the algorithm however, it is desired to transform these equations into the standard form, in which the mean values do not appear explicitly. To this end, take the expectations of equations 3.10 and 3.11, then subtract 3.10 and 3.11 from their expectations. The results are:

$$\epsilon_{i+1}^* = (\phi_i - G_i^* H_i) \epsilon_i^* + \Gamma_i \tau_i - G_i^* \xi_i \tag{3.12}$$

$$\sigma_i^* = H_i \epsilon_i^* + \xi_i \tag{3.13}$$

where

$$\epsilon_i^* = e_i^* - E\left\{e_i^*\right\} \tag{3.14}$$

$$\sigma_i^* = \nu_i^* - E\left\{\nu_i^*\right\} \tag{3.15}$$

$$\tau_i = u_i - \mu_u \tag{3.16}$$

$$\xi_i = v_i - \mu_v \tag{3.17}$$

Note that $\epsilon_i^*, \sigma_i^*, \tau_i$ and $\xi_i$ have the same statistical properties as $e_i^*$, $\nu_i^*$, $u_i$ and $v_i$ respectively, except for the mean values, which are zero instead of $E\left\{e_i^*\right\}$, $E\left\{\nu_i^*\right\}$, $\mu_u$ and $\mu_v$. Note that $\epsilon_i^*$ and $\sigma_i^*$ are now viewed as the predictor error and innovation sequence respectively for the case where the noise means are zero.

In order to obtain the observation vector, $\sigma_i^*$, which contains all information needed to adapt the second order noise statistics, equation 3.15 will be used. The mean values for the innovation sequence can be computed

$$E\left\{\nu_i^*\right\} = \frac{1}{N+1} \sum_{i=1}^{N} \nu_i^* \tag{3.18}$$

which is unbiased, consistent and efficient if the steady state condition of the filter and the ergodicity of the sequence are satisfied. Having ascertained $\sigma_i^*$, the next step is to form the correlation function for $\sigma_i^*$, called the observed correlation function, and the weighted least-squares minimization is performed using the difference (error) between the observed correlation function and the theoretical correlation function, which can be derived as follows:

$$E\left\{\sigma_i^* \sigma_{i-\ell}^{*T}\right\} = H_i E\left\{\epsilon_i^* \epsilon_{i-\ell}^{*T}\right\} H_{i-\ell}^T + H_i E\left\{\epsilon_i^* \xi_{i-\ell}\right\} + R^* \delta_\ell \tag{3.19}$$

To express equation 3.19 in a more desirable form, an explicit function for $\epsilon_i^*$ is required. From equation 3.12:

$$\epsilon_i^* = \psi(i, 0)\, \epsilon_0^* + \sum_{k=0}^{i-1} \psi(i, k+1) \left[\Gamma_k \tau_k - G_k^* \xi_k\right] \tag{3.20}$$

where $\psi(i_1, i_2)$ is the state transition matrix:

$$\psi(i_1, i_2) = A(i_1 - 1)\, A(i_1 - 2) \ldots A(i_2) \qquad \text{for } i_1 > i_2 \tag{3.21}$$

and $A_i$ is defined:

$$A_i = \phi_i - G_i^* H_i \tag{3.22}$$

For a uniformly asymptotically stable filter, the following inequality holds:

$$\|\psi(k, 0)\| < d_1 e^{-d_2 k} \qquad d_1, d_2 > 0 \tag{3.23}$$

Hence the term $\psi(i,0)\, \epsilon_0^*$ in equation 3.20 may be dropped if a sufficiently long time is allowed prior to the processing of the measurement data. Under this condition, equation 3.20 becomes:

$$\epsilon_i^* = \sum_{k=0}^{i-1} \psi(i, k+1) \left\{\Gamma_k \tau_k - G_k^* \xi_k\right\} \tag{3.24}$$

Let $Q^* = \sum_{n=1}^{N} Q_n \alpha_n \qquad R^* = \sum_{n=1}^{N} R_n \alpha_n \qquad S^* = \sum_{n=1}^{N} S_n \alpha_n$

Substitute equation 3.24 into 3.19:

$$E\left\{\sigma_i^* \sigma_{i-\ell}^{*T}\right\} = H_i E\left\{\left(\sum_{k=0}^{i-1} \psi(i, k+1)\left[\Gamma_k \tau_k - G_k^* \xi_k\right]\right) \cdot \right.$$

$$\left. \cdot\left(\sum_{k=0}^{i-\ell-1} \psi(i-\ell, k+1)\left[\Gamma_k \tau_k - G_k^* \xi_k\right]\right)^T\right\} H_{i-\ell}^T$$

$$+ H_i E\left\{\left(\sum_{k=0}^{i-1} \psi(i, k+1)\left[\Gamma_k \tau_k - G_k^* \xi_k\right]\right) \xi_{i-\ell}^T\right\} + R^* \delta_\ell \tag{3.25}$$

After simplification equation 3.25 becomes:

$$E\left\{\sigma_i^* \sigma_{i-\ell}^{*T}\right\} = \sum_{n=1}^{N} F_n(i, \ell)\, \alpha_n \tag{3.26}$$

where

$$F_n(i, \ell) = H_i \left\{\sum_{k=0}^{i-\ell-1} \psi(i, k+1)\left[\Gamma_k Q_n \Gamma_k^T - G_k^* S_n^T \Gamma_k^T - \Gamma_k S_n G_k^{*T}\right.\right.$$

$$\left.\left. + G_k^* R_n G_k^{*T}\right] \psi^T(i-\ell, k+1)\right\} H_{i-\ell}^T + \left[H_i \psi(i, i-\ell+1)\Gamma_{i-\ell}\right] S_n$$

$$- \left[H_i \psi(i, i-\ell+1) G_{i-\ell}^*\right] R_n + R_n \delta_\ell \tag{3.27}$$

In order to apply the conventional least-squares minimization technique, both the observed $\left(\sigma_i \sigma_{i-\ell}^T\right)$ and the the theoretical $\left(\sum_{n=1}^{N} F_n(i, \ell)\alpha_n\right)$ correlation matrices will be transformed into vectors. First, vector $\beta(i, \ell)$ is defined as:

$$\beta(i, \ell) = \mathrm{vec}\left|\sigma_i \sigma_{i-\ell}^T\right| \tag{3.28}$$

For example, if $\left|\sigma_i \sigma_{i-\ell}^T\right|$ is a 2 x 2 matrix:

$$\left| \sigma_i \, \sigma_{i-\ell}^T \right| = \begin{bmatrix} S_{11} & S_{12} \\ S_{21} & S_{22} \end{bmatrix}$$

then

$$\text{vec} \left| \sigma_i \, \sigma_{i-\ell}^T \right| = \begin{bmatrix} S_{11} \\ S_{21} \\ S_{12} \\ S_{22} \end{bmatrix}$$

Similarly, define a set of vectors, vec $\left[F_1 \text{ (i, } \ell)\right]$, vec $\left[F_2 \text{ (i, } \ell)\right]$, ... and vec $\left[F_N \text{ (i, } \ell)\right]$ using matrices $F_1$ (i, $\ell$), $F_2$ (i, $\ell$), ... and $F_N$ (i, $\ell$) for $\ell > 0$. In the case when $\ell = 0$, both vectors vec $\left[F_n \text{ (i, 0)}\right]$ and vec $\left[\sigma_i \, \sigma_i^T\right]$ have repeated entries. That is, in the above example, $S_{12} = S_{21}$ for $\ell = 0$. Repetition of this type will be removed when forming the vectors.

Next, define a matrix:

$$F \text{ (i, } \ell) = \left[\text{vec } F_1 \text{ (i, } \ell) \quad \text{vec } F_2 \text{ (i, } \ell) \ldots \text{vec } F_N \text{ (i, } \ell)\right] \tag{3.29}$$

Clearly, the observation vector defined in equation 3.28 is related to the matrix defined in 3.29:

$$\beta \text{ (i, } \ell) = F \text{ (i, } \ell) \, \alpha + \eta \text{ (i, } \ell) \tag{3.30}$$

in which $\eta$ (i, $\ell$) is the observation noise vector. Equation 3.30 represents the *measurement* equation, in which $\alpha$ is the state vector to be estimated.

The double variables (i, $\ell$) appearing in equation 3.30 may be converted into a single variable using the transformation:

$$t = (i-1)\,(L+1) + \ell \tag{3.31}$$

where L is the maximum value of the shift interger $\ell$.

Hence equation 3.30 can be written as:

$$\beta \text{ (t)} = F \text{ (t)} \, \alpha + \eta \text{ (t)} \tag{3.32}$$

or for simplicity

$$\beta_t = F_t \alpha + \eta_t \tag{3.33}$$

with the understanding that the observations are ordered as follows: $\beta(1,0),\beta(1,1),...\beta(1,L),\beta(2,0),\beta(2,1),...\beta(2,L),\beta(i,0),\beta(i,1),...\beta(i,L),...\beta(L,L)$.

The problem is to find the best estimate of $\alpha_t$, denoted by $\hat{\alpha}_t$, that minimizes the weighted square function:

$$J_t = \eta_t^T W_t^{-1} \eta_t \tag{3.34}$$

in which $W_t^{-1}$ is the weighted function.

Solution to this problem is well documented in Kalman filtering literatures:

$$\hat{\alpha}_{t+1} = \hat{\alpha}_t + P_{t+1} F_{t+1}^T W_{t+1}^{-1} \left[\beta_{t+1} - F_{t+1} \hat{\alpha}_t\right] \tag{3.35}$$

where

$$P_{t+1} = P_t - P_t F_{t+1}^T \left[W_{t+1} + F_{t+1} P_t F_{t+1}^T\right]^{-1} F_{t+1} P_t \tag{3.36}$$

Recursive computation begins with initial values for $\alpha_0$ and $P_0$:

$$\alpha_0 \equiv E\{\alpha\} \quad \text{(a priori estimate)}$$

$$P_0 \equiv E\left\{(\alpha - \alpha_0)(\alpha - \alpha_0)^T\right\}$$

For minimum variance estimation errors, $W_t$ must-be chosen to be

$$W_t = E\left\{\eta_t \eta_t^T\right\}$$

*Part Two — Adaptation of First Order Noise Statistics*

Consider equations 3.1 and 3.2 again, in which the mean values, $\mu_u$ and $\mu_v$, are now the only unknowns to be estimated. Thus, all quantities that do not depend on the noise means are assumed optimal, e.g., $G_i^*$ will be replaced with $G_i$, $Q^*$ will be replaced with Q, etc. The cost function to be minimized is

$$J = \sum_{i=1}^{m} \nu_i^{*T} V_i^{-1} \nu_i^* \tag{3.37}$$

where $\nu_i^*$, is the innovation sequence, m is the terminal time index in the interval of interest and $V_i$ is the covariance of $\nu_i$.
Define:

$$\Delta \mu_v = \mu_v - \mu_v^* \quad \text{and} \quad \Delta \mu_u = \mu_u - \mu_u^* \tag{3.38}$$

From the above definitions, equations 3.11 and 3.10 can be written as:

$$\nu_i^* = H_i e_i^* + \Delta \mu_v + (v_i - \mu_v) \tag{3.39}$$

and

$$e_{i+1}^* = (\phi_i - G_i H_i) e_i^* + \Gamma_i \Delta \mu_u - G_i \Delta \mu_v + \Gamma_i (u_i - \mu_u) - G_i (v_i - \mu_v) \tag{3.40}$$

respectively. In the optimal case ($\Delta\mu_v = \Delta\mu_u = 0$), equations 3.39 and 3.40 become:

$$\nu_i = H_i e_i + v_i - \mu_v \tag{3.41}$$

$$e_{i+1} = (\phi_i - G_i H_i) e_i + \Gamma_i (u_i - \mu_u) - G_i (v_i - \mu_v) \tag{3.42}$$

Define:

$$\Delta \nu_i^* = \nu_i^* - \nu_i \quad \text{and} \quad \Delta e_{i+1}^* = e_{i+1}^* - e_{i+1} \tag{3.43}$$

Combine equations 3.39 and 3.41:

$$\Delta \nu_i^* = H_i \Delta e_i^* + \Delta \mu_v \tag{3.44}$$

Combine equations 3.40 and 3.42:

$$\Delta e_{i+1}^* = (\phi_i - G_i H_i) \Delta e_i^* + \Gamma_i \Delta \mu_u - G_i \Delta \mu_v \tag{3.45}$$

Let

$$\Delta b = \begin{bmatrix} \Delta \mu_u \\ \Delta \mu_v \end{bmatrix} \quad \text{and} \quad b = \begin{bmatrix} \mu_u \\ \mu_v \end{bmatrix} \tag{3.46}$$

and

$$D_{i+1} = (\phi_i - G_i H_i) D_i + \left[\Gamma_i \,\vdots\, -G_i\right] \tag{3.47}$$

Combine equations 3.45, 3.46 and 3.47:

$$\Delta e_i^* = D_i \Delta b \tag{3.48}$$

Hence equation 3.44 can be written as:

$$\Delta \nu_i^* = H_i D_i \Delta b + \Delta \mu_v \tag{3.49}$$

But from equation 3.46:

$$\Delta \mu_v = [0 \vdots I] \Delta b \tag{3.50}$$

Therefore

$$\Delta \nu_i^* = \left\{ H_i D_i + [0 \vdots I] \right\} \Delta b \tag{3.51}$$

Equations 3.47 and 3.51 will be used to implement the estimator, i.e.:

$$\nabla_{b^*} J = -2 \sum_{i=1}^{m} \frac{\Delta \nu_i^{*T}}{\Delta b} V^{-1} \nu_i^* \tag{3.52}$$

The estimation of vector b may be accomplished through the following steps:

1. Initialize the estimator with a starting guess of $b_0^*$, $V_0$, and a predetermined value of m.

2. Use of the observed data $y_i$, the suboptimal means $b_i^*$ and equation 3.26 for updating $V_i$, to compute $\nu_i^*$, $J_i$ and $\Delta J_i / \Delta b_i^*$.

   Since $V_i = E\left\{\nu_i \nu_i^T\right\}$ and $\nu_i$ and $\sigma_i$ have the same statistical properties, $V_i = E\left\{\sigma_i^* \sigma_{i-\ell}^{*T}\right\}$ with the stars removed and set $\ell = 0$.

3. Apply a gradient technique to adjust $b_i^*$ to $b_{i+1}^*$ such that $J_{i+1} < J_i$.

4. Repeat steps 2 and 3 until convergence is obtained.
   Significant remarks relating to this algorithm were given in reference 17.

## IV. SUMMARY AND DISCUSSION

The basic approach to reduce navigation error is to enhance the optimality of the Kalman filter by developing an adaptive scheme for on-line identification of the noise characteristics - specifically the first and second order Gaussian statistics. The solution was presented in two separate stages. The second order noise statistics were estimated first, then results from stage one were used to estimate the first order noise statistics. In both stages, the concept of innovation sequence was applied.

The method of determining the second order noise statistics is based on the special relationship that exists between the covariance matrices and the innovation sequence, i.e., if the system, measurement and correlation covariance matrices can be expressed as linear functions of a set of parameters, then the correlation function of the innovation sequence can also be expressed as a linear function of the *same* set of parameters. This relationship is used to perform the *weighted* least-squares fit to a sequence of measured correlation functions.

The method of determining the first order noise statistics is based on the maximum likelihood criterion and gradient techniques.

A computer program coded in Fortran for the CDC 6600 system was developed to simulate a 7-state INS error dynamic system with latitude and longitude error observations. Data for the simulation was generated by a Monte Carlo technique, rejecting observation errors outside of $\pm 2\sigma$ range. Computer execution time for 500 stages with correlation shift up to 4 was 129 seconds and compilation time was 12 seconds. Memory requirement was 40,100 octal words for a system in which each real/integer variable requires one 60-bit word.

Results indicated that, while an adaptive system will eventually converge on its true statistics, a large number of closely spaced observations are required. Hence the effectiveness of increasing INS accuracy (lengthening the covert operation) depends not only on the efficiency of the algorithm, but also the operations of the boat and associated navigation computer. Recommended future studies are:

- Improved adaptive algorithms so that they will require fewer observations and yet produce better results (rapid convergence).
- Improved INS design and associated navigation computer to ensure effective employment of adaptive algorithms.

## REFERENCES

1. Lainiotis, D.G. "Optimal Adaptive Estimation: Structure and Parameter Adaptation — Part I: Linear Models," Technical Report No. 74, Electronics Research Center, The University of Texas at Austin, Austin, Texas, September 1969

2. Lainiotis, D.G. "Optimal Adaptive Estimation: Structure and Parameter Adaptation," IEEE Transactions Automatic Control, vol. AC-16, no. 2, April 1971, pp 160-170

3. Krause, D.J. and Graupe, D. "Estimation of the Statistical Parameters of the Kalman-Bucy Filter," Proceedings 3rd Symposium on Nonlinear Estimation Theory, San Diego, California, 1972

4. Maybeck, P.S. "Combined State and Parameter Estimation for On-Line Applications," Ph.D. Thesis, Department of Aeronautics and Astronautics, M.I.T., Cambridge, Massachusetts, May 1968

5. Weiss, I.M."A Surveyof Discrete Kalman-Bucy Filtering with Unknown Noise Covariances," Proceedings 1970 Guidance, Control and Flight Mechanics Conference, Santa Barbara, California, August 1970

6. Magill, D.T. "Optimal Adaptive Estimation of Sampled Stochastic Processes," IEEE Transactions Automatic Control, vol. AC-10, October 1965, pp 434-439

7. Parzen, E. "Statistical Inference on Time Series by Hilbert-Space Method I," Technical Report No. 23, Applied Mathematics and Statistics Laboratory, Stanford University, 1959

8. Kashyap, R.L. "Maximum Likelihood Identification of Stochastic Linear Systems," IEEE Transactions Automatic Control, vol. AC-15, February 1970, pp 25-34

9. Mehra, R.K. "Approaches to Adaptive Filtering," IEEE Transactions Automatic Control, vol. AC-17, October 1972, pp 693-698

10. Kailath, T. "An Innovation Approach to Least-Squares Estimation - Part I: Linear Filtering in Additive White Noise," IEEE Transactions Automatic Control, vol. AC-13, December 1968

11. Mehra, R.K. "On the Identification of Variances and Adaptive Kalman Filtering," IEEE Transactions Automatic Control, vol. AC-15, April 1970, pp 175-184

12. Carew, B. and Belanger, P.R. "Identification of Optimum Filter Steady-State Gain for Systems with Unknown Noise Covariances," IEEE Transactions Automatic Control , vol. AC-18, no. 6, December 1973, pp 582-587

13. Belanger, P.R. "Estimation of Noise Covariance Matrices for a Linear Time-Varying Stochastic Process," Automatica, vol. 10, 1974, pp 267-275

14. Neethling, C. and Young, P. "Comments on Identification of Optimum Filter Steady State Gain for Systems with Unknown Noise Covariances," IEEE Transactions Automatic Control, vol. AC-19, no. 5, October 1974, pp 623-625

15. Friedland, B. "Treatment of Bias in Recursive Filtering," IEEE Transactions Automatic Control, vol. AC-14, August 1969, pp 359-367

16. Lin, J.T. and Sage, A.P. "Algorithms for Discrete Sequential Maximum Likelihood Bias Estimation and Associated Error Analysis," IEEE Transactions Systems, Management, and Cybernetics, vol. SMC-1, October 1971, pp 314-324

17. Godbole, S.S. "Kalman Filtering with No A Priori Information about Noise - White Noise Case," IEEE Proceedings Decision and Control Conference, 1973

# ROC PERFORMANCE OF THE ALE AUGMENTED DFT SQUARE LAW DETECTOR

J. Zeidler, Naval Ocean Systems Center, Code 6322, San Diego, CA 92152
M. Shensa, Hydrotronics, Inc., 3420 Kenyon, Suite 111, San Diego, CA 92110
J. T, Rickard, Orincon Corp., 3366 N. Torrey Pines Ct., La Jolla, CA 92037
M. Dentino, Rockwell Int., P. O. Box 3105, Anaheim, CA 92803

Abstract

This paper discusses the comparative Receiver Operating Characteristics (ROC) of Adaptive Line Enhancer (ALE) augmented and conventional Discrete Fourier Transform (DFT) square law detectors for sinusoids in both stationary white Gaussian noise and in nonstationary noise. Probability density functions (pdf's), moments, and ROC curves are derived analytically for sinusoids in stationary white Gaussian noise and verified using extensive Monte Carlo computer simulations.

## 1.0 INTRODUCTION

Several distinct types of data adaptive detection structures have been proposed for use in environments where insufficient a priori information on the signal and noise statistics is available to design fixed optimal detectors. Detector structures which incorporate adaptive mean level detection algorithms, sequential detection, non-parametric statistics, pattern recognition, and decision directed feedback techniques are discussed in numerous references (see 1, 2) for a variety of signal and noise environments in which uncertainty exists. As discussed in References (1–3) the optimal detector for the detection of weak narrowband signals of unknown frequency and random phase in a stationary broadband noise background of known power can be implemented as a bank of narrow-bandpass complex filters whose passbands span the frequency range of interest. The complex output of each filter is sampled at the end of a fixed observation interval and quadratically detected to form the decision statistic for each passband. The above system may be implemented using the DFT to perform the complex filtering operation as described in (4). The decision statistics are then the set of magnitude-squared DFT coefficients corresponding to the frequency range of interest. If the signal duration is greater than the coherent integration time of the DFT, then the detected filter outputs for a given passband may be incoherently integrated to form an improved statistic.[1,2,5]

In recent years, the ALE has been proposed for use in either a prefiltering or a post-filtering role with the conventional DFT square law detector described above.[6–9] The ALE is an adaptive digital transversal filter that is designed to greatly suppress broadband components in its input, while passing narrowband



*D. G. Lainiotis and N. S. Tzannes (eds.), Advances in Control, 378–387.*

components with little attenuation. Extensive analytical treatments of ALE convergence, mean asymptotic behavior of the filter weights, and output statistics for particular classes of inputs have previously been published.[6–14] The purpose of this paper is to derive comparative ROC curves for the ALE augmented and conventional DFT square law detectors for the case of sinusoidal signals of known frequency and unknown phase embedded in white Gaussian noise of known power. The optimal detector for this case is well known[1–3] and provides an upper bound for the ROC performance of both detector implementations considered here. Extensive computer simulations are used to verify key analytical results and to provide a detailed description of comparative performance for cases in which post-detection averaging was employed. Extensive computer simulations of the relative performance of the ALE augmented and conventional DFT square law detectors will also be provided for inputs consisting of a sinusoid in a nonstationary noise background.

## 2.0 SIMULATION CONFIGURATION

A Monte Carlo simulation was made by passing either white noise or a signal and white noise through both the ALE augmented and conventional detection systems pictured in Figure 2. Results included (a) the pdfs of the ALE augmented detector output with and without post-detection integration, (b) comparative ROC curves, (c) the sensitivity in ROC performance to the choice of adaptive time constant, and frequency resolution for the ALE operating over various detection intervals. For the noise only condition, over 250,000 independent test samples were used to generate noise only pdf's and determine the threshold required for a given probability of false alarm ($P_{FA}$). A total of 400 independent samples was made at each input signal to noise ratio (SNR) for the signal plus noise input condition. The signal plus noise simulations were repeated for a range of SNR's. The associated probability of detection ($P_D$) as a function of threshold setting was also obtained for each SNR. A small number of samples (400) for the signal plus noise conditions was adequate for the determination of intermediate values of $P_D$. In all cases the noise bandwidth was 512 Hz, the sampling rate was 1024 Hz, and the signal was bin-centered with respect to the output FFT. The parameters that were varied include input SNR, the number of ALE weights, L, the length of the FFT, K, the ALE resolution ($\beta_1$ = 1024/L Hz), the FFT resolution ($\beta_2$ = 1024/K Hz), the adaptive time constant $\tau$, and the total post-detection integration time T.

## 3.0 ADAPTIVE LINE ENHANCER (ALE) WEIGHT VECTOR MODEL FOR SINUSOIDAL INPUTS IN WHITE NOISE

As illustrated in Figure 1, the ALE consists of an L-weight linear prediction filter in which the filter coefficients, $w_i(k)$, are updated at the input sampling rate, $f_s$. The adaptive filter output r(k) is defined by

$$r(k) = \sum_{i=0}^{L-1} w_i(k)x(k-i-\Delta), \tag{3.1}$$

where $\Delta$ is the prediction distance of the filter measured in iterations. The output r(k) is then subtracted from the input sequence x(k) to form an error sequence $\epsilon(k)$. The error sequence is multiplied by a scalar constant, $\mu$, and fed back to adjust the filter weights according to the Widrow-Hoff LMS algorithm

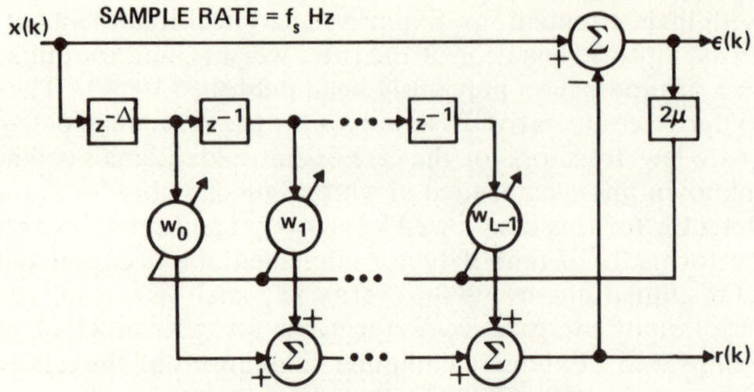


Figure 1. Block Diagram of the ALE

$$w_i(k+1) = w_i(k) + 2\mu\,\epsilon(k)\,x(k-i-\Delta). \qquad i = 0, 1, \ldots, L-1, \tag{3.2}$$

As developed in prior work,(12) a steady-state model for the converged ALE weight vector $\underline{w}(k) = [w_0(k) \ldots w_{L-1}(k)]^T$ may be obtained by decomposing w(k) into two parallel weight vectors $\underline{w}(k) = \underline{w}^* + \underline{\widetilde{w}}(k)$. Here, $\underline{w}^*$ is the optimum Wiener Δ-step predictor weight vector, and $\underline{\widetilde{w}}(k)$ is a very slowly fluctuating vector random process with a mean value of zero. Under the assumptions of relatively slow adaptation, low input SNR, and white Gaussian input noise, the weight vector "misadjustment" $\underline{\widetilde{w}}(k)$ is virtually constant over the typical integration time of a DFT. Thus $\underline{w}(k)$ may be treated as a vector Gaussian random variable (rv) $\underline{w}$, with mean $\underline{w}^*$ and covariance $\sigma_W{}^2 I$, where $\sigma_W{}^2 = \mu\xi_{min}$ and $\xi_{min}$ is the minimum mean squared error associated with Wiener filtering of the input.(6,10–12)

Consider an input to the ALE given by

$$x(k) = A \sin(\omega_0 k + \theta) + n(k), \tag{3.3}$$

where n(k) is a zero-mean white Gaussian noise sequence of variance $\nu^2$, where $\omega_0$ is known a priori. It can then be readily shown that for steady state conditions the mean of the i'th converged weight is given by (11)

$$\overline{w}_i = \frac{2a^*}{L} \cos[\omega_0(i + \Delta)], \qquad 0 \leqslant i \leqslant L-1; \tag{3.4}$$

where

$$a^* = \frac{(L/2)\,\mathrm{SNR}}{1 + (L/2)\,\mathrm{SNR}}, \qquad \text{and} \qquad \mathrm{SNR} = A^2/2\nu^2. \tag{3.5}$$

As discussed in (10), the filter weights converge exponentially to the mean value defined by Equation (3.4) with a time constant $\tau$ given by

$$\tau = \left[ 2\mu\nu^2 f_s \left( 1 + \frac{L}{2}\,\mathrm{SNR} \right) \right]^{-1}. \tag{3.6}$$

The statistical properties of the filter weights are discussed in (6, 13) where it is shown that $\sigma_W{}^2 = \mu\nu^2$ and for a noise only input $\overline{w(k)} = 0$.

## 4.0 PROBABILITY DENSITY OF THE ALE AUGMENTED DFT SQUARE LAW DETECTOR

### 4.1 No-Signal Case: $H_0$ Test Hypothesis

The ALE augmented DFT square law detector system, which will be considered here, is shown in Figure 2. It will be assumed that $L, K \gg 1$, and that the ALE is in steady-state operation. We will evaluate detector performance under two possible test hypotheses, $H_0$ and $H_1$, where $H_0$ implies signal absence and $H_1$ implies signal presence. Under $H_0$, the input x(k) consists of white Gaussian noise with variance $\nu^2$, and $w^* = 0$. The ALE output r(k) is given in Equation (3.1), where the $w_i(k)$ are independent identically distributed (iid) zero-mean Gaussian random variables (rv's) with variance $\sigma_w{}^2$. Given a particular realization of $\underline{w}$, r(k) is a conditional Gaussian process by virtue of the linearity of Equation (3.1). We denote this conditioning by $r(k|\underline{w})$. Now let

$$R_{j|\underline{w}} \triangleq \sum_{k=0}^{K-1} r(k|\underline{w}) \exp\left(-i\frac{2\pi jk}{K}\right) \tag{4.1}$$

be the $j^{th}$ DFT coefficient, where $0 \leqslant j \leqslant K/2 - 1$, and define $y_{j|\underline{w}} \triangleq |R_{j|w}|^2$, which is the basic test statistic of the unintegrated detector output. It is shown in (9) that the pdf of y under $H_0$, $p_0(y)$ is given by

$$p_0(y) = \frac{2a^{n+1}}{\Gamma(n+1)} \left(\frac{y}{a}\right)^{n/2} K_n(2\sqrt{ay}) , \tag{4.2}$$

where

$$a \triangleq KL \left[\nu^2 \sigma_w{}^2 \left(\frac{2K^3L}{3} - \frac{K^4}{6}\right)\right]^{-1} \text{ and } n+1 \triangleq K^2 L^2 \left(\frac{2K^3L}{3} - \frac{K^4}{6}\right)^{-1} ,$$

and $K_n(*)$ denotes the modified Bessel function of the third kind. For $K \gg M$, Equation (4.2) reduces to a $K_o$ distribution in agreement with Burdic's result.(7) For K = M, however,

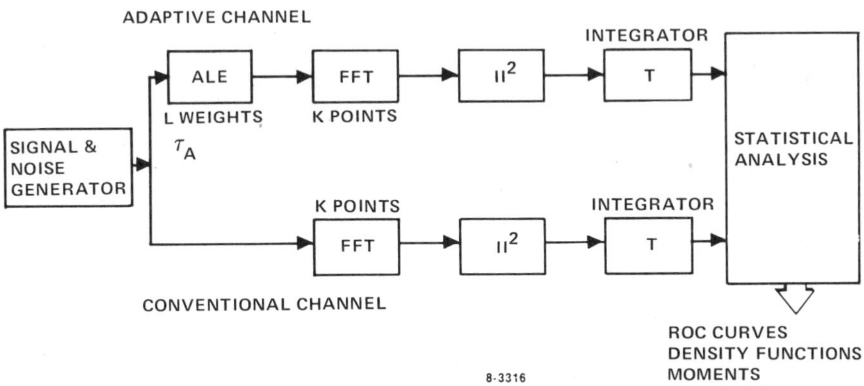


Figure 2. Simulation configuration for the comparison of ALE augmented and conventional FFT square law detectors

$$p_0(y) = \frac{2}{y}\left(\frac{2y}{\sigma^2\sigma_w{}^2KL}\right) K_1 \left(2\sqrt{\frac{2y}{\sigma^2\sigma_w{}^2KL}}\right), \tag{4.3}$$

As shown in Figure 3, for K = L = 1024, both the $K_0$ approximation and the more exact pdf defined by eq. (4.3) give good agreement with the simulation results in the tails of the pdf but substantial deviation occurs in the central part of the pdf. The mean and variance of $p_0(y)$ were derived in (14) and discussed in (8). These results show that the ratio of the squared mean, $m_0{}^2(y)$, to the variance of y under $H_0$, $\sigma_0{}^2(y)$ is given by

$$\frac{m_0{}^2(y)}{\sigma_0{}^2(y)} = \left(3 - \frac{4L}{3K} + \frac{L^2}{3K^2}\right)^{-1} \tag{4.4}$$

A comparison between this theoretical value given by Equation (4.4) and the ratio of the moments measured by Monte Carlo simulation is provided in Table 1. For a $K_0$ distribution the squared mean to variance ratio is 0.333. Table 1 and Figure 3 indicate that the $K_0$ approximation to $p_0(y)$ may be valid for $L \ll K$ but for all other values of L and K, the $K_0$ pdf provides a poor approximation to the first two moments of $p_0(y)$.

TABLE 1

| K | L | $\frac{m_0^2(y)}{\sigma_0^2(y)}$ Measured | $\frac{m_0^2(y)}{\sigma_0^2(y)}$ Theory |
|---|---|---|---|
| 256 | 1024 | .753 | .762 |
| 256 | 512 | .615 | .632 |
| 256 | 256 | .496 | .500 |
| 256 | 128 | .398 | .413 |
| 256 | 64 | .365 | .372 |

From Equation (4.3), for a given threshold, $\alpha$, the $P_{FA}$ for y under the $H_0$ hypothesis is given by

$$P_{FA} = \int_\alpha^\infty p_0(y)\,dy = 2(a\alpha)^{\frac{n+1}{2}} K_{n+1}(2\sqrt{a\alpha}) . \tag{4.5}$$

4.2 Signal-Present Case ($H_1$ Test Hypothesis)

Under $H_0$, the ALE weights are described solely by the random misadjustment filter, $\tilde{w}(k)$ and the output r(k) under $H_0$ is spectrally white, but non-Gaussian noise. For the conventional DFT square law detector the filter weights are deterministic and r(k) has a Gaussian pdf and $y_{j|\underline{w}}$ has a 2 dof gamma distribution.(1,2) When a sinusoidal signal is added to the input, as defined by Equation (3.3), there are four distinct uncorrelated components corresponding to the input signal and input noise being passed through $\tilde{w}$ and $w^*$ as described in (23). This multiplicity of output components leads to intractable analytical problems. An approximate pdf $p_1(y)$ under $H_1$ will be derived below.

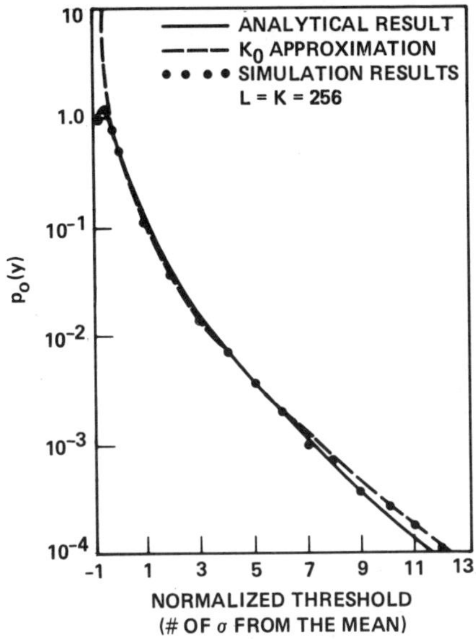


Figure 3. Comparison of the theoretical and approximate pdfs with Monte Carlo simulation results, as a function of the mean-shifted, standard-deviation-normalized, argument plotted in terms of the number of standard deviation from the mean.

The essence of this approximation for $p_1(y)$ is the observation that at the center frequency $\omega_o$, of the input sinusoid, the ALE output spectrum will be dominated by the components of the Weiner filter channel, provided that adequate $SNR_{in}$ exists. Since the noise component of w* is narrowband Gaussian noise (by virtue of the deterministic weights), we approximate the total ALE output noise as being Gaussian in the neighborhood of $\omega_o$, i.e., ignore the non-Gaussianity of the two misadjustment channel output components and lump them with the narrowband Gaussian noise. Under this approximation a K-point magnitude-squared DFT, centered on $\omega_o$ of the ALE output has a non-central gamma pdf with two dof, viz.,

$$p_1(y) = \frac{1}{2\sigma_r^2} \exp\left(-\frac{y+\tilde{A}^2}{2\sigma_r^2}\right) I_o\left(\frac{\tilde{A}}{\sigma_r^2}\sqrt{y}\right), \qquad y \geqslant 0, \tag{4.6}$$

where $\tilde{A} = \frac{a^*AK}{2}$ and $\sigma_r^2$ is the variance of the ALE output, r(k), i.e., (14)

$$\sigma_r^2 = \frac{\mu\nu^4 LK}{2} + \frac{\mu\nu^2 A^2 LK^2}{8} + \frac{2a^{*2}\nu^2 D}{L^2},$$

$$\text{where } D = \begin{cases} \dfrac{K^2L}{4} - \dfrac{K^3}{12} & K \leqslant L \\[2ex] \dfrac{L^2K}{4} - \dfrac{L^3}{12} & K \geqslant L \end{cases} \tag{4.7}$$

For a fixed threshold $\alpha$, the detection probability $P_D$ of the ALE/DFT detector is then given by

$$P_D = \int_{\alpha}^{\infty} dy\, p_1(y) = Q\left(\frac{\widetilde{A}}{\sigma_r}, \frac{\sqrt{\alpha}}{\sigma_r}\right), \tag{4.8}$$

where Q(•,•) denotes the Marcum Q-function. Figure 4 compares the ROC curves obtained by Equations (4.5) and (4.8) to experimental curves derived by computer simulation for the unintegrated ALE augmented square law detector for the case $\beta_1 = \beta_2 = 1$ Hz.

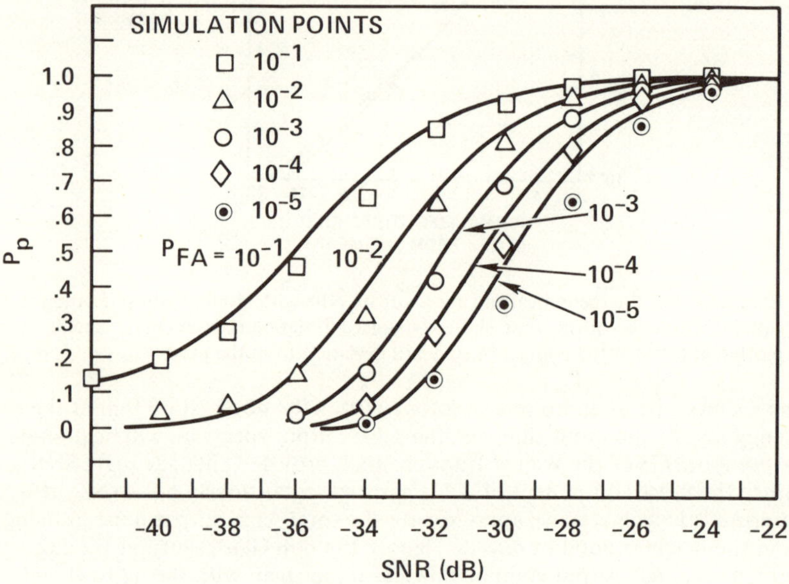


Figure 4. Comparison of theoretical $P_D$ versus input SNR with Monte Carlo simulation results for various $P_{FA}$ for the unintegrated output of an ALE augmented square law detector. K = L = 1024.

## 5.0 COMPARATIVE ROC CURVES FOR ALE AUGMENTED AND CONVENTIONAL SQUARE LAW DETECTORS

### 5.1 Sinusoids in Stationary White Noise

Comparative ROC curves for the conventional and ALE augmented detectors were presented in (8). The maximum gain of the ALE augmented detector observed in these simulations was 1.2 dB when the ALE and DFT frequency resolutions were matched ($\beta_1 = \beta_2$). The simulations provided no overlap or windowing of the FFT coefficients for bin center sinusoids. Consequently, the 1.2 dB gain was well within the bounds of the optimal Fourier receiver which is achieved using maximum time and frequency redundancy of the FFT coefficients as discussed in Reference (5). As discussed in (8), when $\beta_1 < \beta_2$ the noise suppression capability of the ALE does provide improved performance over the conventional detector. When $\beta_2 > \beta_1$, however, the ALE has been shown to degrade the detection performance (8).

## 5.2 Sinusoids in Nonstationary Noise

As discussed in (8), the mean and variance of the ALE augmented detector can be significantly reduced by increasing the ALE time constant for a fixed T and K producing a pdf with almost zero mean and a mean to standard deviation ratio which is significantly reduced in comparison to that of the conventional square law detector. For the case of a stationary noise background of known power this is of no advantage since accurate thresholds can be established for both detector implementations from a knowledge of $p_0(y)$. For the case of a non-stationary noise background, however, the reduced mean to variance ratio of the ALE augmented detector may prove advantageous if alternate means of estimating appropriate thresholds (e.g., 1, 2, 4) either prove ineffective or are impractical to implement. This is illustrated by computer simulation results below.

The type of non-stationary behavior considered in these simulations is restricted to amplitude-level fluctuations, at intervals corresponding to the integration time interval T, of an underlying zero mean, stationary white Gaussian noise process n(k) having variance $\sigma^2$. Let m denote a zero mean Gaussian random variable with variance $\sigma_m{}^2$. Then the input noise n′(k) to the detector during a particular integration interval $[\ell T, (\ell + 1)T]$, where $\ell$ is an integer, is given by

$$n'(k) = (1+m)n(k), \qquad \ell T \leqslant k \leqslant (\ell+1)T \ . \tag{5.1}$$

A new value of m is selected for each integration interval. Over the infinite time interval, the process described by Equation (5.1) is actually a stationary, non-Gaussian process. The false-alarm thresholds for both the conventional and adaptive systems must be determined over the large time period of the simulation, which approximates this latter process. However, with respect to the integration interval T in which a decision must be made, the noise process resembles a non-stationary white Gaussian process whose amplitude correlation time $\tau_m$ equals T.

For all simulations $f_s = 1024$ Hz, L = K = 1024, $T = \tau_A = 128$ sec. The degree of non-stationarity is described by the ratio $\sigma_m/\sigma$ (in percent). Both the ALE filter weights and the integrator output were reset to zero at the start of each integration interval. To illustrate the performance degradation of the two processors as a function of the noise non-stationarity, $P_D$ is plotted versus $(\sigma_m/\sigma)$, for $P_{FA} = 10^{-4}$ and a range of S/N, in Figure 5. Note that the conventional system degrades more rapidly, and at higher S/N values, than the adaptive system.

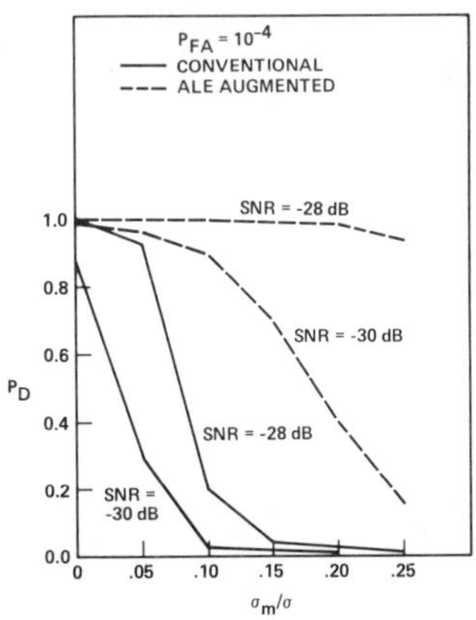


Figure 5. Comparison of the relative degradation in $P_D$ with respect to the degree of non-stationarity in the noise background for specific SNR's for ALE augmented and conventional FFT square law detectors. (Results obtained by Monte Carlo simulation.)

## 6.0 CONCLUSION

A variety of mean level detection algorithms and other related normalization procedures have been devised to design detector structures, which provide a time varying estimate of the mean noise power of the fluctuating noise level in each bin, thereby providing improved detection performance over the conventional FFT square law detector for a non-stationary noise background (e.g., References (1, 2, 4). The ALE attempts to suppress the background noise to a zero mean level, thereby maintaining a more uniform mean output noise level in the presence of noise fluctuations without any attempt to provide explicit estimates of the time varying mean noise power level. A comparison of the performance of ALE augmented detectors with other types of data adaptive detectors for non-stationary inputs remains a topic for future research. As indicated by the complexity of the analytical results presented here for a stationary white Gaussian noise background, however, a detailed treatment of this problem presents formidable analytical difficulties.

## REFERENCES


1. C. W. Helstrom, "Statistical Theory of Signal Detection," Oxford: Pergamon Press, 1968.
2. A. D. Whalen, "Detection of Signals in Noise," New York: Academic Press, 1971.
3. R. Roberts, "On the Detection of a Signal Known Except for Phase," *IEEE Trans. on Information Theory*, Vol IT-11, pp 76–82, Jan 1965.
4. J. T. Rickard and G. M. Dillard, "Adaptive Detection Algorithms for Multiple Target Spectral Data," Proc. of the 1977 IEEE Conf on Decision and Control, New Orleans, La, pp 515–18, 1977.
5. J. Williams and G. Ricker, "Spectrum Analyzer Overlap Requirements and Detectability Using Discrete Fourier Transform and Composite Filters," *J. Acous. Soc. Am.*, Vol 64, pp 815–822, Sep 1978.
6. B. Widrow et al., "Adaptive Noise Cancelling: Principles and Applications," *Proc. IEEE*, Vol 63, pp 1692–1716, Dec 1975.
7. W. S. Burdic, "Detection of Narrowband Signals using Time Domain Adaptive Filters," *IEEE Trans. on Aerospace and Electronics Systems*, Vol AES-14, pp 578–591, July 1978.
8. M. Dentino, M. Shensa, and J. Zeidler, "Detection Performance of an ALE Augmented Detector with Post-Detection Integration," *Proc. of the IEEE Conf. on Decision and Control*, San Diego, CA, pp 1374–1377, Jan 1979.
9. J. T. Rickard, M. Dentino, and J. Zeidler, "Detection Performance of an Adaptive Processor in Non-Stationary Noise," *Proc. of the 1979 IEEE Conf. on ASSP*, Washington, DC, pp 136–139, Apr 1979.
10. J. Treichler, "Transient and Convergent Behavior of the ALE," *IEEE Trans. Acoustics, Speech, and Signal Proc.*, ASSP-27, pp 53–63, Feb 1979.
11. J. R. Zeidler, E. H. Satorius, D. M. Chabries, and H. T. Wexler, "Adaptive Enhancement of Multiple Sinusoids in Uncorrelated Noise," *IEEE Trans. Acoust. Speech, Signal Processing*, Vol ASSP-26, pp 240–254, June 1978.

12. J. T. Rickard and J. R. Zeidler, "Second-Order Output Statistics of the Adaptive Line Enhancer," *IEEE Trans. Acoust., Speech, Signal Processing.* Vol ASSP-27, pp 31–39, Feb 1979.

13. M. Dentino, H. Huey, W. S. Burdic, and J. R. Zeidler, "Statistical Properties of the Adaptive Line Enhanced (ALE) Processor," *Proc. 12th Asilomar Conf. on Circuits, Sys., and Computers*, pp 184–189, Nov 1978.

14. M. Shensa, "A Determination of the First Two Moments of the ALE Output Detector," Naval Ocean Systems Technical Report 330, Nov 1978.

# STATISTICAL ANALYSIS OF A SIGNAL FOR FAULT LOCATION

C. Cresswell, D.R. Towill and J. Hywel Williams

Dynamic Analysis Group, The University of Wales
Institute of Science and Technology, Cardiff, U.K.

For fault detection and location in a large hardware electro-mechanical control system it is required to sift the noise from a corrupted very low frequency sinusoidal message. The noise may be regarded as having pseudo-gaussian characteristics supplemented by an impulse like disturbance which occurs twice per period. In assessing the operational status of the system both message and noise must be recovered to a high degree of accuracy. This is because the response pattern on which operational status is based depends equally on the noise and message characteristics, as distinct from normal measurement techniques where the message has much more significance.

This paper describes various statistical methods used to remove the sinusoidal trend line from a sample test input signal embedded in white noise. The techniques used for 'cleaning' the sine wave include Least Squares Analysis and iterative procedures such as Exponential Smoothing and Maximum Likelihood Estimators. A comparison is made of the efficiency of each technique in removing the cycle trend line. The long term aim is the micro-processor implementation on-line as an automatic status indicator.

## 1. INTRODUCTION

The aim of this paper is to consider methods which will extract the noise from a corrupted, very low frequency sinusoidal message. The noise which has pseudo-gaussian characteristics supplemented by an impulse like disturbance is required for use in a fault detection algorithm.



*D. G. Lainiotis and N. S. Tzannes (eds.), Advances in Control, 388–398.*

Typical pen records of an error channel showing such noise patterns are shown in Fig.1(1).

In assessing the operational status of the system both message and noise must be recovered to a high degree of accuracy. This is because the response pattern on which operational status is based depends equally on the noise and message characteristics, as distinct from normal measurement techniques where the message has much more significance.

This paper describes various statistical methods used to remove the sinusoidal trend line from a sample test input signal embedded in gaussian noise but without any impulse disturbance included. The techniques used for "cleaning" the sine wave are:

(i) Least Squares Analysis
(ii) Exponential Smoothing
(iii) Maximum Likelihood Estimators

A comparison is made of the efficiency of each technique to estimate the cyclic trend line from the noisy signal. The Least Squares method is also used to estimate the trend line when the impulse disturbance is added to the noise.

## 2. STATISTICAL TECHNIQUES

In the Least Squares Analysis, normal equations are derived for fitting a sinusoidal trend line to a set of observations. Explicit relationships are obtained to estimate the amplitude and phase angle parameters. In the other techniques, which are both iterative procedures, established formulae are used to estimate the trend line.

### 2.1. Least Squares Analysis

This technique assumes that we can fit a model or trend line, in this case a sine wave, to a set of observations. From this model normal equations are derived by minimizing the error between the observation and the expected trend line and from these equations estimates are made of the parameters of the models. The main difference between this method and the other procedures discussed later in Sections 2.2 and 2.3 is that in the Least Squares Analysis all the observations available are taken into account when the parameters in the model are estimated. It is however possible to develop an iterative procedure for this Least Squares Analysis but this is not developed here.

We wish to fit to our set of observations a model of the form

$$y_i = B \sin(\omega t_i + \psi) + e_i \tag{2.1}$$

where B = amplitude, $\omega$ = frequency, $\psi$ = phase angle,
$e_i$ = measurement error

It is not possible in this case to find explicit relationships for the estimates of amplitude and phase, B and $\psi$, using normal least squares procedure. However, if we rewrite equation (2.1) in the form

$$y_i = B_1 t_{i1} + B_2 t_{i2} + e_i \tag{2.2}$$

where $B_1 = B \cos \psi$, $B_2 = B \sin \psi$, $t_{i_1} = \sin(\omega t_i)$,

$t_{i_2} = \cos(\omega t_i)$

the new estimates $B_1$ and $B_2$ may be found by minimizing $\sum_{i=1} e_i^2$.

This procedure yields

$$\hat{B}_1 = \frac{\sum_{i=1}^{n} t_{i2}^2 \sum_{i=1}^{n} y_i t_{i1} - \sum_{i=1}^{n} y_i t_{i2} \sum_{i=1}^{n} t_{i1} t_{i2}}{\sum_{i=1}^{n} t_{i1}^2 \sum_{i=1}^{n} t_{i2}^2 - \left( \sum_{i=1}^{n} t_{i1} t_{i2} \right)^2} \tag{2.3}$$

and

$$\hat{B}_2 = \frac{\sum_{i=1}^{n} t_{i1}^2 \sum_{i=1}^{n} y_i t_{i2} - \sum_{i=1}^{n} y_i t_{i1} \sum_{i=1}^{n} t_{i1} t_{i2}}{\sum_{i=1}^{n} t_{i1}^2 \sum_{i=1}^{n} t_{i2}^2 - \left( \sum_{i=1}^{n} t_{i1} t_{i2} \right)^2} \tag{2.4}$$

The original constants B, $\psi$ are then found from equation (2.2) i.e. $B_1 = B \cos \psi$, $B_2 = B \sin \psi$.

$$\hat{B} = \frac{\hat{B}_1}{\cos \psi} , \qquad \hat{\psi} = \tan^{-1}\left(\frac{\hat{B}_2}{\hat{B}_1}\right) \tag{2.5}$$

2.2 Exponential Smoothing

The technique of Exponential Smoothing has been widely used (2,3). It is a method by which a current estimate of a given process is made based on past observations in such a way that more weight is placed on current observations than those obtained a long time ago.

If we have a set of observations $y_n$, n = 1,2......, then the smoothed function of the observation is given by

$$E_n = \Theta y_n + (1 - \Theta) E_{n-1} \qquad 0 < \Theta < 1, \; n = 1, 2, \ldots \tag{2.6}$$

In this analysis the values of the parameter $\theta$ were taken to be 0.2, 0.3. The initial estimate $E_o$ was assigned the value of zero.

## 2.3. Maximum Likelihood Estimators

A similar form to the Exponential Smoothing technique can be developed using basic Maximum Likelihood Estimation(4).

Suppose we have an observation $y_n$, then the next estimate $\hat{x}_{n+1}$ is given by

$$\hat{x}_{n+1} = \hat{x}_n + b\,(y_n - M\,\hat{x}_n) \qquad (2.7)$$

where

$\hat{x}_n$ is the previous estimate.

$b$ is the maximum likelihood weighting factor given by

$$b = \frac{M\,\sigma_x^2}{\sigma_x^2 + \sigma_\varepsilon^2}$$

$\sigma_x^2$ is the variance of the noise
$\sigma_\varepsilon^2$ is the variance of the system
$M$ is a scale factor based on the ratio of the input to output signal

In this case the value of the scale factor M was taken to be 1 and the variance of the system $\sigma_\varepsilon^2$ was taken to be 10% of the noise variance $\sigma_x^2$. The initial value $x_o$ was zero.

## 3. RESULTS

A computer program was written to perform the three statistical techniques described in section 2. Equations (2.3), (2.4), (2.5) were programmed for the Least Squares Analysis, equation (2.6) in the Exponential Smoothing and equation (2.7) for the Maximum Likelihood Estimators. Details of the program can be found in Ref.6.

The program generates a sine wave and normally distributed noise $(N(0,\sigma^2))$ was added to produce the input signal. The standard deviation of the noise was varied (0.1, 0.5) and each technique was tested with the noisy signals. Fig.2 shows the noisy input signals used in all the smoothing methods for the two different levels of noise. Superimposed on the figure is the pure sine wave before any noise was added.

The curves obtained using the Least Squares Analysis are shown in Fig.3. We can see that no significant differences in the Least Squares "best fit" have resulted from increasing the noise levels in the input data. Figs. 4 and 5 show the results obtained using the Exponential Smoothing technique for two standard deviations of noise, namely 0.1 and 0.5 respectively. Each figure has the curves for each of the two smoothing parameters $\theta$ ($\theta$ = 0.2 and 0.3) and the input sine wave without any noise added to it.

From Fig.4 we see that for a small level of noise the Exponential Smoothed curves are good fits for the smooth sine wave except that the amplitude has been reduced and there is a phase shift. As the noise level increases (Fig.5), the Exponential curves are not as "smooth". Fig.6 shows the Maximum Likelihood Estimators for the various noise levels, also shown is the pure sine wave. We can see this method has not produced very good estimates to the sine wave.

Table 1 shows the maximum modular error between the sine wave excluding noise and the smoothed curves using the various statistical techniques. It should be noted that the errors for noise standard deviations other than those shown in the diagrams are included in the Table. These were included in the analysis to test the effectiveness of the techniques, when a range of noisy input signals were used.

Table 1: MAXIMUM ERROR

| Standard Deviation of Noise ($\sigma$) | Least Squares Analysis | Exponential Smoothing $\theta$=0.2 | $\theta$=0.3 | Maximum Likelihood |
|---|---|---|---|---|
| 0.1 | 0.009 | 0.506 | 0.348 | 0.201 |
| 0.2 | 0.018 | 0.481 | 0.393 | 0.386 |
| 0.5 | 0.046 | 0.578 | 0.605 | 0.961 |
| 1.0 | 0.091 | 0.848 | 1.007 | 1.936 |

The maximum modular error for the Least Square Analysis is approximately 9% of the standard deviation of the noise, and in all cases tested this error is very small. For the other techniques the error increases substantially with the increase in noise deviation.

We have seen from the results that the technique which enables us to estimate the underlying trend line, namely the sine wave, from the noisy input data with minimum error is the Least Squares Method.

In the analysis to date no impulse disturbances such as the ones shown in Fig.1, have been included in the noise.

The Least Squares Method was used to estimate the sinusoidal trend line from noisy data when such a disturbance was added. A unit impulse response (5) from a system whose transfer function is given by

$$\frac{1}{1 + \frac{0.4}{\omega_n} s + \frac{s^2}{\omega_n^2}}$$

was added to the input signal (shown in Fig.2) at two points in the period to give the new input signal as shown in Fig.7. Also shown on the figure is the Least Squares "best fit" for the given input signal. We can see that by adding the impulse disturbance to the input signal only a minor increase has occurred in the amplitude in Least Squares estimate. The errors, between the noisy input signal and the Least Squares estimates, also shown in Fig.7 show that the noise and disturbances may be obtained to a fair degree of accuracy which will be required in any subsequent fault detection or location analysis.

## 4. CONCLUSIONS

Of the three methods considered the Least Squares Analysis has clearly produced the best estimate of the sinusoidal trend line from noisy input signals. This will enable the noise to be recovered to a high degree of accuracy to allow further analysis to be carried out.

In the Exponential Smoothing technique, the results were quite good for small levels of noise although a reduction in amplitude and a shift of phase was noted. It would be necessary in this case to develop some form of adaptive procedure to take these factors into account. When noise levels were high and in the Maximum Likelihood Method the output produced was not a good comparison with the pure sine wave.

## 5. ACKNOWLEDGEMENTS

This work has been carried out with the support of the Procurement Executive of the Ministry of Defence.

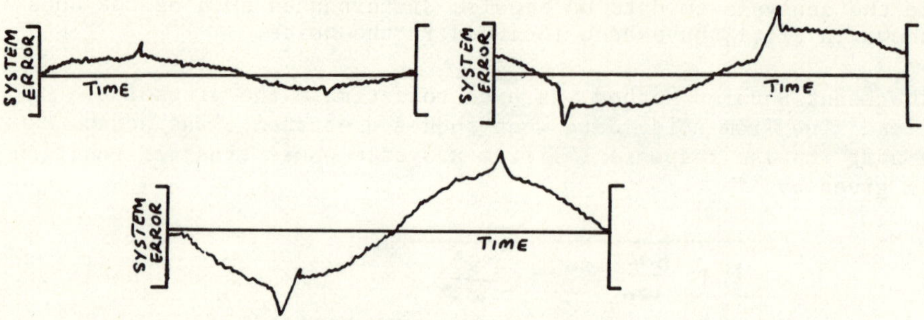


Fig. 1: Records of Error Channel Tests

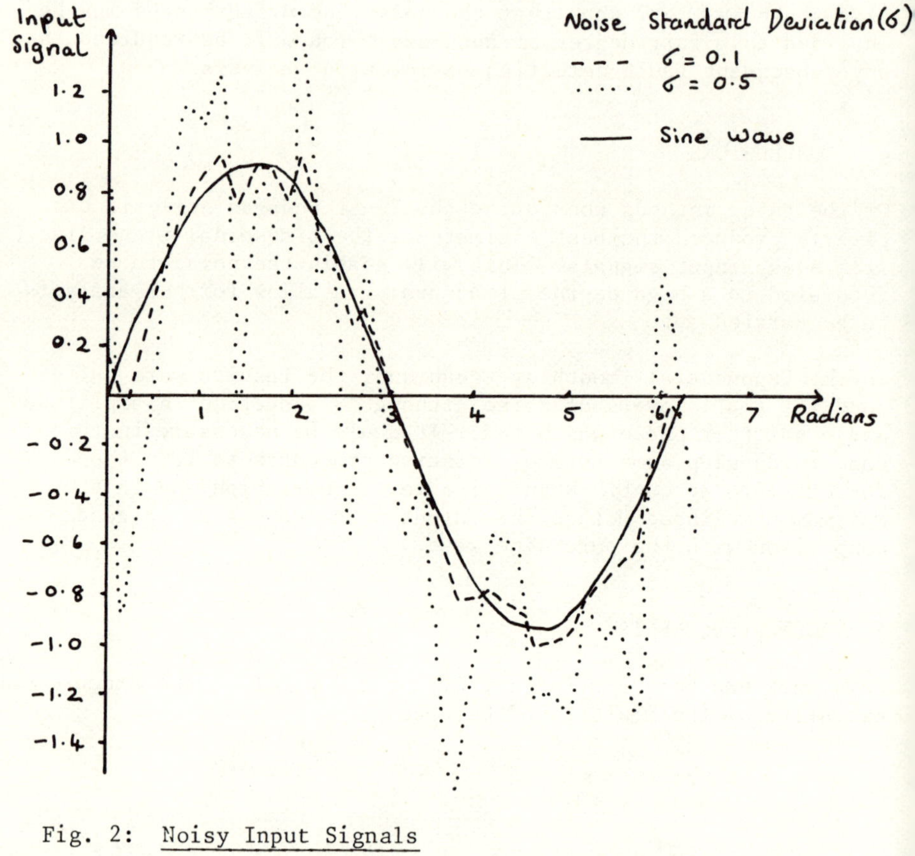


Fig. 2: Noisy Input Signals

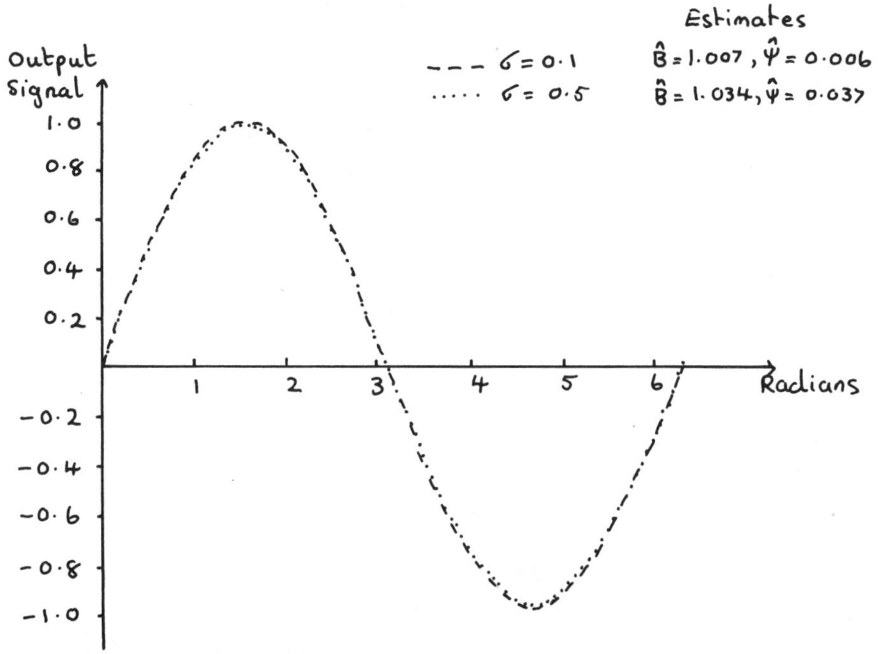


Fig. 3: Least Squares "Best-Fit" Curves

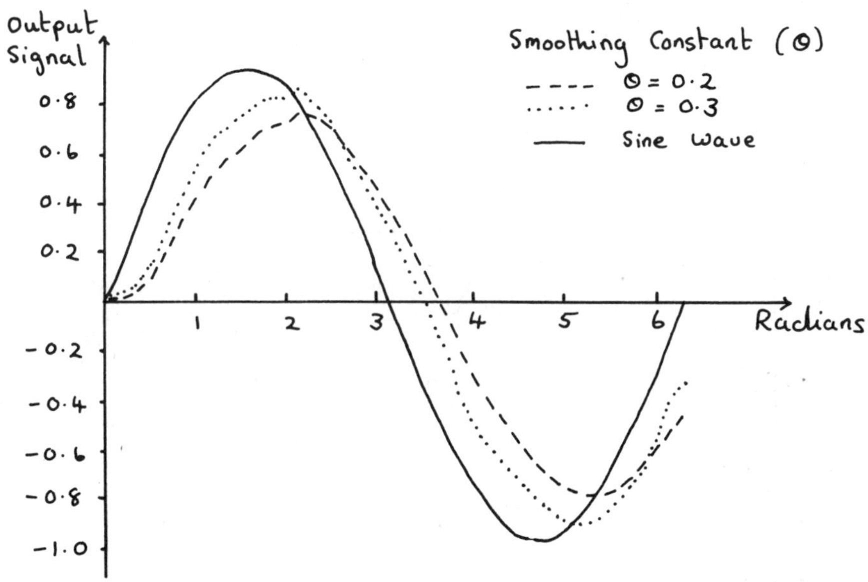


Fig. 4: Exponential Smoothed Curves for Noise Standard Deviation, σ = 0.1

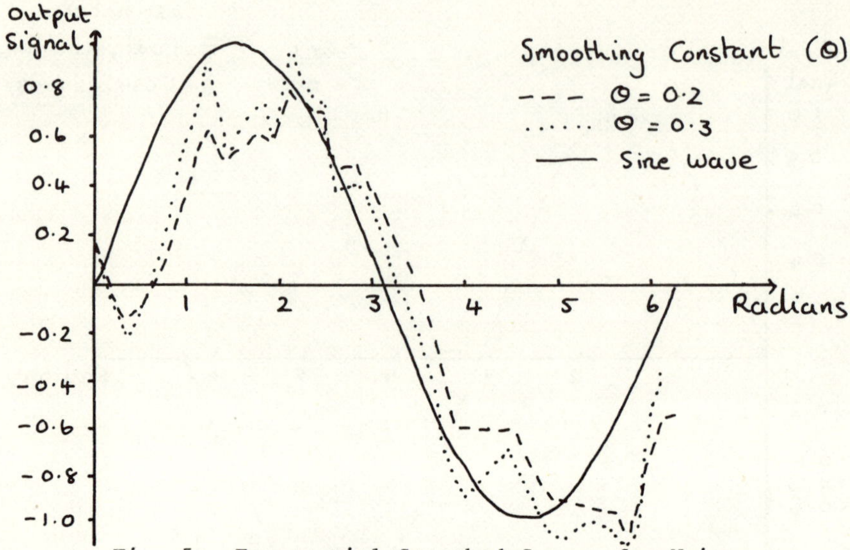


Fig. 5: Exponential Smoothed Curves for Noise Standard Deviation, σ = 0.5

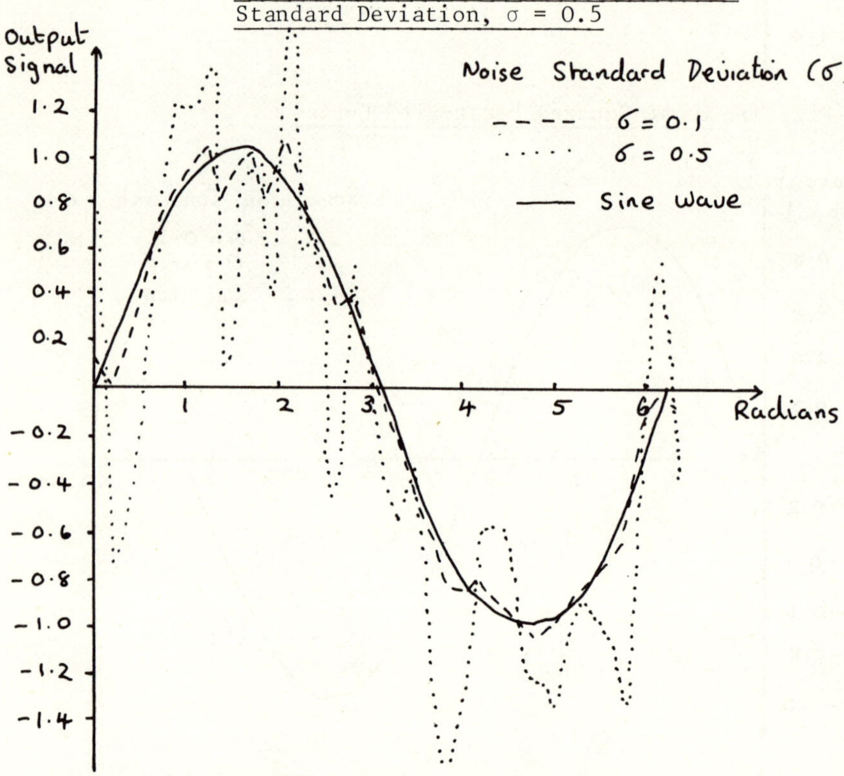


Fig. 6: Maximum Likelihood Estimates

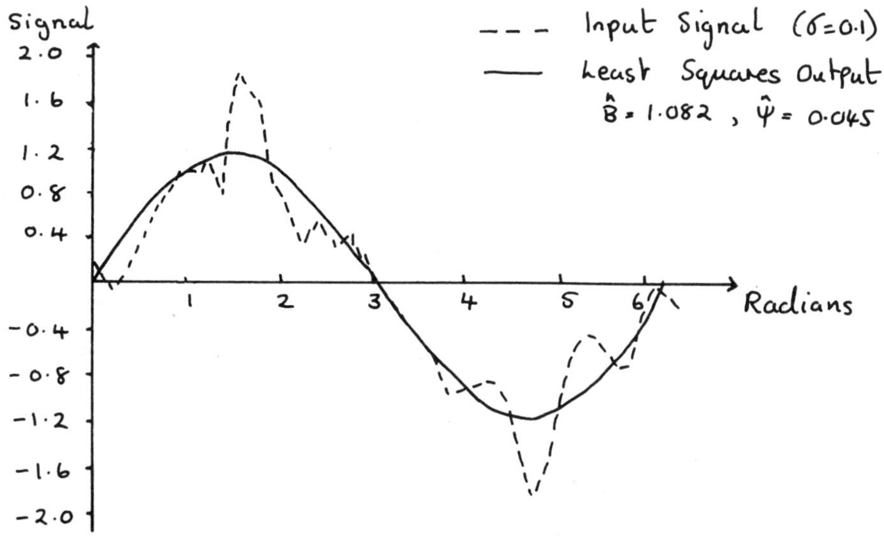


Fig. 7: Least Squares Fit for Input Data with an Impulse Disturbance

REFERENCES

1. Harley, H.V. and Towill, D.R.: "Automation of Large Electro-Mechanical System Testing, ", JNS. Vol.2, No.2, pp 128-136.

2. Brown, R.G.: "Smoothing, Forecasting and Prediction", Prentice Hall, 1962.

3. Williams, J.H.: "Preliminary Note on Data Processing Schemes for Quality Assurance", Technical Note DAG50, UWIST, Cardiff.

4. du Plessis, R.M.: "Poor Man's Explanation of Kalman Filtering", Autonectics Division of North American Rockwell Corporation, June 1967.

5. Dynamic Analysis Group, UWIST, Control Engineering Data Sheets, Issue DRT/1, Sept. 1975.

6. Cresswell, C., Williams, J.H., Towill, D.R.: "Statistical Evaluation of Filtering Techniques Applied to Noisy Low Frequency Waveforms", Technical Note DAG 126, UWIST, Cardiff.

# OPTIMAL CONTROL OF DISTRIBUTED-PARAMETER SYSTEMS VIA THE MATRIX MINIMUM PRINCIPLE

P. Stavroulakis
School of Engrg., Oakland Univ.
Rochester, Michigan

S. Tzafestas
Control Systems Lab.
Patras Univ., Greece

ABSTRACT: In this paper it is shown that the matrix minimum principle can be used to study both deterministic and stochastic control systems described by partial differential equation (PDE) models. It is shown that the matrix formulation of the control problem reduces to the state canonical equations which are identical to the Riccati equations of the conventional state vector representation. The technique employs the powerful optimization techniques of calculus of variations and dynamic programming and lends in a natural way the separation principle for the case of stochastic optimal control problems.

## 1. INTRODUCTION

The purpose of this paper is to provide a statement of the necessary conditions for optimality for distributed parameter control systems which are described by matrix PDEs. A matrix minimum principle for lumped parameter control systems has been studied previously in (1). This study is aiming at extending these results to distributed parameter systems and show that the matrix minimum principle for distributed parameter control systems can be used to derive necessary conditions for optimality for problems which are most conveniently described by matrix equations. The relationship of the canonical equations, thus derived, to the equivalent vector canonical equations is also pointed out. It is shown that the Riccati equation in the vector state formulation becomes one of the canonical equations in the equivalent matrix equation.

The same principle is applied to stochastic systems and it is shown that the separation theorem is derived in a much simpler way.



*D. G. Lainiotis and N. S. Tzannes (eds.), Advances in Control, 399–407.*

## II. PROBLEM FORMULATION AND CANONICAL EQUATIONS

Control problems which are described by matrix PDEs can be thought of as formulations multidimensional systems whose state variables are identical to the elements of the state matrix. The same argument holds for the control variables.

As an example consider the following system which is described by the matrix partial differential equation.

$$\frac{\partial X}{\partial t}(\underline{\eta},t)=\mathcal{L}_\eta\{X\}+BU(\underline{\eta},t) \tag{1}$$

Consider also the cost functional

$$J=\int_0^T\int_\Omega F(X,U)d\eta dt\ , \tag{2}$$

which is to be minimized subject to Eq.(1) where F(.,.) is a scalar valued function of the matrix arguments X,U. Most often we encounter the trace function in place of F(.,.). This is so because of the natural way we define functionals in $L^2$ space and because of the fact that the inner product topology of the space of vectors is preserved under the transformation

$$\underline{x}=[x_{11},x_{12},\ldots,x_{21},x_{22},\ldots,x_{nn}]=\Psi(X) \tag{3}$$

if the inner product of the matrices is defined by:

$$\text{Inner product }(X,Y)=\text{trace }X.Y \tag{4}$$

Similar arguments were used in (1) to prove equivalency between the results of the vector state and matrix state formulations and give a justification of the matrix minimum principle. If we define the Hamiltonian function by

$$H=F(X,U)+\text{trace}(\mathcal{L}_\eta\{X\}+BU)P' \tag{5}$$

then

$$J=\int_0^T\int_\Omega(H-\frac{\partial X}{\partial t}.P')d\eta dt \tag{6}$$

Using procedure similar to that employed in (8) and omitting the details we derive the following set of canonical equations

$$\frac{\partial X}{\partial t}=\mathcal{L}_\eta\{X\}+BU \tag{7}$$

$$-\frac{\partial P}{\partial t}=\frac{\partial F}{\partial X}+\mathcal{L}_\eta^*\{P\} \tag{8}$$

in addition to the condition

$$\partial H/\partial U=0 \tag{9}$$

where $\mathcal{L}_\eta^*$ is the adjoint operator of $\mathcal{L}_\eta(.)$.

We observe that the necessary conditions involve gradient matrices such as

$$\frac{\partial F}{\partial X}, \frac{\partial H}{\partial U} \tag{10}$$

A definition of gradient matrices and some examples can be found in (1) and (2). Boundary conditions have been omitted for simplicity. These canonical equations constitute the necessary conditions for the optimality of the problem stated by Eqs. (1) and (2). Using these necessary conditions, we shall proceed to study linear and stochastic control problems.

## III. DETERMINISTIC LINEAR CONTROL PROBLEMS

Consider the linear control problem described by the following vector partial differential equations

$$\frac{\partial x}{\partial t}(\underline{\eta},t)=\mathcal{L}_{\eta}\{x\}+Bu \tag{11}$$

where $x(\underline{\eta},t)$ is a state vector dependent on the parameters $\underline{\eta}$ and t denoting space and time respectively, $\mathcal{L}_{\eta}(.)$ is a differential matrix linear operator, and B is for simplicity a constant matrix of appropriate domensions.

It is further assumed that the problem defined by Eq. (11) is well-posed in the region $\Omega$ X T defined by the space parameter $\underline{\eta}$ and the time interval $[0,T]$ . Let the cost functional be given by

$$J=\int_0^T\int_\Omega x'(\underline{\eta},t)Qx(\underline{\eta},t)+u(\underline{\eta},t)Ru(\underline{\eta},t)d\underline{\eta}dt \tag{12}$$

We seek to find an optimal control law of the "feedback type" such that

$$\underline{u}^*(\underline{\eta},t)=\int_\Omega M(\underline{\eta},\underline{\eta}',t)x(\underline{\eta}',t)d\underline{\eta}' \tag{13}$$

and the cost functional (12) is minimized subject to Eq. (11).

To apply the matrix minimum principle to this type of problem we shall reformulate Eq. (11) as a matrix equation.

Define

$$X(\underline{\eta},\underline{\eta}',t)=\underline{x}(\underline{\eta},t)\underline{x}'(\underline{\eta}',t) \tag{14}$$

where $\underline{x}'(\underline{\eta}',t)$ denotes the transpose (row) of the vector $x(\underline{\eta},t)$ dependent on a new variable $\underline{\eta}'$.

If we apply the operator $\frac{\partial}{\partial t}(.)$ to Eq. (14) we obtain

$$\frac{\partial X}{\partial t}(\underline{\eta},\underline{\eta}',\tau)=\frac{\partial \underline{x}}{\partial t}(\underline{\eta},t).\underline{x}'(\underline{\eta}',t)+\underline{x}(\eta,t).\frac{\partial \underline{x}'}{\partial t}(\underline{\eta}',t) \tag{15}$$

Substituting Eq. (15) into Eq. (11) we obtain

$$\frac{\partial X}{\partial t} = \mathcal{L}_{\eta}\{x\} + (\mathcal{L}_{\eta'}\{x\})' + B\int_{\Omega} M(\underline{\eta},\underline{\lambda},t)X(\underline{\lambda},\underline{\eta}',t)d\underline{\lambda}$$

$$+\int_{\Omega} X(\underline{\eta},\underline{\sigma},t)M'(\underline{\eta}',\underline{\sigma},t)B'd\underline{\sigma} \tag{16}$$

The cost functional of Eq. (12) can also be written as

$$J=\text{trace}\Big[\int_{0}^{t}\int_{\Omega}\int_{\Omega'} Q\delta(\underline{\eta}-\underline{\eta}')X(\underline{\eta},\underline{\eta}',t)d\underline{\eta}d\underline{\eta}'+$$

$$+\int_{0}^{T}\int_{\Omega}\int_{\Omega}\int_{\Omega}\int_{\Omega} RM(\underline{\eta},\underline{\lambda},t)X(\underline{\lambda},\underline{\sigma},t)M'(\underline{\eta}',\underline{\sigma},t).d\underline{\lambda}d\underline{\sigma}d\underline{\eta}d\underline{\eta}'dt\Big] \tag{17}$$

The problem thus reduces to minimizing the cost functional given by (17) subject to the matrix state equation (16). We observe that the results of the previous section can readily be applied to find necessary conditions for the minimization of (17) subject to (16). The necessary conditions (7), (8), (9) yield the following set of canonical equations:

$$\frac{\partial X}{\partial t} = \mathcal{L}_{\eta}\{X\} + (\mathcal{L}_{\eta'}\{X\})' + B\int_{\Omega} M(\underline{\eta},\underline{\lambda},t)X(\underline{\lambda},\underline{\eta}',t).d\underline{\lambda}$$

$$+\int_{\Omega} X(\underline{\eta},\underline{\sigma},t)M(\underline{\eta}',\underline{\sigma},t)B'(\underline{\eta}',t)d\underline{\sigma} \tag{18}$$

$$-\frac{\partial P}{\partial t} = \mathcal{L}^{*}_{\eta}\{P\} + (\mathcal{L}^{*}_{\eta'}\{P\})' + \int_{\Omega} P(\underline{\eta},\underline{\lambda},t)BM(\underline{\lambda},\underline{\eta}',t)d\underline{\lambda}$$

$$+\int_{\Omega} M'(\underline{\eta},\underline{\lambda},t)B'P(\underline{\lambda},\underline{\eta}',t)d\underline{\lambda} + \int_{\Omega} M'(\underline{\eta},\underline{\lambda},t)R.$$

$$.M(\underline{\lambda},\underline{\eta}',t)d\underline{\lambda} + Q\delta(\underline{\eta}-\underline{\eta}') \tag{19}$$

and

$$M^{*}(\underline{\eta},\underline{\eta}',t) = -R^{-1}B'P(\underline{\eta},\underline{\eta}',t) \tag{20}$$

Using equation (13) we obtain

$$u^{*}(\underline{\eta},t) = -R^{-1}B\int_{\Omega} P(\underline{\eta},\underline{\eta}',t)\underline{x}(\underline{\eta}',t)d\underline{\eta}' \tag{21}$$

If we substitute (21) into (19) we obtain:

$$-\frac{\partial P}{\partial t} = \mathcal{L}^{*}_{\eta}\{P\} + (\mathcal{L}^{*}_{\eta}\{P\})' - \int_{\Omega}\int_{\Omega} P(\underline{\eta},\underline{\sigma},t)BR^{-1}.$$

$$.\ B'P(\underline{\lambda},\underline{\eta}',t)d\underline{\lambda}d\underline{\sigma} + Q\delta(\underline{\eta}-\underline{\eta}') \tag{22}$$

But Eq. (22) is the Riccati equation of the original problem defined by (11) and (12). Hence $P(\underline{\eta},\underline{\eta}',t)$ satisfies the Riccati equation whereas for the matrix formulation it represents the costate matrix of the adjoint system. We observe that the Riccati equation for linear distributed parameter control systems with quadratic cost functionals (the same is true for linear lumped control systems { see reference (1)} obtains a new meaning via the matrix minimum principle. It is the costate matrix of the adjoint system when the original vector control problem is recast an equivalent matrix formulation. The same technique can be extended to filtering and stochastic optimal control of the distributed parameter systems. It is shown in the following sections that as a result of this extension we obtain another statement of the separation theorem which is conceptually easier to understand and much simpler in its derivation than standard techniques such as those presented in (5)-(7).

## IV. STOCHASTIC SYSTEMS

Let the stochastic system

$$\frac{\partial \underline{x}}{\partial t} = \mathcal{L}_\eta\{\underline{x}\} + \underline{\xi}(\underline{\eta},t) \tag{23}$$

$$\underline{z}(\underline{\eta},t) = C_\eta\{\underline{x}\} + \underline{v}(\underline{\eta},t) \tag{24}$$

where $C_\eta\{.\}$ is a linear spatial differential or integral operator, and

$$E\{\underline{v}(\underline{\eta},t)\} = 0 \tag{25}$$

$$E\{\underline{v}(\underline{\eta},t)\underline{v}'(\underline{\eta}',\tau)\} = V(\underline{\eta},t)\delta(\underline{\eta}-\underline{\eta}')\delta(t-\tau) \tag{26}$$

The noise $\underline{v}(\underline{\eta},t)$ doesn't have to be white in space for what follows. It is chosen, however, that way only for convenience. The main objective of this section is to construct a distributed parameter filter to initially estimate the state $\underline{x}(\underline{\eta},t)$. From the practical point of view, the following two questions will be of interest:

i) What is the best dynamical structure of the filter?

e) How good is the filter in estimating the state $\underline{x}(\underline{\eta},t)$?

These are not simple questions if we don't consider them in the light of the existing results applicable to the finite dimensional system (Kalman filter). (4). If we impose, however, the constraints that are present on the Kalman filters, it would seem reasonable, since we have linear system dynamics and Gaussian noises, to choose a linear estimator. We shall therefore constrain the filter to be described by the linear partial integrodifferential equation:

$$\frac{\partial \hat{\underline{x}}}{\partial t} = \hat{\mathcal{L}}_\eta\{\hat{x}\} + \int_\Omega K(\underline{\eta},\underline{\eta}',t)z(\underline{\eta}',t)d\underline{\eta}' \tag{27}$$

Once we choose the structure of the filter, we are now ready to ask the following question. Find $\hat{\mathcal{L}}_\eta\{.\}$ and $K(\underline{\eta},\underline{\eta}',t)$ such that the filter (27) provides and estimate $\hat{\underline{x}}(\underline{\eta},t)$ of $\underline{x}(\underline{\eta},t)$ which will call the minimum variance unbiased estimate if and only if the cost functional

$$J=\int_0^T\int_\Omega \underline{e}'(\underline{\eta},t)R\underline{e}(\underline{\eta},t)d\underline{\eta}dt \tag{28}$$

is minimized, where:

$$\underline{e}(\underline{\eta},t)=\underline{x}(\underline{\eta},t)-\hat{\underline{x}}(\underline{\eta},t) \tag{29}$$

$$E[\underline{e}(\underline{\eta},t)]=\underline{0}. \tag{30}$$

In order to be able to use condition $E|e(\underline{\eta},t)|=\underline{0}$ we formulate the problem as follows: Eqs.(23), (24), (27), and (29) after some manipulation yield:

$$\frac{\partial e}{\partial t}=\mathcal{L}_\eta\{\underline{x}\}-\hat{\mathcal{L}}_\eta\{\hat{\underline{x}}\}+\underline{\xi}(\underline{\eta},t)-\int_\Omega K(\underline{\eta},\underline{\eta}',t)\underline{z}(\underline{\eta}',t)d\underline{\eta}' \tag{31}$$

$$=\{\mathcal{L}_\eta\{.\}-\hat{\mathcal{L}}_\eta\{.\}-\int_\Omega K(\eta,\eta',t)C_\eta\{.\}d\eta'\}\underline{x}(\eta,t)$$

$$+\hat{\mathcal{L}}_\eta\{\underline{e}\}+\underline{\xi}(\underline{\eta},t)-\int_\Omega K(\underline{\eta},\underline{\eta}',t)\underline{v}(\underline{\eta}',t)d\underline{\eta}' \tag{32}$$

The condition (30) gives $E\{\partial\underline{e}(\underline{\eta},t)/\partial t\}=0$,which if applied to (31) yields

$$[\mathcal{L}_\eta\{.\}-\hat{\mathcal{L}}_\eta\{.\}-\int_\Omega K(\underline{\eta},\underline{\eta}',t)C_\eta\{.\}d\underline{\eta}']E[\underline{x}(\underline{\eta},t)]=0 \tag{33}$$

But in general $E[\underline{x}(\underline{\eta},t)]\neq 0$, therefore the structure of $\hat{\mathcal{L}}_\eta\{.\}$ is not completely independent of that of $\mathcal{L}_\eta\{.\}$ but we must have

$$\hat{\mathcal{L}}_\eta\{.\}=\mathcal{L}_\eta\{.\}-\int_\Omega K(\underline{\eta},\underline{\eta}',t)C_\eta\{.\}d\underline{\eta}' \tag{34}$$

What still remains to be found,in order to be able to calculate $\hat{\mathcal{L}}_\eta\{.\}$ completely is $K(\underline{\eta},\underline{\eta}',t)$ in Eq. (34). In order to find $K(\underline{\eta},\underline{\eta}',t)$, we substitute (34) into (27) and use the minimization of (36) as an extra condition subject to (27).
Equation (27) then becomes:

$$\frac{\partial x}{\partial t}=\mathcal{L}_\eta\{\hat{\underline{x}}\}+\int_\Omega K(\underline{\eta},\underline{\eta}',t)\left[\underline{z}(\underline{\eta}',t)-C_\eta\{\hat{x}(\underline{\eta}',t)\}\right].d\eta' \tag{35}$$

and Eq. (31) yields

$$\frac{\partial e}{\partial t}=\mathcal{L}_\eta\{\underline{e}\}-\int_\Omega K(\underline{\eta},\underline{\eta}',t)C_\eta\{\underline{e}(\underline{\eta}',t)\}d\underline{\eta}'+\underline{\xi}(\underline{\eta},t)$$

$$-\int_\Omega K(\underline{\eta},\underline{\eta}',t)\underline{v}(\underline{\eta}',t)d\underline{\eta}' \tag{36}$$

A procedure similar to that used in the previous sections yields:
Let's define

$$\varepsilon(\underline{\eta},\underline{\eta}',t)=E[\ \underline{e}(\underline{\eta},\ )\underline{e}'(\underline{\eta}',t)\ ] \tag{37}$$

Then,using (36) we can easily prove,after some manipulation,that

$$\frac{\partial\varepsilon}{\partial t}(\underline{\eta},\underline{\eta}',t)=[\mathcal{L}_\eta\{.\}-\int_\Omega K(\underline{\eta},\underline{\lambda},t)C_\lambda\{.\}d\lambda]\varepsilon(\underline{\eta},\underline{\eta}',t)+$$

$$+\varepsilon(\underline{\eta},\underline{\eta}',t)\left[\mathcal{L}_\eta\{.\}-\int_\Omega K(\underline{\eta},\underline{\lambda},t)C_\lambda\{.\}d\lambda\right]'$$
$$+\Sigma(\underline{\eta},\underline{\eta}',t)+\int_\Omega K(\underline{\eta},\underline{\xi},t)V(\underline{\xi},t)K^T(\underline{\eta}',\underline{\xi},t)d\underline{\xi} \tag{38}$$

The cost functional given by (28) can similarly be written as

$$J=\text{trace}\left[\int_0^T\int_\Omega R\delta(\underline{\eta}-\underline{\eta}')\varepsilon(\underline{\eta},\underline{\eta}',t)d\underline{\eta}d\underline{\eta}'dt\right] \tag{39}$$

The problem is now reduced to finding a $K(\underline{\eta},\underline{\eta}',t)$ such that (38) is satisfied and (39) minimized. But this problem is in the form we considered in section I. If we use the procedure we used to find $M(\underline{\eta},\underline{\eta}',t)$ in Eq. (20), we obtain

$$K(\underline{\eta},\underline{\eta}',t)=(C_{\eta'}\{\varepsilon(\underline{\eta},\underline{\eta}',t)\})'.\left[V(\underline{\eta}',t)\right]^{-1} \tag{40}$$

We observe that Eq. (40) is identical to the minimum variance estimator gain previously derived using other more complicated approaches. Thus the matrix minimum principle can be used to provide a direct constructive derivation of well established results in distributed systems and when the problem can be most conveniently formulated as a matrix problem this technique can be very useful and powerful.

Moreover, if we substitute (40) into (38) it can be shown that $\varepsilon(\underline{\eta},\underline{\eta}',t)$ satisfies

$$\frac{\partial\varepsilon}{\partial t}(\underline{\eta},\underline{\eta}',t)=\mathcal{L}_\eta\{(\underline{\eta},\underline{\eta}',t)\}+(\mathcal{L}_{\eta'}\ \varepsilon(\underline{\eta},\underline{\eta}',t)\})'+$$
$$+\Sigma(\underline{\eta},\underline{\eta}',t)-\int_\Omega\left[C_\lambda\{\varepsilon(\underline{\eta},\underline{\lambda},t)\}\right]'V^{-1}(\underline{\lambda},t)C_\lambda\{\varepsilon(\underline{\eta}',\underline{\lambda},t)\}d\underline{\lambda} \tag{41}$$

which is the equation of the covariance matrix of the error $\underline{e}(\underline{\eta},t)$.

The same procedure will now be applied to the optimum control of stochastic distributed control systems.
Consider the following stochastic system

$$\frac{\partial\underline{x}}{\partial t}=\mathcal{L}_\eta\{\underline{x}\}+B(\underline{\eta},t)\underline{u}(\underline{\eta},t)+\underline{\xi}(\underline{\eta},t) \tag{42}$$

We shall seek to find the optimal control $\underline{u}(\underline{\eta},t)$ such that the functional of the form

$$J=E\int_0^T\int_\Omega\left[\underline{x}'(\underline{\eta},t)Q\underline{x}(\underline{\eta},t)+\underline{u}'(\underline{\eta},t)R\underline{u}(\underline{\eta},t)\right]d\underline{\eta}dt \tag{43}$$

is minimized. We shall also assume that the measurement vector

$$\underline{z}(\underline{\eta},t)=C_\eta\{\underline{x}\}+\underline{v}(\underline{\eta},t) \tag{44}$$

is available. Again we assume $E\left[\underline{v}(\underline{\eta},t)\underline{v}'(\underline{\eta}',t)\right]=V(\underline{\eta},t)\delta(\underline{\eta}-\underline{\eta}')\delta(t-\tau)$. A procedure similar to that followed to derive (35) leads to

$$\frac{\partial\hat{\underline{x}}}{\partial t}=\mathcal{L}_\eta\{\hat{\underline{x}}\}+B\underline{u}+\int_\Omega K(\eta,\eta',t)\tilde{\underline{z}}(\eta',t)d\underline{\eta}' \tag{45}$$

where $\underline{z}=\tilde{\underline{z}}-C_\eta\{\hat{\underline{x}}\}$. Moreover the functional (43) can be written

$$J = E\Big[\int_0^T\int_\Omega \underline{\hat{x}}'(\underline{\eta},t)Q\underline{\hat{x}}(\underline{\eta},t)+\int_0^T\int_\Omega \underline{u}'(\underline{\eta},t)B\underline{u}(\underline{\eta},t)d\underline{\eta}dt+\int_0^T\int_\Omega \underline{e}'Q\underline{e}d\underline{\eta}dt\Big] \quad (46)$$

where $\underline{e}(\underline{\eta},t)$ satisfies the equation given by (36). We observe that the error function $\underline{e}(\underline{\eta},t)$ is not affected by the control $\underline{u}(\underline{\eta},t)$. Hence, the problem reduces to that of minimizing the cost functional

$$J=E\Big[\int_0^T\int_\Omega\big[\underline{x}'(\underline{\eta},t)Q\underline{x}(\underline{\eta},t)+\underline{u}'(\underline{\eta},t)R\underline{u}(\underline{\eta},t)\big]d\underline{\eta}dt\Big] \quad (47)$$

for a system governed by (45). If we define

$$E\big[\underline{\hat{x}}(\underline{\eta},t)\underline{\hat{x}}'(\underline{\eta}',t)\big]=\hat{X}(\underline{\eta},\underline{\eta}',t) \quad (48)$$

we then seek a $\underline{u}^*(\underline{\eta},t)$ of the form

$$\underline{u}^*(\underline{\eta},t)=\int_\Omega \hat{M}(\underline{\eta},\underline{\eta}',t)\underline{\hat{x}}(\underline{\eta}',t)d\underline{\eta}' \quad (49)$$

Substituting (49) in (45) and (47), we obtain

$$\frac{\partial\underline{\hat{x}}}{\partial t}=\mathcal{L}_\eta\{\underline{\hat{x}}\}+B\int_\Omega \hat{M}(\underline{\eta},\underline{\eta}',t)\underline{\hat{x}}(\underline{\eta}',t)d\underline{\eta}'+\int_\Omega K(\underline{\eta},\underline{\eta}',t)\underline{\tilde{z}}(\underline{\eta}',t)d\underline{\eta}' \quad (50)$$

$$\tilde{J}=\int_0^T\int_\Omega \text{trace}\big[Q\hat{X}(\eta,\eta,t)+\int_\Omega\int_\Omega R\hat{M}(\underline{\eta},\underline{\eta}',t)\hat{X}(\underline{\eta}',\underline{\xi},t).$$

$$.\hat{M}'(\underline{\eta},\underline{\xi},t)d\underline{\xi}\ d\underline{\eta}'\big]d\underline{\eta}dt \quad (51)$$

Assuming also that $\mathcal{L}_\eta\{.\}$ is a generator of a semigroup, it can be shown after some manipulation that

$$E\{\underline{\tilde{z}}(\underline{\eta},t)\underline{\tilde{z}}(\underline{\eta}',\tau)\}=V(\underline{\eta},t)\delta(\underline{\eta}-\underline{\eta}')\delta(t-\tau) \quad (52)$$

If we compare Equations (50) and (51) with (11) (in which (14) is substituted) and (17) and make use of (52) we can easily conclude that $\hat{M}(\underline{\eta},\underline{\eta}',t)$ is identical to that given by (20) which is another statement of the distributed parameter separation theorem. Other types of problems can also be studied using this technique such as those discussed in (9)

## REFERENCES

(1) M. Athans, "The Matrix Minimum Principle, "Information and Control, vol. 11, 1968.

(2) M. Athans and F.C. Schweppe, "Gradient Matrices and Matrix Calculations, MIT Lincoln Lab. Tech., Note 1965-53, Lexington, Mass.

(3) T.L. Johnson and M. Athans, "A Minimum Principle for Smooth First Order Distributed System", IEEE Transactions on Auto Control, pp. 136-139, April 1974.

(4) M. Athans and E.Tse, " A Direct Derivation of the Optimal Linear Filter Using the Maximum Principle", IEEE Trans. Automatic Control AC-12, 1967.

(5) S.G. Tzafestas and J.M. Nightingale, "Optimal Filterin, Smoothing and Prediction in Linear Distributed Parameter Systems", Proceedings of IEE, 115, pp. 1207-1212, 1968.
(6) S.G. Tzafestas and J.M. Nightingale, "Maximum Likelihood Approach to the Optimal Filtering of Distributed-Parameter Systems", Proc. of IEE, vol. 116, No.6, June 1969.
(7) S.G. Tzafestas and J.M. Nightingale, "Optimal Control of a Class of Linear Stochastic Distributed Parameter Systems", Proc. of IEE, vol. 115, No.8, August, 1968.
(8) S.G. Tzafestas, "Optimal Distributed-Parameter Control Using Classical Variational Theory", Int. J. of Control, pp. 593-608, 1970.
(9) P. Stavroulakis and P.E. Sarachik, "Design of Controllers for Distributed Systems Using Finite Dimensional State Observers", IEEE Conference on Decision and Control, Dec. 5-7, 1973, San Diego, California.

# SECTION IV

# ESTIMATION AND IDENTIFICATION

# GENERALIZED WIDE SENSE MARKOV PROCESSES AND QUADRATIC DYNAMICAL DISCRETE SYSTEMS

D. NUALART / J. AGUILAR MARTIN

Universitat de Barcelona / L.A.A.S. (Toulouse)

ABSTRACT

Wide sense Markov processes, as introduced by Doob, are based on semi-group property of orthogonal stochastic projection on the linear manifolds generated by the past values of the process. Replacing the linear manifolds by subspaces generated by nth order polynomials of the past, we generalize the concept of wide sense Markov process. For $n = 2$ an attempt is made to formulate interms of moments up to the fourth order the characteristic property of the semi-group induced; the explicit equations are given for centered and symmetrically distributed stochastic processes. Those results give rise to a quadratic discrete dynamical system perturbed with white independent noise.

For such a system, observed by a linear function with additive white gaussian noise a suboptimal non linear filter can be constructed.

## 1. WIDE SENSE MARKOV PROCESSES

### 1.1. Wide sense p-Markov processes.

Let $\{X_t, t\varepsilon T\}$ , $T \subset R$, be a n-dimensional stochastic process with finite moments up to order 2p, being p a positive integer.

Given k points $t_1 \leq \ldots \leq t_k$ we will denote by $X_{t_1} \otimes \ldots \otimes X_{t_k}$ the $n^k$-dimensional tensor components are

$$\{X_{t_1}^{i_1} \ldots X_{t_k}^{i_k};\ i_1, \ldots, i_k = 1, \ldots, n\}.$$



*D. G. Lainiotis and N. S. Tzannes (eds.), Advances in Control, 411–423.*

Consider the process $X_t$ defined on the probability space $(\Omega, \mathcal{F}, P)$ and for each set of square integrable random variables $\{Z_i\}_{i\varepsilon I}$ denote by $< Z_i, i\varepsilon I>$ the closed linear subspace of $L^2(\Omega, \mathcal{F}, P)$ generated by these variables.

Definition 1.1.

The process $\{X_t, t\varepsilon T\}$ will be called a wide sense p-Markov process if for each $s<t$,

$$\begin{aligned} &\text{proj}\left| \begin{matrix} (X_t \otimes^{m)} \ldots \otimes X_t) \\ <1, X_s \otimes^{k)} \ldots \otimes X_s, k=1,\ldots,p> \end{matrix} \right. \\ = &\text{proj}\left| \begin{matrix} X_t \otimes^{m)} \ldots \otimes X_t) \\ <1, X_{t_1} \otimes \ldots \otimes X_{t_k}, t_1,\ldots,t_k \leq s, k=1,\ldots,p> \end{matrix} \right. \end{aligned} \tag{1.1}$$

for all $m = 1,\ldots,p$.

In this formula the projection of a random tensor is computed by taking the projection of each component.

This definition is equivalent to say that the diference

$$X_t \otimes^{m)} \ldots \otimes X_t - \text{proj}\left| \begin{matrix} X_t \otimes^{m)} \ldots \otimes X_t \\ <1, X_s \otimes^{k)} \ldots \otimes X_s,\ k=1,\ldots,p > \end{matrix} \right. \quad \text{is}$$

centered and orthogonal to each component of

$$X_{t_1} \otimes \ldots \otimes X_{t_k}, t_1,\ldots,t_k \leq s,\ k=1,\ldots,p.$$

For $p = 2$ we will set up equations equivalent to (1.1) in terms of the moments up to fourth order.

To simplify we introduce the following notation :

$E\{X_t\} = m(t) = (m^i(t))_{i=1,\ldots,m} \in R^n,$

$E\{X_s \otimes X_t\} = K(s,t) = (K^{ij}(s,t))_{i,j=1,\ldots,n} \in R^{n^2}$

$E\{X_v \otimes X_s \otimes X_t\} = k(v,s,t) = (K^{ijk}(v,s,t))_{i,j,k=1,\ldots,n} \in R^{n^3}$

$E\{X_u \otimes X_v \otimes X_s \otimes X_t\} = K(u,v,s,t) = (K^{ijkh}(u,v,s,t))_{i,j,k,h=1,\ldots,n} \in R^{n^4}$

These tensors determine the process $X_t$ up to fourth order.

For each $t \in T$ we will put
$K(t)=K(t,t)$
$K_3(t)=K(t,t,t)$
$K_4(t)=K(t,t,t,t)$.

## 1.2. Construction of the quadratic projector

Let $f:R^n \longrightarrow R$ be a measurable function such that $E[f(X_t).X_s^i.X_s^j]$ exists for all $i,j=0,1,\dots,n$ and $s<t$.

The quadratic projector $\widehat{f(X_t)}|_s = \text{proj}\left|\begin{array}{l}(f(X_t))\\ <1,X_s,X_s\otimes X_s>\end{array}\right.$

will be

$$\widehat{f(X_t)}|_s = a + \sum_{i=1}^{m} b^i x_s^i + \sum_{ij=1}^{m} c^{ij} x_s^i x_s^j$$

The coefficients $a,b=(b^i)_{i=1,\dots,n}$, $c=(c^{ij})_{i,j=1,\dots,n}$ can be obtained from the following linear system of equations

$$E\{f(X_t)\} = a + \sum_{i=1}^{n} b^i m^i(s) + \sum_{ij=1}^{n} c^{ij} K^{ij}(s)$$

$$E\{f(X_t).X_s^\alpha\} = a.m^\alpha(s) + \sum_{i=1}^{m} b^i K^{\alpha i}(s) + \sum_{ij=1}^{n} c^{ij} K_3^{\alpha ij}(s) \tag{1.2}$$

$$E\{f(X_t).X_s^\alpha.X_s^\beta\} = a.K^{\alpha\beta}(s) + \sum_{i=1}^{n} b^i K_3^{\alpha\beta i}(s) + \sum_{ij=1}^{n} c^{ij} K_4^{\alpha\beta ij}(s),$$

$\alpha,\beta=1,\dots,n.$

The components of the tensors a,b,c can be properly arranged in a $(1+n+n^2)$ - dimensional vector,

$$\begin{vmatrix} a \\ b \\ c \end{vmatrix} = \begin{vmatrix} a \\ b^i \\ c^{ij} \end{vmatrix}.$$

For each $s\in T$, denote by $M(s)$ the square matrix of order $1+n+n^2$ given by

$$M(s)=\begin{vmatrix} 1 & m^i(s) & K^{ij}(s) \\ m^\alpha(s) & K^{\alpha i}(s) & K_3^{\alpha ij}(s) \\ K^{\alpha\beta}(s) & K_3^{\alpha\beta i}(s) & K_4^{\alpha\beta ij}(s) \end{vmatrix}.$$

Then, the system (1.2) can be written in matrix form

$$\begin{vmatrix} E\ \{f(X_t)\} \\ E\ \{f(X_t).X_s\} \\ E\ \{f(X_t).X_s\otimes X_s\} \end{vmatrix} = M(s).\begin{vmatrix} a \\ b \\ c \end{vmatrix}.$$

1.3. Characterization of wide sense 2-Markov processes

Proposition 1.1.

Let $\{X_t, t\varepsilon T\}$ be a n-dimensional stochastic process, with finite moments up to fourth order. Suppose det $M(s)\neq 0$ for all $s\varepsilon T$. Then, $X_t$ is a wide sense 2-Markov process if and only if for every $s<t$ and $u,v\leq s$ we have

$$\begin{vmatrix} m(t) \\ K(u,t) \\ K(u,v,t) \end{vmatrix} = \begin{vmatrix} 1 & m(s) & K(s) \\ m(u) & K(u,s) & K(u,s,s) \\ m(u,v) & K(u,v,s) & K(u,v,s,s) \end{vmatrix} . M(s)^{-1} . \begin{vmatrix} m(t) \\ K(s,t) \\ K(s,s,t) \end{vmatrix},$$

and

$$\begin{vmatrix} K(t) \\ K(u,t,t) \\ K(u,v,t,t) \end{vmatrix} = \begin{vmatrix} 1 & m(s) & K(s) \\ m(u) & K(u,s) & K(u,s,s) \\ K(u,v) & K(u,v,s) & K(u,v,s,s) \end{vmatrix} . M(s)^{-1} \begin{vmatrix} K(t) \\ K(s,t,t) \\ K(s,s,t,t) \end{vmatrix}.$$

Proof :

Denote by $(a_k\ b_k\ c_k)^t$ the components of $\hat{X}^k_{t|s}$, $k=1,\dots,n$, and by $(a_{kl}\ b_{kl}\ c_{kl})^t$ those of $\widehat{(X^k_t . X^l_t)}_{|s}$, $k=1,\dots,n$.

We know that $X_t$ is a wide sense 2-Markov process if and only if for each $s<t$, $X^k_t - \hat{X}^k_{t|s}$ and $X^k_t X^l_t - \widehat{X^k_t . X^l_t}_{|s}$ are centered random variables orthogonal to $X_u$ and $X_u \otimes X_v$ for all $u,v\leq s$.

That means

$$\begin{vmatrix} m^k(t) \\ K^{\alpha k}(u,t) \\ K^{\alpha\beta k}(u,v,t) \end{vmatrix} = \begin{vmatrix} 1 & m^i(s) & K^{ij}(s) \\ m^\alpha(u) & K^{\alpha i}(u,s) & K^{\alpha ij}(u,s,s) \\ K^{\alpha\beta}(u,v) & K^{\alpha\beta i}(u,v,s) & K^{\alpha\beta ij}(u,v,s,s) \end{vmatrix} . \begin{vmatrix} a_k \\ b^i_k \\ c^{ij}_k \end{vmatrix},$$

and

$$\begin{vmatrix} K^{kl}(t) \\ K^{\alpha kl}(u,t,t) \\ K^{\alpha\beta kl}(u,v,t,t) \end{vmatrix} = \begin{vmatrix} 1 & m^i(s) & K^{ij}(s) \\ m^\alpha(u) & K^{\alpha i}(u,s) & K^{\alpha ij}(u,s,s) \\ K^{\alpha\beta}(u,v) & K^{\alpha\beta i}(u,v,s) & K^{\alpha\beta ij}(u,v,s,s) \end{vmatrix} . \begin{vmatrix} a_{kl} \\ b^i_{kl} \\ c^{ij}_{kl} \end{vmatrix}.$$

Thus using (1.3) the proposition holds.

1.4. Equations for centered symmetric one dimensional process

Let us suppose henceforth that $m(t)=0$ and $K_3(t)=0$ for all $t \in T$. A process $X_t$ with these properties will be called symmetric. This is the case, for instance, if $X_t$ has a symmetric distribution for each t.

If $X_t$ is non centered, we can always take centered moments obtaining the same equations. The hypothesis $K_3(t)=0$ is imposed to simplify computations. Moreover we will place in the onedimensional case (n=1). With these assumptions

$$M(s)^{-1} = \begin{vmatrix} 1 & 0 & K(s) \\ 0 & K(s) & 0 \\ K(s) & 0 & K_4(s) \end{vmatrix}^{-1} = (K(s).Q(s))^{-1} \begin{vmatrix} K(s)K_4(s) & 0 & -K(s)^2 \\ 0 & Q(s) & 0 \\ -K(s)^2 & 0 & K(s) \end{vmatrix}$$

where $Q(s)=K_4(s)-K(s)^2=\sigma^2(X_s^2)$.

Observe that the hypothesis $\det M(s)\neq 0$ for all $s \in T$, holds in this case if and only if $Q(s)\neq 0$ for all $s \in T$ ($X_s^2$ is never deterministic). This condition implies $K(s)\neq 0$ for all $s \in T$.

Then, $X_t$ is a wide sense 2-Markov process if and only if for all $s < t$ and $u,v \leq s$ we have

$$\begin{vmatrix} 0 & K(t) \\ K(u,t) & K(u,t,t) \\ K(u,v,t) & K(u,v,t,t) \end{vmatrix} = (K(s)Q(s))^{-1}. \begin{vmatrix} 1 & 0 & K(s) \\ 0 & K(u,s) & K(u,s,s) \\ K(u,v,s) & K(u,v,s) & K(u,v,s,s) \end{vmatrix}$$

$$\cdot \begin{vmatrix} K(s)K_4(s) & 0 & -K(s)^2 \\ 0 & Q(s) & 0 \\ -K(s)^2 & 0 & K(s) \end{vmatrix} \cdot \begin{vmatrix} 0 & K(t) \\ K(s,t) & K(s,t,t) \\ K(s,s,t) & K(s,s,t,t) \end{vmatrix} .$$

If we put $Q(u,v,s)=K(u,v,s,s)-K(u,v)K(s)=\text{cov.}(X_u.X_v,X_s)$, for all $u,v \leq s$ ; and $Q(s,t)=Q(s,s,t)$ for $s \leq t$, we obtain

$$K(u,t) = \frac{K(u,s)K(s,t)}{K(s)} + \frac{K(u,s,s)K(s,s,t)}{Q(s)}, \quad \text{for all } u \leq s < t \tag{1.4}$$

$$K(u,v,t) = \frac{K(u,v,s)K(s,t)}{K(s)} + \frac{Q(u,v,s,)K(s,s,t)}{Q(s)}, \quad \text{for all } u,v \leq s < t \tag{1.5}$$

$$K(u,t,t) = \frac{K(u,s)K(s,t,t)}{K(s)} + \frac{Q(u,s,s)Q(s,t)}{Q(s)}, \quad \text{for all } u \leq s < t \tag{1.6}$$

$$Q(u,v,t) = \frac{K(u,v,s)K(s,t,t)}{K(s)} + \frac{Q(u,v,s)Q(s,t)}{Q(s)}, \text{ for all } u,v \le s < t. \tag{1.7}$$

1.5. Transition functions

The following functions will play for wide sense symmetric 2-Markov processes the role of transition functions :

(a) $\phi_1(s,t)= \dfrac{K(s,t)}{K(s)}$ , if $s \le t$.

This function verifies :

$\phi_1(s,s)=1$, and

$$\text{proj}_{\langle X_t \rangle} (X_t)=\phi_1(s,t).X_s.$$

Suppose in future that for all $s,t \in T$, $Q(s,t)\neq 0$.

Then, define

(b) $\phi_2(u,v,s) = \dfrac{K(u,v,s)}{Q(u,v)}$, if $u,v \le s$.

This function verifies

$\phi_2(s,s,s)=0$

$$\text{proj}_{\langle 1,X_u.X_v \rangle} (X_s) =\phi_2(u,v,s) \frac{Q(u,v)}{\sigma^2(X_u.X_v)} (X_u.X_v-K(u,v)), \text{ if } \sigma^2(X_u.X_v)\neq 0$$

For $s \le t$ let $\phi_2(s,t)=\phi_2(s,s,t)= \dfrac{K(s,s,t)}{Q(s)}$ ; then,

$$\text{proj}_{\langle 1,X_s,X_s^2 \rangle} (X_t) =\phi_1(s,t)X_s +\phi_2(s,t) (X_s^2-K(s)). \tag{1.8}$$

From (1.8) we will deduce a quadratic dynamical model for symmetric wide sense 2-Markov processes.

(c) $\phi_3(s,t)= \dfrac{K(s,t,t)}{K(s)}$ , if $s \le t$.

This function verifies

$\phi_3(s,s)=0$,

$$\text{proj}_{\langle X_s \rangle} (X_t^2)=\phi_3(s,t).X_s.$$

(d) $\phi_4(u,v,s)=\dfrac{Q(u,v,s)}{Q(u,v)}$ , if $u,v \leq s$.

$\phi_4(s,s,s)=1$

$$\underset{<1,X_u\cdot X_v>}{\text{proj}}\Big|(X_s^2-K(s))= (u,v,s)\ \frac{Q(u,v)}{\sigma^2(X_u\cdot X_v)}\ \{X_u\cdot X_v-K(u,v)\}\ ,\qquad \text{if}\quad \sigma^2(X_u\cdot X_v)\neq 0$$

For $s \leq t$ let $\phi_4(s,t)=\phi_4(s,s,t)$; then

$$\underset{<1,X_s,X_s^2>}{\text{proj}}\Big|(X_t^2)=K(t)+\phi_3(s,t)X_s+\phi_4(s,t)\{X_s-K(s)\}\ . \tag{1.9}$$

From (1.9) we can deduce a quadratic dynamical model for $X_t^2$.

Proposition 1.2.

Let $X_t$, $t \in T$ be a onedimensional stochastic process with finite moments up to fourth order, and such that $m(t)=0, K_3(t)=0$, and $Q(s,t)\neq 0$ for all $s \leq t$ in T.

Then $X_t$ is a wide sense 2-Markov process if and only if

$$\phi_1(u,t)=\phi_1(u,s)\phi_1(s,t)+\phi_3(u,s)\phi_2(s,t) \tag{1.10}$$

$$\phi_2(u,v,t)=\phi_2(u,v,s)\phi_1(s,t)+\phi_4(u,v,s)\phi_2(s,t) \tag{1.11}$$

$$\phi_3(u,t)=\phi_1(u,s)\phi_3(s,t)+\phi_3(u,s)\phi_4(s,t) \tag{1.12}$$

$$\phi_4(u,v,t)=\phi_2(u,v,s)\phi_3(s,t)+\phi_4(u,v,s)\phi_4(s,t) \tag{1.13}$$

It is say, in matrix form

$$\begin{vmatrix}\phi_1(u,t) & \phi_3(u,t)\\ \phi_2(u,v,t) & \phi_4(u,v,t)\end{vmatrix} = \begin{vmatrix}\phi_1(u,s) & \phi_3(u,s)\\ \phi_2(u,v,s) & \phi_4(u,v,s)\end{vmatrix}\cdot\begin{vmatrix}\phi_1(s,t) & \phi_3(s,t)\\ \phi_2(s,t) & \phi_4(s,t)\end{vmatrix}$$

for all $u,v \leq s < t$.

## 1.6. Partial differential equations for transition functions

In this section T will be an open interval of R. Suppose that the following limits exist and are continuous functions of t :

$$\lim_{h\to 0}\frac{\phi_1(s,s+h)-1}{h}=\frac{\partial\phi_1(s,t)}{\partial t}\Big|_{t=s}=\alpha_1(s)\qquad ,$$

$$\lim_{h\to 0} \frac{\phi_2(s,s+h)}{h} = \left.\frac{\partial \phi_2(s,t)}{\partial t}\right|_{t=s} = \alpha_2(s),$$

$$\lim_{h\to 0} \frac{\phi_3(s,s+h)}{h} = \left.\frac{\partial \phi_3(s,t)}{\partial t}\right|_{t=s} = \alpha_3(s),$$

$$\lim_{h\to 0} \frac{\phi_4(s,s+h)-1}{h} = \left.\frac{\partial \phi_4(s,t)}{\partial t}\right|_{t=s} = \alpha_4(s) \quad , \quad \text{for all} \quad s \in T.$$

We suppose also that all moments up to fourth order are continuous.

Then, the relations (1.10) to (1.13) imply

$$\frac{\partial}{\partial t}\phi_1(u,t)=\phi_1(u,t)\alpha_1(t)+\phi_3(u,t)\alpha_2(t),$$

$$\frac{\partial}{\partial t}\phi_2(u,v,t)=\phi_2(u,v,t)\alpha_1(t)+\phi_4(u,v,t)\alpha_2(t),$$

$$\frac{\partial}{\partial t}\phi_3(u,t)=\phi_1(u,t)\alpha_3(t)+\phi_3(u,t)\alpha_4(t),$$

$$\frac{\partial}{\partial t}\phi_4(u,v,t)=\phi_2(u,v,t)\alpha_3(t)+\phi_4(u,v,t)\alpha_4(t),$$

Or, in matrix form

$$\frac{\partial}{\partial t}\begin{vmatrix} \phi_1(u,t) & \phi_3(u,t) \\ \phi_2(u,v,t) & \phi_4(u,v,t) \end{vmatrix} = \begin{vmatrix} \phi_1(u,t) & \phi_3(u,t) \\ \phi_2(u,v,t) & \phi_4(u,v,t) \end{vmatrix} \cdot \begin{vmatrix} \alpha_1(t) & \alpha_3(t) \\ \alpha_2(t) & \alpha_4(t) \end{vmatrix} .$$

Thus, to construct the process up to fourth order it suffices to know the functions $\alpha_i(t)$, $i=1,2,3,4$, the positive function $K(t)$ and the covariance $Q(s,t)$. However, if we start with these functions and we find solutions $\phi_1$, $\phi_2$, $\phi_3$, $\phi_4$ of (1.10) to (1.13) we can not assert, in general, the existence of a symmetric wide sense 2-Markov process having those transition functions.

1.7. Recursive model

Let $\{X_t,\ t \in T\}$ be a wide sense 2-Markov process ve rifying the hypotheses of proposition 1.2. For each $s < t$ in T, define

$$\varepsilon_{st}=X_t-\mathrm{proj}\Big|_{<1,X_s,X_s^2>}(X_t)=X_t-\phi_1(s,t)X_s-\phi_2(s,t)\{X_s^2-K(s)\}\ ,$$

$$\delta_{st} = X_t^2 - \text{proj}_{|< 1, X_s, X_s^2>}(X_t^2) = X_t^2 - K(t) - \phi_3(s,t)\, X_s - \phi_4(s,t)\{ X_s^2 - K(s)\} .$$

The random variables $\varepsilon_{st}$ and $\delta_{st}$ are centered and orthogonal to $X_u$ and $X_u . X_v$ for all $u,v \leq s$. This a consequence of (1.8) and (1.9).

Thus, in the discrete case $\{ T= (0,1,\dots )\}$ our process $\{ X_k,\ k \geq 0\}$ admits the following recursive model

$$X_{k+1} = \phi_1(k)X_k + \phi_2(k)\ X_k^2 - K(k)\ + \varepsilon_k \quad , \tag{1.14}$$
$$X_{k+1}^2 = K(k+1) + \phi_3(k)X_k + \phi_4(k)\{ X_k^2 - K(k)\} + \delta_k \quad ,$$

where we have put $\phi_i(k) = \phi_i(k,k+1)$ for $k \geq 0$, and

$$\varepsilon_k = \varepsilon_{k,k+1},\ \delta_k = \delta_{k,k+1} .$$

Moreover, we have, for all $j < k$

$$X_k = \phi_1(j,k)X_j + \phi_2(j,k)\{ X_j^2 - K(j)\} + \varepsilon_{jk}$$
$$X_k^2 = K(k) + \phi_3(j,k)X_j + \phi_4(j,k)\{ X_j^2 - K(j)\} + \delta_{jk} \tag{1.15}$$

Observe that given a sequence $\varepsilon_k$, $k \geq 0$ of independent and centered random variables, functions $\phi_1(k)$, $\phi_2(k)$ and a random variable $X_o$ ; the process $X_k$ constructed from (1.4) will not be, in general, wide sense 2-Markov because we do not know if the process $X_k^2$ admits a quadratic dynamical model too, and neither if (1.15) holds.

## 1.8. Particular case of Gaussian processes

Consider a Gaussian process $\{X_t, t \in T\}$ with $E\ (X_t)=0$ and $K(t) \neq 0$.

Then it is easy to verify that

$E\ \{X_v . X_s . X_t\} = 0$ for all $v,s,t \in T$; and, therefore we have the following property.

Proposition 1.3.

A centered Gaussian process $X_t$ is a wide sense 2-Markov process if and only if it is a Markov process

Proof :

We know that $X_t$ is Markov if and only if

$$K(u,t)=\frac{K(u,s)K(s,t)}{K(s)} \quad , \text{ for all } u \le s \le t \text{ in } T.$$

Then, we have only to prove that in this case

$$\phi_4(u,v,t)=\phi_4(u,v,s).\phi_4(s,t) \quad \text{holds for all } u,v \le s \le t.$$

That means, the sequence $\{X_u . X_v, X_s^2, X_t^2\}$ is Markovian; and this is a consequence of

$$K(u,v,s,t)=K(u,s)K(s,t)+K(u,v)K(s,t)+K(u,t)K(v,s),$$

for all u,v,s,t in T.

## 2. PREDICTION AND FILTERING OF WIDE SENSE 2-MARKOV PROCESSES

### 2.1. Recursive prediction

Let $\{X_k, k \ge 0\}$ be a discrete symmetric wide sense 2-Markov process, which has the model (1.14) and (1.15).

Then, we have

$$X_{k|j} = \phi_1(j,k)X_j + \phi_2(j,k)\{X_j^2-K(j)\}$$

$$X^2_{k|j} = \phi_3(j,k)X_j + \phi_4(j,k)\{X_j^2-K(j)\}$$

$$X_{k+1|j} = \phi_1(j,k+1)X_j + \phi_2(j,k+1)\{X_j^2-K(j)\}$$

$$X^2_{k+1|j} = \phi_3(j,k+1)X_j + \phi_4(j,k+1)\{X_j^2-K(j)\} + K(k+1).$$

Therefore,

$$\begin{vmatrix} X_{k|j} & \phi_1(j,k) & \phi_2(j,k) \\ X^2_{k|j}-K(k) & \phi_3(j,k) & \phi_4(j,k) \\ X_{k+1|j} & \phi_1(j,k+1) & \phi_2(j,k+1) \end{vmatrix} = 0$$

, and

$$\begin{vmatrix} X_{k|j} & \phi_1(j,k) & \phi_2(j,k) \\ X_{k|j}-K(k) & \phi_3(j,k) & \phi_4(j,k) \\ X_{k+1|j}-K(k+1) & \phi_3(j,k+1) & \phi_4(j,k+1) \end{vmatrix} = 0 \; ;$$

which implies

$$\begin{aligned} X_{k+1|j} &= A_1(j,k)X_{k|j} + B_1(j,k)\{X^2_{k|j} - K(k)\} \\ X^2_{k+1|j} &= K(k+1) + A_2(j,k)X_{k|j} + B_2(j,k)\{X^2_{k|j} - K(k)\} \end{aligned} \tag{2.1}$$

where

$$A_1(j,k) = \begin{vmatrix} \phi_3(j,k) & \phi_4(j,k) \\ \phi_1(j,k+1) & \phi_2(j,k+1) \end{vmatrix} \cdot \begin{vmatrix} \phi_1(j,k) & \phi_2(j,k) \\ \phi_3(j,k) & \phi_4(j,k) \end{vmatrix}^{-1}$$

$$B_1(j,k) = -\begin{vmatrix} \phi_1(j,k) & \phi_2(j,k) \\ \phi_1(j,k+1) & \phi_2(j,k+1) \end{vmatrix} \cdot \begin{vmatrix} \phi_1(j,k) & \phi_2(j,k) \\ \phi_3(j,k) & \phi_4(j,k) \end{vmatrix}^{-1}$$

$$A_2(j,k) = \begin{vmatrix} \phi_3(j,k) & \phi_4(j,k) \\ \phi_3(j,k+1) & \phi_4(j,k+1) \end{vmatrix} \cdot \begin{vmatrix} \phi_1(j,k) & \phi_2(j,k) \\ \phi_3(j,k) & \phi_4(j,k) \end{vmatrix}^{-1}$$

$$B_2(j,k) = \begin{vmatrix} \phi_1(j,k) & \phi_2(j,k) \\ \phi_3(j,k+1) & \phi_4(j,k+1) \end{vmatrix} \cdot \begin{vmatrix} \phi_1(j,k) & \phi_2(j,k) \\ \phi_3(j,k) & \phi_4(j,k) \end{vmatrix}^{-1} .$$

## 2.2. Suboptimal non linear filter

Given a sysmmetric wide sense 2-Markov process $\{X_k, \ k \geq 0\}$, we observe the process $y_k$ given by

$$y_k = h_k \cdot X_k + \eta_k , \quad k > 0 \tag{2.2}$$

where $\{\eta_k, \ k > 0\}$ is a sequence of independent centered random variables, which are independent of the process $X_k$, and such that

$$E\{\eta_k^2\} = R_k , \quad \sigma^2\{\eta_k^2\} = P_k .$$

Using the recursive model (1.14) we can compute $\hat{X}_{k+1|k}$ and $\widehat{X^2}_{k+1|k}$ from $\hat{X}_{k|k}$ and $\widehat{X^2}_{k|k}$, where for each $k \geq 0$ these projections are taken over $< 1, y_o, \ldots, y_k;\ y_i y_{i'}, 0 \leq i, i' \leq k-1 >$ :

$$\begin{aligned} \hat{X}_{k+1|k} &= \phi_1(k)\hat{X}_{k|k} + \phi_2(k)\{\widehat{X^2}_{k|k} - K(k)\} \\ \widehat{X^2}_{k+1|k} &= \phi_3(k)\hat{X}_{k|k} + \phi_4(k)\{\widehat{X^2}_{k|k} - K(k)\} + K(k+1) . \end{aligned} \tag{2.3}$$

The observation model (2.2) allows us to form a suboptimal filter as it follows,

$$\hat{X}_{k|k}=\hat{X}_{k|k-1}+\begin{vmatrix}P_{xy} & P_{xy^2}\end{vmatrix}\begin{vmatrix}P_{yy} & P_{yy^2}\\ P_{y^2y} & P_{y^2y^2}\end{vmatrix}^{-1}\cdot\begin{vmatrix}y_k-h_k\cdot\hat{X}_{k|k-1}\\ y_k^2-h_k^2\cdot\widehat{X^2}_{k|k-1}-R_k\end{vmatrix}$$

$$\widehat{X^2}_{k|k}=\widehat{X^2}_{k|k-1}+\begin{vmatrix}P_{x^2y} & P_{x^2y^2}\end{vmatrix}\begin{vmatrix}P_{yy} & P_{yy^2}\\ P_{y^2y} & P_{y^2y^2}\end{vmatrix}^{-1}\cdot\begin{vmatrix}y_k-h_k\cdot\hat{X}_{k|k-1}\\ y_k^2-h_k^2\cdot\widehat{X^2}_{k|k-1}-R_k\end{vmatrix},$$

where $P_{x^iy^j}=\mathrm{cov}\,(X_k^i,y_k^j)$.

That means,

$$P_{xy}=h_k\cdot K(k)$$

$$P_{xy^2}=R_k$$

$$P_{x^2y}=0$$

$$P_{x^2y^2}=h_k^2\cdot Q(k)$$

$$P_{yy}=R_k+h_k^2\cdot K(k)$$

$$P_{yy^2}=0$$

$$P_{y^2y^2}=h_k^4\cdot Q(k)+4h_k^2K(k)R_k+P_k\ .$$

Therefore, we obtain the following equations,

$$\begin{aligned}\hat{X}_{k|k}&=\hat{X}_{k|k-1}+h_kK(k)\{R_k+h_k^2K(k)\}^{-1}(y_k-h_k\hat{X}_{k|k-1})+\\ &\quad+R_k\{h_k^4Q(k)+4h_k^2K(k)R_k+P_k\}^{-1}(y_k^2-h_k^2{}_{k|k-1}-R_k)\ ,\qquad(2.4)\\ \widehat{X^2}_{k|k}&=\widehat{X^2}_{k|k-1}+h_k^2Q(k)\{h_k^4Q(k)+4h_k^2K(k)R_k+P_k\}^{-1}(y_k^2-h_k^2\widehat{X^2}_{k|k-1}-R_k)\ .\end{aligned}$$

This filter is suboptimal because we do not have taken in account innovations like

$$y_iy_h-h_ih_k\widehat{(X_iX_k)}_{|k-1}\quad,\quad 1\le i\le k-1.$$

To introduce these innovations we should dispose of a quadratic dynamical model for $X_iX_k$ similar to that of $X_k^2$, for i=0,...,k-1.

This would be true if we had used a stronger Markov condition;

but, in this case, the dimension of the filter would not be bounded.

## REFERENCES

1 DOOB, J.L. "Stochastic Processes". Wiley and Sons, 1953.

2 NUALART, D. and AGUILAR MARTIN, J. "Estimation optimale en puissances de degré N". C.R.A.S., 284, A81-A83, (1977).

# USE OF POSSIBILITY MEASURES FOR MODELING AN A PRIORI KNOWLEDGE IN PARAMETRIC ESTIMATION

G. BANON

L.A.A.S.-C.N.R.S. - 7, Av. du Colonel Roche
31400 TOULOUSE - France

ABSTRACT

The main interest in this note is to consider the nature and the role of an a priori knowledge in parametric estimation.

The usual approach consists to assume that an a priori knowledge is probabilistic in nature, in contrast our concern is to assume that it is possibilistic in nature. This assumption seems to be more realistic because it is less restrictive in the sense that the additivity of the usual probability measure is replaced by what is sometimes called fuzzy-additivity.

The very simple problem of estimation of the mean of an independent sequence of real random variables is considered.

An alternative approach to the classical linear mean squares estimate is given which is based on the assumption that the initial guess is consistent with a possibility measure.

## 1. INTRODUCTION

In the field of parametric estimation subjective a priori informations or knowledges are often available. The usual way of modeling this information is to assume that the involved uncertainty is probabilistic in nature. For instance this is usually done in linear filtering where before solving the equations of the filter the initial conditions on the mean and the covariance of the state are chosen assuming implicitly that the initial uncertainty behaves like a random variable.



*D. G. Lainiotis and N. S. Tzannes (eds.), Advances in Control, 424–430.*

But is such an assumption convenient in the case of subjective uncertainty ?

Comparing probabilistic and possibilistic models for a priori information one will see that the latter seems more convenient.

The issue of such a study is very important for the case of small sample size, where, in contrast to the infinite case, the estimates are closely dependent on the initial guess.

In section 2, two classical estimates of the mean are recalled.

In section 3, two important remarks are made concerning the role and the nature of the a priori information.

Finally, in section 4, a new estimate for the mean is proposed.

## 2. ESTIMATION OF THE MEAN OF A RANDOM VARIABLE

Let $(\Omega, \mathcal{A}, P)$ be a probability space and V a real random variable (r.r.v.) defined on $(\Omega, \mathcal{A}, P)$ and whose probability law is known (in particular it is assumed that EV = 0).

Let $1_A$ be the r.r.v. defined on $(\Omega, \mathcal{A})$ by

$$\omega \in \Omega \longmapsto 1_A(\omega) = \begin{cases} 1 \text{ if } \omega \in A \\ 0 \text{ if not} \end{cases}$$

for $A \in \mathcal{A}$.
For $x \in R$, let $\{Y_i\}_{i=1}^n$ be the sequence of r.r.v. defined by

$$Y_i = x\, 1_\Omega + V_i$$

where $\{V_i\}_{i=1}^n$ is a sequence of independent r.r.v. defined on $(\Omega, \mathcal{A}, P)$ and having the same law as the r.r.v. V.

The problem is to estimate the unknown parameter x (the common mean of the $Y_i$'s) from a sequence $\{Y_i(\omega)\}_{i=1}^{i}$ $(\omega \in \Omega)$ using or not an a priori information on x.

i) <u>The sample mean</u>

Since $x = EY_i$, i=1,...,n, a natural estimate for x is the sample mean

$$\frac{1}{n} \sum_{i=1}^{n} Y_i \tag{2-1}$$

denoted here by $\overline{Y}_n$.

This is the non-bayesian point of view, no a priori information is included in $\overline{Y}_n$.

ii) <u>The least squares estimate</u>

If, in contrast, x can be interpreted as the value taken by a r.r.v. X defined on a probability space $(\Omega', \mathcal{A}', P')$, and having a density function $\pi$, one may include that initial information in the estimation scheme of x.

This is the case of the least squares estimate which can be written (if V has a density)

$$\frac{\int_R u\mathcal{L}(u, Y_1, \ldots, Y_n)\,\pi(u)\,du}{\int_R \mathcal{L}(u, Y_1, \ldots, Y_n)\,\pi(u)\,du}, \tag{2-2}$$

where $\mathcal{L}$ stands for likelihood function, $\mathcal{L}(u, y_1, \ldots, y_n)$ being simply the value of the joint density function of $Y_1, \ldots, Y_n$ at point $y_1, \ldots, y_n$ given the parameter u.

The least squares estimate corresponds to the bayesian point of view.

From the above two estimates for the mean two remarks can be made.

## 3. INTRODUCTION OF A NON PROBABILISTIC MODEL FOR SUBJECTIVE A PRIORI INFORMATION

i) <u>Remark 1</u> (complete ignorance)

If X is no longer taken as a r.r.v. and if $\pi$ is chosen as a constant function on R, the expression (2-2) of the least squares estimate becomes

$$\frac{\int_R u\mathcal{L}(u, Y_1, \ldots, Y_n)\,du}{\int_R \mathcal{L}(u, Y_1, \ldots, Y_n)\,du} \tag{3-1}$$

If in addition, V is taken as a gaussian random variable, it is known that (3-1) is nothing more than the sample mean $\overline{Y}_n$.

In other words if X is supposed to be something else than a r.r.v. expression (2-2) could include the case where no a priori information is avalaible, that is to say the case of complete ignorance about the value x.

At this stage, another point of view (5) is to consider (3-1) as the limit of bayesian estimates of type (2-2), but then the problem of modeling a priori information is eluded.

ii) Remark 2 (wrong knowledge)

If the r.r.v. V is still assumed to be gaussian, say $N(0,\sigma^2)$, it can be shown that

$$E(\overline{Y}_n - x\,1_\Omega)^2 = \frac{\sigma^2}{n}$$

Let $\hat{X}_n$ be the least squares estimate given by (2-2), if the r.r.v. X is in addition gaussian, say $N(0,\sigma_0^2)$, then

$$E(\hat{X}_n - x\,1_\Omega)^2 = \frac{x^2\sigma^4 + n\sigma_0^4\sigma^2}{(n\sigma_0^2 + \sigma^2)^2}$$

Hence, the mean square error is a way to compare both estimates (here the expectations are computed with respect to P).

It can be seen from these that if the actual value x is such that

$$|x| > \sqrt{2\sigma_0^2 + \frac{\sigma^2}{n}}$$

then $\quad E(\hat{X}_n - x\,1_\Omega)^2 > E(\overline{Y}_n - x\,1_\Omega)^2$

In other words having a wrong item of knowledge (e.g. assuming x is zero when it is in fact far from zero) could lead to a very bad performance.

The above remarks have shown limitations of the probabilistic model for a priori information, but they also suggest to introduce a new model which should include the notion of complete ignorance and should not be risky for wrong guesses.

Actually, if $\pi$ in expression (2-2) is chosen as a possibility distribution of Zadeh (6), those exigencies can be satisfied.

At this stage, to say that $\pi$ is a possibility distribution only means that $\pi$ is bounded on R ( $\pi$ is not necessarily a density).

The following class of possibility distributions is suggested

$$\pi(u) = \begin{cases} 1 & \text{if } u \in I \\ t & \text{if not} \end{cases},$$

where I is an interval of R and $t \in ]0, 1[$.

This can be interpreted by saying that the unknown value x is assumed to belong to I but it is not excluded with a degree t that x is outside this interval.

To such possibility distributions it corresponds the t-1 possibility measures focused on I (1)

$$A \subset R \longmapsto \begin{cases} 1 \text{ if } A \cap I \neq \emptyset \\ t \text{ if } A \cap I = \emptyset \\ 0 \text{ if } A = \emptyset . \end{cases}$$

Instead of the additivity property of the probability measures, such possibility measures have the F-additivity property :

$$\forall A, B \subset R \qquad \mu(A \cup B) = \max(\mu(A), \mu(B)),$$

which appears to be less restrictive.

## 4. INTRODUCTION OF A NEW ESTIMATE FOR THE MEAN

The proposed estimate for the mean is given by expression (2-2) where $\pi$ is chosen among the above class of possibility distributions. Let note this estimate by $X_n$.

The following properties for $X_n$ have been proved in (2).

Theorem 1 ($X_n$ is not risky)
$E(X_n - x1_{\Omega})^2$ is bounded on R and the upper bound is $\frac{\sigma^2}{n}$ for $x = \pm\infty$.

Theorem 2 ($X_n$ can be better than $Y_n$)
If x is "in the middle" of I then

$$E(X_n - x\, 1_\Omega)^2 < \frac{\sigma^2}{n} .$$

Figure 1 shows the values of the mean square error as a function of x for the four following a priori models, $u \in R$,

i) $\pi(u) = 1$,

ii) $\pi(u) = \begin{cases} 1 \text{ if } u \in ]-1, 1[ \\ 1/2 \text{ if not}, \end{cases}$

iii) $\pi(u) = \begin{cases} 1 \text{ if } u \in ]-1, 1[ \\ 1/4 \text{ if not}, \end{cases}$

iv) $\pi(u) = \sqrt{\frac{3}{2\pi}}\; e^{-3/2\, u^2} .$

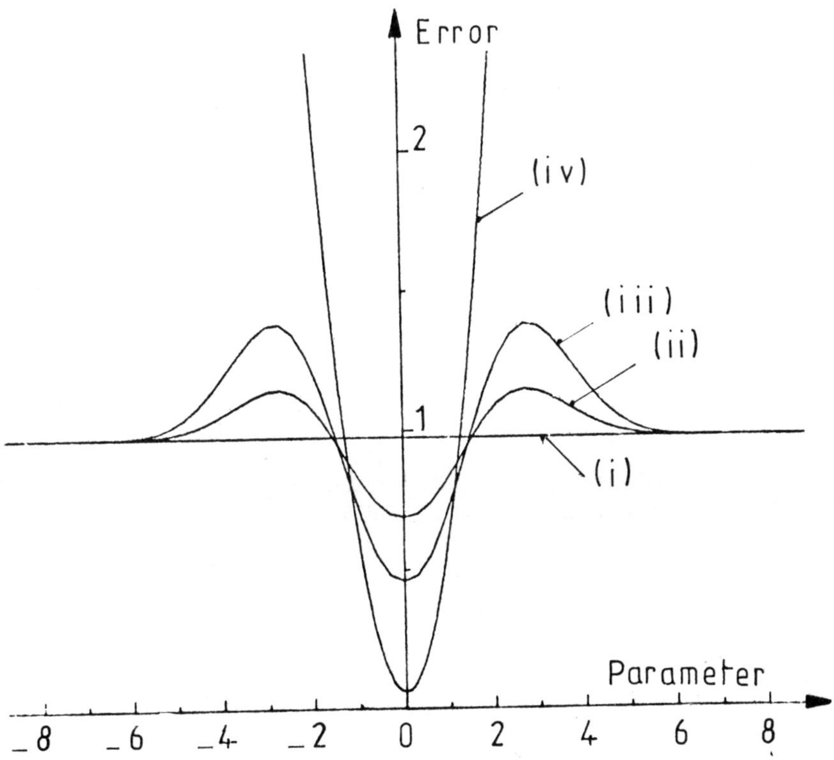


Figure 1. The mean square error as a function of x for the models (i)-(iv)

One can see that the mean square error corresponding to the possibility model behaves like the mean square error corresponding to the gaussian model inside the interval $]-1, 1[$ while it behaves like the mean square error corresponding to the model of complete ignorance outside this interval, which seems satisfactory.

In this approach x can be seen as the value taken by a fuzzy variable X having a t-1 possibility measure.

Finally the construction of the new estimate could be seen as obtained by application of Dempster rule of combination (3)-(4), instead of the classical Bayes rule.

## 5. CONCLUSION

In this note it has been shown that the probabilistic models and in particular the gaussian models are not satisfactory to represent subjective a priori information in the estimation of the mean. A more convenient model has been proposed which is a class of Zadeh's possibility measures.

## REFERENCES


(1) Banon, G. : "Distribution between several subsets of fuzzy measures", Submitted for publication at International Journal of Fuzzy-Sets and Systems (April 1979).

(2) Banon, G. : "Utilisation de familles de mesures de possibilité comme modèles d'information a priori subjective dans l'estimation paramétrique", June 1979, Note Interne, LAAS 79 I 33.

(3) Dempster, A.P. : "Upper and lower probabilities induced by multivalued mapping", 1967, Ann. Math. Statist. 38, pp. 325-339.

(4) Shafer, G. : "A mathematical theory of evidence", 1976, Princeton University Press, Princeton, New Jersey.

(5) Zacks, S. : "The theory of statistical inference", 1971, John Wiley and Sons, New York.

(6) Zadeh, L.A. : "Fuzzy sets as a basis for a theory of possibility", 1978, International Journal of Fuzzy Sets and Systems 1 (1), pp. 3-28.

# AN APPROXIMATE METHOD FOR NONLINEAR FILTERING PROBLEMS VIA MAXIMAL LIKELIHOOD PRINCIPLE

Yoshifumi Sunahara* and Yoji Morita**

* Professor, Kyoto Institute of Technology
** Research Fellow, Kyoto Institute of Technology
Matsugasaki, Kyoto, 606 Japan

From the information theoretic viewpoint, this paper presents the best approximated filter dynamics for state estimates of nonlinear dynamical systems. The basic notion is to introduce the Kullback's information criterion which minimizes the information loss due to an advent of linearization techniques. First, the linearization technique is performed so as to minimize the information loss between the true nonlinear system and the linearized one. Secondly, an important aspect is shown that the minimization of the Kullback's information quantity corresponds to maximizing the likelihood function.

The remainder of this paper is devoted to show the linearization technique based on the criterion of the maximal likelihood function. A numerical example is also given, including comparative discussions with another approximated filter dynamics using a familiar method of Taylor series expansion.

## 1. INTRODUCTION

Questions of interest in the performance of an approximated filter dynamics are
(Q-1) how close the approximated filter is to the optimal one?
(Q-2) as approximation techniques adopted are not unique but ad hoc, different approximated filter dynamics may be obtained for the same nonlinear systems under noisy observations. How can we establish comparative standards for various kinds of approximated filter dynamics?

In spite of the universality of these questions, there exist no generally accepted evaluation methods of standards for their judgement.



*D. G. Lainiotis and N. S. Tzannes (eds.), Advances in Control, 431–442.*

The purpose of this paper is to present a method of deriving the best approximated filter dynamics from the information theoretical point of view. Use is made of the Kullback's information criterion which minimizes the information loss due to an advent of linearization techniques. Furthermore, we show that the linearization technique presented here answer (Q-1) and (Q-2).

The general problem of nonlinear estimation theory is concerned with an n-dimensional state vector x and an m(≤n)-dimensiona vector which is the observation made at the system output. It is assumed that observations are additively contaminated by random noise. The dynamical equations that are normally used to formulate the problem are,

$$\left.\begin{aligned} dx_t &= f(x_t)dt + G_t dw_t,\ x(0) = x_0,\ t \in [0,T] \quad (1.1a)\\ dy_t &= h(x_t)dt + R_t dv_t,\ y(0) = 0 \quad (1.1b)\end{aligned}\right\}\Sigma_1$$

where $x_t$, $y_t$ are respectively the sample processes of x(t), y(t), $w_t$, $v_t$ are respectively n and m dimensional vectors of Brownian motion processes, all of whose components are stochastically independent of one another, f, h are n and m dimensional vector valued nonlinear functions respectively and $G_t$, $R_t$ are known n×n and m×m matrices respectively whose components depend on time

For convenience of the present descriptions, symbolic conventions are stated as follows. Throughout this paper, vector and matrix notations follow the usual manner. The transpose is denoted by a prime. The Euclidean norm is expressed by $||x||^2=x'x$. Let μ and ν be respectively measures on a proper measurable space, μ>>ν denotes the absolute continuity of ν with respect to μ, and μ∿ν denotes the mutually absolute continuity, i.e., μ>>ν and ν>>μ. The symbol, $\mathcal{Y}_t$ denotes the smallest σ-algebra of ω sets with respect to which the random variables y(τ) with τ≤t are measurable, where ω is a generic point of the probability space Ω. The mathematical expectation is denoted by $E$. The conditional expectation of a random variable conditioned by $\mathcal{Y}_t$ is simply expressed by "$\hat{\ }$" such that $\hat{x}_t \triangleq E\{x_t|\mathcal{Y}_t\}$.

## 2. PRELIMINARIES AND PROBLEM STATEMENT

Let (Ω,$\mathcal{F}$,P) be the probability triple and $\mathcal{F}_t$ be a nondecreasing right continuous family of sub σ-algebra. In order to guarantee the existence and uniqueness of the solution process to Eq.(1.1), the following assumptions are required:

(A-1) There exists a positive constant K such that

$$||f(u_1) - f(u_2)|| + ||h(u_1) - h(u_2)|| \le K||u_1 - u_2|| \tag{2.1}$$

(A-2) For the initial value of Eq.(1.1a), $E\{||x_0||^4\} < \infty$.

Although the minimal variance estimate $\hat{x}_t$ is formally determined by the solution to the nonlinear filtering equation, (1), the estimator dynamics generating $\hat{x}_t$ may not be realized due to the existence of nonlinear functions f(x) and h(x). Consequently, both the nonlinear functions f and h may respectively be expanded into

$$f(x_t) = a_t + B_t x_t + \varepsilon_1;\ h(x_t) = c_t + D_t x_t + \varepsilon_2 \tag{2.2}$$

where $a_t$, $c_t$ are n and m dimensional vectors, $B_t$ and $D_t$ are n×n, m×n matrices, $\varepsilon_1$, $\varepsilon_2$ are the collections of n and m dimensional vector error terms, respectively.

Using the approximations to (2.2), i.e.,

$$f(x_t) \cong a_t + B_t x_t,\ h(x_t) \cong c_t + D_t x_t, \tag{2.3}$$

for Eq.(1.1a), we write

$$dx_t^* = (a_t + B_t x_t^*)dt + G_t dw_t,\ x_0^* = x_0 \tag{2.4a}$$

and for Eq.(1.1b),

$$dy_t^* = (c_t + D_t x_t^*)dt + R_t dv_t,\ y_0^* = 0 \tag{2.4b}$$

$\Big\}\ \Sigma_2$

where a set of new symbols $x_t^*$ and $y_t^*$ have been used, denoting the approximated state variable and observations. It should be noted that, although the true observations $y_t$ are directly available for us, due to the existence of the nonlinear function h, observations are obliged to use the approximations obtained by (2.4b). An assumption is made on Eqs. (2.4a) and (2.4b) as follows:

(A-3) There exists a unique solution $(x_t^*, y_t^*)$ satisfying, the following inequality:

$$\int_0^T \{\|a_t + B_t x_t^*\|^2 + \|c_t + D_t x_t^*\|^2\}dt < \infty, \text{ a.s.} \tag{2.5}$$

In order to derive the approximated filter dynamics based on the approximated state variables $x_t^*$ and observations $y_t^*$, the coefficients $a_t$, $B_t$, $c_t$ and $D_t$ should be determined through the best possible way. To do this, we shall invite here the Kullback's information criterion which is interpreted as a measure of the difference between the systems $\Sigma_1$ and $\Sigma_2$, and is given by the following definition:

(*Definition*) (Kullback's information ) (2) Let $\nu_1$ and $\nu_2$ be measures respectively and $\nu_1 \sim \nu_2$. Then, the Kullback's information is defined by

$$I_{1,2} \triangleq E_{\nu 1}\{\log(d\nu_1/d\nu_2)\} \tag{2.6}$$

where $E_{\nu 1}\{\cdot\}$ denotes the mathematical expectation with respect to $\nu_1$-measure.

*Remark*: It is an easy exercise to show that $I_{1,2} \geq 0$ and the equality holds if and only if $\nu_1 = \nu_2$.

## 3. RADON-NIKODYM DERIVATIVES AND KULLBACK'S INFORMATION FOR $\Sigma_1$ AND $\Sigma_2$

Let $\mu_{X,Y}$, $\mu_{X^*,Y^*}$ be measures in the space $(C_T, \mathcal{B}_T)$, corresponding to the systems $\Sigma_1$ and $\Sigma_2$ respectively, where $C_T$ is the space of continuous functions $u=\{u_s, s\le T\}$ and $\beta_T$ is the σ-algebra generated by $\{u_s, s\le T\}$. The following assumption is basic to derive the Radon-Nikodym derivative, $d\mu_{X,Y}/d\mu_{X^*,Y^*}$.

(A-4) The coefficients $a_t$, $B_t$, $c_t$ and $D_t$ are functionals of $\{y^*_\tau, 0\le\tau\le t\}$, i.e., $a_t=a(y^*_\tau, 0\le\tau\le t)$, $B_t=B(y^*_\tau, 0\le\tau\le t)$ and so on. The assumption (A-4) implies that these coefficients are assumed to be $\beta_T$-measurable.

For $0\le t\le T$, consider the stochastic processes $\bar{x}_t$ and $\bar{y}_t$ with the respective stochastic differential

$$\Sigma_0 : d\bar{x}_t = G_t dw_t,\ d\bar{y}_t = R_t dv_t \tag{3.1}$$

Let $\mu_{\bar{X},\bar{Y}}$ be the measure in $(C_T,\beta_T)$ induced by the system $\Sigma_0$. From (A-1), (A-2) and (A-3), it is apparent that $\mu_{X,Y}\sim\mu_{\bar{X},\bar{Y}}$ and $\mu_{X^*,Y^*}\sim\mu_{\bar{X},\bar{Y}}$. Hence, we have $\mu_{X,Y}\sim\mu_{X^*,Y^*}$ and this allows us to compute the Radon-Nikodym derivative as state in the following lemma.

[*Lemma*-1] With the assumption (A-1) to (A-4), the Radon-Nikodym derivative of $\mu_{X,Y}$ with respect to $\mu_{X^*,Y^*}$ is given by

$$\begin{aligned}\frac{d\mu_{X,Y}}{d\mu_{X^*,Y^*}} = \exp[&-\int_0^t(a_s+B_sx_s-f(x_s))'G_s^{-1}dw_s - \int_0^t(c_s+D_sx_s-h(x_s))'R_s^{-1}dv_s\\ &+ \frac{1}{2}\int_0^t(a_s+B_sx_s-f(x_s))'(G_sG_s')^{-1}(a_s+B_sx_s-f(x_s))ds\\ &+ \frac{1}{2}\int_0^t(c_s+D_sx_s-h(x_s))'(R_sR_s')^{-1}(c_s+D_sx_s-h(x_s))ds]\end{aligned} \tag{3.2}$$

*Proof*: Since *Lemma*-1 is proved as a direct consequence of the application of Girsanov's theorem, (3) the procedure is outlined in the sequel. For economy of notations, we introduce new symbols $z_t$ and $z^*_t$ augmented respectively by $z_t=(x_t', y_t')'$ and $z^*_t=(x^{*\prime}_t, y^{*\prime}_t)'$. Furthermore, set as

$$f_1(z_t) = \begin{pmatrix} f(x_t)\\ h(x_t)\end{pmatrix},\quad f_2(z^*_t) = \begin{pmatrix} a_t+B_tx^*_t\\ c_t+D_tx^*_t\end{pmatrix},\quad Q_t = \begin{pmatrix} G_t & 0\\ 0 & R_t\end{pmatrix} \text{ and } B_t = \begin{pmatrix} w_t\\ v_t\end{pmatrix} \tag{3.3}$$

Then, the systems $\Sigma_1$ and $\Sigma_2$ can be described as

$$\Sigma_1 : dz_t = f_1(z_t)dt + Q_t dB_t \tag{3.4}$$

$$\Sigma_2 : dz^*_t = f_2(z^*_t)dt + Q_t dB_t \tag{3.5}$$

with the initial condition, $z_0=z^*_0$. From Eq.(3.4), it follows that

$$\begin{aligned} dz_t &= f_2(z_t)dt + Q_t\{dB_t - Q_t^{-1}(f_2(z_t) - f_1(z_t))dt\}\\ &= f_2(z_t)dt + Q_t d\tilde{B}_t\end{aligned} \tag{3.6}$$

where

$$d\tilde{B}_t = dB_t - \phi_t dt \tag{3.7}$$

and

$$\phi_t = Q_t^{-1}\{f_2(z_t) - f_1(z_t)\}. \tag{3.8}$$

It is well-known by Girsanov's theorem that the measure induced by $z_t$ in Eq.(3.4) is transformed into that by Eq.(3.6) with the Radon-Nikodym derivative,

$$\exp\{\int_0^t \phi_s' dB_s - \frac{1}{2}\int_0^t \|\phi\|^2 ds\} \tag{3.9}$$

and that $\tilde{B}_t$ in Eq.(3.7) becomes a Brownian motion process with respect to the transformed measure by Eq.(3.9). Then, from Eqs.(3.5) and (3.6), both $z_t$ and $z_t^*$ are the solutions to the same type of differential equation driven by the different Brownian motion process. Hence, noting that, with the initial condition, $z_0=z_0^*$, $z_t^*$ by Eq.(3.5) and $z_t$ by Eq.(3.6) induce the same probability measure in $(C_T, B_T)$, we obtain

$$\frac{d\mu_{X^*,Y^*}}{d\mu_{X,Y}} = \frac{d\mu_{Z^*}}{d\mu_Z} = \exp[\int_0^t \{f_2(z_s) - f_1(z_s)\}' Q_s^{-1} dB_s - \frac{1}{2}\int_0^t \{f_2(z_s) - f_1(z_s)\}'(Q_s Q_s')^{-1}\{f_2(z_s) - f_1(z_s)\} ds] \tag{3.10}$$

Hence, computing $d\mu_{X,Y}/d\mu_{X^*,Y^*}$ and using (3.3), the proof is completed.

From (2.6) and Lemma-1, for the systems $\Sigma_1$ and $\Sigma_2$, Kullback's information is defined by

$$\begin{aligned} I_\Sigma &= E_{\mu X,Y}\{\log(d\mu_{X,Y}/d\mu_{X^*Y^*})\} \\ &= \frac{1}{2}\int_0^t E_{\mu_{X,Y}}\{(a_s+B_s x_s-f(x_s))'(G_s G_s')^{-1}(a_s+B_s x_s-f(x_s)) \\ &\quad + (c_s+D_s x_s-h(x_s))'(R_s R_s')^{-1}(c_s+D_s x_s-h(x_s))\} ds \end{aligned} \tag{3.11}$$

## 4. MAIN THEOREM

The coefficients $a_s$, $B_s$, $c_s$ and $D_s$ are, in this section, determined in such a way that the Kullback's information $I_\Sigma$ becomes minimal.

*Theorem*-1: With (A-2) to (A-4), under the Kullback's information criterion, the coefficients $a_s$, $B_s$, $c_s$ and $D_s$ are determined by

$$a_s = \hat{f}(x_s) - B_s\hat{x}_s,\quad B_s = (\widehat{f(x_s)x_s'} - \hat{f}(x_s)\hat{x}_s')P_s^{-1} \tag{4.1}$$

and

$$c_s = \hat{h}(x_s) - D_s\hat{x}_s,\quad D_s = (\widehat{h(x_s)x_s'} - \hat{h}(x_s)\hat{x}_s')P_s^{-1} \tag{4.2}$$

where, in the sequel, the symbol "$\hat{\cdot}$" denotes $E_{\mu_{X,Y}}\{\cdot|\mathcal{Y}_s\}$ and $P_s = \text{cov.}\{x_s|\mathcal{Y}_s\}$

*Proof*: Since $E_{\mu_{X,Y}}\{\cdot\}=E_{\mu_{X,Y}}\{E_{\mu_{X,Y}}\{\cdot|\mathcal{Y}_s\}\}$, the minimization of the Kullback's information given by (3.11) is equivalent to minimizing the conditional expectation $\Psi_s$ defined by

$$\Psi_s = E_{\mu_{X,Y}}\{(a_s+B_sx_s-f(x_s))'(G_sG_s')^{-1}(a_s+B_sx_s-f(x_s))$$

$$+ (c_s+D_sx_s-h(x_s))'(R_sR_s')^{-1}(c_s+D_sx_s-h(x_s))|\mathcal{Y}_s\}. \quad (4.3)$$

The condition $\partial\Psi_s/\partial a_i = 0$ $(i = 1,2,\cdots,n)$ gives

$$(0,\cdots,\underset{(i)}{1},0,\cdots,0)'(G_sG_s')^{-1}(a_s+B_s\hat{x}_s-\hat{f}(x_s)) = 0 \quad (4.4)$$

Hence, we have $a_s$ in (4.1). For the (i,j)-element of the matrix $B_s$, the condition $\partial\Psi_s/\partial b_{ij}=0$ is written by

$$a_s'(G_sG_s')^{-1}B_s^{(ij)}\hat{x}_s + \widehat{(B_sx_s-f)'(G_sG_s')^{-1}B^{(ij)}x_s} = 0 \quad (4.5)$$

where $B_s^{(ij)} = \partial B_s/\partial b_{ij}$. From (4.5), it follows that

$$(a_s'(G_sG_s')^{-1})_i\hat{x}_{sj} = -\widehat{((B_sx_s-f)'(G_sG_s')^{-1})_ix_{sj}} \quad (4.6a)$$

or, in a vector form

$$(a_s'(G_sG_s')^{-1})'\hat{x}_s' = -\widehat{((B_sx_s-f)'(G_sG_s')^{-1})'x_s'} \quad (4.6b)$$

where $(\cdot)_i$ is the i-th component of the vector "$\cdot$" and $x_{sj}$ is the j-th component of $x_s$. With (4.1), (4.6b) becomes

$$(G_sG_s')^{-1}(\hat{f}\hat{x}_s' - B_s\hat{x}_s\hat{x}_s' + B_s\widehat{x_sx_s'} - \widehat{fx_s'}) = 0 \quad (4.7)$$

Noting that $P_s = \widehat{x_sx_s'} - \hat{x}_s\hat{x}_s'$, from (4.7), we have (4.1). Equation (4.2) can be obtained by a similar method to (4.1), respectively. Thus, the proof has been completed. It should be emphasized that Theorem 1 gives the same result as shown in Ref.(*4*).

## 5. MINIMAL KULLBACK'S INFORMATION AND MAXIMAL LIKELIHOOD FUNCTION

Let $\mu_Y$, $\mu_{Y*}$ and $\mu_{\bar{Y}}$ be measures induced by the processes $y_t$, $y_t^*$ and $\bar{y}_t$ defined by Eqs.(1.1b), (2.4b) and (3.1). It is apparent that $\mu_Y \sim \mu_{\bar{Y}}$ and $\mu_{Y*} \sim \mu_{\bar{Y}}$. This implies that there exists the Radon-Nikodym derivative of $\mu_{Y*}$ with respect to $\mu_Y$ and that the likelihood function is defined by $\Lambda_t = d\mu_{Y*}/d\mu_Y$.

We shall consider the conditional expectation of the logarithmic likelihood function $E_{\mu Y}\{\log(d\mu_{Y*}/d\mu_Y)\}$. By the definition (2.6), it can be seen that

$$E_{\mu_Y}\{\log(d\mu_{Y*}/d\mu_Y)\} = - E_{\mu_Y}\{\log(d\mu_Y/d\mu_{Y*})\} = - I_{Y,Y*} \quad (5.1)$$

Before describing a theorem stating the relation between the Kullback's information and the likelihood function, we need the following lemma.

[*Lemma*-2] (5) Let $\mu_{X\otimes\bar{Y}}$ and $\mu_{X*\otimes\bar{Y}}$ be product measures respectively associated with $(x,\bar{y})$ and $(x^*,\bar{y})$. Then, the Radon-Nikodym derivatives with respect to processes $(y_t,\bar{y}_t)$ and $(y_t^*,\bar{y}_t)$ are respectively given by

$$\frac{d\mu_Y}{d\mu_{\bar{Y}}} = E_{\mu_{X\otimes\bar{Y}}}\{(d\mu_{X,Y}/d\mu_{X\otimes\bar{Y}})|\mathcal{B}_t\} \quad (5.2)$$

and

$$\frac{d\mu_{Y*}}{d\mu_{\bar{Y}}} = E_{\mu_{X*\otimes\bar{Y}}}\{(d\mu_{X*,Y*}/d\mu_{X*\otimes\bar{Y}}|\mathcal{B}_t\} \quad (5.3)$$

where $\mathcal{B}_t$ is the σ-algebra generated by continuous mapping from $[0,t]$ to the observation space $E^{(m)}$.

*Theorem*-2: The following inequality holds

$$0 \geq E_{\mu_Y}\{\log\Lambda_t\} = - I_{Y,Y*} \geq - I_\Sigma \quad (5.4)$$

*Proof*: From Remark described in Section 2, it is obvious that $I_{Y,Y*}\geq 0$ and $I_{Y,Y*}=0$, if and only if $\mu_Y=\mu_{Y*}$. Using a measure $\mu_{\bar{Y}}$, the likelihood function $\Lambda_t$ is

$$\Lambda_t = \frac{d\mu_{Y*}}{d\mu_{\bar{Y}}} \Big/ \frac{d\mu_Y}{d\mu_{\bar{Y}}} \quad (5.5)$$

Substituting (5.2) and (5.3) into (5.5), we obtain

$$\Lambda_t = E_{\mu_{X*\otimes\bar{Y}}}\{\frac{d\mu_{X*,Y*}}{d\mu_{X*\otimes\bar{Y}}}|\mathcal{B}_t\} \Big/ E_{\mu_{X\otimes\bar{Y}}}\{\frac{d\mu_{X,Y}}{d\mu_{X\otimes\bar{Y}}}|\mathcal{B}_t\} \quad (5.6)$$

Since

$$d\mu_{X*,Y*}/d\mu_{X*\otimes\bar{Y}} = d\mu_{X*,Y*}/d\mu_{X,Y}\cdot d\mu_{X,Y}/d\mu_{X*\otimes\bar{Y}} \quad (5.7)$$

using the property of Radon-Nikodym derivative,

$$E_{\mu_{X*\otimes\bar{Y}}}\{\frac{d\mu_{X*,Y*}}{d\mu_{X*\otimes\bar{Y}}}|\mathcal{B}_t\} = E_{\mu_{X,Y}}\{\frac{d\mu_{X*,Y*}}{d\mu_{X,Y}}|\mathcal{B}_t\}E_{\mu_{X*\otimes\bar{Y}}}\{\frac{d\mu_{X,Y}}{d\mu_{X*\otimes\bar{Y}}}|\mathcal{B}_t\} \quad (5.8)$$

Similarly, in the denominator of (5.6), the following equality can be obtained,

$$E_{\mu_{X\otimes\bar{Y}}}\{\frac{d\mu_{X,Y}}{d\mu_{X\otimes\bar{Y}}}|\mathcal{B}_t\} = E_{\mu_{X^*\otimes\bar{Y}}}\{\frac{d\mu_{X,Y}}{d\mu_{X^*\otimes\bar{Y}}}|\mathcal{B}_t\}E_{\mu_{X\otimes\bar{Y}}}\{\frac{d\mu_{X^*\otimes\bar{Y}}}{d\mu_{X\otimes\bar{Y}}}|\mathcal{B}_t\}. \quad (5.9)$$

From (5.8) and (5.9), the likelihood function $\Lambda_t$ given by (5.6) is computed to be

$$\Lambda_t = E_{\mu_{X,Y}}\{\frac{d\mu_{X^*,Y^*}}{d\mu_{X,Y}}|\mathcal{B}_t\}/E_{\mu_{X\otimes\bar{Y}}}\{\frac{d\mu_{X^*\otimes\bar{Y}}}{d\mu_{X\otimes Y}}|\mathcal{B}_t\}. \quad (5.10)$$

Bearing in mind the fact that $d\mu_{X^*\otimes\bar{Y}}/d\mu_{X\otimes\bar{Y}}=d\mu_{X^*}/d\mu_X$, it is easy to see that

$$E_{\mu_{X\otimes\bar{Y}}}\{(d\mu_{X^*\otimes\bar{Y}}/d\mu_{X\otimes\bar{Y}})|\mathcal{B}_t\}=\int(d\mu_{X^*}/d\mu_X)\mu_X(dX)\int\mu_{\bar{Y}}(d\bar{Y}) = 1. \quad (5.11)$$

From (5.11), (5.10) is simply expressed by

$$\Lambda_t = E_{\mu_{X,Y}}\{(d\mu_{X^*,Y^*}/d\mu_{X,Y})|\mathcal{B}_t\}. \quad (5.12)$$

Since the maximization of $\Lambda_t$ is equivalent to that of $\log\Lambda_t$, from now on let us consider $\log\Lambda_t$. Furthermore, operating the conditional expectation with respect to $\mu_Y$-measure on $\log\Lambda_t$, it follows that

$$E_{\mu_Y}\{\log\Lambda_t\} = E_{\mu_Y}[\log\{E_{\mu_{X,Y}}\{(d\mu_{X^*,Y^*}/d\mu_{X,Y})|\mathcal{B}_t\}\}] \quad (5.13)$$

Noting that the function $-\log(u)$ is a convex function(downward), Jensen's inequality can be applied to be $E\{-\log(u)\} \geq -\log E\{u\}$. Combining this with (5.13),

$$E_{\mu_Y}\{\log\Lambda_t\} \geq E_{\mu_Y}\{E_{\mu_{X,Y}}\{\log(d\mu_{X^*,Y^*}/d\mu_{X,Y})|\mathcal{B}_t\}\}. \quad (5.14)$$

The right hand side of (5.14) becomes

$$E_{\mu_Y}\{E_{\mu_{X,Y}}\{\cdot|\mathcal{B}_t\}\} = E_{\mu_{X,Y}}\{E_{\mu_{X,Y}}\{\cdot|\mathcal{B}_t\}\} = E_{\mu_{X,Y}}\{\cdot\}. \quad (5.15)$$

From (5.14) and (5.15), we have

$$E_{\mu_Y}\{\log\Lambda_t\} \geq - E_{\mu_{X,Y}}\{\log(d\mu_{X,Y}/d\mu_{X^*,Y^*})\} = - I_\Sigma. \quad (5.16)$$

Thus, we have the inequality (5.4) and the proof has been completed. Theorem-2 reveals us that the stochastic linearization by using (4.1) and (4.2) based on the minimal Kullback's information criterion maximizes simultaneously the lower bound of the expected value of logarithmic likelihood function.

## 6. STOCHASTIC LINEARIZATION VIA MAXIMAL LIKELIHOOD PRINCIPLE

The results obtained in the previous section suggest a possibility of replacing the minimal Kullback's information criterion by the maximal likelihood principle. In this section, the computational aspect of stochastic linearization for practical uses is shown, based on the maximal likelihood estimate. The system dynamics and observation mechanism are the same as given by $\Sigma_1$, where all elements of both matrices $G_t$ and $R_t$ are considered to be time-independent, and both matrices are denoted by G and R respectively.

We shall start with the approximated dynamics of the system and observation mechanism labeled by $\Sigma_2$ (see Eq.(2.4)), where, for simplicity of descriptions, the symbols $x_t^*$ and $y_t^*$ are simply denoted by $x_t$ and $y_t$, respectively. Furthermore, in the sequel, we use the same symbols for the true stochastic process as for the linearized stochastic process because no confusions will result. Define the likelihood (ratio) function $\lambda_t$ by $\lambda_t \triangleq d\mu_Y/d\mu_{\bar{Y}}$. Under suitable conditions for nonlinear functions $f(x_t)$ and $h(x_t)$, it is known that $\lambda_t$ is of the form (6).

$$\lambda_t = \exp\{\int_0^t \hat{h}'(x_s)(RR')^{-1}dy_s - \frac{1}{2}\int_0^t \hat{h}'(x_s)(RR')^{-1}\hat{h}(x_s)ds\} \qquad (6.1)$$

where $\hat{}$ means, in this section, the conditional expectation $E\{\cdot|Y_t\}$. For convenience of discussions, $\log\lambda_t$ will be taken into consideration, instead of $\lambda_t$ itself.

In this section, two cases will be considered

Case 1: The system dynamics is considered to be linear and, on the other hand, observations are obtained through the nonlinear dynamics. Thus, the system $\Sigma_2$ given by Eqs.(2.4a) and (2.4b) are, in Case 1, described by

$$dx_t = A_t x_t dt + Gdw_t, \quad x(t_0) = x_0 \qquad (6.2a)$$

$$dy_t = (c_t + D_t x_t)dt + Rdv_t, \quad y(t_0) = y_0 = 0. \qquad (6.2b)$$

In order to apply the maximal likelihood principle, for Eq.(6.2b), we shall write

$$dy_t = (c + Dx_t)dt + Rdv_t, \quad y(t_0) = y_0 = 0 \qquad (6.2c)$$

where c and D are respectively an m-dimensional constant vector and an m×n constant matrix. Our purpose is to estimate the unknown coefficients c and D so as to maximize the logarithmic likelihood function $\log\lambda_t$. The following form of $\log\lambda_t$ is a direct consequence from Eq.(6.1):

$$\log\lambda_t = \int_0^t (c + D\hat{x}_s)'(RR')^{-1}dy_s - \frac{1}{2}\int_0^t (c + D\hat{x}_s)'(RR')^{-1}(c + D\hat{x}_s))ds \qquad (6.3)$$

Let $\tilde{c}(t)$ and $\tilde{D}(t)$ be respectively the maximal likelihood estimator. Both $\tilde{c}(t)$ and $\tilde{D}(t)$ may be obtained by the conditions

$$\left.\frac{\partial \log\lambda_t}{\partial c_i}\right|_{\text{at } c_i = \tilde{c}_i(t)} \text{ and } \left.\frac{\partial \log\lambda_t}{\partial d_{ij}}\right|_{d_{ij} = \tilde{d}_{ij}(t)} = 0 \qquad (6.4)$$

After somewhat tedious calculations, we have

$$\tilde{c}(t) = \frac{1}{t}[\int_0^t dy_s - \tilde{D}(t)\int_0^t \hat{x}_s ds] \qquad (6.5)$$

and

$$\tilde{D}(t) = [\int_0^t \hat{x}_s dy_s' - \frac{1}{t}(\int_0^t \hat{x}_s ds)(\int_0^t \hat{x}_s ds)']' \times [\int_0^t \hat{x}_s \hat{x}_s' ds - \frac{1}{t}(\int_0^t \hat{x}_s ds)(\int_0^t \hat{x}_s ds)']^{-1} \tag{6.6}$$

With $\tilde{c}(t)$ and $\tilde{D}(t)$ obtained by (6.5) and (6.6), the nonlinear function is approximated as

$$h(x_t) \cong \tilde{c}(t) + \tilde{D}(t)x_t \tag{6.7}$$

Furthermore, it is a simple exercise to show that the approximated filter dynamics is given by

$$d\hat{x}_t = A_t \hat{x}_t dt + P_t \tilde{D}_t'(RR')^{-1}[dy_t - (\tilde{c}_t + \tilde{D}_t \hat{x}_t)dt] \tag{6.8a}$$

$$\hat{x}_0 = E\{x_0\} \tag{6.8b}$$

with the associated error covariance equation,

$$\frac{dP_t}{dt} = A_t P_t + P_t A_t' + GG' - P_t \tilde{D}_t'(RR')^{-1}\tilde{D}_t P_t \tag{6.9a}$$

$$P_0 = \text{cov.}[x_0] \tag{6.9b}$$

A sample run of the approximated nonlinear filter dynamics is obtained by solving simultaneously Eqs.(6.8) and (6.9) with (6.5) and (6.6). The aspect of numerical computation will be described with respect to the second case. Before proceeding with the second example, it is a topic of some interest to compare the result obtained above with that by the stochastic linearization (4). From (6.5), (6.6) and (6.7), we have

$$h(x_t) \cong \frac{1}{t}\int_0^t dy_s + \tilde{D}(t)[x_t - \frac{1}{t}\int_0^t \hat{x}_s ds] \tag{6.10}$$

On the other hand, the stochastic linearization technique gives

$$h(x_t) \cong \hat{h}(x_t) + [\widehat{h(x_t)x_t'} - \hat{h}(x_t)\hat{x}_t']P_t^{-1}(x_t - \hat{x}_t) \tag{6.11}$$

Let $\tilde{w}_t$ be defined by

$$d\tilde{w}(t) = dy_t - \hat{h}(x_t)dt. \tag{6.12}$$

Then, it is apparent that $\tilde{w}_t$ is the innovation process and that this is equivalent to the Wiener process on $\mathcal{Y}_t = \sigma\{y_s, s \le t\}$. Hence,

$$\frac{1}{t}\int_0^t dy_s = \frac{1}{t}\int_0^t \{dy_s - \hat{h}(x_s)ds\} + \int_0^t \hat{h}(x_s)ds = \frac{1}{t}\int_0^t d\tilde{w}_s + \frac{1}{t}\int_0^t \hat{h}(x_s)ds. \tag{6.13}$$

Taking into account the property of Wiener process, the following identity holds:

$$\lim_{t\to\infty} \frac{1}{t}\int_0^t d\tilde{w}_s = \lim_{t\to\infty} \frac{\tilde{w}_t}{t} = 0 \tag{6.14}$$

Hence,

$$\lim_{t\to\infty} \frac{1}{t}\int_0^t dy_s = \lim_{t\to\infty} \frac{1}{t}\int_0^t \hat{h}(x_s)ds \tag{6.15}$$

Assume that the filtering process $\hat{h}(x_s)$ has an invariant measure $\nu$. Then

$$\lim_{t\to\infty}\frac{1}{t}\int_0^t \hat{h}(x_s)ds = E_\nu\{\hat{h}(x_t)\}, \tag{6.15}$$

which corresponds to $\hat{h}(x_t)$ in the stochastic linearization.

Case 2: For our better understanding an exemplary scalar system is considered with a numerical experiment. A system dynamics is nonlinear and, on the other hand, the observation mechanism is linear:

$$dx_t = -\tan^{-1}x_t dt + gdw_t, \quad dy_t = x_t dt + rdv_t \tag{6.16}$$

The nonlinear function $-\tan^{-1}x_t$ is expanded into a linear function $a+bx_t$ where a, b are unknown constants. These are estimated through the logarithmic likelihood function $\log\lambda_t$:

$$\left.\frac{\partial\log\lambda_t}{\partial a}\right|_{a=\tilde{a}(t)} = 0, \quad \left.\frac{\partial\log\lambda_t}{\partial b}\right|_{b=\tilde{b}(t)} = 0 \tag{6.17}$$

By the maximal likelihood estimators $\tilde{a}(t)$ and $\tilde{b}(t)$, a linearization of $-\tan^{-1}x_t$ is

$$-\tan^{-1}x_t \cong \tilde{a}(t) + \tilde{b}(t)x_t \tag{6.18}$$

Hence, the estimator dynamics are given by

$$d\hat{x}_t = \{\tilde{a}(t) + \tilde{b}(t)\hat{x}_t\}dt + p_0(t)r^{-2}\{dy_t - \hat{x}_t dt\}, \tag{6.19}$$

$$\frac{dp_0(t)}{dt} = 2\tilde{b}(t)p_0(t) + g^2 - p_0^{\,2}(t)r^{-2}, \tag{6.20}$$

where

$$\tilde{a}(t) = \frac{3}{t^3}\{\int_0^t sdy_s - \int_0^t\hat{\eta}_s sds - \tilde{b}(t)\int_0^t\hat{\rho}_s ds\} \tag{6.21a}$$

$$\tilde{b}(t) = [\int_0^t\hat{\rho}_s dy_s - \int_0^t\hat{\rho}_s\hat{\eta}_s ds - \frac{3}{t^3}\int_0^t\hat{\rho}_s sds\int_0^t sdy_s$$
$$+ \frac{3}{t^3}\int_0^t\hat{\rho}_s sds\int_0^t\hat{\eta}_s sds][\int_0^t\hat{\rho}_s^{\,2}ds - \frac{3}{t^3}\{\int_0^t\hat{\rho}_s sds\}^2]^{-1} \tag{6.21b}$$

and where $\hat{\rho}_s \triangleq E\{\int_0^s x_\tau d\tau|\mathcal{Y}_s\}$ and $\hat{\eta}_s \triangleq E\{x_0+\int_0^s gdw_\tau|\mathcal{Y}_s\}$ are determined by a set of equations:

$$d\hat{\rho}_s = p_1(s)\{dy_s - \hat{x}_s ds\} + \hat{x}_s ds, \tag{6.22}$$

$$d\hat{\eta}_s = p_2(s)\{dy_s - \hat{x}_s ds\}, \tag{6.23}$$

$$\frac{dp_1(s)}{ds} = p_0(s) + \tilde{b}(s)p_1(s), \quad p_1(0) = 0 \tag{6.24}$$

and

$$\frac{dp_2(s)}{ds} = g^2 + \tilde{b}(s)p_2(s), \quad p_2(0) = p_0(0). \tag{6.25}$$

The $\hat{x}_t$-run is given by the simulataneous treatments of approximated structure of state estimator dynamics and of parameter estimation $\tilde{a}(t)$ and $\tilde{b}(t)$ with running evaluation of the error covariance $p_0(t)$.

For comparative discussions, an approximated filter dynamics derived by Taylor series expansion (7) also calculated as shown in Fig.1.

Equations (6.19) to (6.25) are simulated on a digital computer. Figure 1 shows one of representative results of simulation experiments.

It should be noted that the algorithm of estimator dynamics presented here may be efficient after some amount of the observation data is obtained, because both $\tilde{a}(t)$ and $\tilde{b}(t)$ in Eq.(6.21) are numerically meaningless at the initial time t=0. With the values of $t_\delta$, $\tilde{a}(t)=a_0$ and $\tilde{b}(t)=b_0$ suitably chosen, the $\hat{x}(t)$ and $p_0(t)$ runs start and after $t_\delta$, the estimator and the associated error covariance algorithms are effecitive. In Fig.1, the followings were set as $t_\delta=2.3$, $a_0=0$ and $b_0=-0.6$.

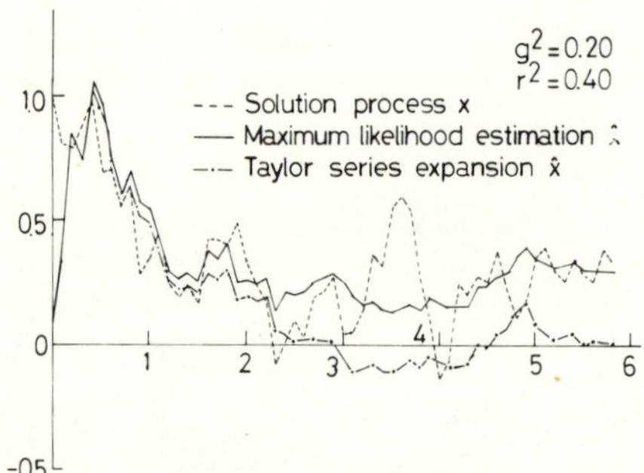


A sample run of the filter

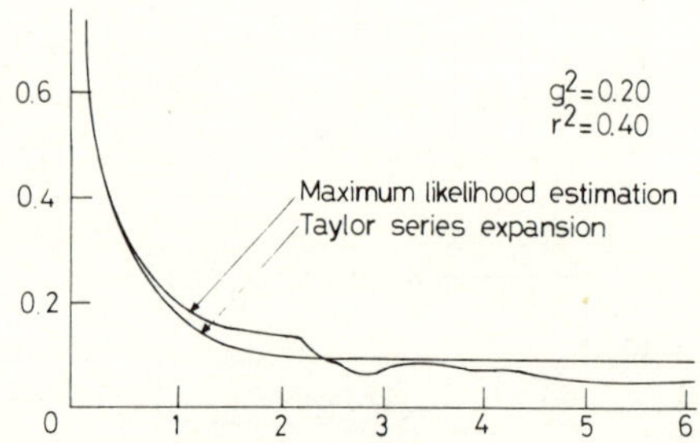


A sample run of the covariance

## 7. CONCLUSIONS

The key notion in this paper is to introduce the Kullback's information as the information criterion which measures the difference between the original nonlinear system and an approximated linear one. Thus, the stochastic linearization technique is now valid from the information theoretic viewpoint.

REFERENCES

1. Kushner, H.J.: 1967, J. Diff. Eqs.3, pp.179-190.
2. Kullback, S.: Information Theory and Statistics, 1959, John Wiley, New York.
3. Girsanov, I.V.: 1960, Theo. Prob. Appl.5, pp.285-301.
4. Sunahara, Y.: 1970, Trans. ASME, J. of Basic Eng.92, pp.385-393.
5. Liptser, R.S. and Shiryaev, A.N.: Statistics of Random Processes 1978, Springer Verlag, New York.
6. Kailath, T.: 1969, IEEE Trans. IT-15, pp.350-361.
7. Schwartz, L.: 1966, Hughes Aerospace Tech. Res. Rept., SSD 60472R

# CONSISTENT ESTIMATION OF CONTINUOUS-TIME SIGNALS FROM QUANTIZED NOISY SAMPLES

Elias Masry and Stamatis Cambanis

U. California, La Jolla; U. North Carolina, Chapel Hill

## SUMMARY

In general a signal s(t) cannot be reconstructed from its hardlimited version, i.e. from its sign. The question arises whether it may be possible to reconstruct the signal from the hardlimited version of an appropriate transformation of the signal. The answer to this question turns out to be positive in the precise sense described below. An appropriate transformation consists in first sampling the signal periodically and then adding noise to the samples prior to hardlimiting; this is a random transformation and it is clear that no nonrandom transformation could do the job. The signal can then be estimated consistently from its hardlimited noisy samples as the sampling rate tends to infinity. The estimates, which are generally nonlinear, can be made linear by a proper choice of the noise distribution. This rather unexpected result holds for all bounded and uniformly continuous signals.

In describing the results we restrict attention to continuous signals s(t) defined on the unit interval $0 \le t \le 1$ with a known bound b: $|s(t)| \le b$ for all $0 \le t \le 1$. The general problem can be modelled as a transmitter/receiver (with no channel noise) with structure depicted in Figure 1. The signal s is sampled at the equally spaced points $\{k/W, k=0,1,\ldots,W\}$ where the sampling rate W is an integer. The samples $\{s(k/W)\}$ are then deliberately corrupted by additive noise $\{X_k\}$, where the random variables $X_0, X_1, \ldots, X_W$ are independent and identically distributed with distribution specified below. The noisy samples $Y_{W,k} = s(k/W) + X_k$ are passed through a hardlimiter whose binary output



*D. G. Lainiotis and N. S. Tzannes (eds.), Advances in Control, 443–446.*

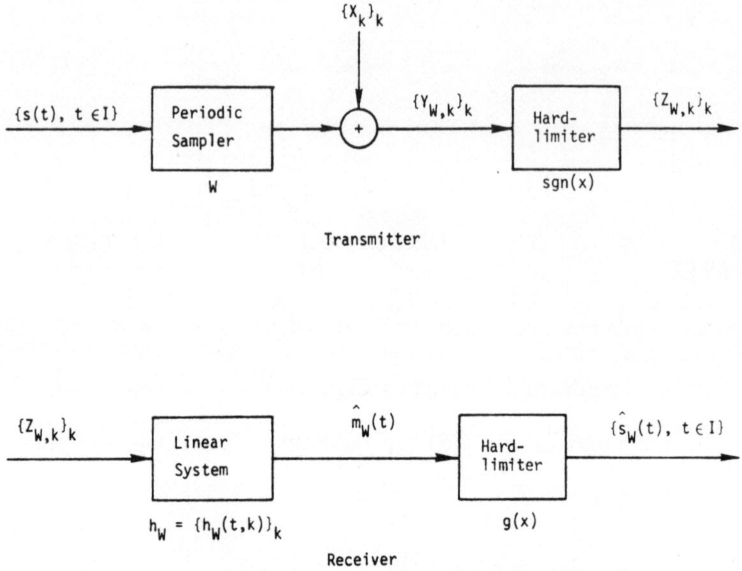


Figure 1. The structure of the transmitter/receiver model.

$$Z_{W,k} = \operatorname{sgn}[Y_{W,k}] = \operatorname{sgn}[s(k/W) + X_k] \ , \ k=0,1,\ldots,W \ ,$$

is transmitted. The receiver uses the binary sequence $\{Z_{W,k}\}$ to obtain an estimate $\hat{s}_W(t)$ of $s(t)$ as follows. The receiver or estimator consists of the linear system

$$h_W = \{h_W(t,k) = \binom{W}{k} t^k (1-t)^{W-k} \ , \ k=0,1,\ldots,W,\ 0 \le t \le 1\}$$

cascaded with a memoryless nonlinearity $g(x)$. The output $\hat{m}_W(t)$ of the linear system $h_W$ is given by

$$\hat{m}_W(t) = \sum_{k=0}^{W} h_W(t,k) Z_{W,k} \ , \ 0 \le t \le 1 \ ,$$

and the estimate $\hat{s}_W(t)$ of $s(t)$ is defined by

$$\hat{s}_W(t) = g[\hat{m}_W(t)] \ , \ 0 \le t \le 1 \ .$$

The choice of the memoryless nonlinearity $g(x)$ depends on the choice of the noise distribution. When the distribution of the noise is Gaussian with mean zero and variance $\sigma^2$ we choose

$$g(x) = \begin{cases} \mu^{-1}(x) , & |x| \le \mu(c) \\ 0 , & |x| > \mu(c) \end{cases} , \quad c = b+\varepsilon, \ \varepsilon>0 ,$$

where

$$\mu(x) = (2/\pi)^{\frac{1}{2}} \int_0^{s/\sigma} \exp(-u^2/2)du , \quad -\infty<x<\infty .$$

When the distribution of the noise is uniform over $(-b,b)$ we choose

$$g(x) = \begin{cases} bx , & |x| \le 1 \\ 0 , & |x| > 1 \end{cases}$$

so that in this case the estimator or receiver is linear in the binary data $\{Z_{W,k}\}$ (since $|\hat{m}_W(t)| \le 1$).

The first result gives the mean-square consistency of the estimate $\hat{s}_W(t)$ and provides bounds on the rate of convergence, expressed in terms of the modulus of continuity of s

$$\omega(s;\delta) = \sup\{|s(t)-s(t')|: |t-t'| \le \delta, 0\le t, t'\le 1\}.$$

Theorem 1. For every $0<t<1$ the estimate $\hat{s}_W(t)$ converges in the mean-square sense to $s(t)$ as the sampling rate $W \to \infty$, and

$$E[\hat{s}_W(t)-s(t)]^2 \le k_1\omega^2(s;\sqrt{\frac{t(1-t)}{W}}) + k_2\frac{1+o(1)}{\sqrt{\pi t(1-t)W}}$$

where the constants are given by

$$k_1 = (16/\pi\sigma^2)k_2 , \quad k_2 = (\pi\sigma^2/4)[1+(b/\varepsilon)^2]\exp[(b+\varepsilon)/\sigma]^2$$

when the noise is Gaussian, and when the noise is uniform by

$$k_1 = 4 , \quad k_2 = b^2/2 .$$

For somewhat smoother signals the next result shows that the estimate $\hat{s}_W(t)$ is strongly consistent so that $s(t)$ can be reconstructed from almost every realization of the binary data $\{\mathrm{sgn}[s(k/W)+X_k]\}$.

Theorem 2. If $s(t)$ is Lip $\gamma$, $0<\gamma\leq 1$, then for each fixed $0<t<1$ we have with probability one

$$W^{\alpha}|\hat{s}_W(t)-s(t)| \to 0 \quad \text{as } W \to \infty$$

where $\alpha$ is any constant satisfying $0<\alpha<\frac{1}{2}\min(\gamma,\frac{1}{2})$.

The proofs of these results can be found in [1], along with further results concerning the asymptotic normality of the estimate, the case of an infinite interval, and other memoryless nonlinearities than the hardlimiter.

## ACKNOWLEDGEMENTS

The work of E. Masry was supported by the Office of Naval Research under Contract N00014-75-C-0652. The work of S. Cambanis was supported by the Air Force Office of Scientific Research under Grant AFOSR-75-2796.

## REFERENCE

1. Masry, E., and Cambanis, S. Consistent estimation of continuous-time signals from nonlinear transformations of noisy samples, 1979, manuscript submitted for publication.

# OBSERVERS AND NOISE: PERFORMANCE CHARACTERIZATION

Nicholas Kyriakopoulos and Chester W. Wong

The George Washington University
Washington, D.C. 20052, USA

In this paper, the behavior of deterministic observers in the presence of noise is examined. For small system and measurement noises it might be advantageous to design a deterministic observer instead of a minimum variance estimator, and accept a certain estimation error. Such an error is, of course, a function of the plant and measurement noises.

The error covariance equation for an observer has been formulated and, for the case of a constant system, solved. A bound has also been defined for the steady state error covariance matrix as a function of the noise variances and the observer dynamics.

For general second order systems the error variance has been computed as a function of the system and measurement noise variances as well as the system parameters. These results are compared to the Kalman filter error variance.

## INTRODUCTION

Deterministic observers as developed by Luenberger [1], [2], [3], are designed without reference to either plant or measurement noise; for constant systems the error between the actual states and their estimates is an exponentially decreasing function whose time constant can be controlled by the location of the poles of the observer. Besides the error control, another advantage of the deterministic observers is their relative ease of implementation in either analog or discrete form.



*D. G. Lainiotis and N. S. Tzannes (eds.), Advances in Control, 447–455.*

In reality, however, there are no noise-free systems; furthermore, some information is always either available or assumed about the noise characteristics. Kalman and Bucy [4] have utilized this additional information to develop a linear recursive estimator which is optimal in the minimum error variance sense. Because of its recursive nature this estimator requires a certain computational power even for constant systems; in addition, it fails in the absence of noise because of the singularity of the error covariance matrix.

In situations, however, where the plant and measurement noises are small the simplicity of a noisy observer might outweigh the optimality of the computationally complex estimator. It is therefore desirable to examine the behavior of observers in noisy environments. Observers for stochastic systems have been examined [5], [6], [7], [8]; in all cases the observer gains are obtained by rather elaborate recursive procedures.

Another approach to the design of observers operating in a noisy environment is taken here. Since the observer output is a stochastic process, the designer establishes bounds for both the speed of covergence of the observer and the output error power. These two parameters are affected in a contradictory manner by the observer gain which is, in turn, affected by the location of the observer poles. Thus, the design problem may be formulated as a constrained optimization problem, in which the locations of the poles are chosen to satisfy both the error bound and the convergence rate.

In this paper the observer error is expressed in terms of the plant and measurement noise covariance matrices as well as the observer gain matrix. The norm of the error covariance matrix as well as the sample variance both taken at steady state are used as measures of performance. An observer for a second order system has been designed and its behavior has been examined as a function of the plant and measurement noises. The performance of the observer has been compared to that of the corresponding Kalman filter.

## PROBLEM FORMULATION

A linear discrete time-variant system corrupted by noise is characterized by

$$x(k+1) = Ax(k)+Bu(k)+Gw(k) \tag{1}$$

where u(k) is the deterministic input vector and w(k) is the plant noise vector. The corresponding measurement equations are

$$y(k) = Cx(k) + \Gamma v(k) \tag{2}$$

where $v(k)$ is the measurement noise vector.

For the plant described by (1) the observer is given by

$$\hat{x}(k+1) = A\hat{x}(k) + Bu(k) + M(y(k) - \hat{y}(k)) \tag{3}$$

where $M$ is the observer gain matrix and

$$\hat{y}(k) = C\hat{x}(k) \tag{4}$$

From (1), (2), (3), and (4), the estimation error

$\tilde{x}(k) = x(k) - \hat{x}(k)$ is given by

$$\tilde{x}(k+1) = [A-MC]\tilde{x}(k) + Gw(k) - M\Gamma v(k) \tag{5}$$

The last two terms in (5) reflect the contribution of the plant and measurement noises to the behavior of the observer. For the noise-free case the estimation error $\tilde{x}(k)$ approaches zero exponentially, the rate of decrease being controlled by the choice of the observer gain M. If it is therefore assumed that an observer has been designed to estimate the state of a system under noise-free conditions and with a given time constant, it would be of interest to determine whether such an observer could still be used as a state estimator in a noisy environment. Thus, it must first be ascertained that the output of the observer will approach some steady state and then the observer behavior at steady state must be characterized.

Inspection of (5) reveals that for G, M and $\Gamma$ constant, and $x(k)$, $v(k)$ bounded, the solution will also be bounded. It remains therefore to determine the statistics of $\tilde{x}(k)$ in terms of M as well as in terms of the statistics of $v(k)$ and $w(k)$.

Under the assumption that the expected values of $v(k)$ and $w(k)$ are zero, then, from (5), the expected value of $x(k)$ approaches zero asymptotically. The error variance $P(k)$ is determined from

$$P(k+1) = E[\tilde{x}(k+1)\tilde{x}(k+1)^T] \tag{6}$$

by substituting (5) into (6) and assuming that $x(k)$, $w(k)$ and $v(k)$ are uncorrelated; then

$$P(k+1) = [A-MC]P(k)[A-MC]^T + GR_w(k)G^T + M\Gamma R_v(k)\Gamma^T M^T \tag{7}$$

where $R_w(k)$ and $R_v(k)$ are the covariance matrices of $w(k)$ and $v(k)$, respectively.

For simplicity, it will subsequently be assumed that $R_w(k)=R_w$ and $R_v(k)= R_v$.

By substitution (7) yields

$$P(k) = [A-MC]^k P_o ([A-MC]^T)^k + \sum_{i=1}^{k} [A-MC]^{i-1} R([A-MC]^T)^{i-1} \quad (8)$$

where

$$R \overset{\Delta}{=} GR_w G^T + M \Gamma R_v \Gamma^T M^T \quad (9)$$

The behavior of the observer can be examined by considering the steady state value of P(k) defined as

$$\bar{P} = \lim_{k\to\infty} P(k) \quad (10)$$

or

$$\bar{P} = \lim_{k\to\infty} [A-MC]^k P_o ([A-MC]^T)^k + \sum_{i=1}^{\infty} [A-MC]^{i-1} R([A-MC]^T)^{i-1} \quad (11)$$

Under the assumption that the eigenvalues of [A-MC] have negative real parts, the first part of (11) goes to zero and

$$\bar{P} = \sum_{i=1}^{\infty} [A-MC]^{i-1} R([A-MC]^T)^{i-1} \quad (12)$$

As it is expected, the steady state error covariance depends not only on the plant and measurement noise convariances but also on the observer gain. A bound on $\bar{P}$ can be determined by taking the norm of (12),

$$||\bar{P}|| = ||\sum_{i=1}^{\infty} [A-MC]^{i-1} R([A-MC]^T)^{i-1}||$$

$$\leq \sum_{i=1}^{\infty} ||A-MC||^{2(i-1)} \cdot ||R|| \tag{13}$$

where

$$||R|| \leq ||G||^2 \cdot ||R_v|| + ||M||^2 \cdot ||\Gamma||^2 \cdot ||R_v||$$

One can then perceive $||\bar{P}||$ as estimation error power and $||R||$ as estimation noise power; thus, an upper bound of the error-to-noise power ratio is given by

$$\frac{||\bar{P}||}{||\bar{R}||} \leq \sum_{i=1}^{\infty} ||A-MC||^{2(i-1)} \tag{14}$$

For the corresponding continuous case the steady state error covariance matrix is found [9] to satisfy the equation

$$[A-MC]\bar{P} + \bar{P}\,[A-MC]^T + R = 0 \tag{15}$$

Equations (12), (14), (15), suggest that the observer gain could be selected so that the error-to-noise power ratio is minimized. At this time, no easy solution to this problem has been found.

To examine, however, the effects of the observer pole locations on the estimation error, an observer was designed for a second order system and its performance was examined for various values of plant and measurement noises. Although the steady state performance is defined as $t \to \infty$, a commonly used value is $t = 4\zeta$ where $\zeta$ is the system time constant; thus, one can use the sample variance at that instant (Figure 1) as an alternative measure performance since the output of the observer can be taken as a reasonable estimate of the state.

EXAMPLE

A system defined by

$$H(s) = K\frac{s+a}{s^2+Qs+1} \tag{15}$$

as well as the measurements were corrupted by white gaussian noise; the observable realization by (15) is

$$\begin{pmatrix} x_1 \\ x_2 \end{pmatrix} = \begin{bmatrix} -Q & 1 \\ -1 & 0 \end{bmatrix} + \begin{bmatrix} w_1 \\ w_2 \end{bmatrix} \tag{16}$$

$$y = \begin{bmatrix} 1 & 0 \end{bmatrix} \begin{pmatrix} x_1 \\ x_2 \end{pmatrix} + v \tag{17}$$

where $v = N(0,R_v)$, $w_1$, $w_2 = N(0,R_w)$

On the basis of (16) and (17), a deterministic observer was designed. Since the observer response should be much faster than the system response, the eigenvalues of the observer were chosen to times larger than those of the system. Three types of systems were examined; underdamped (Q=1), critically damped (Q=2), overdamped (Q=3). The corresponding observer poles were placed at s=-5, s=-10, and s=-26.

The system given by (16), along with the Kalman filter and the Luenberger observer were simulated on a Honeywell 6060 computer, the estimation error sample variances are given in Figures 2 through 7. In all cases the observer error sample variance was one order of magnitude greater than that of the Kalman filter.

## SUMMARY

A deterministic observer operating in the presence of plant and measurement noises has been shown to converge to a steady state as long as the noise covariance matrices are bounded. The expected value as well as the covariance matrix of the estimation error have also been computed.

Since the observer gain affects both the rate of convergence as well as the estimation error variance in a contradictory manner, two measurements of performance incorporating both of these parameters have been proposed. The simplest one is the sample variance at steady state; using that as a basis for comparison, the performance of an observer for a second order system was compared with that of a Kalman filter. Another performance index is the ratio of the norm of the error covariance matrix at steady state to the norm of an "estimation nose" matrix; such a ratio is shown to be bounded, the bound being a function of the observer gain. The proper pole locations for the observer could thus be obtained by minimizing the ratio of the estimation error power-to-the "estimation noise" power. There has not yet been found an easy way to perform this optimization.

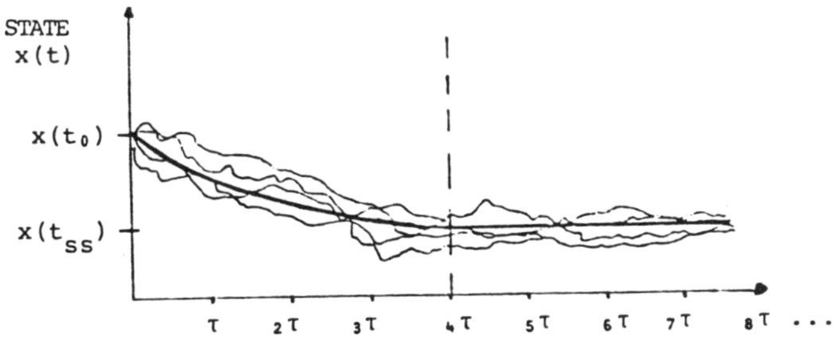


Figure 1. Sample variance at steady state.

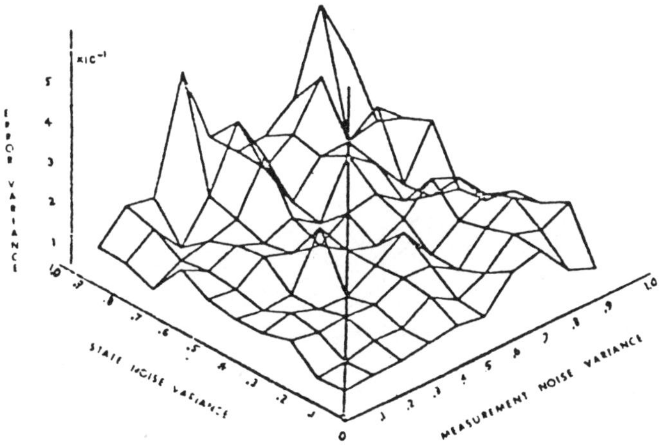


Figure 2. Observer error variance for Q=1.

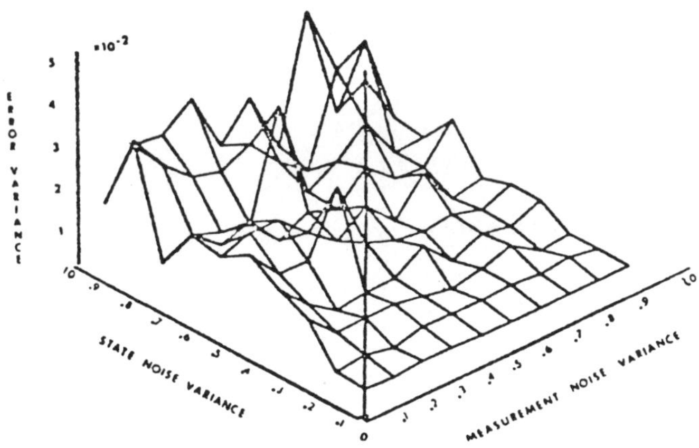


Figure 3. Kalman filter error variance for Q=1.

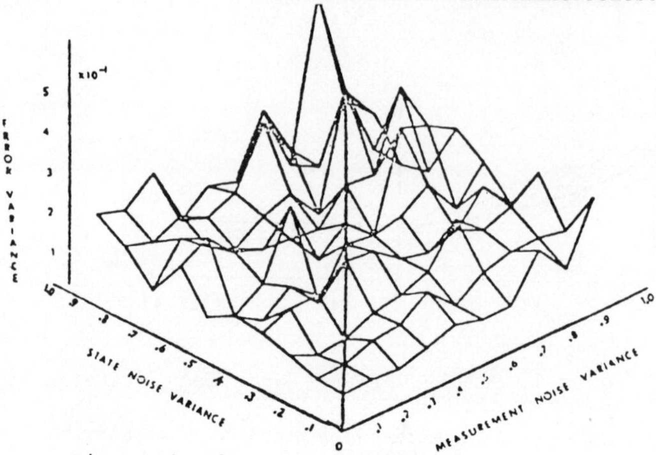


Figure 4. Observer error variance for Q=2.

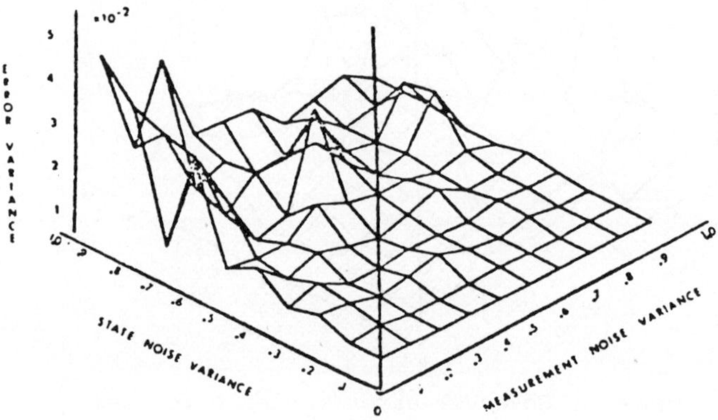


Figure 5. Kalman filter error variance for Q=2.

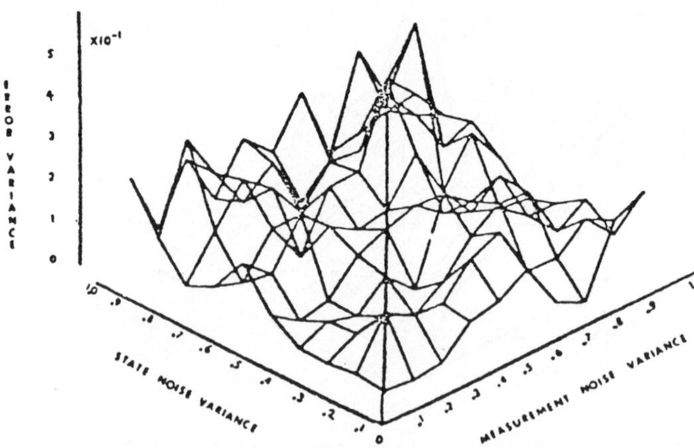


Figure 6. Observer error variance for Q=3.

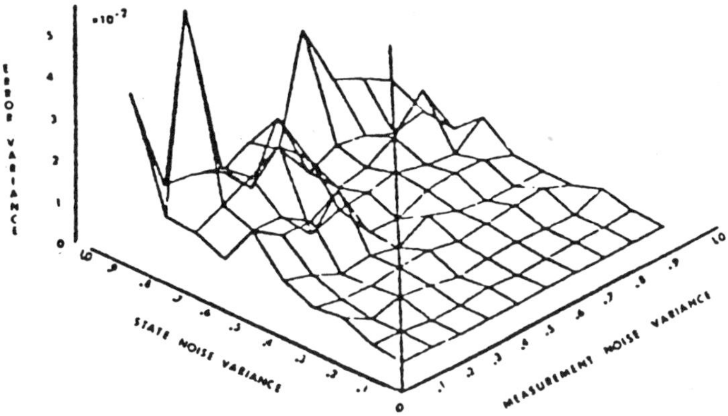

Figure 7. Kalman filter error variance for Q=3.

REFERENCES

1. D. G. Luenberger, An Introduction to Observers, IEEE Trans. on Automatic Control, 16 (1971), 596-602.

2. ________________, Observing the state of a linear system, IEEE Trans. Military Electronics, 8 (1964), 74-80.

3. ________________, Observers for multivariable systems, IEEE Trans. on Automatic Control, 11 (1966), 190-197.

4. R. E. Kalman and R. Bucy, New results in linear filtering and prediction theory, Journal of Basic Engineering (ASME), 83 (1961), 95-108.

5. C. T. Leondes and C. M. Novak, Optimal minimal-order observers for discrete-time systems - A unified theory, Automatica, 8 (1972), 379-387.

6. K. K. Sundareswaren, P. J. McLane, and M. M. Bayoumi, Observers for linear systems with arbitrary plant disturbances, IEEE Trans. on Automatic Control, 22 (1977), 870-871.

7. H. Weiss, On the structure of the Luenberger Observer in discrete-time stochastic systems, IEEE Trans. on Automatic Control, 22 (1977), 871-873.

8. M. Aoki and J. R. Huddle, Estimation of the state vector on a linear stochastic system with constrained estimator, IEEE Trans. on Automatic Control, 12 (1967), 432-433.

9. C. Wong, Considerations for Observers in Noise, Master's Thesis, The George Washington University, 1980.

# BUILDING OF ROBUST "ORTHOGONAL" ESTIMATORS

Robert BADARD and J. C. COMMERÇON

Laboratoire d'Informatique Appliquée 502
I.N.S.A. 20, avenue Einstein
69621 - VILLEURBANNE FRANCE

We have developped a method for building estimators about linear input-output relations, with different a priori informations on the process. These estimators we call "orthogonal's", are defined by a geometric approach in a determinist environment. By this way it is easy to see what a priori information and hypotheses are needed to ensure the identifiability of the system. After this step we extend these estimation procedures to stochastic cases and obtain estimators which are, under very weak hypotheses, strongly consistent. We give applications to solve problems such that :

- Estimation or the parameters or A.R.M.A. model, and of the covariance matrix of the noise, this in a stationary case, and without replicated experiments.
- An extension of the preceding problem to non stationary cases, when for instance the initial conditions are not known, and have a non negligible effect.

In identification it is generally a good assumption to suppose that the discrepancies have no reason not to be distributed in a stationary and non preferential way. But it is sometimes not clear that these discrepancies come from a stochastic process of which we know the distribution function to within a few parameters. There are situations in which even if the assumptions about the distribution are not true, there is a class of distributions for which the results are yet satisfactory. For this reason estimates are often done with normality assumption and then follows a study of their robustness. We will try to build estimators based simply on independence and stationary assumptions. For this purpose we will have to make the discrepancies as close as possible to these assumptions. We could use a robust test of



*D. G. Lainiotis and N. S. Tzannes (eds.), Advances in Control, 456–465.*

independence (built especially for the case where no information is known about the distribution), but in fact in order to obtain simple and practical results we will only study the residual correlations. The estimators we obtain can be called orthogonal and are built in a determinist situation in which we assume that the discrepancies exactly verify the orthogonality conditions.

We have applied this technique more especially to the estimation of the parameters of an Auto Regressive Moving Average (A.R.M.A.) filter, because of their practical interest, and we sometimes obtain estimations of the discrepancy distribution as a subproduct. By the building of these estimators in determinist situations, it is easy to see what the conditions for identifiability will be, and this is a fondamental point. So, the part A explains the buiding of the orthogonal estimators for the parameters of a linear relation. The part B shows how this technique can be translated to deal with estimation of the parameters of an A.R.M.A. filter, and this, depending on the fact that the covariance matrix of the noises and the initial conditions are known or not. Finally, the part C is concerned with the asymptotic repartition of these estimators. For space constraint all the proves will be omitted ; they can be found in the report (3).

## A. BUILDING OF THE ORTHOGONAL ESTIMATORS FOR A LINEAR RELATION

Let $Z = (z_i^j)$, $1 \leqslant i \leqslant n$, $1 \leqslant j \leqslant k$ be an array of real numbers, with n rows and k columns, so that $n \geqslant k$. Z is the sum ($Z = X + B$) of an array X of unobservable measures, and an array of errors B. Every row of X verifies linear relation : $\sum_{j=1}^{k} x_j^i \bar{\Theta}_j = 0$, so we will write $X\bar{\Theta} = 0$, and we remark that only the direction of the vector $\bar{\Theta}$ is important.
We will try to solve the problem : can we possibly find $\bar{\Theta}$ with a few more assumptions. An answer is given by :
<u>Proposition 1</u> : if ${}^tXB = 0$ and if we know $C = k\,{}^tBB$, every solutions $\hat{\Theta}$ of :

$$\min_{\Theta} {}^t\Theta\,{}^tZZ\Theta, \text{ with } {}^t\Theta C\Theta = 1, \text{ satisfies } {}^t\hat{\Theta}\,{}^tXX\hat{\Theta} = 0$$

If ${}^tXX$ is of rank $k - 1$ (Identifiability assumption) then $\hat{\Theta} = \bar{\Theta}$ is the only solution. We extend the preceding result to the stochastic case in the following way. Let $x_1, x_2, \ldots, x_k$ be variables (not necessarly random) which verify
$x_1\bar{\Theta}_1 + x_2\bar{\Theta}_2 + \ldots x_k\bar{\Theta}_k = 0$, and let $b_1, \ldots, b_k$ be the corresponding random errors. We do n experiments and oberve $z_1^i, z_2^i, \ldots, z_k^i$ where i is the experiment index, with $z_j^i = x_j^i + b_j^i$, and we know, to within one positive multiplicative factor, the variance covariance matrix of the $b_j$, say C. Let $\hat{\Theta}$ be an estimator of $\bar{\Theta}$, built in the following way : As in the preceding section we compute ${}^tZZ$, then $\hat{\Theta}$ is the eigenvector associated with the minimal eigenvalue of ${}^tZZ\Theta = \lambda C\Theta$, then we have :

Proposition 2 : We assume that :

- $\frac{1}{n} \sum_{i=1}^{n} x_j^i x_q^i \xrightarrow[n \to \infty]{} s_{jq}$ the convergence being the usual one if $x_j$ are not random variables and almost sure if $x_j$ are random. $S^j = (s_{jq}) \neq 0$ is of rank k-1.

- $\frac{1}{n} \sum_{i=1}^{n} b_j^i b_q^i \xrightarrow[n \to \infty]{a.\ s} E(b_j b_q)$, convergence being almost sure, $V = (E(b_j b_q))$.

- $\frac{1}{n} \sum_{i=1}^{n} x_j^i b_q^i \xrightarrow[n \to \infty]{a.\ s} 0$.

Then $\hat{\theta}$ is strongly consistent, $\hat{\theta} \xrightarrow[n \to \infty]{a.\ s} \bar{\theta}$.

Let us now recall that this estimator is the maximum likelihood when the $b_j$ have a Laplace-Gauss distribution, and when the experiments are made independently. It is also known that in this case this estimator can be inconsistent, because there are many incidental parameters (if we do not know the matrix C).

## B. IDENTIFICATION OF AN A.R.M.A. FILTER

### B.1. THE NOISE OF THE COVARIANCE MATRIX C IS KNOWN

We will consider a system with one input and one output, this especially to avoid too heavy notations, but the extension would be straight-forward. We will suppose in the sequel that the input-output relation is of the form : $x_k = a_1 x_{k-1} + a_2 x_{k-2} + \ldots + a_p x_{k-p} + b_1 u_{k-1} + \ldots + b_p u_{k-p}$. The 2p parameters $a_1, a_2, \ldots, a_p, b_1, b_2 \ldots, b_p$ are associated with the vector :

$${}^t\theta = (\theta_1, \theta_2, \ldots, \theta_{2p+1}) = -\theta_1 \ (-1, a_1, a_2, \ldots, a_p, b_1, b_2, \ldots b_p)$$

and let T be the set of vectors $\theta$ so that at least one $b_j$ is non-zero (causality). As before we note $\Theta$ as the quotient of T by the relation $\theta_1 \sim \theta_2$ if $\theta_1 = k\,\theta_2$ with $k \neq 0$. But even in this case we can have $\theta_1 \not\sim \theta_2$, which are associated with the same system, this when, in the "z" transfert function, there are common factors in the numerator and the denominator. So we assume that for $\bar{\theta}$ there are no such common factors ($\bar{\theta}$ is then said a minimal representation). We will also note $\Theta_s$ the subset of $\Theta$ for which the corresponding systems are stable (the roots of the polynomial $z^p - a_1 z^{p-1} - \ldots - a_p z^0 = 0$ are so that $|\rho_i| < 1$).

We will frequently us the notations :

$$- X^n(\Theta,x_0,x_1,\ldots) = \begin{bmatrix} x_p, & x_{p-1},\ldots\ldots, & x_0, u_{p-1}, u_{p-2},\ldots\ldots, & u_0 \\ x_{p+1}, & , x_1, u_p, & , u_1 \\ . & & . \\ . & & . \\ x_{p+n-1}, & , u_{p+n-2}, & , u_{n-1} \end{bmatrix}$$

where $u_k$ is the exact value of the input at time k, and $x_k$ is the exact output at time k, computed by the use of $\Theta$ and of the corresponding initial conditions.

$$B^n = \begin{bmatrix} \varepsilon_p, & \varepsilon_{p-1},\ldots\ldots, & \varepsilon_0, \gamma_{p-1}, \gamma_{p-2},\ldots\ldots, & \gamma_0 \\ \varepsilon_{p+1}, & , \varepsilon_1, \gamma_p, & , \gamma_1 \\ . & & \\ . & & \\ \varepsilon_{p+n-1}, & & , \gamma_{n-1} \end{bmatrix}$$

where $\varepsilon_k$ and $\gamma_k$ are the errors at time k on the output and the input respectively. $Z^n = X^n(\bar{\Theta}, \bar{x}_0,\ldots) + B^n$ is the array built with the observed input and output at different times, where $\bar{\Theta}$, $\bar{x}_0,\ldots$ are the true parameters and initial conditions. We also note $Y^n(\Theta)$ for $X^n(\Theta, 0, 0,\ldots)$ and $A_i^n(\Theta)$ the array built as $X^n$ which is associated to a free response with an initial condition $x_i = 1$.

<u>Study of a periodic case</u> : Let the input u be periodic, of period n, and we note : ${}^t u = (u_0,u_1,\ldots,u_{n-1})$, ${}^t x(\Theta) = (.,x_i(\Theta).)$

for the exact periodic input and the exact periodic output if it exists, ${}^t\varepsilon = (\varepsilon_0,\ldots, \varepsilon_{n-1})$ and ${}^t\gamma = (\gamma_0, \ldots, \gamma_{n-1})$ for the output and input errors that we also assume to be periodic. In the first step we recall some results about identifiability when the input-output sequences are known without error. Let D be the matrix associated with the delay operator, so that $(x_1, x_2,\ldots, x_{n-1}, x_0) = {}^t x {}^t D$. If we want to compute the periodic output $x(\Theta)$ for the parameter $\Theta$, we have to ask : Is $(I - a_1 D^{-1} - \ldots - a_p D^{-p})$ invertible, and if this were not always the case in what conditions this would be true ? The answer is given by :

<u>Lemma 1</u>: The matrix $(I-a_1 D^{-1}-\ldots-a_p D^{-p})$ is invertible if and only if the polynomial $\lambda^p - a_1\lambda^{p-1} - \ldots - a_p$ has not roots of the type $\exp(i2k\pi/n)$ with $k \in \{0, 1, \ldots\}$ and n being the period lenght. About the identifiability in the periodic case we have the following result.

<u>Lemma 2</u> : Let $\bar{\Theta} \in \Theta_s$ and $\bar{\Theta}$ be a minimal representation and we know periodic sequences of the input and of the output (of period n) then $\bar{\Theta}$ is identifiable if :

$$U^n = \begin{bmatrix} u_{2p-1},\ldots\ldots, & u_1, u_0 \\ & \ldots\ldots \\ & u_0, u_{n-1} \end{bmatrix} \text{ is of rank } 2p.$$

If we note $X^n_s(\theta) = X^n_s(\theta, x^s_0, \ldots)$ the array with initial conditions corresponding to a periodic output (if there is one) then $X^n_s(\bar{\theta})$ is of rank 2p.

We now go to the problem of identifiability in the presence of noise in a periodic determinist case. We will suppose that $\bar{\theta} \in \Theta_s$ and is a minimal representation. We also know a periodic sequence of period n of the input and of the output, but this with unknown periodic additive errors.

Proposition 3 : Let $\bar{\theta} \in \Theta_s$ be a minimal representation. We know periodic sequences of the input and output, say

${}^t v = {}^t u + {}^t \gamma$ and ${}^t z = {}^t x(\bar{\theta}) + {}^t \varepsilon$ .

We assume that $\varepsilon$ and $\gamma$ are periodic of period n and orthogonal to u in the sense that ${}^t\varepsilon D^k u = 0$ and ${}^t\gamma D^k u = 0$ for every k. We assume that $U^n$ is at least of rank 2p and that we know the matrix ${}^tB^nB^n$, say C, to within one multiplicative factor. The system is then identifiable and $\bar{\theta}$ is the only eigenvector associated with the minimal eigenvalue of ${}^tZ^nZ^n\theta = \lambda C\theta$.

Stochastic case : We now extend this result, obtained in a periodic case, to a very broad class of situations, that is to say, when the noises $\varepsilon$ and $\gamma$ are stochastic processes. We first assume that the input is "sufficiently exciting" in order to have identifiability (Aström, 2), so we have the result :

Proposition 4 : The estimator $\hat{\theta}$ built as the eigenvector associated with the minimal eigenvalue of $\frac{1}{n}{}^tZ^nZ^n\theta = \lambda C\theta$, where C is the matrix $E\left\{{}^t(\varepsilon_{p+i}, \ldots, \varepsilon_i, \gamma_{p+i-1}, \gamma_i)(\varepsilon_{p+i}, \ldots, \gamma_{p+i-1}, \ldots, \gamma_i)\right\}$ known to within one multiplicative factor, is strongly consistent if the following conditions are verified :

- the noises $\varepsilon$ and $\gamma$ are zero mean S.A.I.4 processes, independent of the input.
- $\bar{\theta} \in \Theta_s$ and is a minimal representation.
- If u is determinist and if $\lim_{n\to\infty} {}^tU^nU^n$ exists and is of rank 2p.
- If u is a S.A.I.4 process and if the matrix :

$$E\left\{{}^t(u_{2p+i-1}, \ldots u_i)(u_{2p+i-1}, \ldots, u_i)\right\} \text{ is of rank } 2p.$$

Aoki has shown that when the noises $\varepsilon$ and $\gamma$ are independent gaussian white noises and when identifiability assumptions are fulfiled, this estimator (obtained with C = I), known as Levin's (6), is not the maximum likelihood one but it is a good approximation and it has a good asymptotic efficiency (1).

## B.2. THE COVARIANCE MATRIX C IS UNKNOWN

As in the preceding paragraph we assume that the data are obtained sequentially, but in particulary cases the knowledge of C is not necessary. We treat a special case, that is when we assume that there is no error on the input ($\gamma=0$). let us first built the orthogonal estimator in the periodic determinist situation.

Periodic case : We again suppose that $u, \varepsilon$ and the true output $x(\bar{\theta})$ are all periodic with period n. The method is still the same : We search for a $\theta$ for which we have the orthogonality ${}^tX_s^n(\bar{\theta})\ (Z^n - X_s^n(\bar{\theta})) = 0$. But it is complicated to use this formula so we will show that we can adopt an iterative process : Take a $\theta$ and compute $x(\theta)$, we thus obtain $\hat{\varepsilon}(\theta)$ and we can estimate the variance-covariance matrix C. We then compute a new approximation of $\bar{\theta}$ by the preceding method. We then have :

Proposition 5 : Let $\bar{\theta} \in \Theta_s$, be a minimal representation, then $\bar{\theta}$ is identifiable if the following assumptions are fulfiled :
- $U^n$ is of the rank 2p
- The errors $\varepsilon$ are orthogonal to u in the sense that : ${}^t\varepsilon D^k u = 0$, for every k.

Then $\bar{\theta}$ is the only $\hat{\theta}$ in $\Theta_s$ which verifies

$$\frac{{}^t\hat{\theta}\,{}^t(Z^n - X_s^n(\hat{\theta}))\ (Z^n - X_s^n(\hat{\theta}))\hat{\theta}}{{}^t\hat{\theta}\,{}^tZ^nZ^n\hat{\theta}} \geqslant \frac{{}^t\theta\,{}^t(Z^n - X_s^n(\hat{\theta}))\ (Z^n - X_s^n(\hat{\theta}))\theta}{{}^t\theta\,{}^tZ^nZ^n\theta}$$

Stochastic case : The principle of the method is a straightforward extension of the method used by Levin (6), because, with the sequentiality of the data, we can estimate for every $\theta$ the covariance matrix of the output errors. This method has been used, for instance, by Grosjean (5), but we think that its consistence and its asymptotic properties have never been correctly studied.

With the samples we build the array $Z^n$ and search for $\hat{\theta}$, for wich :

$$\frac{{}^t\theta\,{}^t(Z^n - X^n(\hat{\theta}, x_0,..))\ (Z^n - X^n(\hat{\theta}, x_0..))\theta}{{}^t\theta\,{}^tZ^nZ^n\theta}$$

is maximal in $\theta$ for $\theta = \hat{\theta}$, if the initial conditions are known. On the contrary, we hope that the error we make in taking $x_0 = z_0, \ldots$ as an instance will be negligible. In pratice, this estimator is obtained by the iterative procedure we described.

Proposition 6 : The estimator $\hat{\theta}$ is strongly consistent if the following conditions are fulfiled
- $\bar{\theta} \in \Theta_s$ and is a minimal representation
- The input u is determinist so that $\lim_{n\to\infty} \frac{1}{n}\,{}^tU^nU^n$ exists and is of rank 2p, or u is a S.A.I.4 process, and the matrix $E({}^t(u_{2p+i-1},..,u_i)\ (u_{2p+i-1},..,u_i))$ is of rank 2p
- $\varepsilon$ is zero mean S.A.I.4 process independent of the input.

Convergence of the iterative procedure : To build $\hat{\theta}$ we use the algorithm :

Let $\theta_i$ be an approximation of $\hat{\theta}$, we then compute $X^n(\theta_i, x_0, \ldots)$ and ${}^t\tilde{B}^n(\theta_i)\tilde{B}^n(\theta_i) = {}^t(Z^n - X^n(\theta_i, x_o,..))\ (Z^n - X^n(\theta_i, x_o))$

From this we compute a new approximation $\theta_{i+1}$ of $\hat{\theta}$ by taking the

eigenvector associated with the minimal eigenvalue of : ${}^tZ^nZ^n\Theta = \lambda\, {}^t\tilde{B}^n(\Theta_i)\tilde{B}^n(\Theta_i)\Theta$. We then ask ourselves if this algorithm converges towards $\hat{\Theta}$?
First we assume that the orthogonality is fulfiled, then if $\Theta_i$ is in a sufficiently small neighbourhood of $\bar{\Theta}$ we find by first order approximation : ${}^t\tilde{B}^n(\Theta_i)\tilde{B}^n(\Theta_i) \approx {}^tB^nB^n$ and then if u is a sufficiently exciting input, we have $\Theta_{i+1} = \bar{\Theta}$ so the procedure is locally stable. It is clear by continuity that if orthogonality is not exactly verified, we still have local stability for $\hat{\Theta}$.

## B.3. THE COVARIANCE MATRIX C AND THE INITIAL CONDITIONS ARE UNKNOWN

Let $\bar{\Theta}\in \Theta_s$ be a minimal representation and we suppose that the initial conditions are so that ${}^tX^n(\bar{\Theta},\bar{x}_o,\ldots)B^n \neq 0$. But we assume that there are initial conditions, say $x_o^s x_1^s,..$ so that ${}^tX^n(\bar{\Theta},x_o^s,\ldots)B^n = 0$. We can again use the preceding method after we have made the following remarks :

${}^t\Theta\,{}^t(Z^n - \sum_{i=1}^p (\bar{x}_i - x_i^s)A_i^n(\bar{\Theta}))\;(Z^n - \sum_{i=1}^p (\bar{x}_i - x_i^s)A_i^n(\bar{\Theta}))\Theta$ is minimal with the contraint ${}^t\Theta\,{}^tB^nB^n\Theta = 1$ for $\Theta = \bar{\Theta}$. The matrix $B^n$ being computed by : $B^n = Z^n - X^n(\bar{\Theta}, x_o^s,\ldots) - \sum_{i=1}^p (\bar{x}_i - x_i^s)A_i(\bar{\Theta})$.

So the process will be of the same type as before but we must, on each iteration, compute the terms $(\bar{x}_i - x_i^s)$ in order to make ${}^tX^n(\Theta,x_o^s..)B^n$ as small as possible.
We think that here again it is convenient to procede as in the periodic situation. Let $u_o, u_1, \ldots, u_{p+n-2}$ and $z_o, z_1 \ldots, z_{p+n-2}$, $z_{p+n-1}$ be the known input output sequences (the input being exactly known). We build the matrix $Z^n$, and we can see the p input columns as p different periodic inputs with period n. If we suppose that the noise is also periodic so that :

$$ {}^t\begin{vmatrix} u_{p-1}, \ldots\ldots, u_1, u_o \\ u_1 \\ \\ u_{p+n-2}, \ldots\ldots, u_{n-1} \end{vmatrix} D^k \begin{vmatrix} \varepsilon_o \\ \varepsilon_1 \\ \\ \varepsilon_{n-1} \end{vmatrix} = 0 $$

for every k, then we can ensure that for every $\Theta$ for which a periodic output exists we have ${}^tX_s^n(\Theta)B^n = 0$.

It is in this way that we compute the $x_i^s(\Theta)$. For the estimation of the $\bar{x}_i(\Theta)$, we will try to make

${}^tX_s^n(\Theta)\;(Z^n - X_s^n(\Theta) - \sum_{i=1}^p (\bar{x}_i(\Theta) - x_i^s(\Theta))A_i^n(\Theta)$ as small as possible,

this for instance by minimizing the sum of the square of the corresponding elements.

We then have the algorithm :
- Compute the $x^s_i(\Theta_q)$.
- Compute the $\bar{x}_i(\Theta_q)$ as those for which the sum of the square of the elements of the matrix ${}^tX^n_s(\Theta_q)\ (Z^n - X^n_s(\Theta_q) - \sum_{i=1}^p (\bar{x}_i(\Theta_q) - x^s_i(\Theta_q))A^n_i(\Theta_q)$ is minimal.
- Compute $\tilde{B}^n(\Theta_q) = Z^n - X^n_s(\Theta_q) - R^n(\Theta_q)$, with $R^n(\Theta_q) = \sum_{i=1}^p (\bar{x}_i(\Theta_q) - x^i_s(\Theta_q))\ A^n_i(\Theta_q)$
- Compute $\Theta_{q+1}$ as the eigenvector associated with the minimal eigenvalue of :
  ${}^t(Z^n - R^n(\Theta_q))\ (Z^n - R^n(\Theta_q))\Theta = \lambda {}^t\tilde{B}^n(\Theta_q)\tilde{B}^n(\Theta_q)\Theta$ .

It is obvious that $\bar{\Theta}$ is a fixed point of this iterative process, but unfortunately this algorithm can be unstable when the initial conditions are very great even if all the orthogonality assumptions are fulfiled. But with the many tests we have done we remark that for "reasonable" unknown initial conditions in respect of the sequence lenght, it gives good results in practice.

## C. ASYMPTOTIC PROPERTIES OF THESE ESTIMATORS :

In all situations we have studied (case A, B.1, B.2), we remark that if the orthogonality is exactly verified and if we have identifiability, for instance in the periodic case, then we find exactly the true values of the parameters. The repartition of the estimates is closely related to the repartition of the statistics they use.
For the case A and B.1 we can compute the asymptotic estimator repartition by using perturbation theory, this, in the case where it is consistent (identifiability assumption) because in this situation the smallest eigenvalue is unique. So, we have :

$$\frac{1}{n}\ ({}^tX^n(\bar{\Theta},..)X^n(\bar{\Theta},..) + {}^tX^n(\bar{\Theta},..)B^n + {}^tB^nX^n(\bar{\Theta},..) + {}^tB^nB^n) = \lambda C\,\Theta.$$

So, the perturbation matrix of the left member is :
$\frac{1}{n}\ ({}^tX^n(\bar{\Theta},..)B^n + {}^tB^nX^n(\bar{\Theta},..) + {}^tB^nB^n) - kC$ and it is clear that for S.A.I.4 processes, its elements are asymptotically normal with zero mean. So, if we norm $\hat{\Theta}$ by ${}^t\hat{\Theta}C\hat{\Theta} = 1$, then asymptotically $\hat{\Theta}$ is normal centered around $\bar{\Theta}$, more precisely.

<u>Proposition 7</u> : Under the assumption of the proposition 4, we can assert that the estimator $\hat{\Theta}$, normalized by ${}^t\hat{\Theta}C\hat{\Theta}$= constant, is asymptotically normal, centered around $\bar{\Theta}$.
For the sase B.2, the result is stronger because in this case

by a perturbation tool we easily show that asymptotically the only perturbation term is $\frac{1}{n}({}^tX(\bar{\theta},..)B^n + {}^tB^nX^n(\bar{\theta},...))$.

Proposition 8 : Under the asumption of the proposition 6 we can assert that the estimator $\hat{\theta}$, normalized by ${}^t\hat{\theta}C\hat{\theta}$= constant where $C = \lim_{n\to\infty} \frac{1}{n}\,{}^tB^nB^n$, is asymptotically normal, centered around $\bar{\theta}$. About the robustness, we can say that these estimators are robust depending upon the manner with which the statistics they use tend to normality. For this reason, the case B.2 is certainly the most robust because only the perturbation term $\frac{1}{n}({}^tX^n(\bar{\theta},..)B^n + {}^tB^nX^n(\bar{\theta},..))$ is to be taken into account. The tendance to normality of these term can be very fast, if for instance, the process $\varepsilon$ has asymetric distribution around zero and if the the distribution of $\varepsilon u$ has not too slowly decreasing tails. In the contrary, say if $\varepsilon u$ has slowly decreasing tails, it would be better to use robust test of independance ; the method would be of the same type because we would have to choose $\theta$ such that the residual errors seem as independent as possible of the input.

## D. CONCLUSIONS

The estimators we have described are very useful in practice and they give good results if we take care, for instance on the sampling frequency when we treat continuous time systems (4). They are a reasonable compromise between efficiency and complexity of computation. In fact they principally use the Jacobi algorithm which is a particularly simple and fast one. But the point which can be criticized is certainly that we assume some independence properties and in fact we only use their correlation consequences. We think that if we suppose that the distributions of errors are symetric and have continuous densities with tails which are slowly decreasing relative to the normal case then our computation of correlations can not be perfectly adapted. In this case it would be best to use another robust test of independence. The procedure would be of the same type but at each step we would have to choose $\theta$ for which the independence seems the best satisfied. Unfortunly in this case we could not use the Jacobi algorithm other than as a first approximation. It would certainly be interesting to test such algorithms.

## APPENDIX

We define what we call a Stationary Asymptotically Independent process of order k(S.A.I.k), because we think that it is a sufficiently general type of stochastic process which has nice properties.

Definition : A stochastic process $X_i \mid_{i \in N}$ is said a S.A.I. k

process (with values in $R^p$) if it verifies :
- It is a stationary process.
- It is an asymptotically independent process.
- The moment of order k exists ( $E[|X_i|^k] < \infty$)

For a S.A.I.k process with k=2q, we have the following properties
- $\frac{1}{n}\sum_{i=1}^{n} X_i^{q_1} \xrightarrow[n\to\infty]{as} E\, X_i^{q_1}$ when $q_1 \in N$ and $q_1 \leqslant q$
- $\frac{1}{n}\sum_{i=1}^{n} X_i^{q_1} X_{i+r}^{q_2} \xrightarrow[n\to\infty]{as} E\left[X_i^{q_1} X_{i+r}^{q_2}\right]$ with $q_1, q_2 \in N$ and $q_1 + q_2 \leqslant q$.
- These sums are asymptotically normal, this last point is a consequence of the fact that the weak convergence of distributions correspond to the pointwise convergence of the corresponding characteristic functions, Prohorov (7).

For instance a white gaussian process filtered by a stable linear invariant system is a S.A.I.k process for every value of the integer k.

## BIBLIOGRAPHY


(1) AOKI M, and YUE P C "On a Priori Estimates of Some Identification Methods" IEEE Transactions on Automatic Control, AC-15, n° 5, 1970.

(2) ASTROM K J, and BOHLIN T, and WENSMARK S "Automatic construction of linear stochastic dynamic models for stationary industrial processes with random disturbances using operating records. I.B.M. Nordic Lab. Sweden, Tech. Rep. T.P. 18.150, June 1965.

(3) BADARD R, and COMMERÇON J C, "Building orthogonal estimators" Report n° 100579 Dpt Informatique Appliquée INSA LYON.

(4) COMMERÇON J C, and BADARD R, "A Methodology for Graphic Identification" International Congress of Patras 1979.

(5) GROSJEAN A "Contribution à l'identification statistique de procédés dynamiques linéaires échantillonnés" Thèse, Grenoble, 1976.

(6) LEVIN J M "Estimation of a system pulse transfer function in the presence of noise" IEEE Transactions on Automatic Control, AC-9, 1964.

(7) PROHOROV Y V, ROZANOV Y A "Probability Theory" Springer-Verlag, 1969.

NEW UNKNOWN-BUT-BOUNDED ALGORITHMS WITH APPLICATION TO SYSTEM IDENTIFICATION AND RADAR TRACKING

R.J. Evans
F. Barker
Department of Electrical Engineering, University of Newcastle, New South Wales, 2308, Australia
Plessey Radar Ltd., Oakcraft Road, Chessington, Surrey, United Kingdom, KT9, 1QZ.

## 1. INTRODUCTION

In this paper we consider several aspects of the problem of system identification with unknown-but-bounded noise disturbances on the input-output observations. For the first case discussed the input to the real system is observed exactly and the output is corrupted by unknown-but-bounded errors. We show that the best linear approximation to the real system, which is consistent with the observations, can be determined adaptively. Also some new persistently exciting conditions are presented and several open questions are discussed. For the second case we further assume that measurements made on the input to the real system are also corrupted by unknown-but-bounded noise. An algorithm for determining the best linear model consistent with the input-output observation is presented.

Much still remains to be done on the problems discussed above, however, the fact that real observation errors are nearly always unknown-but-bounded, and the fact that it is possible to achieve asymptotically unbiased rapidly converging estimates, despite correlated, biased observation errors is very attractive, and worth persuing.

## 2. OUTPUT OBSERVATION ERRORS

Referring to Figure 1 it is clear that if we are given an observed output $y_i$ then the actual output of the real system $x_i$ must satisfy

$$|x_i - y_i| \leq \varepsilon_i.$$



*D. G. Lainiotis and N. S. Tzannes (eds.), Advances in Control, 466–474.*

Now if our linear model, represented by a parameter vector $\underline{c}$, is to be consistent with the observations $(s,y)$ and our knowledge of the noise error bounds $\varepsilon$, we must have

$$|\langle \underline{s}_i, \hat{\underline{c}}\rangle - y_i| \le \varepsilon_i \;\forall\, i.$$

The <u>best</u> linear model $\hat{\underline{c}}^*$ can now be defined in a variety of ways but the most obvious method is to choose $\hat{\underline{c}}^*$ to

$$\text{minimize} \;\; \Sigma[\langle \underline{s}_i, \hat{\underline{c}}\rangle - y_i]^2$$

where, of course, $\hat{\underline{c}}^*$ must also satisfy the data consistency conditions above.

Given a set of input-output data $\{y_i, \underline{s}_i\}_{i=1}^N$ an iterative algorithm for determining $\hat{\underline{c}}^*$ is given by

$$\hat{\underline{c}}(k+1) = \hat{\underline{c}}(k) - 2\alpha(k)[\sum_{i=1}^{N} \underline{s}_i(\langle s_i, \hat{\underline{c}}(k)\rangle - y_i + \tfrac{1}{2}\lambda_i(k)]$$

$$\lambda_i(k+1) = \lambda_i(k) + \alpha(k)\ell_i(k), \; i = 1, \ldots, N$$

where

$$\ell_i(k) = \begin{cases} \langle \underline{s}_i, \hat{\underline{c}}(k)\rangle - y_i - \varepsilon_i & \text{if } \lambda_i(k) > 0 \\ & \text{or if } \lambda_i(k) = 0 \text{ and } \langle \underline{s}_i, \hat{c}(k)\rangle - y_i > \varepsilon_i \\ \langle s_i, \hat{\underline{c}}(k)\rangle - y_i + \varepsilon_i & \text{if } \lambda_i(k) < 0 \\ & \text{or if } \lambda_i(k) = 0 \text{ and } \langle s_i, \hat{\underline{c}}(k)\rangle - y_i < \varepsilon_i \\ 0 \text{ if } \lambda_i(k) = 0 & \text{and } |\langle s_i \hat{\underline{c}}(k) - y_i\rangle \le \varepsilon_i \end{cases}$$

and it can be shown that when $\sum_k \alpha(k) = \infty$, and $\alpha^k(k) \to 0$,

$$\hat{\underline{c}}(k) \xrightarrow[k\to\infty]{} \hat{\underline{c}}^*$$

or for a fixed step size $\alpha(k) = \alpha$

$$\limsup_{k\to\infty} ||\hat{\underline{c}}(k) - \underline{c}^*||^2 \le V(\alpha) \xrightarrow[\alpha\to 0]{} 0.$$

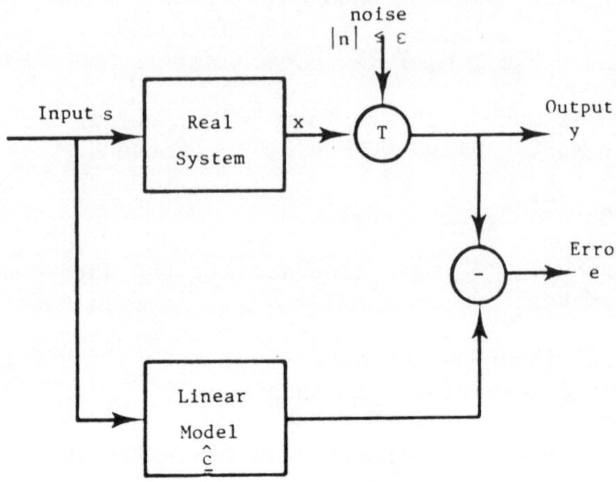


FIGURE 1

This algorithm can be implemented off-line on one set of data, or on-line using repeated $\{\underline{s}_i\}_{i=1}^{N}$. It can be shown that if $\underline{s}_i$ is corrupted by finite variance unbiased noise with $\underline{n}_i$ and $\underline{n}_{i+1}$ uncorrelated then for a step sequence $\{\alpha(k)\}$ satisfying $\Sigma\alpha(k) = \infty$ and $\Sigma[\{(k)]^2 < \infty$, $\hat{\underline{c}}(k)$ converges to $\hat{\underline{c}}*$ in mean-square and with probability one, and if a fixed step is used $c(k)$ stays within a mean-square neighbourhood of $\hat{\underline{c}}*$ as $k \to \infty$.

Above we have been concerned with finding a model of the system which is consistent with the observed data. It is also interesting to determine what conditions the input signal must satisfy in order to guarantee that the true system parameters $\underline{c}^0$ will be found, if the real system is linear. Such conditions are called persistently exciting conditions and they ensure that the exciting signal is rich enough to ensure that the observed output contains sufficient information to identify the system. For our case the system is uniquely identified if there is only one element consistent with the observed data, i.e. $\underline{c}^0$. If the real system is linear then

$$y_i = \langle s_i, \underline{c}^0 \rangle + n_i \quad \text{hence}$$

$$|\langle \underline{s}_i, \hat{\underline{c}} - \underline{c}^0 \rangle - n_i| \leq \varepsilon_i$$

Defining $\underline{e} = \hat{\underline{c}} - \underline{c}^0$ we say that a signal is <u>persistently exciting</u> if the conditions

$$\left.\begin{array}{l} \langle \underline{s}_i, \underline{e} \rangle \leq \varepsilon_i + n_i \\ \langle \underline{s}_i, \underline{e} \rangle \leq \varepsilon_i - n_i \end{array}\right\} \quad i = 1, \ldots, N$$

imply $\underline{e} = 0$. Thus the system has been identified after a finite number of observations. We have the following result.

<u>Theorem</u>: Given a set of signals $\{s_i^+\}$ for which the noise achieves its upper bound, $n_i = \varepsilon_i$, and a set $\{s_i^-\}$ for when $n_i = -\varepsilon_i$, then the $M$ parameters of the true linear system are identified uniquely if the $M \times M$ matrix $\underline{S}$ whose rows $\{s_i\}_1^M$, are any selection of $M$ signals from the sets $\{s_i^+\}$ or $\{-\underline{s}_i^-\}$ has full rank, and there exists a further signal either $\underline{s}^-$ such that

$$\underline{s}^- = \sum_{i=1}^{M} \lambda_i \underline{s}_i, \quad \lambda_i \geq 0,$$

or

$$\underline{s}^+ = \sum_{i=1}^{M} \lambda_i \underline{s}_i, \quad \lambda_i \leq 0.$$

There are clearly several open questions associated with the above technique. For example what happens if the noise bounds are incorrectly chosen either too big or too small. If the bounds are too big then the set of parameters which are consistent with the data $\hat{C}$ always contains the true parameter set $\hat{C}^*$ thus the best element in $\hat{C}$ may not be in $\hat{C}^*$. However, if the noise is reasonable in the sense that there are both positive and negative errors $\hat{C}$ approaches $\hat{C}^*$ monotonically and $\hat{C}^*$ approaches $\hat{\underline{c}}^*$ monotically. It has been our experience in real radar tracking problems that a choice of noise bound which is too big is perfectly acceptable and convergence is almost as fast as with correct bounds. If the bounds are too small is it possible for $\hat{C}$ to be smaller than $\hat{C}^*$. It turns out however that the algorithm still finds a sensible parameter and converges to the standard least-square solution. A further advantage of our approach is that if one is confident about the noise error bounds chosen, then the above approach is capable of finding bad data points - or detecting changes in the system being modelled.

## 3. INPUT-OUTPUT ERRORS

The case we consider here is when measurements on the real system input and output are corrupted by unknown-but-bounded observation errors as shown in Figure 2. The input to the real system cannot be measured exactly and only $s$ is available. However, since the noise is bounded we know that $|i-s| \leq \theta$ . Thus for the output we know that any linear model $\hat{\underline{c}}$ for which

$$|<\underline{i}_i, \hat{\underline{c}}> - y_i| \leq \varepsilon_i \;\; \forall\, \underline{i}_i : |\underline{i}_i - \underline{s}_i| \leq \underline{\theta}_i$$

will be consistent with observed data and again we can choose a unique linear model by

$$\text{minimize} \;\; \Sigma[<s_i, \hat{\underline{c}}> - y_i]^2$$

subject to the consistency constraint.

The trouble is that the constraint is in a form which is difficult to handle. It can be shown however that the constraint is exactly equivalent to the more tractible constraint

$$|<\underline{s}_i, \hat{\underline{c}}> - y_i| + <|\hat{\underline{c}}|, \theta_i> \leq \varepsilon_i$$

This leads to a nondifferentiable optimization problem which can be solved by the following convergent algorithm,

$$\hat{\underline{c}}(k+1) = \hat{\underline{c}}(k) - \alpha(k)[2[\underline{s}_i(<\underline{s}_i, \hat{\underline{c}}(k)> - y_i) + (\lambda_i(k) - \gamma_i(k))\, \underline{s}_i + \underline{f}(k)]\}$$

$$\lambda_i(k+1) = \max\{0, \lambda_i(k) + \alpha(k)[(<s_i, \hat{\underline{c}}(k)> - y_i) + <|\hat{c}(k) , \underline{\theta}_i> - \varepsilon_i]\}$$

$$\gamma_i(k+1) = \max\{0, \gamma_i(k) - \alpha(k)[(<\underline{s}_i, \hat{\underline{c}}(k)> - y_i) - <|\hat{\underline{c}}(k)|, \underline{\theta}_i> + \varepsilon_i]\}$$

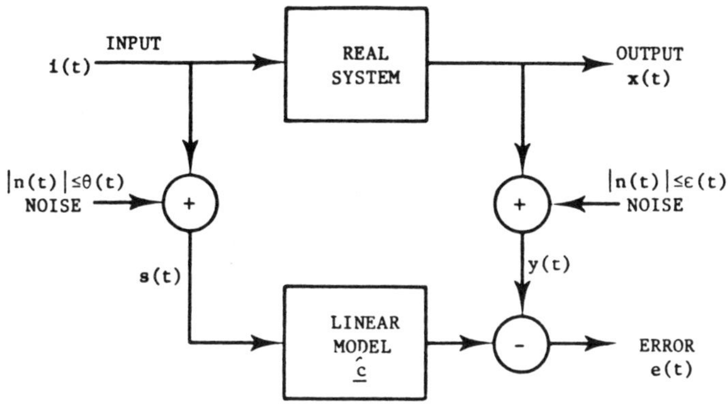


FIGURE 2

where

$$[f(k)]_j \triangleq \begin{cases} [\sum \underline{\theta}_i(\lambda_i(k) - \gamma_i(k)]_j \quad \text{if} \quad [\hat{\underline{c}}(k)]_j > 0 \\ \qquad \text{or if} \quad [\hat{\underline{c}}(k)]_j = 0 \quad \text{and} \\ \qquad [\sum \underline{s}_i(<\underline{s}_i, \hat{\underline{c}}(k)> - y_i + [\lambda_i(k)-\gamma_i(k)]\underline{s}_i \\ \qquad + [\lambda_i(k)+\gamma_i(k)]\underline{\theta}_i]_j > 0 \\ 0 \text{ if } [\hat{\underline{c}}(k)]_j = 0 \quad \text{and} \quad |\sum \underline{s}_i(<\underline{s}_i, \hat{\underline{c}}(k)> -_i + \lambda_i(k) \\ \qquad -\gamma_i(k))|_j \\ \qquad \leq \{\sum(\gamma_i(k)+\lambda_i(k))\theta_i\} \\ -[\sum \underline{\theta}_i(\lambda_i(k)-\gamma_i(k)]_j \quad \text{if} \quad [\underline{c}(k)]_j < 0 \\ \qquad \text{or if} \quad [c(k)]_j = 0 \quad \text{and} \\ \qquad [\sum \underline{s}_i(<\underline{s}_i, \hat{\underline{c}}(k)> - y_i \\ \qquad + \lambda_i(k)-\gamma_i(k)]+[\lambda_i(k)+\gamma_i(k)]\underline{\theta}_i]_j < 0 \end{cases}$$

This algorithm can be shown to satisfy similar converge properties to the algorithm described earlier.

## 4. RECURSIVE SOLUTION AND OCCAM FILTER

We have been unable to obtain recursive solutions to either of the problems discussed above. For the output noise case however it is possible to devise useful recursive procedures which approximately solve the problem. These algorithms differ from standard recursive least-squares algorithms, like the Kalman Filter, in that the gains are nonlinearly dependent on the data. Schweppe [2] has also considered recursive solutions to unknown-but-bounded type problems, and has developed fairly general elliptic bounding algorithms with data dependent gains. To illustrate we consider a simple radar tracking problem. Assume that an aircraft is flying with a constant speed and we are trying to determine this speed from noisy range measurements. The aircraft model is simply

$$\underline{x}(k+1) = \begin{bmatrix} r \\ v \end{bmatrix}_{k+1} = \begin{bmatrix} 1 & T \\ 0 & 1 \end{bmatrix} \begin{bmatrix} v \\ v \end{bmatrix}_k = \underline{A}\,\underline{x}(k)$$

$$\underline{x}(0) = \underline{x}_0$$

and the range observation is

$$z(k) = [1, \ 0]\underline{x}(k) + n(k)$$

$$= \underline{h}'\underline{x}(k) + n(k)$$

with noise bounds

$$\varepsilon_\ell(k) \le n(k) \le \varepsilon_u(k)$$

Now defining a new state vector

$$\underline{s}(k+1) = \begin{bmatrix} \underline{x}_u(k+1) \\ \hline \underline{x}_\ell(k+1) \end{bmatrix} = \begin{bmatrix} r_u \\ v_u \\ r_\ell \\ v_\ell \end{bmatrix}$$

$$\underline{p}(k) = \begin{bmatrix} z_u(k) \\ z_\ell(k) \end{bmatrix} = \begin{bmatrix} z(k) + \varepsilon_u(k) \\ z(k) + \varepsilon_\ell(k) \end{bmatrix}$$

where the subscripts $u$ and $\ell$ refer to upper and lower bounds. The simplified Occam filter for determining the aircraft's speed is now given by

$$\underline{\hat{s}}(k+1) = \underline{\tilde{A}}\,\underline{\hat{s}}(k) + \underline{K}(k)(\underline{p}(k+1) - \underline{P}\,\underline{\hat{s}}(k))$$

where

$$\underline{K}(k) \triangleq \left[\begin{array}{cc} K_{ru} & 0 \\ K_{vu} & 0 \\ \hline 0 & K_{r\ell} \\ 0 & K_{v\ell} \end{array}\right]_k$$

and

$$K_{ru}(k) = \begin{cases} 0 & \text{if} \quad z_u(k) \geq \underline{h}'\underline{Ax}_u(k) \\ 1 & \text{if} \quad z_u(k) < \underline{h}'\underline{Ax}_u(k) \end{cases}$$

$$K_{r\ell}(k) = \begin{cases} 0 & \text{if} \quad z_\ell(k) \leq \underline{h}'\underline{Ax}_\ell(k) \\ 1 & \text{if} \quad z_\ell(k) > \underline{h}'\underline{Ax}_\ell(k) \end{cases}$$

$$K_{vu}(k) = \begin{cases} 0 & \text{if} \quad z_u(k) \geq \underline{h}'\underline{Ax}_u(k) \\ \frac{1}{kT} & \text{if} \quad z_u(k) < \underline{h}'\underline{Ax}_u(k) \end{cases}$$

$$K_{v\ell}(k) = \begin{cases} 0 & \text{if} \quad z_\ell(k) \quad \underline{h}'\underline{Ax}_\ell(k) \\ \frac{1}{kT} & \text{if} \quad z_\ell(k) \quad \underline{h}'\underline{Ax}_\ell(k) \end{cases}$$

and

$$\underline{P} \triangleq \left[\begin{array}{cc} \underline{h}'\underline{A} & 0 \\ \hline \underline{0} & \underline{h}'\underline{A} \end{array}\right]$$

$$\underline{\tilde{A}} = \left[\begin{array}{cc} \underline{A} & \underline{0} \\ \hline \underline{0} & \underline{A} \end{array}\right]$$

<u>Remarks</u>: (1) The simplified Occam filter is stable and converges to the true speed. Moreover, the upper and lower bounds converge monotically.

(2) The algorithm has been the basis of a tracking algorithm in a real system for several years and has performed extremely well.

(3) Simulation studies using white Gaussian noise have shown that its error variance converges only marginally slower than the optimum recursive least-squares procedure.

(4) In the radar tracking application it has been possible to confidently detect manoeuvers by simply noting when observations lie outside the predicted upper/lower bound interval.

## 5. CONCLUSION

In this paper we have briefly reported on some techniques which appear to be of value in certain system modelling and estimation problems.

## REFERENCES

[1] F. Barker, "A Technical Introduction to the Occam Filter", Plessey Radar Limited, Company Research Report, October 1971.

[2] F. Schweppe, "Recursive State Estimation: Unknown-but-Bounded Errors and System Inputs", IEEE Trans. on Automatic Control, Vol. AC-13, No. 1, Feb. 1968, pp. 22-28.

[3] R. Evans, A. Cantoni and T.E. Fortmann, "Envelope-Constrained Filters III: Uncertain Input" submitted for publication

[4] P. Eykoff, <u>System Identification</u>, van Nostrand, 1974.

# ESTIMATION AND IDENTIFICATION METHODS FOR EUROPEAN SATELLITE GEOS*

P. Y. Willems, Professor
G. Campion, Chercheur qualifié FNRS
D. A. Johnson, Senior Research Associate
J. C. Samin, Chargé de Recherches FNRS

Université Catholique de Louvain

This paper presents the results obtained in state vector estimation and parameter identification of a linear model of European satellite GEOS. Two methods are propsed : the extended Kalman filter and the sequential decision method based on a binary quantization procedure. These methods are tested with simulated data as well as with in-flight data provided by telemetry.

## INTRODUCTION

This paper summarizes results obtained at the University of Louvain for the modelling, state vector estimation and parameter identification of European satellite GEOS. As proposed by the European Space Agency (ESA) the problem can be described as follows : GEOS is a geostationnary satellite consisting of a rigid central body carrying short antennas and two long flexible booms with tip masses (experimental devices). ESA is interested in the study of the attitude motion of GEOS, i.e. in the evolution of the orientation of the satellite with respect to some reference frame. This orientation is important for telecommunication antennas solar pannels, instrumental devices... In order to perform this study, it is necessary to elaborate a mathematical model for the attitude motion as well as methods for estimating the state and identifying the parameters of this model. The knowledge or the monitoring of these parameters, which have a physical significance, is important in itself, in order to check the stability of the attitude motion during configuration modifications (boom deployment...). The estimation and

* This research was partially supported by ESA under contract 2163/74.



D. G. Lainiotis and N. S. Tzannes (eds.), Advances in Control, 475–481.

the identification must be achieved by processing telemetry data provided by two accelerometers and three magnetometers located on the central body.

Theoretically this satellite should be modelled as a continuum, resulting in partial differential equations. Practically a reduced order model has to be elaborated for estimation and identification purposes. This finite dimensional model is obtained by mode truncation. It has been shown that the first six modes are dominant and that the effect of the neglected modes on the accelerometer and magnetometer outputs is not significant (no observation spillover). On the other hand this reduced model can be linearized and can therefore be described by a set of 12 first-order linear differential equations. The elaboration and properties of this model are discussed in section I.

With this linear model the estimation problem is easily solved by use of the Kalman filter, provided by dynamical parameters of the model are perfectly known. As uncertainty on these parameters occurs( due, for example, to fuel consumption, incomplete boom deployment, thermal effects,...), it is necessary to implement an adaptive estimation procedure. As the knowledge of these dynamical parameters is important in itself we have chosen methods providing their explicit identification in order to follow the evolution of the dynamical structure of the satellite.

Two identification methods were implemented. The first uses state vector augmentation by the unknown parameters with the resulting non linear estimation problem treated by means of the well-known Extended Kalman Filter (EKF). The second method is the sequential decision technique (SDT), and more precisely the binary quantization procedure proposed by Lainiotis. These methods are briefly described in sections II and III.

These methods were tested successfully, first with measurements provided by simulations based on the reduced order linear model, and, in a second step, with real data provided by telemetry. The results of these systematic studies are given in section IV.

## I. Model Description.

GEOS can be described as a main body with several appendages (antennas, radial booms and two 20 meter radial cables carrying instrumental devices). A representation of GEOS is given in fig. 1. A dynamical analysis of this system has beeen performed by means of the "multibody formalism" ([1]) based on a modelisation of this satellite as a set of interconnected bodies. Because of the sensor location it is reasonable to neglect the internal degrees of freedom which are not observable from the central body. It has been shown in fact, that the dynamical behaviour of the main body is almost unaffected by the high frequency modes induced by the vibrations of the short booms and the bending of the cables ([2]). For estimation and identification purposes it is therefore sufficient to consider a reduced order model representing GEOS as a main rigid body with two attached rigid cables. The relative motion allowed for the cables is a pendulous motion with respect to the main body.

The attitude motion is passively stabilized by spinning the satellite at about 25 r.p.m. around its axis of symmetry. Without perturbation this nominal motion is defined as the equilibrium state, which, because of physical asymmetry, can be tilted, i.e. be different from the nominal reference equilibrium The actual configuration of GEOS is described by seven angles : 3 angles characterizing the position of the main body with respect to the equilibrium position, and for each cable, 2 deflection angles characterizing the pendulum motion in the meridional and equatorial planes. As one of the three angles describing the position of the main body can be eliminated by means of an integral of motion, the configuration of the satellite is completely characterized by six angles only.

For small displacements around the equilibrium configuration the equations of motion may be linearized under the form

$$M\ddot{y} + G\dot{y} + Ky = 0$$

where y represents the six defined angles and M, G and K are three (6 x 6) matrices depending on the physical characteristics of the satellite. The system may also be described by a set of 12 first order differential equations

$$\dot{x}(t) = A(\theta)x(t) \quad , \qquad (2)$$

where

$$x^T = [y^T \dot{y}^T]^T \quad ,$$

or, in the discrete-time version, by

$$x(k+1) = \Phi(\theta)x(k) \quad , \qquad (3)$$

where $\Phi(\theta)$ is the state transition matrix corresponding, for a given time interval, to $A(\theta)$. $\theta$ represents the vector of the physical parameters characterizing the model. The sensor outputs (accelerometers and magnetometers) may also be expressed as linear functions of the state vector x. Details may be found in ([2], [3] , [4]).

In order to provide a flexible tool for the estimation / identification algorithm, a numerical programme has been elaborated to generate, for any set of physical parameters characterizing GEOS, the equilibrium configuration, the dynamical and the transition matrices of the system (A or $\Phi$), corresponding the the motion around this equilibrium and the characteristic elements of the measurement equation.

## II. Extended Kalman Filter (EKF) Method.

The first identification method is well known and consists in augmenting the state vector by the physical parameters to be identified :

$$X^T = [x^T \ \theta^T] \ , \tag{4}$$

with the following additional dynamical equation

$$\dot{\theta} = 0 \ . \tag{5}$$

The dynamical equations as well as the output equations become non-linear in the variable X, so the estimation of X becomes a non-linear estimation problem.

A suboptimal solution is given by the so-called "Linearized Kalman Filter", resulting from a linearization of the dynamical equations and measurement equations around a nominal reference trajectory, in this case in linearizing these equations with respect to the unknown physical parameters around their nominal values, and in applying the Kalman Filter algorithm to this linearized system. This linearization is performed numerically on the basis of the dynamical programme. A more sophisticated solution is given by the Extended Kalman Filter. In this algorithm the linearization is performed, at each step, around the best available estimate of the physical parameters, which means that the reference trajectory is actualized at each step. As this linearized is computer-time consumming an intermediate solution has been chosen : the reference values of the parameters are actualized (and the system equations are relinearized) only when the difference between the last reference value and the best available estimate reaches somes prespecified level. It has been observed that the performance of the algorithm is significantly improved by means of this relinearization. The implementation of this method is described in detail in [2] and [5].

## III. Sequential Decision (SD) Method.

The second method is an adaptive estimation algorithm deduced from Lainiotis's partition theorem. Consider a descretization of the parameter space in M possible values of $\theta$, say $\theta_i$.

The conditional probability density function of the state is given, following Bayes'rule,

$$p[x(k)|Z_\circ^k] = \sum_{i=1}^{M} p[x(k)/Z_\circ^k, \theta_i]Pr[\theta_i/Z_\circ^k] \qquad (6)$$

where $Z_\circ^k$ represent the past measurements up to time k. As for a given $\theta_i$ the system is linear, $p[x(k)/Z_\circ^k,\theta_i]$ is given directly by the Kalman filter based on $\theta_i$. On the other hand $Pr[\theta_i/Z_\circ^k]$ can be computed recursively using the properties of the innovation process ([6]). The resulting structure is a bank of M Kalman filters working in parallel. If many parameters have to be identified, on-line processing is not possible. For this reason we chose a simplified procedure based on a binary quantization of the parameter space ([7]). Basically this technique consists in decomposing the original problem into a sequence of elementary choices between two possible values of $\theta$. A particular mechanism of construction of the elementary alternatives ensures convergence of the algorithm. The resulting structure is a bank of only two Kalman filters working in parallel and based on values of $\theta$ which are reactualized at each elementary decision. The implementation of this method is discussed in detail in [2], [3] and [8].

The advantage of this technique is that it preserves the linear structure of the system and does not complicate it artificially. The corresponding computer programme is easy to implement and is self-contained : it does not need auxiliary programmes such as linearization of the system equations with respect to the physical parameters.

## IV. Numerical results.

Numerical results were obtained in two separate stages

a) As this study started some time before GEOS was actually launched, the measurements had at first to be generated by means of the dynamic simulation programme. This had the advantage of providing a check on the validity and the accuracy of the algorithms : a number of offset parameters were fed into the simulator which cranked out series of measurements at the prescribed sample rate. The estimation/identification algorithms then processed these data to reconstruct x and $\theta$ . An appreciation of the results was then a matter of direct comparison with the values in the simulation programme. The conclusion of these systematic investigations is that the moments of inertia (directly related to the stability of the motion), the products of inertia (related to the offset of the actual equilibrium with respect to the nominal equilibrium), and the lengths of the cables are identifiable with good precision and that the results of the two methods (EKF and SD) are quite comparable. Numerical values may be found in [2].

b) Having tested these algorithms we were ready to apply them to in-flight data from the real-life GEOS(effectively launched in April 1977). Different records of data were investigated corresponding to characteristic phases in GEOS's mission (boom deployment...). From one phase to the other the physical parameters are modified and the European Space Agency was interested in the identification of these parameters during the quiescent periods following these configuration modifications. Unfortunately only accelerometers outputs are available but an interesting point is that the satellite is in a pure nutation phase so that all information related to the unknown physical parameters is translated into 5 characteristics of the two sinusoïdal outputs of the accelerometers (signal frequency, 2 offsets of the signals, ratio of amplitudes and relative phase). This property provides a possibility to check the results by feeding the estimated parameters into the dynamics simulation programme and comparing the characteristics of the outputs with those of the real measurements. The conclusion is that the agreement is excellent for both methods.

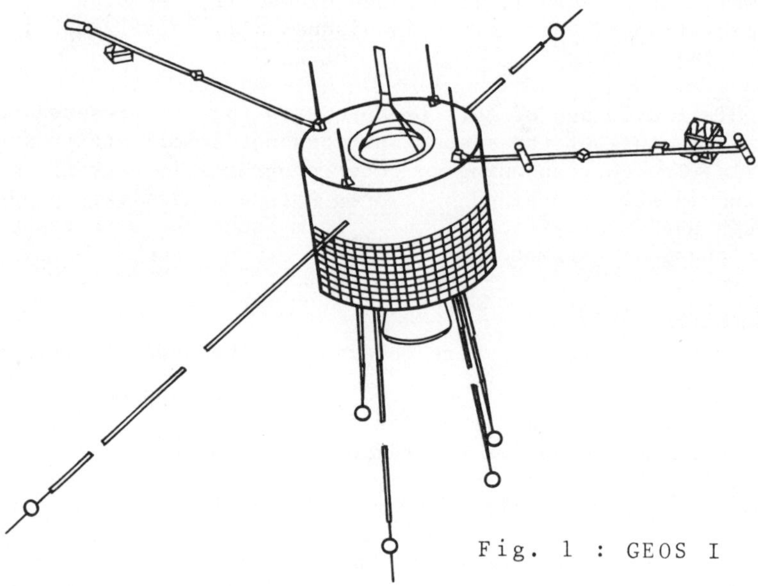

Fig. 1 : GEOS I

## Conclusions

1) From the results obtained with simulated data as well as with in-flight data we feel secure in concluding as to the feasibility of estimation/identification for spacecraft and, in particular, as to the power of Extended Kalman Filtering and Sequential Decision Technique in that respect.

2) An interesting parallel may be drawn between the performances of EKF and SD techniques. As concerns their convergence properties, SD is markedly superior when the parameters are far from their nominal values; conversely EKF produces a higher final accuracy when the uncertaintly on the parameters is small, requiring only a short series of measurements (of the order of 100) compared with SD (which, however processes each step much faster, so the total CPU time is comparable). The two techniques are therefore complementary : ideally one should implement SD first to give a rough estimate of the parameters, and then EKF to increase accuracy.

References


[1] J.C. Samin, P.Y. Willems, Ph. Boland
"Stability analysis of interconnected deformable bodies in topological tree"
AIAA Journal, vol. 12, Aug. 1974.

[2] P.Y. Willems, G. Campion, D. Johnson, J.C. Samin
"Study of state estimation and parameter identification applied to spinning flexible satellite"
ESA-CR-868 Sept. 1977.
Part I : Dynamics
Part II : Estimation and Identification
Part III : User's guide.

[3] G. Campion, J.C. Samin
"Partitioning Estimation and Identification algorithms : Application to European satellite GEOS"
Proc. 1979 JACC Conference Denver, June 1979.

[4] J.C. Samin
"Further Extension of Multi-Body Formalism"
Proc. AIAA Symposium on dynamics and control of large flexible Spacecraft"
Blacksburg, June 1979.

[5] D.A. Johnson
"Spacecraft Identification through Kalman filtering"
Proc. AIAA Symposium on dynamics and control of large flexible spacecraft"
Blacksburg, June 1979.

[6] D.G. Lainiotis
"Optimal adaptive Estimation : Structure and parameter adaptation"
IEEE-AC-16 , n° 2, April 1977.

[7] R.L. Senbush, D.G. Lainiotis
"Simplified parameter quantization procedure for adaptive estimation"
IEEE Trans. Autom. Contr. AC-14, August 1969.

[8] G. Campion
"Spacecraft Identification by Sequential Decision Technique"
Proc. AIAA Symposium on dynamics and control of large flexible spacecraft, Blacksburg, June 1979.

# MULTI-MODEL PARALLEL CONTROL STRUCTURE - METHOD AND APPLICATION

Z. Binder

Laboratoire d'Automatique de Grenoble,
E.N.S.I.E.G. - I.N.P.G. , Grenoble, France

ABSTRACT

Limitations imposed by linear control theory for practical applications have lead to the search for alternative methods for modelling systems. In this general trend, multi-model control technique helps overcome the limits of linearisation and still ensures the continuity with existing methods so that existing algorithms or existing knowledge can still be employed. Multi-model technique have very reasonable justifications. In fact, there are several situations where multiple model representation is useful or inevitable. In the case of nonlinear or time varying systems with time varying characteristics, a single model can hardly represent the system for different operating points. Another example can be related to changes of the physical structure of the system or the need for different representations according to different methods or different hierarchical levels.

The multi-model control technique proposed, exhibits a strong parallelism between its two parts : location of the models with reference to the system and the synthesis of the control law. In addition to this, one of these functions requires processing the different models in parallel. *Location*, consists of two stages : positioning the models and their classification. The criteria used for the synthesis of the algorithms for these two stages are usually identical. These algorithms can be derived from the optimal quadratic input-output tracking or from an optimal filter. Due to this location, we position the models "close" to the system. The classification consists of comparing the results of this positioning for choosing the best models. The synthesis of the *control* variables to be applied to the system is obtained as a weighted combination of the basic control law of each model.



*D. G. Lainiotis and N. S. Tzannes (eds.), Advances in Control, 482–493.*

## 1. INTRODUCTION

The recent progress achieved simultaneously in the fields of control theory and microprocessor technology has created several new ideas in the control of complex processes which have enabled to overcome the limitations of existing methods. Classical control methods which employ a linearized representation of systems do not give desirable results. The behaviour of actual systems do not always follow a linear law. The actual nature of the system, external influencing factors, the consequence of control actions do not generally justify a unique representation, much less a single linear model representation. Naturally, neglecting these facts does simplify analysis but also leads to failures in the case of practical applications. Attempts to overcome this difficulty have resulted in the use of multiple representation of systems.

LANIOTIS has introduced in 1969 (10) the theorem called " Partitioning theorem " for estimation which leads to multiple representation (11) (12). In our case, since 1972 (15) we have been studying the behaviour of physical systems under different operating conditions and operating points which lead us to propose a variable structure control algorithm (5) and the multi-model control technique (7) (8) (16). Simultaneously other publications have appeared (1), motivated either by the theoretical interest (13) or the practical implications of this class of problems (2).

## 2. MULTIPLE REPRESENTATION

There could be many reasons for an interest in multi-model representation of a system.

Evolution of a nonlinear or variable parameter system can be approximated by a set of linear models which are valid about some nominal operating point. The partitionning (12) of the complete model provides the set of models. The difficulty of any practical application of this approach is dependent on the construction and identification of the complete model. This is why identification (3) of the actual system is used very often for obtaining linear models directly. On the other hand, the choice of the nominal operating points for identification is defined by the nature and the specifications of the system represented by a complete model (16).

Change in the structure of the system can be very difficult to account for by any other means except by using multiple - models, where each model represents the system for a given structure. The change of system structure could be related to the incidence of interactions within the system as well as changes in the operating strategy of the system. These changes modify considerably the activation or suppression of the different parts of the system leading to a model with a different structure.

Different representations of a system could be envisaged as a function of the end use of the models. The same physical system is represented by different models. They could differ in terms of the simplifications or reductions involved as in the case of hierarchical control (19), or in terms of the characteristics which are neglected like in the case of static or dynamic models or models with different dynamic characteristics. Finally, models could be complimentary i.e. each one representing only one part of the complete system. The same arguments apply to the case of the same system being modeled for different operating conditions e.g. start up, shut-down, regulation, constrained control etc., the representation being decided as a function of the control methods implemented during these operating conditions.

We shall restrict ourselves to the multi-model technique discussed in the first paragraph where the models are obtained by identification or partionning and are intended for controlling the system.

## 3. STRUCTURE OF THE MULTI-MODEL CONTROL

The availability of several models for a system introduces an additional degree of freedom in the control of the system. The problem of control synthesis is augmented by the problem of model selection which we call location (16). Thus, the structure of a multi-model control algorithm consists of the following steps :

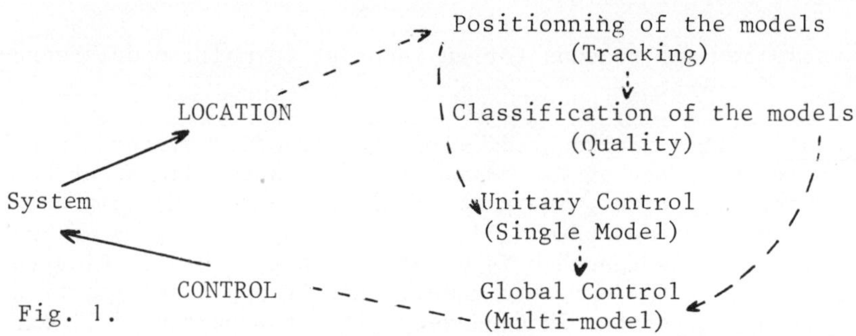


Fig. 1.

It must be noted that a strong parallelism exists in the execution of the above operations simultaneously on each model both during the location as well as the control synthesis stages.

## 4. LOCATION

The aim of this operation is to be able to compare the different models and select those which approximate best the system. Two problems arise : the first problem involves fixing the best position of each model with respect to the system observed. The second problem involves the choice of the criterion for evaluating the quality of the models so that they can be classified accordingly.

### 4.1. Positioning the models

For dynamic systems the positioning can be ensured by tracking the system by the models which are in competition to approximate the system behaviour. Because of the tracking approach adopted (6), the models approach the system and their position depends principally on the system evolution, on the form of the criterion adopted and the control action applied for modifying the evolution of the models.

Input - Output Tracking. A quadratic criterion J which expresses the deviation between the input and the output variables of the system $(u_s, y_s)$ and between the variables corresponding to the model $(u_m, y_m)$ is chosen. This criterion characterises the tracking of the system by the model : (4)

$$J = \sum_{t=o}^{T} [(u_s(t-1) - u_m(t-1))^T R(u_s(t-1)-u_m(t-1)) + (y_s(t) - y_m(t))^T Q(y_s(t) - y_m(t)]$$

Minimisation of this criterion helps determine the input $u_m$ of the model as a function of its states $x_m$ and the observed system variables $(u_s, y_s)$.

$$u_m = f(x_m , u_s , y_s)$$

We observe that for determining the optimal tracking, i.e. the optimal input $u_m$ of the model we must predict the evolution of the system for the optimising horizon T. Nevertheless certaine hypothesis on the nature of the system and on the class of its input permit developing an algorithm applicable in real-time ( 16 ).

Optimal filter. An other way of positioning the models consists of filtering (10) the output signals of the system $(y_s)$ on the basis of each candidate model. This is another possibility of tracking the system by the models. The corresponding criterion can possibly by written in the following form :

$$J_{FILT} = \{(y_s(t)- y_m(t)) (y_s(t) - y_m(t))^T | \{y_s(t),...y_s(1), y_s(0)\}\}$$

The solution of the problem consists of calculating the best estimate of the output of the model $y_m$. For linear state variable models the problem is resolved by the Kalman filter which gives the estimated state of the model.

$$x_m(t+1) = f(x_m(t), y_s(t+1))$$

### 4.2. Classification of the models

The ultimate aim of the location of the system with respect to the set of its models is to classify the models according to their quality, i.e. according to their possibility of approximating the system during a given period or at the instant of classification. The quality of models could be judged by comparing with the system, its evolution or its variables. In practical applications, only the external variables of the system are accessible and can be compared with the corresponding variables of the models.

In general, the criteria chosen for positioning the models (4.1) could be utilised for measuring the quality of these models. For example,

$$Q = f\,[(u_s - u_m),\ (y_s - y_m)]$$

The observation horizon (T), for calculating the index of quality (Q) could be more or less long. We often reduce it to one single step.

The results of the model classification are used in the synthesis of the control action. It would sometimes be useful to determine the normalised values of the quality index ($q_m$) for the N models, i.e.

$$\sum_{m=1}^{N} q_m = 1$$

The index ($q_m$) could be interpreted as the a posteriori probability of the model m (12) (13) in case of filtering is used for positioning the models.

## 5. CONTROL

Multiple representation is often interesting in the context of automatique control. In fact changes in operating zones or in the operating points of a system require the models to adapt themselves fast to the system.

Classical adaptive control can hardly answer to these conditions, on the other-hand, multi-model control can respond rapidly to changes in the system parameters under the condition that the system can be approximated by one or a combination of the models available. The synthesis of the control action applied thus depends on the quality of these models where each one serves to define the basic control law.

### 5.1. Basic control law (single-model)

The computation of the basic control $u_m$ consists of calculating the control sequence which when applied to the system (well represented by the model m), can fulfil the desired objective.

In the case of a change in the operating point, this objective could be the tracking of the reference variables which describe the desired trajectory in space. In the case where we track simultaneously trajectories desired for the input ($z_i$) and trajectories desired for the output ($z_o$), the criterion can be expressed as follows :

$$J_c = \sum_{t=o}^{T} [(z_i(t-1) - u_m(t-1))^T R((z_i(t-1) - u_m(t-1))$$

$$+ (z_o(t) - y_m(t))^T Q(z_o(t) - y_m(t))]$$

where $u_m$ and $y_m$ are the input and the output of the system as represented by the model m.

The minimisation of this generalised criterion enables determining the control input $u_m$

$$u_m = f(x_m , z_i , z_o)$$

Certain hypotheses on the class of desired trajectories and model linearities enable calculating $u_m$ on-line (16).

We observe that the states $x_m$, necessary for this calculation characterise the internal behaviour of the system as described by the states of the model m. Often, the states of the system are not accessible or do not correspond to any physical state. In which case, for the above computations, we can make use of the states estimated from the model m, for example by using the state vector $x_m$ calculated during the positioning phase (para. 4.1.).

5.2. Global control law (multi-model)

The synthesis of the control action in its final form, is a function of the different strategies concerning the number of models considered for the synthesis.

Only one model is used at each instant for approximating the system. In this case, we apply to the system, the basic control law $u_m$ (5.1) associated with the classification of the best model $m$ (4.2).

$$u = u_m \quad \text{for} \quad Q_m < Q_j \quad j = 1, \ldots N$$

Normally, each change of the classification leads to changes in the basic control law at the instant when this change appears. We remark that the change of this approach consists of the calculation of a sequence of control variables in which several of the first best models are utilised. The sharing of time between different basic control laws is a function of the quality of the corresponding models.

Progressive transition between models (m) (n) which approximate the system can be resolved by using generalised tracking. Thanks to the hypotheses on the period and the form of the transition (T), the global control action applied to the system can be computed on line as a function of the following variables (8):

$$u = f(u_m , u_n , x_m , x_n , T)$$

We note that the solution is obtained by varying the weighting matrices and can be used for all the cases of transition between models for a given period.

Several models approximate the system. This case is the most commonly employed for practical applications. The solution consists of finding a weighted combination of the basic control law for the N available models.

$$u = \sum_{m=1}^{N} q_m \cdot u_m$$

This type of control can be advantageous due to the moderation which results since we calculate an average value of the control action whose violent variations are attennated. On the other hand, the effect of the influence of the past evolution during the location stage, is felt due to the index q in the calculations.

Consequently, the basic control law calculated by the best model is "corrected" using the control law obtained by other models as long as their quality index does not become zero. This fact poses some difficulty in comparison with the control law derived from the single best model. An intermediate solution can be obtained, say, by using the control laws of a few very good models.

## 6. SIMULATION EXAMPLE

The two approaches presented above for multimodel control have been tested on a nonlinear simulation example. The positioning of the candidate models is achieved by a quadratic tracking of the system (para. 4.1). (7) (9).

The first approach which we have termed deterministic consists of using one model at a time.

In the second approach, called probabilistic, the positioning and the classification of models is based on Kalman filters (para. 4.1). The global control law is calculated as a combination of the basic control laws of all the models (para. 5.2).

The nonlinear system we have taken is of the form :

$$\frac{dx}{dt} = -2x + xu + 2u$$

$$z = x + v$$

where v is a white, gaussian noise with zero mean and known covariance $\sigma^2$, $v = 0$ in deterministic case.

The static characteristics $(\frac{dx}{dt} = 0)$ of the system are given by :

$$x = \frac{2u}{2-u}$$

We have taken three static operating points on this characteristic and have obtained three models linearised about these points, which can be expressed in the discrete form as follows :

$$M_1 \ (x=0,\ u=0) : x(t+1) = 0.82x(t) + 0.2u(t)$$

$$M_2 \ (x=1,\ u=2/3) : x(t+1) = 0.87x(t) + 0.3u(t)$$

$$M_3 \ (x=2,\ u=1) : x(t+1) = 0.9x(t) + 0.4u(t)$$

(discretisation interval 0.1 sec.).

We assign the following aim for the control of the system : to lead the system from the point (w=0, u=0) to (x=2, u=1), by tracking the reference variables zi , zo). The reference variable zi and zo are generated by two second order linear reference models.

We have compared these two approaches to multimodel control, with the optimal control solution obtained by non linear programming (method of gradients). The figure 2a gives the global control u(t) to be applied to the system as given by each approach and figure 2b represents the output of the system given by each approach. The following table gives an idea of the performance of each approach and the utility of introducing multiple models for control (with $\sigma = 0.001$). (9).

| Models used | $M_1$ | $M_2$ | $M_3$ | $M_1+M_3$ | $M_1+M_2+M_3$ |
|---|---|---|---|---|---|
| Deterministic | 4296 | 102 | 35.8 | 36.2 | 36.8 |
| Stochastic | 4326 | 76.28 | 35.42 | 36.4 | 34.14 |
| Optimal (NLP) | | | 26 | | |

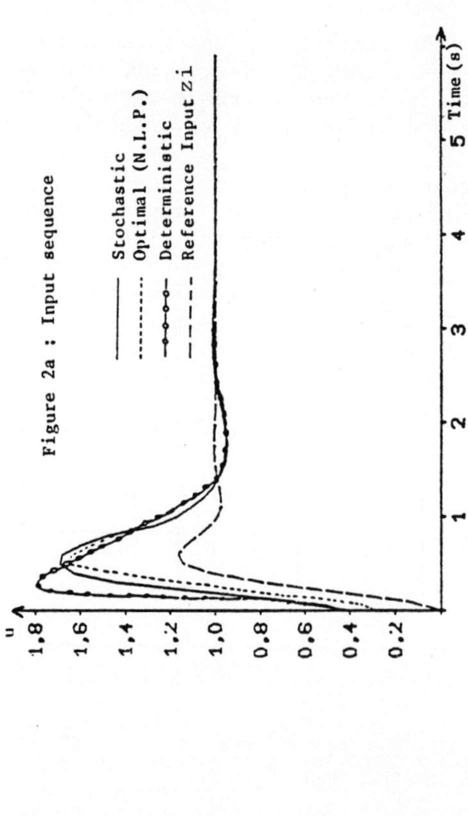


Figure 2a : Input sequence

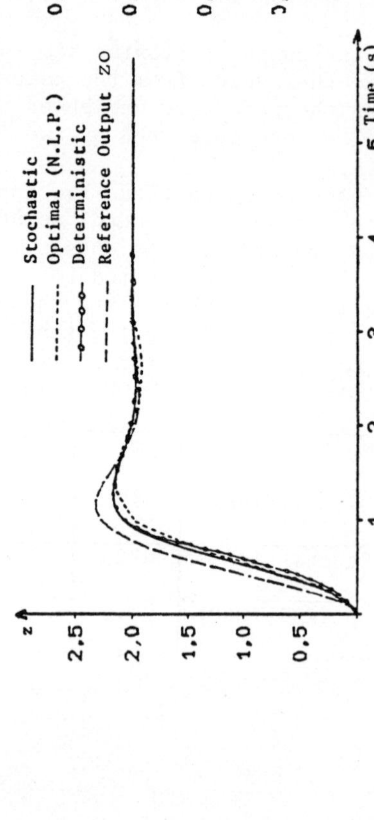


Fig. 2b : Output Sequence

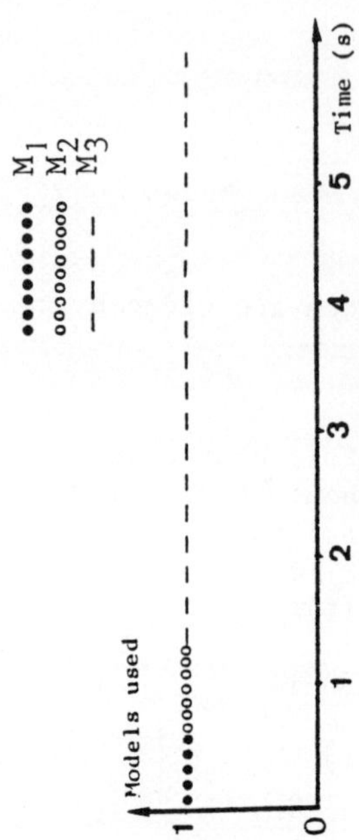


Fig. 2c : Deterministic case

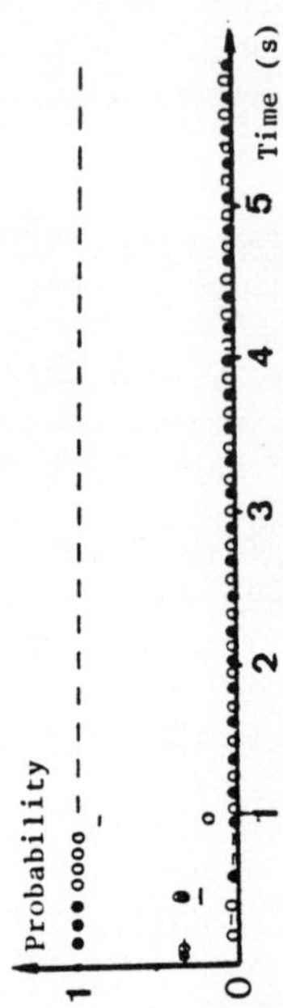


Fig. 2c : Stochastic case (σ = 0.001)

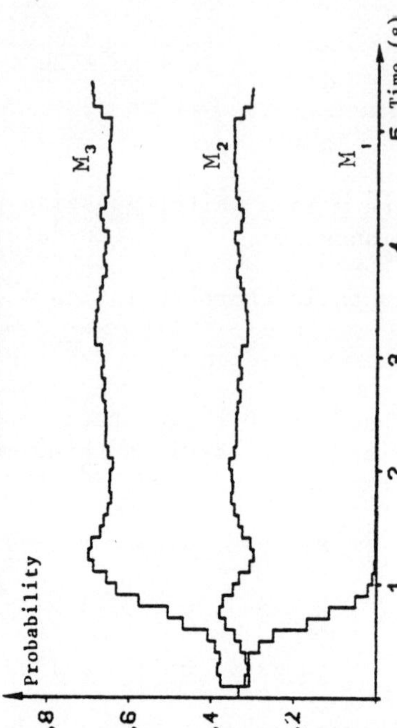


Fig. 2d : Stochastic case (σ = 0.1)

So as to compare the different approaches developped (fig. 2a, 2b, 2c), we have taken first a weak noise term ($\sigma = 0.001$) on the system output. We observe that in the case where the noise level increases $\sigma = 0.1$ (fig. 2d), the commutation between models takes place at a slower rate and the convergence to a single model is not assured. This is explained by the fact that a non linear system perturbed by noise fluctuates violently from its operating point and a model otherwise good would not be sufficient to represent the system or to control it.

The two approaches, deterministic as well as stochastic, have been observed to give highly satisfactory results as verified by the non-linear programming approach. In addition to this, they have proved feasible to be implemented with reasonable costs.

## CONCLUSION

The multi-model technique presented above offers several advantages. This method permits to overcome the limitations of linear control theory for non-linear, or variable parameter or incompletely specified systems. It could also be seen as an extension of the results obtained for linear systems by the use of numerous algorithms available in this field.

From the point of view of methodology, the method brings in a contribution for progressive automation. Thus, the number and the quality of models can be modified along with improvements in the knowledge of the system. Continuous comparison of models helps to test new representations without any risk of serious consequences on the system.

Multi-model technique has a very vast area of application for systems with different natures, or for the different hierarchical levels of a system as well as for testing different control structures. The multi-model approach leads to algorithms with parallel structure. Asynchronism in the parallel processing introduces new constraints in the conception and operation of the processing system chosen as a support for this technique. This fact leads to new approaches for the architecture of such systems wherein certain multi-microprocessors can be employed (17) (18). Their development and use can introduce multi-model techniques for applications which we cannot imagine well at present.

Although the first attempts are about ten years old, multi-model technique is at the beginning of a very major development. There is still a lot of work to be done, both on the theoretical as well as the application aspects. We are however convinced that this method although new, adapts well to practical applications, which increases its value.

BIBLIOGRAPHY

(1) Athans, M., and Varaiya, P.P. : 1975, *A survey of adaptive stochastic control methods*, in Proc. Eng. Foundation Conf. on Syst. Eng., New England College, Henniker, NH.

(2) Athans, M., and all. 1977, *The stochastic control of the F-8C Aircraft using a multiple model adaptive control (MMAC)*, IEEE Trans. Automat. Contr., vol. AC-22, n°5, pp. 768-780.

(3) Barraud, A. : 1975, *Réalisation minimale et approximation optimale des systèmes dynamiques linéaires invariants*, Thesis Docteur-es-Sciences.

(4) Binder, Z. : 1971, *Commande optimale dynamique d'un procédé industriel multivariable* , Thesis Docteur-Ingénieur, I.N.P.G.

(5) Binder, Z., and Nawarecki, E. : 1974, *A computer algorithm of variable structure for the control of complex objects*, Proc. IFAC/IFIP Int. Conf. on Digital Computer Application, Zurich.

(6) Binder, Z., Badr, O., and Perret, R. : 1975, *Tracking approach to the control problems*, Congress IFAC Boston, ed : ISA Pitsburg 1975, part IIIA - 12.4., pp. 1-8.

(7) Binder, Z., Jawhari, S., and Monnier, B. : 1978, *Multi-model techniques in the simulation and control of nonlinear systems*, Proceeding Int. Conf. IASTED, Montreal, ed. ACTA PRESS.

(8) Janex, A. : 1979, *Changement de modèles dans la commande à plusieurs niveaux*, Thesis Docteur en Automatique, I.N.P.G.

(9) Jawhari, S. : 1978, *Sur la commande stochastique adaptative multi-modèles*, Thesis Docteur en Automatique, I.N.P.G.

(10) Lainiotis, D.G. : 1969, *Optimal adaptive estimation, structure and parameter adaptation*, Proceedings of the 1969 IEEE Adapt. Processes Symp. (IEEE Catalog. n°69 C 52-AC).

(11) Lainiotis, D.G., Deshpande, J.C., and Upadhyay, T.N. : 1972, *Optimal adaptive control : a nonlinear separation theorem*, Int. J. Control, vol. 15, n°5, pp. 877-888.

(12) Lainiotis, D.G. : 1976, *Partitionning : a unifying framework for adaptive system, I : estimation, II : control*, Proceedings of the IEEE, vol. 64, n°8.

(13) Lee, A.Y., and Sims, C.S. : 1974, *Adaptive estimation and stochastic control for uncertain models*, Int. J. Control, vol.19, n°3, pp. 625-639.

(14) Magill, T.D. : 1965, *Optimal adaptive estimation of sampled stochastic process*, IEEE Trans. Automat. Contr., vol. AC-10, pp. 434-439.

(15) Monnier, B. : 1973, *Algorithme à structure variable*, Projet Section Spéciale Automatique, I.N.P.G.

(16) Monnier, B. : 1977, *Contributions à la commande d'une classe de procédés dynamiques industriels dans de grands domaines de fonctionnement*, Thesis Docteur-Ingénieur, I.N.P.G.

(17) Olaiwan, Z., and Binder, Z. : 1976, *Algorithm processing unit for automatic control*, Proceeding IEEE Computer Society, Int. Conf. on Parallel Processing, Bellaire (IEEE n°78 CH 1321-9C, pp. 139-145).

(18) Olaiwan, Z. : 1979, *Système multimicroprocesseur pour la commande automatique*, Thesis Docteur-Ingénieur, I.N.P.G.

(19) Rey, D. : 1978, *Sur la commande décentralisée coordonnée - application à un procédé pilote de distillation*, Thesis Docteur-Ingénieur I.N.P.G.

(20) Wittenmark, B. : 1975, *Stochastic adaptive control methods : a survey*, Int. J. Control, vol. 21, n°5, 705-730.

# OPTIMAL NON-QUADRATIC COST CONTROL AND ITS RELATIONSHIP WITH NON-GAUSSIAN INITIAL STATE ESTIMATE

Guy Campion
Chercheur Qualifié FNRS, Université Catholique de Louvain.

The optimal control problem for linear systems with non quadratic final cost function admits a partitioned solution dual of the partitioned solution of the estimation problem for linear systems with non-gaussian initial state. In order to make possible the implementation of this method a general solution of the fixed end-point problem is provided. This result appears as a particular case of transposition in the control domain of the partition theorem stated by Lainiotis for the estimation problem.

## INTRODUCTION

This paper presents an original solution for the optimal control problem of linear systems with non-quadratic final cost function. The structure of this solution is inspired by the concept of duality between the estimation problem and the optimal control problem. Kalman was the first to point out the similarity of the solution structures for these problems in the linear case ([1], [2]). It is well known that the solutions of these two problems are completely characterized by two functions : the conditional probability density function (p.d.f.) of the state, given the past measurements, for the estimation problem, and the minimum cost function for the optimal control problem. In the linear case this conditional p.d.f. is gaussian with mean and covariance matrix given by the Kalman filter, while the minimum cost function is a quadratic form, if the terminal cost and the integral cost are quadratic. The covariance matrix and the weighting matrix of the quadratic form are given as the solutions of Riccati equations (forward equation in estimation and backward equation in control) which are dual of each other. Accordingly to this duality rela-



D. G. Lainiotis and N. S. Tzannes (eds.), Advances in Control, 494–504.

tionship the a priori p.d.f. of the initial state corresponds to the terminal cost function, the noise covariance matrices to the weighting matrices of the integral cost and the output (observation) matrix to the input (control) matrix.

The similarity between the solution structures of the two problems has been exploited by Campion and Dajani ([3] ,[4]) in order to generate new methods of optimal control by transposition of known estimation methods. In a first approach a suboptimal solution of the non linear control problem has been obtained by transposition of the Gaussian sum approximation method introduced by Sorenson and Alspach in the estimation problem ([5]). In the present paper it is shown how to obtain the solution of the optimal control problem of a linear system with quadratic integral cost but non-quadratic final cost function by transposition of a known solution of the dual estimation problem. Accordingly to the described duality relationship this dual problem is nothing but the estimation of a linear system with gaussian additive noises but non gaussian a priori p.d.f. of the initial state. The solution of this estimation problem has been given by LAINIOTIS, as a particular case of its well-known "partition theorem ([6],[7],[8]). The proposed solution of the control problem, directly inspired by the method of Lainiotis, presents the same partitioning structure and can therefore be considered as a particular case of transposition in the optimal control domain of the "partition theorem".

In section I the partitioned solution of the control problem is presented and discussed, while in section II an original solution of the fixed end-point optimal control problem is proposed. Section III describes the practical implementation of the method.

## I. PARTITIONED SOLUTION

Consider a dynamical system described by the following linear differential equation

$$\dot{x} = A(t)x + B(t)u \quad , \tag{I.1}$$

where x is a n-vector (state vector), u a m-vector (input vector). Consider, in addition, a performance index J of the following form

$$J(c,t_o,t_f) = \theta[x(t_f))] + \frac{1}{2}\int_{t_o}^{t} x^T(\tau)Q(\tau)x(\tau) + u^T(\tau)R(\tau)u(\tau)d\tau \quad , \tag{I.2}$$

where the time interval $[t_o,t_f]$ is given, $\theta(\cdot)$ is the terminal cost function (non negative and convex), Q and R are respectively a (n x n) semi-definite positive symmetric and a (m x m) definite positive symmetric matrix.

The optimal control problem consists, for a given initial state c, in the elaboration of a control policy (i.e. $u(\tau)$, $t_o \leq \tau \leq t_f$) minimizing $J(c,t_o,t_f)$ subject to the constraint (I.1). The admissible control laws $u(\tau)$ lie in the Hilbert space $L_2(t_o,t_f)$, that is

$$\int_{t_o}^{t_f} ||u(\tau)^2|| dt < \infty \ . \tag{3}$$

Define now $E(c,t_o,t_f)$ as the set of all possible trajectoires defined between $t_o$ and $t_f$, starting at c, and corresponding to all the control laws satisfying (I.1) and (I.3). It is well known (see for example [9]) that this problem admits a solution completely characterized by the knowledge of the minimum cost function $V(x,t)$

$$V(x,t) = \min_{E(x,t,t_f)} J(x,t,t_f) \ . \tag{I.4}$$

This function satisfies the Hamilton-Jacobi equation

$$\frac{\partial V}{\partial t} + \min_{u} H\left[x,\frac{\partial V}{\partial x}, u, t\right] = 0 \quad , \tag{I.5}$$

where $H(x,\lambda,u,t)$ is the Hamiltonian defined by

$$H(x,\lambda,u,t) \triangleq \frac{1}{2} x^T(t)\, Q(t)\, x(t) + \frac{1}{2} u^T(t)\, R(t)\, u(t) + \lambda^T[A(t)x + B(t)u] \ , \tag{I.6}$$

and with the terminal condition

$$V(x,t_f) = \theta(x) \tag{I.7}$$

Suppose that $V(x,t)$ is available, then the optimal control at time $t$ is given by the minimization of the Hamiltonian $H[x,\frac{\partial V}{\partial x},u,t]$. In the particular case of quadratic final cost function $\theta$, it is easy to verify that $V(x,t)$ is itself a quadratic form. The original problem (I.1), (I.2) reduces therefore to the construction of this $V(x,t)$ function. A method proposed by Liang([10]) is based on the variational approach of Caratheodory while the present paper gives another method inspired by the duality between control and estimation.

As said before the dual of this problem is the estimation problem for a linear system, with white gaussian additive noises but non gaussian a priori p.d.f. for the initial state. The solution of this estimation problem is completely characterized by the conditional p.d.f. of the state corresponding to the minimum cost function characterizing the solution of the optimal control problem. Using the partition theorem the construction of the estimation problem solution is decomposed by Lainiotis ([8]) into two subproblems. The first one consists in the non adaptive estimation of the state with fixed initial state and is solved by use of the Kalman filter. It is important to see that the solution of this first subproblem is independent of the particular form of the a priori p.d.f. of the initial state. The second subproblem is non linear and consists in the adaptive a posteriori estimation of the initial state. The conditional p.d.f. of the state is then obtained by connecting the solutions of the subproblems by use of Bayes' rule. It is therefore, expected that the solution structure of the optimal control problem defined by (I.1) and (I.2) can be partitioned in the same way in two subproblems : the first one is the optimal control problem with fixed terminal state, independent of the particular terminal cost function, and the second one is the optimal choice of the terminal state considered as a parameter. The minimum cost function $V(x,t)$ has then to be obtained by some combination of the solutions of the two subproblems.

Define the set of attainability at time $t_f$ corresponding to the state $c$ at time $t_o$ :

$$K(c,t_o,t_f) \text{ .} \qquad \text{(I.8)}$$

For each $\sigma$ belonging to $K(c,t_o,t_f)$ it is possible to define the fixed end-point problem corresponding to (I.1) and (I.2) by the additional constraint

$$x(t_f) = \sigma \qquad \text{(I.9)}$$

Define $E_\sigma(c,t_o,t_f)$ as the set of all possible trajectories, obtained by $u(t) \in L_2(t_o,t_f)$, with the constraints (I.1), $x(t_o) = c$ and $x(t_f) = \sigma$ and let $V^*(\sigma,c,t_o)$ be the minimum cost for the problem with fixed end-point $\sigma$. Therefore, for a given $\sigma$ ,

$$V^*(\sigma,c,t_o) = \min_{E_\sigma(c,t_o,t_f)} J(c,t_o,t_f). \qquad \text{(I.10)}$$

It is clear that

$$E(c,t_o,t_f) = \bigcup_{\sigma \in K(c,t_o,t_f)} E_\sigma(c,t_o,t_f) \text{ ,} \qquad \text{(I.11)}$$

with the consequence that

$$\min_{E(c,t_o,t_f)} J(c,t_o,t_f) = \min_{\sigma \in K(c,t_o,t_f)} \{\min_{E_\sigma(c,t_o,t_f)} J(c,t_o,t_f\} \tag{I.12}$$

or $$V(c,t_o) = \min_{\sigma \in K(c,t_o,t_f)} V^*(\sigma,c,t_o)$$

or, in general,

$$V(x,t) = \min_{\sigma \in K(x,t,t_f)} V^*(\sigma,x,t) \tag{I.13}$$

From expression (I.13) it can be seen that the construction of $V(x,t)$ is achieved by considering two subproblems : the first one is the fixed end-point problem and the second one is the optimal choice of the terminal state by minimizing $V^*(\sigma,x,t)$ for $\sigma \in K(x,t,t_f)$.

The optimal trajectory of the original problem is the optimal trajectory for the particular fixed end-point problem corresponding to $\sigma$ minimizing $V^*(\sigma,c,t_o)$. It must be noted that the optimal $\sigma$ does not change along this trajectory. Let $\sigma_o$ the optimal $\sigma$ corresponding to the couple $(c,t_o)$ and assume that for a particular $(x,t)$ belonging to the corresponding trajectory the optimal $\sigma$ is $\sigma_1 \neq \sigma_o$. This would imply that there exists a trajectory $(c,t_o) \to (\overset{o}{x},t) \to (\sigma_1,t_f)$ with a "better" cost than the original trajectory $(c,t_o) \to (\sigma_o,t_f)$, which is in contradiction with the fact that $\sigma_o$ is the optimal $\sigma$ corresponding to $(c,t_o)$.

The implementation of this method needs a general analytical expression of $V^*(\sigma,x,t)$, the minimum cost corresponding to the fixed end-point problem. An original solution of this problem is presented in the next section.

## II. The fixed end-point problem

The description of the corresponding fixed end-point problem is the same as (I.1), (I.2) with the additional constraint

$$x(t_f) = \sigma \tag{II.1}$$

It is well-known that the solution of this problem is obtained by solving the canonical equations

$$\dot{x} = A(t)\, x(t) - B(t)\, R^{-1}\, B^T\, \lambda(t) \tag{II.2}$$
$$\dot{\lambda} = -\, Q(t)\, x(t) - A^T(t)\, \lambda(t)$$

with the terminal conditions $x(t_o) = c$ and $x(t_f) = \sigma$. In these equations $\lambda(t)$ is a n-vector, the costate vector. Defining the (2n x 2n) transition matrix $\psi(t,t_f)$

$$\psi(t,t_f) = \begin{bmatrix} \psi_{11}(t) & \psi_{12}(t) \\ \psi_{21}(t) & \psi_{22}(t) \end{bmatrix} \tag{II.3}$$

by the corresponding equations

$$\dot{\psi}(t,t_f) = \begin{bmatrix} A(t) & -B(t)R^{-1}(t)B^T(t) \\ -Q(t) & -A^T(t) \end{bmatrix} \psi(t,t_f) , \tag{II.4}$$

with the terminal condition

$$\psi(t_f,t_f) = I \tag{II.5}$$

the solution of (II.2) can be expressed as

$$\begin{bmatrix} x(t) \\ \lambda(t) \end{bmatrix} = \psi(t,t_f) \begin{bmatrix} \sigma \\ \lambda(t_f) \end{bmatrix} \tag{II.6}$$

with the constraint $x(t_o) = c$.

For a given $\sigma$, the existence of a solution for (II.2) or (II.6) is equivalent to the fact that $\sigma$ belongs to the set of attainability $K(x,t,t_f)$. On the other hand a solution of (II.2) or (II.6) defines an extremal trajectory. It is known (see for example [11]) that two extremal trajectories with the same end-point (and, of course, same initial state) coïncide almost everywhere. A solution of the canonical equations defines therefore, if it exists, the unique optimal solution of the fixed end-point problem. We now discuss the existence and the construction of solution of the canonical equations.

The properties of the canonical system are well known. It can be shown particularly, that the (n x n) submatrix $\psi_{22}(t,t_f)$ can be considered itself as a transition matrix and is therefore regular. The 2 equations (II.6) can be combined to produce

$$\psi_{12}(t)\,\psi_{22}^{-1}(t)\,\lambda(t) = x(t)-[\psi_{11}(t)-\psi_{12}(t)\psi_{22}^{-1}(t)\psi_{21}(t)]\sigma \tag{II.7}$$

Defining 2 (n x n) matrices $\Pi_o(t)$ and $\psi_o(t)$ by

$$\Pi_o(t) = \psi_{12}(t)\,\psi_{22}^{-1}(t) , \tag{II.8}$$

and

$$\psi_o(t) = \psi_{11}(t)-\psi_{12}(t)\psi_{22}^{-1}(t)\psi_{21}(t) , \tag{II.9}$$

the existence of a solution of the canonical equations is equivalent to the existence of $\lambda(t)$ such that

$$\Pi_o(t)\ \lambda(t) = x(t) - \psi_o(t)\ \sigma \tag{II.10}$$

It must be noted that $\Pi_o(t)$ as defined in (II.8) is nothing but the solution of the inverse Riccati equation :

$$\dot{\Pi}_o(t) = A(t)\ \Pi_o(t)+\Pi_o(t)\ A^T(t)+\Pi_o(t)\ Q(t)\ \Pi_o(t)-B(t)R^{-1}(t)\ B^T(t)$$

$$\text{with}\quad \Pi_o(t_f) = 0\ , \tag{II.11}$$

and that $\psi_o(t)$ is the transition matrix corresponding to the dynamical matrix $[A(t) + \Pi_o(t)\ Q(t)]$ :

$$\dot{\psi}(t) = [A(t) + \Pi_o(t)\ Q(t)]\psi_o(t) \tag{II.12}$$

$$\text{with}\quad \psi_o(t_f) = I$$

If $\Pi_o(t)$ is a regular matrix the solution of (II.10) exists and is unique

$$\lambda(t) = \Pi_o^{-1}(t)[x(t)-\psi_o(t)\sigma] \tag{II.13}.$$

Consider, on the other hand, the case of a singular matrix $\Pi_o(t)$. Define $N[\Pi_o(t)]$ as the nullspace of $\Pi_o(t)$, i.e.

$$N[\Pi_o(t)] = \{y : \Pi_o(t)y = 0\}\ , \tag{II.14}$$

and $R[\Pi_o(t)]$ as the row-space of $\Pi_o(t)$, i.e. the space spanned by the rows of $\Pi_o(t)$. It is well-known (see, for example, [12]) that $N[\Pi_o(t)]$ is the orthogonal complement of $R[\Pi_o^T(t)]$ or $R[\Pi_o(t)]$, as $\Pi_o(t)$ is symmetric.
The equation (II.10) admits a solution if and only if $[x(t) - \psi_o(t)\ \sigma]$ is orthogonal to $N[\Pi_o(t)]$, i.e. if this vector belongs to $R[\Pi_o(t)]$. If this vector does not belong to $R[\Pi_o(t)]$, there is no solution for equation (II.10), that means that $\sigma$ does not belong to $K(x,t,t_f)$. If this vector belongs to $R[\Pi_o(t)]$, the general solution has the following form :

$$\lambda(t) = \lambda_r(t) + \lambda_n(t) \tag{II.15}$$

where $\lambda_r(t)$ belongs to $R[\Pi_o(t)]$ and is given by

$$\lambda_r(t) = \Pi_o^+(t)[x-\psi_o(t)\sigma]\ , \tag{II.16}$$

with $\Pi_o^+(t)$ denoting the pseudo-inverse of $\Pi_o(t)$ and $\lambda_n(t)$ is any vector belonging to $N[\Pi_o(t)]$.
It will now be shown that the component $\lambda_n(t)$ has no influence on the corresponding extremal trajectory and that is therefore suffi-

cient to consider as solution the component belonging to $R[\Pi_o(t)]$, i.e.

$$\lambda(t) = \Pi_o^+(t)\ [x-\psi_o(t)\sigma] \qquad (II.17)$$

Consider a decomposition of the solution (II.6) in two vectors :

$$\begin{bmatrix} x(t) \\ \lambda(t) \end{bmatrix} = \begin{bmatrix} x_r(t) \\ \lambda_r(t) \end{bmatrix} + \begin{bmatrix} x_r(t) \\ \lambda_n(t) \end{bmatrix} \qquad (II.18)$$

where $x_r(t)$ corresponds to $\lambda_r(t) \in R[\Pi_o(t)]$ and $x_n(t)$ to $\lambda_n(t) \in N[\Pi_o(t)]$ , with the terminal conditions

$$x_r(t_f) = \sigma \quad \text{and} \quad x_n(t_f) = 0 \ .$$

This decomposition is given by

$$x_r(t) = \psi_{11}(t)\sigma + \psi_{12}(t)\ \lambda_r(t_f) \qquad (II.19)$$

$$\lambda_r(t) = \psi_{21}(t)\sigma + \psi_{22}(t)\ \lambda_r(t_f)$$

and

$$x_n(t) = \psi_{12}(t)\ \lambda_n(t_f) \qquad (II.20)$$

$$\lambda_n(t) = \psi_{22}(t)\ \lambda_n(t_f)$$

The system (II.20) reduces to

$$x_n(t) = \psi_{12}(t)\ \psi_{22}^{-1}(t)\lambda_n(t) = \Pi_o(t)\lambda_n(t) = 0.$$

This last result shows that an extremal trajectory is characterized only by the component of $\lambda(t)$ belonging to $R[\Pi_o(t)]$ which is the desired result.

With the expressions (II.13) or (II.17) for $\lambda(t)$, it is straight forward to verify that the corresponding minimum cost $V(\sigma,x,t)$ is given, for $[x-\psi_o(t)\sigma] \in R[\Pi_o(t)]$ , by

$$V^*(\sigma,x,t) = \frac{1}{2}\ [x-\psi_o(t)\sigma]^T\ \Pi_o^+(t)[x-\psi_o(t)\sigma] + \beta(\sigma,t)\ , \qquad (II.21)$$

where $\beta(\sigma,t)$ is defined by

$$\dot{\beta}(\sigma,t) = -\frac{1}{2}\ \sigma^T\ \psi_o^T(t)\ Q(t)\ \psi_o(t)\ \sigma \qquad (II.22)$$

$$\text{with } \beta(\sigma,t_f) = \theta(\sigma)\ , \qquad (II.23)$$

or, equivalently, by

$$\beta(\sigma,t) = \theta(\sigma) + \frac{1}{2}\ \sigma^T\ \Lambda(t,t_f)\sigma \qquad (II.24)$$

where

$$\Lambda(t,t_f) = \int_t^{t_f} \psi_o^T(\tau,t_f)\ Q(\tau)\ \psi_o(\tau,t_f)\ d\tau \qquad (II.25).$$

As conclusion of this section, this method of solution of the fixed end-point problem can be summarized as follows :

a) Integrate first backward from $t_f$ to $t_o$ the equations, (II.11), (II.12) and (II.22) in order to generate ${}^o\Pi_o(t_o)$, $\psi_o(t_o)$ and $\beta(\sigma,t_o)$

b) Consider now the vector $[c-\psi_o(t_o)\sigma]$. If this vector belongs to $R[\Pi_o(t_o)]$, the problem admits a solution that means that $\sigma \in K(c,t_o,t_f)$. Of course, if $\Pi_o(t_o)$ is invertible $R[\Pi_o(t_o)]=R^n$ and the system is controlable

c) If $[x-\psi_o(t_o)\sigma] \in R[\Pi_o(t_o)]$, then the minimum cost function $V^*(\sigma,x,t)$ is given by (II.21) and the optimal control $u^o(t)$ has the form :

$$U^o(t) = - R^{-1}(t)B^T(t)\Pi_o^+(t)[x-\psi_o(t)\sigma] \tag{II.26}$$

As the general solution for the fixed and point problem is now available it is possible to present the practical implementation of the partitioned solution of the original problem (I.1) and (I.2).

## III. Implementation of the solution

The partitioned solution is implemented in three steps :

a) Backward integration from $t_f$ to $t_o$ of equations (II.11), (II.12) and (II.24), in order to generate $\Pi_o(t_o)$, $\psi_o(t_o)$ and $\beta(\sigma,t_o)$. It must be noted that this integration can be achieved independently of the particular form of the terminal cost $\theta(x)$. This step produces the general expression of $V^*(\sigma,x,t)$ (II.21)

b) Optimal choice of the terminal point
Accordingly to the result (I.13) the minimum cost function of the original problem is obtained by minimizing $V(\sigma,c,t_o)$, with respect to $\sigma$, for $\sigma$ belonging to $K(c,t_o,t_f)$, i.e. for all $\sigma$ such that $[c-\psi_o(t_o)\sigma]$ belongs to $N[\Pi_o(t_o)]$. Let $\sigma^*(c,t_o)$ be the optimal value corresponding to $(c,t_o)$.

c) The optimal trajectory is obtained as the optimal trajectory of the fixed end-point problem corresponding to the particular $\sigma^*(c,t_o)$. The optimal control is given as in (II.26) by

$$u^o(t) = -R^{-1}(t)B^T(t)\Pi_o^+(t)[x-\psi_o(t)\sigma^*(c,t_o)] \tag{III.2}$$

If, in addition, the terminal point of the trajectory is constrained to belong to some target set $G \subset R^n$, then $V^*(\sigma,c,t_o)$ has only to be minimized with respect to $\sigma$ on the subset of $K(c,t_o,t_f)$ belonging to G, i.e. on the intersection

$$\{K(c,t_o,t_f) \cap G\} .$$

## CONCLUSIONS

1) The partitioned solution of the optimal control problem of linear system with non quadratic terminal cost function appears as dual of the partitioned solution of the estimation problem for a linear system with non gaussian a priori p.d.f. of the initial state. This solution structure is very flexible because the major part of the work (namely step a) of the implementation) can be achieved independently of the terminal cost function.

2) This paper proposes a method of construction of the set of attainability. An original solution of the fixed end-point problem is also exposed.

3) As this solution is dual of the non gaussian initial state estimation problem it is possible to deduce the results derived by Lainiotis concerning the Riccati equation solution ([13],[14]). It can be shown that the solution of the control Riccati equation with arbitrary terminal condition can be obtained algebraically from the solution of the inverse Riccati equation with zero terminal condition : this result can be exploited, as in Lainiotis' work, to generate fast Riccati equation solution algorithms.

4) This partitioned solution has to be considered as a particular case of a more general effort to transpose in the optimal control domain the "partition theorem" stated by Lainiotis for the estimation problem.

## REFERENCES

[1] KALMAN, R.E., "On the general theory of control systems", Proc. 1st IFAC Congress, Moscow 1960.

[2] KALMAN, R.E., "A new approach to linear filtering and prediction problems", J. of Basic Engineering, March 1960.

[3] CAMPION, G., "Estimation et commande optimale", Ph.D. Thesis, University of Louvain, 1974.

[4] M.Z. DAJANI, and G. CAMPION, "Closed-loop control design for non-linear non-quadratic systems", Proc. of the IEEE Symposium on Decision and Control, 1973, San Diego.

[5] ALSPACH, D.L., and SORENSON H.N., "Non-linear Bayesian estimation using gaussian sum approximation", IEEE AC 17, N° 4, 1972.

[6] LAINIOTIS D.G., "Partitioned estimation algorithms I : Non-linear estimation", J. of Information Sciences, 7, 1974.

[7] LAINIOTIS D.G., "Optimal non-linear estimation",Int. Journal of Control, 1971, Vol. 14, N° 6.

[8] LAINIOTIS D.G., S.K. PARK and R. KRISHNAIAH, "Optimal state vector estimation for non-Gaussian initial state vector", IEEE-AC-16 N° 2, 1971.

[9] SAGE, A.P., "Optimum systems control, Prentice Hall, Englewood Cliffs, 1968.

[10] LIANG, A.C.C., "Non-linear controllability and involutory actuator orbits", AFOSR, Scientific Report, AFOSR2200 TR, 1970.

[11] LEE, E.B. and L. MARKUS, "Foundations of optimal control theory, John Wiley, New-York, 1967.

[12] STRANG, G., "Linear algebra and its applications, Academic Press, New-York, 1976.

[13] LAINIOTIS, D.G., "Fast Riccati Equation Solutions Partitioned Algorithms", Comput. and Electr. Engng, Vol.2, Pergamon Press, 1975.

[14] LAINIOTIS, D.G.,"Partitioned Riccati Algorithms", Proc. of the 1975 IEEE Decision and Control Conference.

RECURSIVE MULTIVARIABLE LEAST SQUARES IDENTIFICATION WITH BIAS ESTIMATION

W.C. Amaral, L. Gimeno

UNICAMP, Department Electrical Engineering
CP 1170 13.100-CAMPINAS-S.P. BRASIL

In this paper a new algorithm for recursive identification is presented. This algorithm is an extension and modification of Hsia's work as it considers the multivariable situation and an ARMA model for the noise. A recursive multivariable LS algorithm is utilized for identification of process parameters. Residuals are utilized to estimate noise parameters utilizing an extended LS procedure.

When compared with the Generalized Multivariable LS method, this algorithm has the following advantages: i) it considers a more general noise model and ii) when considering the recursive version this algorithm avoids the matrix inversion problem of the recursive GLS procedure. Experimental results are included to illustrate the algorithm performance.

## 1. INTRODUCTION

In this work a recursive algorithm for parameter estimation in multivariable stochastic difference equations is presented. The algorithm is based in bias estimation of model parameters, it is an extension and modification of the algorithm presented by Hsia (1) as it considers the multivariable situation and an ARMA model for the noise.

Let the system be represented by (1)

$$A(z^{-1})\underline{y}_k = B(z^{-1})\underline{u}_k + \underline{v}_k \qquad \underline{y},\underline{v},\underline{e} \in R^n \quad ; \ u \in R^m \tag{1}$$

and the noise $\underline{v}_k$ by (2)



*D. G. Lainiotis and N. S. Tzannes (eds.), Advances in Control, 505–512.*

$$|I+C(z^{-1})|\underline{v}_k = |I+D(z^{-1})|\underline{e}_k \tag{2}$$

where $\underline{e}_k$ is a vector of zero-mean white noise gaussian processes with covariance matrix V. Matrices A,B,C and D are polynomial matrices defined in (3)

$$\begin{aligned} A(z^{-1}) &= I+A_1 z^{-1}+\ldots+A_{n_a} z^{-n_a} \\ B(z^{-1}) &= B_1 z^{-1}+\ldots+B_{n_b} z^{-n_b} \\ C(z^{-1}) &= C_1 z^{-1}+\ldots+C_{n_c} z^{-n_c} \\ D(z^{-1}) &= D_1 z^{-1}+\ldots+D_{n_d} z^{-n_d} \end{aligned} \tag{3}$$

It is well known that ordinary least squares algortihm leads to non consistent estimates when the sequence $\underline{v}_k$ is not white noise. In Section 2 a recursive algorithm based in LS procedure is presented, it introduces a bias correction term obtained through the estimation of noise model parameters in (2). Section III contains some simulated examples where the bias correction algorithm is compared with the generalized least squares algorithm.

## 2. DERIVATION OF THE BIAS CORRECTION ALGORITHM

In (1) a bias correction is obtained for SISO models utilizing the multivariable situation and a more general model for the noise (equation 2).

The algorithm is obtained by minimizing (4)

$$\min_{\underline{\theta}} J_N = \min_{\underline{\theta}} \sum_{k=1}^{N} \underline{w}_k^T V^{-1} \underline{w}_k \tag{4}$$

where $\underline{w}_k$ is the prediction error and N is the number of data available. The vector $\underline{\theta}$ includes all the parameters to be estimated.

The prediction error is given by (5)

$$\underline{w}_k(\underline{\theta}) = |A(z^{-1})\underline{y}_k - B(Z^{-1})\underline{u}_k + C(z^{-1})\underline{v}_k - D(z^{-1})\underline{e}_k| \tag{5}$$

which also can be rewritten as in (6)

$$\underline{w}_k(\underline{\theta}) = \underline{y}_k - \phi_k^P \underline{\theta}^P - \phi_k^N \underline{\theta}^N \tag{6}$$

where: $\underline{\theta}^{NT} = |\underline{\theta}_1^{NT} \ldots \underline{\theta}_n^{NT}|$ ; $\underline{\theta}_i^{NT} = |\underline{c}_i^T \ \underline{d}_i^T|$

$\underline{\theta}^T = |\underline{\theta}^{PT} \vdots \underline{\theta}^{NT}|$; $\underline{\theta}^{PT} = |\underline{\theta}_1^{PT} \ldots \underline{\theta}_n^{PT}|$; $\underline{\theta}_i^{PT} = |\underline{a}_i^T \ \underline{b}_i^T|$

$\underline{a}_i^T$ : i-th row of $A(z^{-1})$; $\underline{b}_1^T$ : i-th row of $B(z^{-1})$

$$\phi_k^P = \text{block diagonal } |\underbrace{\underline{\psi}_k^{PT}}_{-1-} \ \underbrace{\underline{\psi}_k^{PT}}_{-2-} \ldots \underbrace{\underline{\psi}_k^{PT}}_{-n-}|$$

$$\underline{\psi}_k^{PT} = |-\underline{y}_{k-1}^T \ldots -\underline{y}_{k-n_a}^T \ \underline{u}_{k-1}^T \ldots \underline{u}_{k-n_b}^T|$$

$$\phi_k^N = \text{block diagonal } |\underline{\psi}_k^{NT} \ \underline{\psi}_k^{NT} \ldots \underline{\psi}_k^{NT}|$$

$$\underline{\psi}_k^{NT} = |-\underline{v}_{k-1}^T \ldots -\underline{v}_{k-n_c}^T \ \underline{e}_{k-1}^T \ldots \underline{e}_{k-n_d}^T|$$

Defining $\phi_k$ and $\underline{\psi}_k$ as in (7)

$$\phi_k = |\phi_k^P \vdots \phi_k^N| \ ; \quad \underline{\psi}_k^T = |\underline{\psi}_k^{PT} \vdots \underline{\psi}_k^{NT}| \qquad (7)$$

equation (6) can be rewritten as

$$\underline{w}_k(\underline{\theta}) = \underline{y}_k - \phi_k \underline{\theta} \qquad (8)$$

Minimization in equations 4 leads to

$$\hat{\underline{\theta}} = |\sum_{k=1}^{N} \phi_k^T V^{-1} \phi_k|^{-1} \ |\sum_{k=1}^{N} \phi_k^T V^{-1} \underline{y}_k|$$

Due to the combination of blocks diagonal structure of $\phi_k$, the estimate $\hat{\underline{\theta}}$ can be rewritten separatedly for each $\hat{\theta}_i$ as follows:

$$\hat{\underline{\theta}}_i = |\sum_{k=1}^{N} \underline{\psi}_k \underline{\psi}_k^T|^{-1} \ |\sum_{k=1}^{N} \underline{\psi}_k y_k^i| \quad i = 1,2,\ldots,n \qquad (9)$$

where $y_k^i$ is the i-th measurement at stage k.

Matrix V does not enter the estimate equations due to the special structure of $\phi_k$. This result also holds for other estimation methods as has been shown in (2).

Defining matrix $\hat{\Theta}$ as in (10)

$$\hat{\Theta} = |\underline{\hat{\theta}}_1 \underline{\hat{\theta}}_2 \ldots \underline{\hat{\theta}}_n| \tag{10}$$

we have

$$\hat{\Theta} = \left| \sum_{k=1}^{N} \underline{\psi}_k \underline{\psi}_k^T \right|^{-1} \left| \sum_{k=1}^{N} \underline{\psi}_k \underline{y}_k^T \right| \tag{11}$$

Partitioning of $\hat{\Theta}$ as $\hat{\Theta}^T = |\hat{\Theta}^{PT} \vdots \hat{\Theta}^{NT}|$ leads to the following equation:

$$\begin{vmatrix} \hat{\Theta}^P \\ \cdots \\ \hat{\Theta}^N \end{vmatrix} = \begin{vmatrix} \sum_{k=1}^{N} \underline{\psi}_k^P \underline{\psi}_k^{PT} & \sum_{k=1}^{N} \underline{\psi}_k^P \underline{\psi}_k^{NT} \\ \sum_{k=1}^{N} \underline{\psi}_k^N \underline{\psi}_k^{PT} & \sum_{k=1}^{N} \underline{\psi}_k^N \underline{\psi}_k^{NT} \end{vmatrix}^{-1} \begin{vmatrix} \sum_{k=1}^{N} \underline{\psi}_k^P \underline{y}_k^T \\ \sum_{k=1}^{N} \underline{\psi}_k^N \underline{y}_k^T \end{vmatrix} \tag{12}$$

Utilizing well known results of matrix calculus equation (12) results in:

$$\hat{\Theta}^P = \left| \sum_{k=1}^{N} \underline{\psi}_k^P \underline{\psi}_k^{PT} \right|^{-1} \sum_{k=1}^{N} \underline{\psi}_k^P \underline{y}_k^T - \left| \sum_{k=1}^{N} \underline{\psi}_k^P \underline{\psi}_k^{PT} \right|^{-1} \sum_{k=1}^{N} \underline{\psi}_k^P \underline{\psi}_k^{NT} \hat{\Theta}^N \tag{13}$$

where the second term is the bias which would be obtained by an ordinary least square algorithm. The bias correction algorithm is an extension of the LS algorithm where the bias is corrected through an estimation of $\Theta^N$. This estimation is done as follows: First equation (2) is written as

$$\underline{v}_k = \Theta^{NT} \underline{\psi}_k^N + \underline{e}_k$$

The estimate $\hat{\Theta}^N$ of $\Theta^N$ is obtained by minimizing (4) where the prediction error $\underline{w}_k$ is now given by:

$$\underline{w}_k(\Theta) = \underline{v}_k - \Theta^{NT}\,\underline{\psi}_k^N$$

The value $\underline{v}_k$ is not known and is estimated by $\hat{\underline{v}}_k$

$$\hat{\underline{v}}_k = \underline{y}_k - \hat{\Theta}^{PT}\,\underline{\psi}_k^P$$

The estimate $\hat{\Theta}^N$ will be given by equation (14)

$$\hat{\Theta}^N = \left| \sum_{k=1}^{N} \underline{\psi}_k^N \psi_k^{NT} \right|^{-1} \left| \sum_{k=1}^{N} \underline{\psi}_k^N \hat{\underline{v}}_k^T \right| \tag{14}$$

The recursive version of this algorithm is then given by equations (15)

$$\begin{aligned}
\hat{\theta}^{PLS}_{k+1} &= \hat{\theta}^{PLS}_k + P_k\underline{\psi}^P_{k+1}(1+\underline{\psi}^{PT}_{k+1}P_k\underline{\psi}^P_{k+1})^{-1}\left|\underline{y}^T_{k+1} - \underline{\psi}^{PT}_{k+1}\hat{\theta}^{PLS}_k\right| \\
P_{k+1} &= P_k - P_k\underline{\psi}^P_{k+1}(1+\underline{\psi}^{PT}_{k+1}P_k\underline{\psi}^P_{k+1})^{-1}\underline{\psi}^{PT}_{k+1}P_k \\
\hat{\Theta}^N_{k+1} &= \hat{\Theta}^N_k + Q_k\underline{\psi}^N_{k+1}(1+\underline{\psi}^{NT}_{k+1}Q_k\underline{\psi}^N_{k+1})^{-1}\left|\underline{v}^T_{k+1} - \underline{\psi}^{NT}_{k+1}\hat{\Theta}^N_k\right| \\
Q_{k+1} &= Q_k - Q_k\underline{\psi}^N_{k+1}(1+\underline{\psi}^{NT}_{k+1}Q_k\underline{\psi}^N_{k+1})^{-1}\underline{\psi}^{NT}_{k+1}Q_k
\end{aligned} \tag{15}$$

The final estimate is given by

$$\hat{\Theta}^P_{k+1} = \hat{\Theta}^{PLS}_{k+1} - \hat{\Theta}^{PB}_{k+1} \tag{16}$$

Where the bias correction term $\hat{\Theta}^{PB}_k$ is given by equation (17):

$$\hat{\Theta}^{PB}_{k+1} = \hat{\Theta}^{PB}_k + P_k\underline{\psi}^P_{k+1}(1+\underline{\psi}^{PT}_{k+1}P_k\underline{\psi}^P_{k+1})^{-1}\left|\underline{\psi}^{NT}_{k+1}\hat{\Theta}^N_{k+1} - \underline{\psi}^{PT}_{k+1}\hat{\Theta}^{PB}_k\right| \tag{17}$$

At each stage the algorithm: i) obtains the LS estimate of process and noise parameters: $\hat{\theta}^{PLS}_k$, $\hat{\theta}^N_k$, ii) calculates the bias correction term $\hat{\theta}^{PB}_k$ by equation 17 and finally delivers the final estimate $\hat{\theta}^P_k$ (equation 13).

Hsia (1) has analized the particular situation where n=1 and $D(q^{-1}) = 0$. In this case it is proved (2) that this algorithm converges to an unique solution when the signal to noise ratio is sufficiently high, this means an analogous convergence to the generalized least squares algorithm. Algorithm convergence in the MIMO situation has not been studied.

When $D(z^{-1}) = 0$ this algorithm can be compared with the generalized least squares method (GLS) which has been presented in its multivariable version by Payne (3,4) and Keviczky (5). The GLS algorithm does not allow to estimate separatedly each $\underline{\theta}_i$ (equation 9) because the filtered values of the input and output lead to a matrix $\phi^P$ (equation 6) which has not a block diagonal structure. Due to this fact the bias correction algorithm will represent a great reduction in computational time. In the next section the quality of the estimates obtained with these two algorithms will be compared.

## 3. SIMULATION RESULTS

In this section the results obtained in one simulated example are presented. Bias correction (BC) and GLS algorithm will be compared.

Let the system be given by:

$$\underline{y}_k + A_1\underline{y}_{k-1} = B_1\underline{u}_{k-1} + \underline{v}_k \; ; \quad \underline{v}_k + C_1\underline{v}_{k-1} = \underline{e}_k$$

$$A_1 = \begin{vmatrix} 0.8 & -2.9 \\ 0.4 & -1.4 \end{vmatrix} ; B_1 = \begin{vmatrix} 1 & 0 \\ 0 & 1 \end{vmatrix} ; \; C_1 = \begin{vmatrix} -0.2 & -0.4 \\ 0.2 & -0.8 \end{vmatrix} ; \; V = \begin{vmatrix} 1 & 1 \\ 1 & 2 \end{vmatrix}$$

where $\underline{e}_k$ is a gaussian white noise vector with mean value zero and covariance matrix V. A model is considered with polynomial matrices of the same degree as the system. For comparison purposes the following criteria are utilized:

$$\delta^2 = \sum_{i=1}^{p+n} \left| \frac{\Delta\theta_i}{\theta_i} \right|^2 \; ; \qquad \delta_p^2 = \sum_{i=1}^{p} \left| \frac{\Delta\theta_i}{\theta_i} \right|^2 \tag{18}$$

where:
p : number of process parameters (A,B) to be estimated
n : number of noise parameters(C) to be estimated
$\theta_i$: true value of parameter
$\Delta\theta_i$: estimate error in parameter $\theta_i$

The comparison of input-output characteristics is done by evaluating the variance of each component of the prediction error given in equation 5. Figures 1, 2 and 3 contain the result from a Monte-Carlo simulation of 50 runs.

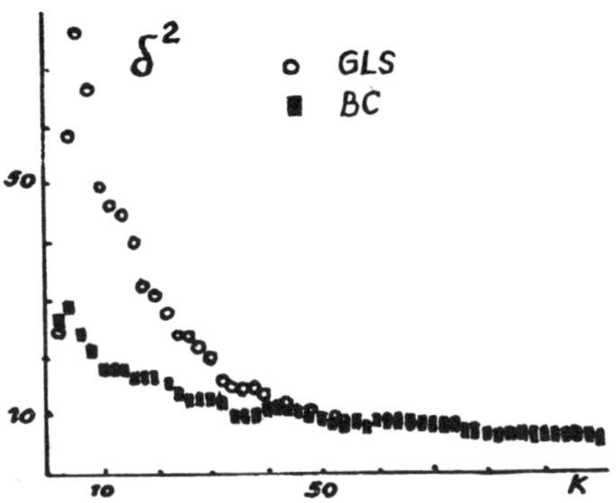


fig.1 - $\delta^2$ evolution

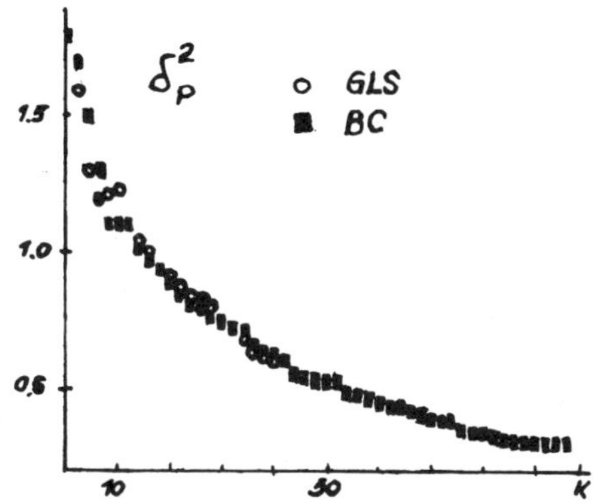


fig.2 - $\delta_p^2$ evolution

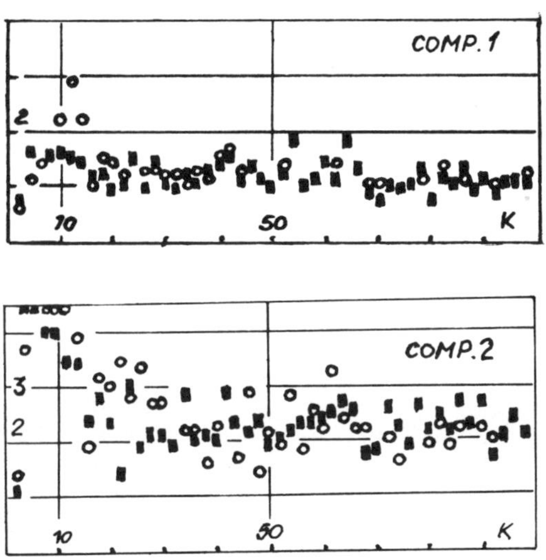


fig.3 Prediction error components variance
■ BC algorithm
o GLS algorithm

In figures 1 and 2, it can be seen that BC performs better than GLS in estimating the noise parameters. It is well known the slow convergence pattern of noise parameters for GLS. As far as only input-output characteristics are significant, it is shown in figure 3 that the two algorithms have a similar behaviour. In this example CPU time for BC is 25% of the time needed by GLS; this decrease is even greater when order and number of parameters increase.

## 4. CONCLUSIONS

In this work is presented the multivariable extension of the bias correction algorithm presented in (1). A general ARMA model for the noise is considered. The recursive version of the algorithm is presented and it is shown that the estimate equations do not depend on noise covariance matrix. The special structure of the estimate equations allows to estimate separatedly each row of the polynomial matrices thus leading to a lower computational time when compared with the generalized least squares algorithm. Simulations results show a similar behaviour of the two algorithms as far as input-output characteristics are considered.

## REFERENCES

(1) Hsia, T.C., *On Least Squares for System Parameter Identification* IEEE Trans. Automat. Contr., February 1976.

(2) Amaral, W.C., *Multivariable Estimation in Discrete Time Systems.* Ph.D. Dissertation. Universidade Estadual de Campinas, Campinas, S.P., Brasil, March 1980.

(3) Goodwin, G.C., Payne, R.L. and Kabaila, P.V., *On canonical forms and algorithms for multivariable time series analysis*, 4th IFAC Symp. Ident. Syst. Param. Estimation, Tbilisi USSR (1976)

(4) Goodwin, G.C., Payne, R.L., *Dynamic System Identification: Experiment Design and Data Analysis*. Ac. Press 1977.

(5) Keviczky, L.K., Banyasz, Cs.M., *Some new results of multiple input-multiple output identification methods*. 4th IFAC Symp. Ident. Syst. Param. Estimation, Tbilisi, USSR (1976).

# A NOTE ON THE JOINT-PROCESS IDENTIFICATION METHOD

A. Frosini, P. Piazzesi, A. Tiano[(†)] and G. Zappa

Istituto di Elettronica, Università di Firenze.
(†) Laboratorio per l'Automazione Navale del CNR
Genova, Italy.

Abstract. Given the input process u and the output process y of a dynamical system S, the joint-process identification method consists in identifying an innovation representation for the joint-process $z = [y' \; u']'$ , from which an input-output model can be derived. The main advantage of the method lies in the simplification of the structure selection procedure. On the other hand loss of accuracy must be expected due to the effort of identifying the feedback too. In this paper an identification algorithm based on the joint process, recently proposed by Akaike, is revisited and the effects of it straightforward implementation on parameter estimates and structure selection is considered. Then a modification of the above method is presented, which, acting differently on input and output, gives better performances with lower computational burden.

## 1. INTRODUCTION

The task of identifying a multivariable system from input-output data is far from being trivial since it involves the selection of an adequate structure for the model representing the system. The usual identification procedure can be schematically described as follows. First a class of models is chosen based on some preliminary analysis of the data and the a priori information about the system. Then the free parameters of each model are estimated according to some fitting criterion and finally a comparison among the various calibrated models is performed. Despite some recent effort [1] , it is still difficult to perform these operations in an interactive way, so that the outlined procedure, though simple in principle, turns out to be numerically difficult and time- consuming.



*D. G. Lainiotis and N. S. Tzannes (eds.), Advances in Control, 513–520.*

For these reasons remarkable interest arose in the last few years around the so-called joint-process identification method: viz., given an input process u(t) and an output process y(t), identify an innovation representation for the joint process z(t) = $= [y(t)' \vdots u(t)']'$, from which, by algebraic manipulations, an input-output model can be derived. In fact structure selection is accomplished here through the factorization of a Hankel matrix of the covariance function and does not require a preliminary parameter estimation.

In this context a preminent role is played by an identification package developed by the japanese statistician Akaike [2] and succesfully applied by his coworkers. Here the identification procedure is accomplished in three steps:

a) preliminary identification of an Auto-Regressive (A.R.) model;
b) Selection of a structure of a state-space model with the help of canonical correlation analysis (c.c.a.) and an information criterion;
c) maximum likelihood estimation of the free parameters of the selected structure exploiting as initial guesses for the parameter value the estimates obtained in the previous steps.

Point b) is the most critical since the maximization of the likelihood function of several variables can be efficiently performed only if a good parametrization and a good starting value for the parameters are available. According to Akaike's quotation "the canonical correlation analysis constitute a significant step toward the practical use of markovian models in time-series analysis" [3].

Unfortunately this, like others joint-process based identification methods, has the disadvantage, with respect to direct identification methods, that also the feedback loop must be estimated, which in most cases is unnecessary. Therefore more severe assumption on the regulator must be adopted and an inferior performance must be expected, regarding both the adequacy of the selected structure and the accuracy of the parameter estimate.

Aim of this paper is to analyze such problems and to propose a modification of Akaike's structure selection algorithm, which, concerning only a model for the forward loop and exploiting some a-priory available information about the feedback, has lower computational complexity and better performances. The paper has a rather informal and preliminary character. Nevertheless, even if a theoretical comparison between the two approaches has not yet been developed, we think that such problems are of some importance and worth of discussion and further analysis.

The paper is organized as follows. In Sec. 2 some definitions and preliminary considerations are reported. In Sec. 3 Akaike's struc ture selection algorithm is summarized and criticized. In Sec. 4 the new algorithm is presented while in Sec. 5 some comments end the paper.

## 2. STATEMENT OF THE PROBLEM

In the following we assume that our identification experiment is based on the observations of two discrete-time stochastic processes u(t) and y(t) which are described by the model of Figure 1

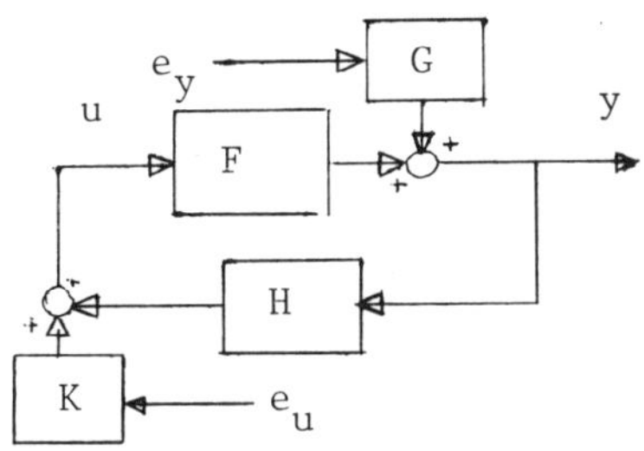


Figure 1. Closed loop system.

where $y(t) \in R^p$ is the output, $u(t) \in R^m$ is the input, $e_y(t) \in R^p$, $e_u(t) \in R^m$, and the couple $[e_y \ e_u]$ is a zero mean,white,gaussian stochastic process with covariance matrix positive definite. F, H, G and K are rational causal transfer function matrices of appropriate dimensions, describing respectively the process dynamics, the feedback dynamics and the filters generating the noise on the process and the feedback. Moreover we shall make the following two additional assumptions:

a) the joint process $z(t) = [y(t)' \mid u(t)']'$ is a full rank, stationary stochastic process;

b) there is a delay both in the process and in the feedback dynamics, that is:

$$\lim_{z\to\infty} F(z) = 0 \quad \text{and} \quad \lim_{z\to\infty} H(z) = 0.$$

According to a), z(t) can also be described by one of its spectral factors, that is a causal rational,stable and invertible transfer function matrix $\Phi$ of dimension $(p+m)\times(p+m)$, relating z(t) with its innovation process $e_z(t)$.(see Figure 2).

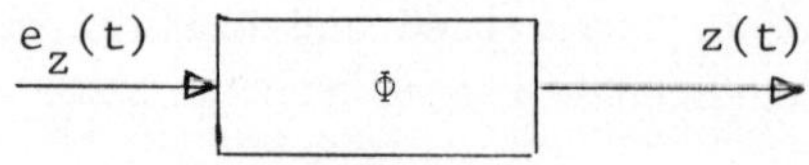


Figure 2. Innovation representation for z(t)

There is a 1 - 1 correspondence between $\Phi$ and the matrices F,G,H and K [4]. Since there are many spectral factors, condition b) assures identifiability properties for both the forward and feedback loops if we choose $\Phi(z)$ such that $\lim_{z\to\infty} \Phi(z) = I$ [5].

It should be stressed that the model of Figure 1 is not the most general linear, time-invariant stochastic system relating y and u. For istance a causal dynamical dependence between $e_y$ and $e_u$ can also be considered. However this more general structure leads to some difficult identifiability problems [6] which we desire to skip here.

Though model of Figure 1 is the usual description for a closed loop system, algorithms presented in this paperwill concern the identification of a state-space model for the forward loop:

$$\begin{cases} x_1(t+1) = A_1x_1(t) + B_1u(t) + K_1e_y(t) \\ y(t) = C_1x_1(t) + e_y(t) \end{cases} \tag{2.1}$$

with (A,C) observable, $(A,[B \vdots K])$ reachable and $e_y(t)$ white and uncorrelated with u(t). Clearly (2.1) is equivalent to the couple of transfer functions (F,G). Nevertheless we prefer the state-space model (2.1) since it is more suitable for control purposes and it is a more compact description of the process (in fact F and G could share common poles).

An analogue description for the feedback loop ($1\leftrightarrow 2$, $y\leftrightarrow u$) leads to a state-space model for the joint process

$$\begin{cases} \begin{bmatrix} x_1(t+1) \\ x_2(t+1) \end{bmatrix} = \begin{bmatrix} A_1 & B_1C_2 \\ B_2C_1 & A_2 \end{bmatrix} \begin{bmatrix} x_1(t) \\ x_2(t) \end{bmatrix} + \begin{bmatrix} K_1 & I \\ I & K_2 \end{bmatrix} \begin{bmatrix} e_y(t) \\ e_u(t) \end{bmatrix} \\ z(t) = \begin{bmatrix} y(t) \\ u(t) \end{bmatrix} = \begin{bmatrix} C_1 & 0 \\ 0 & C_2 \end{bmatrix} \begin{bmatrix} x_1(t) \\ x_2(t) \end{bmatrix} + \begin{bmatrix} e_y(t) \\ e_u(t) \end{bmatrix} \end{cases} \tag{2.2}$$

We shall focus our attention essentially on the selection of a structure for (2.1) and on the estimates for the A and C matrices. In fact rough estimates of other parameters can be easily refined by a subsequent direct identification algorithm if good values for A and C matrices have already been obtained.

## 3. AKAIKE'S STRUCTURE SELECTION PROCEDURE

We first consider the case where no input is present. Therefore let z(t) be a discrete-time, zero-mean, stationary, gaussian stochastic process of full rank with covariance function

$$R_z(k) = E[z(t+k)z(t)]$$

and let

$$Z^t = \text{span}\{z(\tau), \quad t \leqslant \tau\}$$

$$Z_t = \text{span}\{z(\tau), \quad t > \tau\}$$

denote the Hilbert spaces generated by future and past values z. Then, according to [7], the state x(t) of a innovations representation of z:

$$\begin{cases} x(t+1) = Ax(t) + Ke(t) \\ z(t) = Cx(t) + e(t) \end{cases} \tag{3.1}$$

is a collection of random variables belonging to $Z_t$ such that

$$E[z(t+k)|Z_t] = E[z(t+k)|x(t)] \qquad k \geqslant 0.$$

Hence x(t) is a basis in $Z_t$ for the predictors of the future of z upon its past and the selection of a structure for (3.1) reduces to the chóice of a set of linearly independent predictors such all other predictors are a linear combination of them.

Let us define the random vectors

$$v = [z(t)' \vdots \; . \; . \; . \; . \; \vdots z(t+s)']'$$

$$w = [z(t-1)' \vdots \; . \; . \; . \; . \; . \; \vdots z(t-r)']' \qquad r > s.$$

Then, for a sufficiently high s, this linear dependence analysis of predictors is performed through canonical correlation analysis (c.c.a.) between v and w [3] . Numerically it requires the Cholewsky decomposition of $R_v = E[vv']$ and $R_w = E[ww']$ and singular value decomposition of $R_{vw} = E[vw']$. If only estimates of $R_z$ are available, a statistical test based on AIC is adopted.

Besides the structure of (3.1), c.c.a. provides estimates of A and C. K and the covariance of e(t) can be obtained either solving a Riccati equation [8], or from the impulse response of (3.1) estimated from a A.R. model.

If an imput u(t) is present, under the assumptions of Sec. 2, we can apply Akaike's procedure in a straightforward way to the joint process, obtaining an innovation representation like (3.1). Finally, partitioning K and C matrices like:

$$K = \begin{bmatrix} K_1 & K_2 \end{bmatrix} \quad \text{and} \quad C' = \begin{bmatrix} C_1' & C_2' \end{bmatrix}$$

$$\dim K_1 = \dim C_1' = n \times p \ , \ \dim K_2 = \dim C_2' = n \times m$$

we get the desidered state-space model (2.1)

$$\begin{cases} x_1(t+1) = (A_1 - K_2C_2)x_1(t) + K_1e_y(t) + C_2u(t) \\ y(t) = C_1x_1(t) + e_y(t) \end{cases}$$

which, of course, is in general unobservable, hence non minimal. Note that c.c.a. is now performed between

$$v(t) = \left[v_y(t)' \ v_u(t)'\right]' = \left[y(t)', \cdots, y(t+s)', u(t)', \cdots, u(t+s)'\right]'$$

and

$$w(t) = \left[w_y(t)' \ w_u(t)'\right]' = \left[y(t-1)', \cdots y(t-r)', u(t-1)', \cdots u(t-r)'\right]'$$

The main disadvantage of this procedure is the computational burden due to the higher dimensionality of v and w. However algorithm performances will also suffer from the following drawbacks:

a) The state -space model is not directly estimated, nor it is obtained in canonical forms.
b) Stationarity, linearity and finite-dimensionality could not hold exactly for the regulator dynamics.
c) No a-priori information upon the feedback link between y and u can be exploited to improve identification results.

As a consequence, both a poorer estimate of the parameters and the selection of an inadequate structure can be expected. Such conjecture has been confirmed by the results of identification experimentes on ship and hydrofoil dynamics recently performed. See [10] for a detailed description.

## 4. ALTERNATIVE APPROACH

Since (2.1) is a predictive model for the output only, we could perform c.c.a. between $v_y$ and w (that is between the future of the output and the past). But, following the previous outlined analysis, we obtain:

$$\begin{cases} x_1(t+1) = A_1x_1(t) + K_1e_y(t) + K_2e_u(t) \\ y(t) = C_1x_1(t) + e_y(t) \end{cases}$$

and, unfortunately, this model does not compute the innovations process $e_u(t)$ from u(t). So we are forced to identify also a state-space model for u(t) having y(t) as input. Therefore,combining them we go back to the joint-process model and no improvement is obtained.

On the other hand, by (2.1), the following relations hold:

$$E\left[y(t+k)\,|Z_t,v_u\right] = E\left[y(t+k)\,|x(t),v_u\right] \quad k \leq s, \Longrightarrow$$

$$E\left[y(t+k)\,|Z_t^{\perp}{}_{|v_u}\right] = E\left[y(t+k)\,|x(t)^{\perp}{}_{|v_u}\right]$$

where $^{\perp}|v_u$ denotes the residual of the orthogonal projection on $v_u$. Hence the structure of (2.1) is provided by the linear depencence analysis among the predictors of the future output upon $Z^{\perp}_{|v_u}$, that is the subspace of $Z_t$ uncorrelated with future inputs. We therefore propose the following algorithm:

1) Compute tha rank of $R_{v_uw}$

2) Find an orthogonal trasformation Q such that

$$Qw = \tilde{w} = \begin{bmatrix} \tilde{w}_y \\ \tilde{w}_u \end{bmatrix} \quad \text{with } \dim \tilde{w}_u = \text{rank } R_{v_uw} \text{ and } E\left[v_u\tilde{w}_y'\right] = 0.$$

3) Perform c.c.a. between $v_y$ and $\tilde{w}_y$ .

Note that if u(t) is white the two approaches coincide. If u(t) is feedback free than rank $R_{v_uw}$ = rank $R_{v_uw_u}$ and point 1) is simplified. The computational burden is lowered, due to the reduced dimensionality of $v_y$ and $w_y$ and since Q can be easily obtained through a QR decomposition of $R_{v_uw}$.

In this way we get a structure for (2.1) and estimates of $A_1$ and $C_1$ matrices. As before, B1 and $K_1$ can be obtained by an estimate of the impulse response of (2.1) obtained from an A.R. model. Now no particular assumption on the feedback is necessary, provided that good estimates for $R_z$ are available. A similar algorithm has been derived in a different context also in [9] .

## 5. CONCLUSIONS

The Akaike's algorithm for structure selection and preliminary identification of a state-space model has been revisited, and a modification proposed when an observable input is present. In this case the feedback loop is not estimated, thus reducing computing time and weakening the assumptions on the regulator itself. Despite some preliminary successfull applications, further analysis is necessary, in our opinion, on the following items:

a) identifiability conditions analogue to those derived in [6];
b) optimal input design;
c) analytical comparison of the outlined approaches, involving, for istance, estimate variances, probabilities of selecting a wrong structure, conditioning of matrices.

## REFERENCES

[1] van Overbeek, A.J.M. and Ljung, L., "On Line Structure Selection Multivariable State Space Models", Preprints 5th IFAC Symp. on Identification, Darmstadt, 1979, pp.387-396.

[2] Akaike, H., Arahata, E. and Ozaki, T., "TIMSAC 74", Computer Science Monograph, n° 5, Annals Inst. Statis. Mathem., Tokyo 1975.

[3] Akaike, H., "Canonical Correlation Analysis and an Information Criterion" in System Identification: Advançes and Case Studies (Mehra and Lainiotis eds.), Acad. Press, New York,1976,pp.27-96

[4] Ng, T.S., Goodwin, G.C. and Anderson, B.D.O., "Identifiability of MIMO Dynamical Systems", 1977, vol. 13, pp. 477-485.

[5] Gevers, M.R., "On the identification of Feedback Systems", Preprints 4th IFAC Symp. on Identification, Tbilisi, USSR, 1976.

[6] Anderson, B.D.O. and Gevers, M.R., "Identifiability of Closed Loop Systems", Preprints 5th IFAC Symp. on Identification, Darmstadt, 1979, pp. 645- 652.

[7] Akaike, H., "Markovian Representation of Stochastic Processes", by Canonical Variables", SIAM J. Control, 13, 1975, pp.162-173.

[8] Faurre, P.L., "Stochastic Realization Algorithms" in System Identification: Advances and Case Studies" (Mehra and Lainiotis eds.), Acad. Press, New York, 1976, pp. 1-25.

[9] Canuto, E. and Menga, G., "Identifiability of Multivariable Stochastic Systems from Second Order Input-Output Statistics" Preprints 5th IFAC Symp. on Identification, Darmstadt, 1979, pp. 703-710.

[10] Tiano, A. et al., "Multivariable Model Identification of Ship Dynamics", Preprints 5th IFAC Symp. on Identification, Darmstadt, 1979, pp.185-196.

# IDENTIFICATION OF INDEPENDENT TIME-DEPENDENT COMPONENTS IN MULTIPLE TIME SERIES DATA

J.G. Crookes, K. Croston, P. Sypsas*

University of Lancaster, U.K.
*also University of Patras, Greece

## ABSTRACT

A technique for analysing multiple time series has been developed. Its efficiency on a number of type of series has been tested. The method involves seeking time dependent components analogous to principal components but using criteria based on time dependent functions of the resulting components. The functions are those used in the Box-Jenkins forecasting system and results in components ideally suited to forecasting using that system. The initial results are extremely encouraging and the paper consists of a description of the method together with a summary of the results for a wide range of series types. The system is offered as a diagnostic tool useful for those who need to identify the underlying sources of variation present in large volumes of data.

## INTRODUCTION

A commonly occurring problem is that of understanding a set of related variables developing together over time. Both variable to variable correlations and time to time correlations need to be understood. Problems of this nature arise both in Simulation, where the task is to design a sampling scheme for such series, and in Planning, where the requirement is for forecasts of the series.

It is generally desirable to have a quantitative understanding of the true relationship between the variables and between the values of one variable at different times, so that independent



*D. G. Lainiotis and N. S. Tzannes (eds.), Advances in Control, 521–528.*

sources of variation can be recognized and, for Simulation, represented by suitable sampling processes. When this understanding is not available a method of analysing the multiple-time series data with a view to discovering such sources of variation is desirable. This paper describes a method which we believe will be a useful addition to the analyst's armoury.

## THE METHOD

The proposed system is a combination of an extension of the method of Principal Components, called Process Components, together with single series Time Series Analysis. We assume we wish to sample ( or forecast) a number of series

$$\{X_i(t)\} = X_i, \; i=1,2,\ldots,n$$

which are correlated. To use Principal Component Analysis we would ignore the time dependence of the series and obtain a new set of series

$$\{Y_i(t)\} = Y_i, \; i=1,2,\ldots,n$$

each uncorrelated with the others and with criterion of maximum variance, being used to sequentially choose the series.

We believe that, in a situation where understanding of time series is the objective, it is misguided to ignore the time dependence and in adopting maximum variance as a criterion for choosing components, we can not avoid the noise being included in the early components, making the estimation task more difficult.

Using Process Components we propose retaining the examination of linear combinations of the $X_i$'s as the basic method for forming the $Y_i$'s but replace the maximum variance criterion with a criterion based on functions describing the time dependent behaviour of the resulting $Y_i$'s. As in Principal Components we confine ourselves to axis rotations without scale changes.

The criteria we have used are based on the diagnostic methods of the Box-Jenkins system and use the Autocovariance and the Partial Autocovariance function of the $Y_i$'s (output series).

In contrast with the Principal Component Method the criterion may be changed as we extract each component.

The Box-Jenkins method for a non-seasonal, stationary series, regards the series as arising from either a moving average of order p, MA(p), process, an autoregressive of order q, AR(q), or a combination of the two, ARMA(q,p). The autocovariance and partial autocovariance functions take characteristic shapes depending on the particular process and there exists a duality in the behaviour of the two functions for MA(p) and AR(p) processes i.e. autocovariance function of a MA(p) process looks like the partial autocovariance function of an AR(p) process and vice-versa.

Our method inspects these functions for linear combinations of the $X_i$'s and chooses the linear combination which has the function's shape closest to chosen "target" shapes. This search uses a numerical optimization method.

When a satisfactory component has been identified, the chosen component becomes the new Y and a new orthogonal axis system is found including Y as one of the axis. The data expressed in terms of these axes, but excluding the new Y (i.e. one less series to analyse) are now subjected to the same routine, though, possibly, with a different criterion being adopted. See figure 1 for a flow diagram of the method.

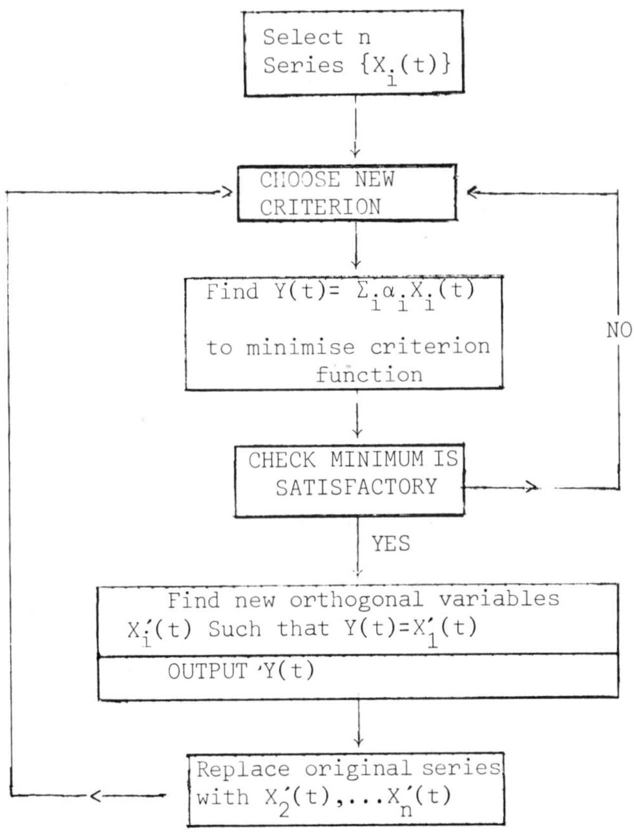


Fig.1 BLOCK DIAGRAM OF EXTENDED METHOD

TESTING PROCESS FOR SEARCHING DIRECTIONS

We have produced an interactive FORTRAN system for carrying out the component seeking method described above and have tested the concept using simulated series of known structure.

This system includes the following steps:

1. Simulate $\underset{\sim}{X}'(t)=(X_1'(t), X_2'(t),\ldots,X_n'(t))$, series of known structure, independent and with unit variance.

2. Scramble them with an axis rotation produce

$\underset{\sim}{X}(t)=A.\underset{\sim}{X}'(t)$ where $A=\{\alpha_{ij}\}$ $i=1,\ldots,n$, $j=1,\ldots,n$

subject to $\sum_{j=1}^{n} \alpha_{ij}^2=1$, $i=1,\ldots,n$.

3. Search for direction ($\alpha$'s as cosine directions) which produce the first "target" series $Y_1(t)$.

4. Project to n-1 space producing n-1 $Z_i$'s orthogonal to $Y_1(t)$.

5. Choose new criterion to search for the second "target" series.

6. Finish the job, by having extracted n orthogonal series and having found the structure that causes the cross correlation.

7. Compare the extracted "OUTPUT" series to "INPUT" series $\underset{\sim}{X}'(t)$.

SEARCHING PROCESS

We recommend a search strategy of first seeking constants. When this search fails, then zero order processes are sought (i.e. noise) next first order processes and so on.

In this way by using the autocovariance function to identify MA(p) processes and the partial autocovariance function to identify AR(q) processes we avoid actually specifying the magnitude of the effects we seek but rather specify that higher order terms shall be close to zero which makes for a more economical search.

So, the searching steps are as follow:

1. Look for constants or white noise with very low variance, using Principal Component analysis to eliminate the components that explain almost nothing of the total variance of the system.

2. Look for white noise by minimizing the sums of the squares of the autocorrelations from lag 1 onwards. i.e. find

$$\underset{\sim}{\alpha}^T=(\alpha_1,\ldots,\alpha_n) \text{ s.t. } Z_1(t)=\underset{\sim}{\alpha}^T\underset{\sim}{Y}(t), \text{ is a noise series}$$

by minimizing function F, where

$$F = \sum_{k\geq 1}^{\text{upperlimit}} \left(\frac{\underset{\sim}{\alpha}^T S_k \underset{\sim}{\alpha}}{\underset{\sim}{\alpha}^T S_0 \underset{\sim}{\alpha}}\right)^2$$

$S_k$ being the cross-covariance matrix of $Y_i$'s, i=1,...,n, at lag k.

3. Look for MA(1) by minimizing F for $k\geq 2$.

Generally look for MA(p), using the same function F but summing over $k\geq p+1$.

4. Look for AR(1) by minimizing function $\Phi$, where

$$\Phi = \sum_{k\geq 2}^{\text{upperlimit}} \phi_{kk}^2$$

where $\hat{\phi}_{kk}$'s are the partial autocorrelations of $Y_i$'s, i=1,...,n at lag k, estimated using Durbin's formula.

For the minimization process we have used numerical methods not based on derivatives in particular Powell's routine.

5. If seasonality of period S is expected, then the terms

k=S,2S,3S... etc., would be omitted from the summation.

6. If non-stationary series were to be analysed, Box-Jenkins differencing procedure seems to be sensible as a first step.

## CONCLUSIONS

A system for the analysis and hence either sampling or forecasting, of multiple time series has been described. Given the basic plausibility of the two processes, evidenced by the general success of the Box-Jenkins system and given the success of the linear combination approach in general multivariate applications it is reasonable to expect that our recommended procedure will prove of value in a wide range of applications.

Tests of the system have shown it capable of detecting those structures it is designed to seek.

In Appendix 2 we give tables of the correlation coefficients between INPUT and OUTPUT series. In general these correlations are good indicating the system's above mentioned capability.

A simpler version of the system has proved useful in practical situations.

We conclude that the system merits consideration as a useful approach when faced with multiple Time Series problems.

## A1. PROJECTING THE n-SPACE ONTO AN (n-1)-SPACE TO REMOVE A CHOSEN COMPONENT

Given a Time Series $\{X_i(t)\}=X_i$, $i=1,2,\ldots,n$, with unit variance, we choose the linear combination

$$Y_1(t) = \sum_{i=1}^{n} \alpha_i X_i(t), \text{ subject to } \sum_{i=1}^{n} \alpha_i^2=1,$$

as the "BEST" component for our purposes and wish to obtain (n-1) orthogonal variables, orthogonal also to $Y_1(t)$.

A modified version of Gram-Schmidt's orthogonalization theorem gives:

$$Y_2(t)=k_2(\alpha_1 X_1(t) - \frac{\alpha_1^2}{\alpha_2} X_2(t))$$

$$Y_3(t)=k_3(\alpha_1 X_1(t)+\alpha_2 X_2(t) - \frac{\alpha_1^2+\alpha_2^2}{\alpha_3} X_3(t))$$

$$Y_n(t)=k_n(\alpha_1 X_1(t)+\ldots+\alpha_{n-1} X_{n-1}(t) - \frac{1-\alpha_n^2}{\alpha_n} X_n(t))$$

where

$$k_r^2 \left\{\alpha_1^2+\alpha_2^2+\ldots+\alpha_{r-1}^2 + \left(\frac{\sum_{i=1}^{r-1} \alpha_i^2}{\alpha_r}\right)^2\right\} = 1.$$

## A2. TABLES OF INPUT-OUTPUT CORRELATIONS

C : Constant
N : Noise
MA p : Moving Average process of order p.
AR q : Autoregressive process of order q.

In cases where constants are present a preliminary Principal Component Analysis has been carried out to eliminate them and reduce the dimension.

CORRELATION COEFFICIENTS

Series Types

| | | | | | |
|---|---|---|---|---|---|
| N | .99 | .89 | .99 | .88 | .83 |
| MA1 | .78 | .79 | .94 | .68 | .60 |
| AR1 | .71 | .74 | .77 | .72 | .66 |

| | | | | | | | | |
|---|---|---|---|---|---|---|---|---|
| C | 1.0 | .95 | .99 | .99 | .99 | .95 | .97 | .93 |
| MA1 | .94 | .98 | .93 | .86 | .99 | .70 | .93 | .98 |
| AR1 | .83 | .80 | .80 | .83 | .78 | .94 | .83 | .80 |

| | | |
|---|---|---|
| N | .89 | .95 |
| MA2 | .80 | .77 |
| AR2 | .96 | .75 |

| | |
|---|---|
| N | .85 |
| MA1 | .6 |
| MA2 | .65 |

| | |
|---|---|
| N | .93 |
| AR1 | .65 |
| AR2 | .62 |

| | | | |
|---|---|---|---|
| C | .97 | .97 | .97 |
| MA1 | .80 | .81 | .81 |
| MA1 | .77 | .76 | .78 |
| AR1 | .88 | .87 | .88 |

REFERENCES

1. G.E.P. Box and G.M. Jenkins (1970). "Time Series Analysis, Forecasting and Control". Holden Day.

2. D.F. Morrison (1967). "Multivariate Statistical Methods". McGraw-Hill.

3. J. Durbin (1960). "The Fitting of Time Series models". Rev. Int. Inst. Stat., 28, 233.

4. O.D. Anderson (1967). "Time Series Analysis and Forecasting". Butterworths.

# AUTOPACK - A NUMERICAL SOFTWARE FOR AUTOMATIC CONTROL

A.Y. Barraud

Laboratoire d'Automatique E.N.S.I.E.G.
BP 46 Saint Martin d'Hères 38402 FRANCE

Abstract - Considerable efforts have been made, for almost ten years, in the development of high quality software. Needs and availability of general purpose libraries increasingly answering this criterion are now well understood. In a more specialized area such as automatic control, it seems that it has not yet been attempted to produce widely available good routines except for some particular subjects such as Kalman filtering. The aim of this paper is to present the first efforts made to fill in this gap, through a self contained package, called AUTOPACK, whose development began almost three years ago at Nantes'University and is now continuing at Grenoble's University.

## 1. INTRODUCTION

Needs and development of high quality mathematical software are progressively more and more well understood. The increasing number of specialized books, journals and papers shows clearly the today trends in this particular area where data processing numerical analysis and various other mathematical sciences are merged. In the last book edited by Rice in 1977 (1) the paper "Mathematical Software Production" by Crowell and Fosdick is quite representative of the last ten years trends that we have tried to follow in the development of AUTOPACK. Our paper is organized as follows : In a first part, the overall design of the library through an appropriate modular structure is considered. A second paragraph is devoted to the efforts made around the basic modules used throughout the package. Lastly, some insights are given about the developments and implementations of algorithms to solve some specifically difficult problems such as matrix exponential , Lyapunov and Riccati equations.



*D. G. Lainiotis and N. S. Tzannes (eds.), Advances in Control, 529–534.*

## 2. AUTOPACK STRUCTURE

The AUTOPACK project has been built around some basic constrains such as: easiness of use, robustness, efficiency and portability. Briefly, the library has been written and documented in such a way that a user could easily find the routine which performs the required function and how to use it. Robustness is potentially ensured by an appropriate choice of the best available algorithms, and, when necessary, an error handling facility has been embodied, as in PORT (2). Efficiency is considered here as a compromise between a high execution speed, a core memory as small as possible and an algorithm conception easiness. Finally, portability is only limited by the FORTRAN IV one, according to the 1966 standards (3) and the various existing compiler versions; in fact, the FORTRAN used in AUTOPACK is a subset of the FORTRAN 77 standards (4).

With these aims in mind, AUTOPACK has been developed according to a deep modular structure. Although this approach is more difficult to realise than a monolithic coding and is slightly more time consuming, it has been chosen because of its global efficiency. More precisely, each module can be coded and tested independently; furthermore already certified routines can be quickly used to test new modules of higher levels, and modifications inside a module are without any consequences for the others. Finally, new routines can be easily introduced in the library or updated.Following these arguments a hierarchical structure with three levels has been adopted :

Level 1 : automatic control routines
Level 2 : automatic control sub-modules
Level 3 : general purpose modules.

### 2.1. Names coding

In order to manage more efficiently this set of modules, the name of each routine has been coded according to its hierarchical level (1, 2 or 3) and its purpose in this level.

The level one modules begin with a letter specifying the system type, upon which they operate : C stands for continuous systems, D discrete ones, M multivariable transformations, S monovariable transformations. The second letter is an S for state space representation and T for transfer function models. The last character is a letter which is an I if all the input parameters are integers, S if all the arguments which are not of the integer type are single precision real parameters, and D for a double precision version. The other characters are alphanumeric ones and, generally, more or less mnemonic.

The level two modules begin with the first same letter as the level 1 module(s) which call(s) them, and the second letter is always U . The other characters follow the same conventions as explained before.

The level three modules begin with a G and the last character is I, S or D according to the previously given definitions. The other characters are usually mnemonic. However, the second character is M for matrix operations, V for vector operations, P for polynomial calculations, S for systems resolutions and O for optimization.

## 2.2. Global description

The general purpose modules solves classical numerical problems. A first set deals with elementary computations on vectors and matrices, some of them being more specific. For example :

| | |
|---|---|
| GMTBAS | computes $C = A^T B A$ , with B sym. $\geqslant 0$ |
| GMPD2S | computes $B = A B$ |
| GMPD3S | computes $B = B A$ |
| GMTR1S | performs $A = A^T$ in situ, A square or not |

A second set deals with linear algebra problems and a last one is attached to the non linear numerical problems. Most of these routines are adapted from Harwell library (5) or Nats projects (6)

Another class of routines is the automatic control modules which constitute the core of AUTOPACK. The main sub-classes are: signals simulation, linear dynamic systems time responses (continuous and discrete), frequency responses, linear filtering (some of the excellent JPL routines are included (7) ), identification with various off line and on line techniques, and optimal control via the well known LQG formulation (Linear Quadratic Gaussian) including Lyapunov and Riccati problems.

## 3.BASIC MODULES AND ACCURACY

Some of the lower level routines, such as inner products, play a fundamental role versus speed, accuracy and robustness for most of the higher level routines. A detailed study of this specific problem has been done in (8) in the same spirit as the Basic Linear Algebra Subroutines (9) used in the LINPACK Nats project. Two techniques have been coded: The classical double length accumulation which is the method run by default, and the Malcolm's algorithm which guarantees an accurate result within the machine precision. This second possibility is offered to the user via a command "CALL GFACUR" which corresponds to the accurate version of AUTOPACK; with a CALL GFAST the user can come back to the first method which corresponds to the fast version of AUTOPACK. In fact this change can be done anywhere, each time speed or accuracy becomes of primary importance.

## 4. SOME DIFFICULT NUMERICAL PROBLEMS

Now we will consider some specific and difficult numerical problems in automatic control, and how they have been treated.

### 4.1. The matrix exponential

This kind of computation appears during the simulation of continuous systems, but also in optimal control equations both directly or indirctly as integrands.For example we have the following result (10) :

$$Q(\alpha) = \int_0^{\alpha} e^{A^T t} \, V \, e^{At} \, dt$$

and

$$\exp\left\{ \alpha \begin{bmatrix} -A^T & V \\ 0 & A \end{bmatrix} \right\} \triangleq \begin{bmatrix} X(\alpha) & Y(\alpha) \\ 0 & Z(\alpha) \end{bmatrix}$$

$$Q(\alpha) = Z^T(\alpha).Y(\alpha)$$

The computation of $e^A$ is basically done in several steps: reduction of $\|A\|$ (balancing, scaling), computation of a Padé approximant of the reduced matrix, and back transformations of the first step. For stiff systems, A is firstly decomposed in block diagonal form each block having eigenvalues of the same order (11). The coding of the whole method is relatively important but the routine is quite efficient.

### 4.2. Lyapunov equations

These equations are solved through a technique which transforms the initial matrix A of:

$$\begin{cases} A^T X + X A - Q = 0 & \text{continuous case} \\ A^T X A - X - Q = 0 & \text{discrete case} \end{cases}$$

into the real Schur form. In these conditions X can be computed through a generalized back substitution process similar to the well known gaussian elimination associated with the classical linear systems (12),(13). Clearly iterative refinement and/or balancing can be efficiently obtained. The algorithm can operate in situ, in some $n^3$ multiplications, whatever the stability of A may be.

### 4.3. Riccati equations

These equations are the to many problems in filtering and optimal control. For discrete system the equations are:

$$A^TPA - P - A^TPB(B^TPB + R)^{-1}B^TPA + Q = 0$$

With the classical assumptions (A,B) stabilizable, (C,A) detectable ($C^TC = Q$), $R > 0$ and $Q \geq 0$, we have to solve the equations for the unique symmetric non negative definite matrix P. Many methods exist, essentially two have been included in AUTOPACK, a fast square root algorithm via Chandrasekhar formulation and an algorithm operating upon the bilinear transformation of the symplectic matrix associated with the Riccati equation via matrix sign function (14).

For continuous systems, we have to solve the following equation, assumptions and conclusion about P beeing the same:

$$PA + A^TP - PBR^{-1}B^TP + Q = 0$$

The solution is obtained here by means of the computation of the hamiltonian matrix sign function (15) or through an equivalent discrete system determined by bilinear transformation.

## 5. CONCLUSION

The AUTOPACK project started in 1977 (16) has reached today a certain degree of maturity. Although its development is far from being finished, AUTOPACK has come into operation with approximately 200 FORTRAN modules. Most of the level 3 routines have so far been written. They are also in AUTOPACK the necessary basic routines (levels 1-2) to simulate discrete and continuous systems under various representations. Several routines have been included in order to perform identification of ARMAX models using the maximum likelihood method. In a very near future there will be two new sets of routines implementing multivariable identification from arbitrary input/output data through a canonical state space model using, both, on line approach (square root filtering ) and off line approach (non linear programming). The main numerical tools to perform optimal control are already in the package. Our efforts are now directed towards the implementation of easy-to-use routines answering the practical problems of optimal control such as performance of the closed loop systems under various conditions.

REFERENCES

(1) J.R.Rice *Mathematical Software 3* Academic Press, 1977.
(2) P.A.Fox *The PORT mathematical subroutine library* 2° ed. 1977, Bell laboratories, Murray Hill,U.S.A.
(3) USA Standard Inst. *ANSI FORTRAN X 3.9 - 1966*
(4) USA Standard Inst. *ANSI FORTRAN X 3.9 - 1978*
(5) M.J.Hopper *Harwell subroutine library - a catalogue of subroutines* AERE- R 9185, UKAEA, Harwell
(6) B.T.Smith *The Nats Projects, a National Activity to Test Software* SHARE - SSD, oct. 1972
(7) G.J.Bierman,M.W.Nead *A Parameter Estimation Subroutine Package* JPL Report 77-26, revision 2, oct. 1978, California Institute of Technology,Pasadena
(8) J.Hanen,A.Y.Barraud *Accurate Basic Modules in Linear Algebra* Texas Conf. Math. Soft., March 1978, Austin, USA
(9) C.L.Lawson et al. *Basic Linear Algebra Subprograms for FORTRAN Usage* T.O.M.S. Vol 5-3, sept. 1979, pp 308-323
(10) C.F.Van Loan *Computing integrals involving the matrix exponential* I.E.E.E. Tr.Aut.Cont. Vol AC-23 June 1978
(11) C.A.Bavely,G.W.Stewart *An algorithm for computing reducing subspaces by block diagonilazation* SIAM J. Num.Anal. Vol 16-2, April 1979, pp359-367
(12) R.H.Bartel,G.W.Stewart *Solution of the matrix eqution AX+XB =C* C.A.C.M. Vol 15,Sept 1972, pp821-826
(13) A.Y.Barraud *A numerical algorithm to solve $A^TXA-X=Q$* I.E.E.E. Tr.Aut.Cont. Vol AC-22, Oct.1977,pp883-885
(14) A.Y.Barraud *Produit etoile et fonction signe de matrice - application à l'éqhation de Riccati dans le cas discret* RAIRO Automatique n°1, 1980, à paraitre
(15) A.Y.Barraud *Investigations autour de la fonction signe d'une matrice-application à l'équation de riccati* RAIRO Automatique vol 13 n°4, 1979
(16) A.Y.Barraud,J.Hanen *Design and development of an advanced numerical package in automatic control* Internat. Conf. System Science IV Wroclaw, Pologne,Sept.1977

# PRACTICAL EXPERIMENTAL TECHNIQUES TO IDENTIFY THE NATURAL FREQUENCY OF SECOND-ORDER SYSTEMS

Jonnathan Yung-chang Chen

Department of Electrical & Power Engineering
Tsing Hua University, Hsin-chu, Taiwan, R.O.C.

## ABSTRACT

Two practical techniques using the frequency response method or the correlation function method to identify the natural frequency of second-order systems are presented. The results of applying these techniques to a magnetic suspension system are also shown.

## 1. INTRODUCTION

The transfer function of a system or substystem in the engineering world can often be approximated in the second-order form. The real-time identification of its instantaneous natural frequency is frequently desired.

The known experimental approaches to identifying a system are normal input method; step, impulse and frequency response methods; correlation function techniques,.....etc. For our purposes, two simple techniques using either the frequency response method or the correlation function method will be introduced.

## 2. FREQUENCY RESPONSE IDENTIFICATION

The measurement of $G(j\omega)$ for obtaining a frequency-response can be performed by applying sine-wave inputs $A\sin\omega t$ for various $\omega$ and recording the respective outputs $B\sin(\omega t+\phi)$. The values of $B/A$ and $\phi$ are thus recorded for each frequency $\omega$ that is considered, to yield the required frequency response. This procedure is tedious and time-consuming.



*D. G. Lainiotis and N. S. Tzannes (eds.), Advances in Control, 535–540.*

For a second-order system, the aforementioned method can be greatly simplified and automated.
The transfer function of a second-order system is given by:

$$G(s)= \frac{\omega_n}{s^2 + 2\xi\omega_n s + \omega_n^2}$$

where $\xi$ is damping ratio and $\omega_n$ is natural angular frequency.
At the resonant frequency, the gain function becomes:

$$G(j\omega_n)= -j \frac{1}{2\xi\omega_n}$$

The output-phase lags the input-phase by $90^o$.
The frequency at which $90^o$ phase-lag occurs can be detected automatically. As shown in Fig.1, the system to be identified is excited by a voltage-controlled oscillator(VCO). The excitation signal is multiplied with the output signal. The product is then averaged and fed through an integrator to adjust the frequency of VCO. The frequency of VCO will be varied until the average value of the product vanishes. The steering voltage of the VCO is then a measure for the natural frequency of the second-order system.

## 3. THE NATURAL FREQUENCY OF A MAGNETIC SUSPENSION BALANCE

The magnetic suspension balance (Gast-Sartorius)[1], which is now in use in a number of laboratories, measures weight changes of a sample in a closed container. Inside the container, vacuum or controlled atmospheres are maintained. The balance proper is totally separated from the interior of the container and therefore protected from corrosion or contamination by gases or vapours surrounding the sample. Likewise, the interior of the container is proof against outgasing from the balance. The necessary coupling between the balance pan and the balance consists of a controlled magnetic suspension (Fig.2).
Mechanically the controlled magnetic suspension is a mass-spring-damper system. Its natural angular frequency is given by:

$$\omega_n = \sqrt{\frac{k}{m_s}}$$

where k is the spring constant and $m_s$ is the mass of suspended magnet. The arrangement shown in Fig.1 has been applied to identify the natural frequency of the magnetic suspension system.
The results are encouraging. Nevertheless the suspension status

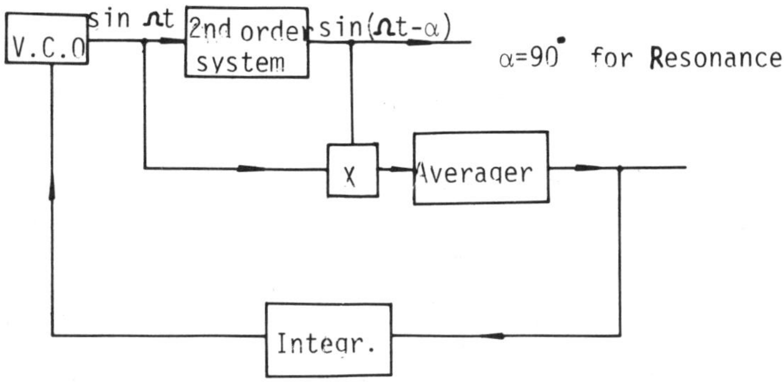


Fig.1 : Identification with sinusoidal inputs

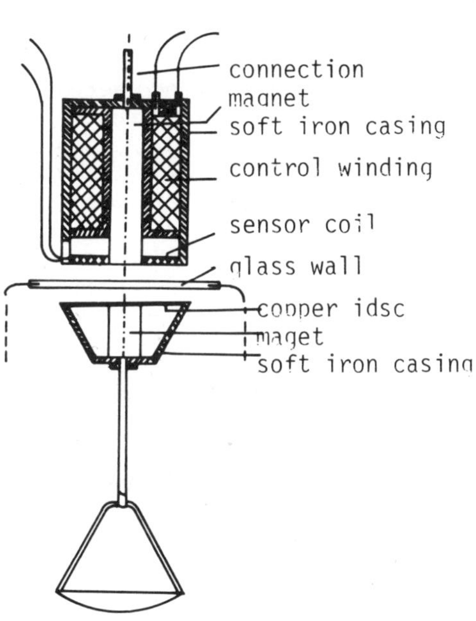


Fig.2 : Magnetic Suspension

was disturbed somewhat by the added sinusoidal excitation during the identification process.

## 4. CORRELATION FUNCTION METHOD

The correlation function approach is based on applying a white noise input to the system to be identified. The energy of a white noise is distributed over a wide frequency band. If this noise is of a sufficiently low amplitude, it may be superimposed on the normal-operation input to the system without affecting performance, and can therefore be applied on-line. Furthermore, the normal-operation input usually does not affect the identification procedure.

It is well known that the cross-correlation function $\phi_{xy}(\tau)$ of the system with the white noise input is proportional to the impulse response of the tested system. The time required to set up the cross-correlation function can be reduced by ultilizing pseudo-noises. The most convenient class of pseudo-noises to consider, for our purposes, is the class of maximum-length pseudo-random binary sequences (m-sequences for short).

### 4.1 Recognition of the first zero-crossing of the impulse response

The eigenfrequency of an underdamped second-order system can be determined by recognizing its first zero-crossing $\tau_1$. A suitable arrangement is shown in Fig.3. The delay time $\tau$ for the correlation function will be automatically adjusted through an I-controller until the cross-correlation function $\phi_{xy}(\tau)$ becomes zero at $\tau=\tau_1$, which is the first zero-crossing time. At this moment the control voltage of the VCO is a measure for the unknown eigenfrequency.

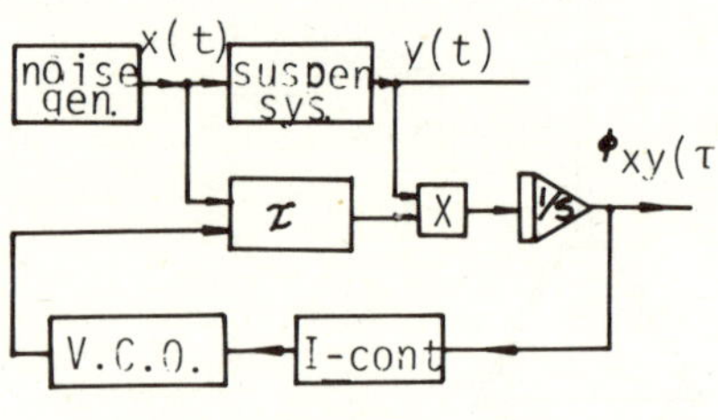


Fig. 3 : Arrangement for recognizing $\tau_1$

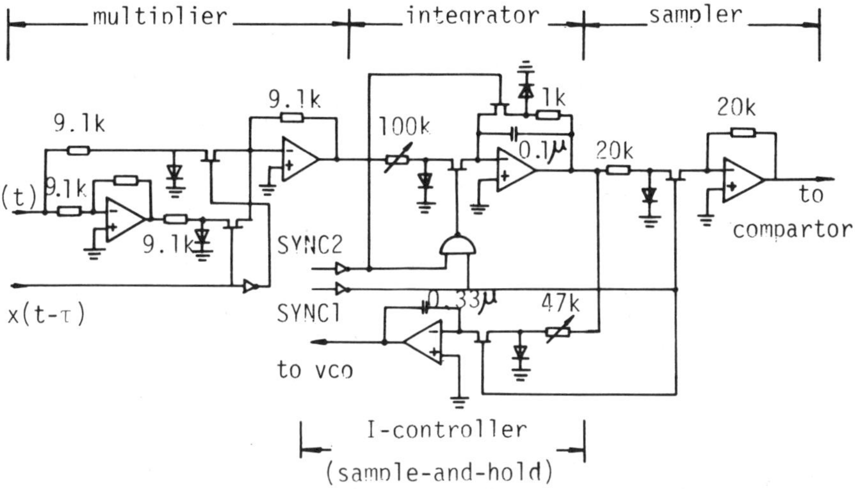


Fig. 4 : Main parts of correlator

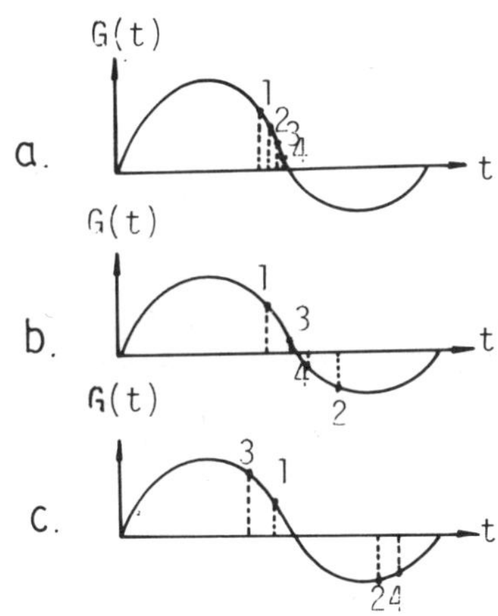


Fig.5 : Possible idnetification-processes

### 4.2 Application to the magnetic suspension balance

An 8-stage shiftregister is ultilized to generate an m-sequence of the length of 255 bits. Stages 1,2,7 and 8 are fed back through modulo-two-addition. Automatic start and synchronization signals are also provided.
A 40-stage shiftregister is used to delay the binary sequences. The clock signal of this shiftregister comes from a VCO, which is controlled by the output voltage of the I-controller in loop. The multiplication is very simple. The output signal of the tested system is allowed to pass according to the logic state of the delayed binary sequence (Fig.4). The time constant of the following integrator must be adjusted so that its output signal is, without oversteering, as large as possible. This signal, which represents the value of the cross-correlation function, is sent to the I-controller.
According to the open-loop gain of the identification loop, the identification can proceed in three different ways (Fig.5). In case (a) the identification converges "aperiodically". In case (b) it converges "periodically". Case (c) is to be avoided, since it is unstable.
The clock rate of the m-sequence generator was set at 800Hz. With the proper choice of the time constant of the I-controller, the identification process was completed in 2 seconds. The suspension status was scarcely disturbed during this identification process.

## 5. CONCLUSION

To identify the natural frequency of second-order systems, two practical techniques, using the frequency response method and the correlation function method respectively, are proposed.
The frequency at which $90^{o}$ phase-lag occurs can be detected automatically by using a feedback loop with phase-sensitive rectifier, integrator and voltage-controlled oscillator.
Maximum-length pseudo-random binary sequences are used as input signals for the correlation function method. The natural frequency can be indirectly determined by identifying the first zero-crossing of the I/O cross-correlation function.
Both techniques have been successfully implemented on a magnetic suspension system.

## REFERENCES

[1] Gast, Th. : Waagen mit freischwebender magnetischer Aufhangung. Naturwissenschaften 56(1969), pp.434-438

[2] Davies, W.D.T. : System Identification for Self-Adaptive Control. Wiley-Interscience 1970

# DIGITAL COMPUTER SIMULATION OF A POWER THREE-PHASE INVERTER AND ESTIMATION OF ITS HARMONIC CONTENT

Kl. Giavassoglou*, G. Athanassiadis*, A. Safacas**

*Research Assistant, **Prof.Dr.-Ing., Electromechanical Energy Conversion Laboratory, University of Patras,Greece

ABSTRACT

A three phase power inverter with an adjustable frequency and rms value output voltage is used supplied from a costant d.c. source. A general digital simulation method is used covering every operation mode of variable topology circuits. The branch voltages and currents are examined by the internal logic of the program at the beginning of each computation step and the state of each semiconductor element is recognized. The appropriate differential equations system is constructed and its solution gives the branch voltages and currents which are analysed according to Fourier analysis. The high harmonics content of the above waveforms is dependent on the nature of the load, the firing angle and the triggering method.

## 1. INTRODUCTION

As it is know the induction motor, because of its nature, lacks of an economical and technologically effective regulation of its speed and consequently of its mechanical power. However we can obtain considerable effects using the induction motor connected to a self - commutated, variable rms - value and frequency output voltage converter.

This electromechanical system shows almost perfect ability in adaptation to various load demands, which might require, a wide range speed and power adjustement.



*D. G. Lainiotis and N. S. Tzannes (eds.), Advances in Control, 541–549.*

Many converters of this kind have been already used for industrial purposes and generally everywhere we need variable rms - value and frequency voltages. Usually this is achieved by conversion of a d.c. voltage to a an a.c. one by a power electronic inverter. The waveforms of the output voltage and current of such a kind of device are not of sinusoidal shape and consequently they include harmonics of high order which are undesirable in the power systems and we are thus obliged to device methods of analyzing and neutralizing them.

A three - phase inverter, the most suitable for practical use, is shown in figure 1.

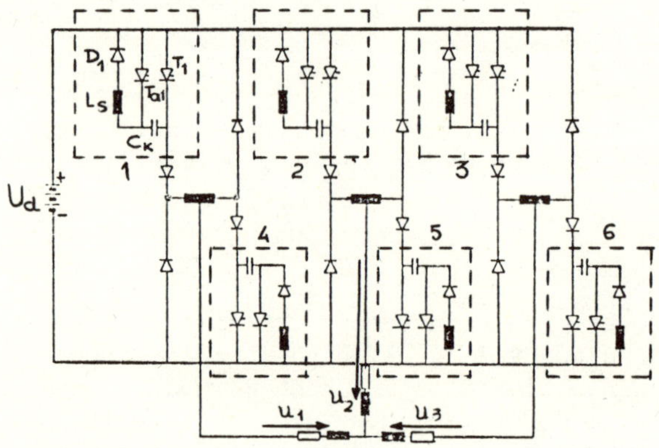


Figure 1: Three phase Inverter

## 2. SYSTEM DESCRIPTION

The above converter, consisting of six d.c. -choppers, inverts the d.c. input voltage E to a three - phase with adjustable frequency and rms - value output voltage.

Two thyristors, are included in each chopper, a main one $T_i(i=1,2,\ldots,6)$ which connects the input voltage to the load and an auxiliary one $T_{ai}(i=1\ldots6)$ with the aid of which we turn off the main thyristor.

This inverter operates only if two choppers at least conduct, one from the above three and one from the bellow three. Additionally each d.c. chopper can be conducting up to the half of the operating period T, where T is the desirable sinusoidal output period and depends only on the triggering frequency $f_1$.

The conducting time t of the main thyristor of a chopper, depends on the time interval between the triggering pulses to the main thyristor and to the auxiliary one. Each main thyristor $T_i$ can be turned off only by triggering its auxiliary one $T_{ai}$.

In relation with the conducting time t of each main thyristor $T_i$ we have the following conducting situations.

| | |
|---|---|
| $T_1/3 < t < T_1/2$ | Two or three choppers conduct |
| $T_1/6 < t < T_1/3$ | One or two choppers conduct |
| $0 < t < T_1/6$ | The inverter doesn't operate |

The variance of the conducting time t affects the rms - value of the output voltage.

In figure 2 the theoretical waveformes of a phase and line voltage are shown with the conducting time as parameter.

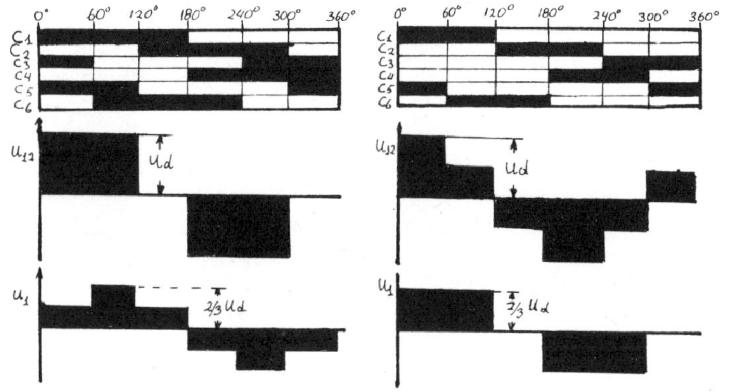


Figure 2: Theoretical waveforms of output voltages.

## 3. SYSTEM ANALYSIS

We connect for convinience a complex load $Z_L$ to the inverter, instead of an a.c. motor, in order to analyze the competition of the above mentioned three phase inverter (figure 1). The graph of the circuit of figure 1 is shown in figure 3. This graph has b=46 branches and $n_t$=24 nodes. In the same picture an arbitrary tree is shown with dark lines with $n=n_t-1=23$ tree branches and $l=b-n=23$ links.

For this prime graph we can write by inspection the following equation:

$$(E_S) = (Z)(I) \tag{1}$$

where: $(E_S)$ is the loop voltage source vector of order l=23
$(Z)$ is the loop impedance matrix of order lxl=23x23
$(I)$ is the loop independent current vector of order l=23

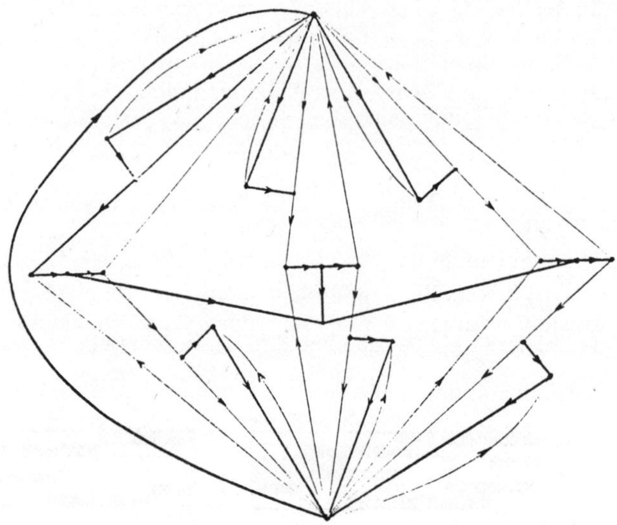

Figure 3: Graph of the three phase Inverter

The solution of the equation (1), gives the independent current (I) in case that the inverter circuit would be a time - invariant circuit. But in our case this circuit is time - varying. So the topology of the system varies and the above prime graph isn't constant. We express this variance mathemetically, using a variable order matrix $(C_n)$ called "incidence" matrix of such a kind that

$$(I) = (C_n)(I_n) \tag{2}$$

where

$(I_n)$ is the loop independent current vector at each moment in accordance with the reduced graph.

Additioanally we introduce a matrix (X) of order 6x5, called "state matrix" of the semiconducting elements. Each element $x_{i,j}$ of this matrix takes the values "1" or "0" in accordance with "on" or "of" condition of the corresponding semiconductor.

The matrix $(C_n)$ is constructed in relation with the values "1" or "0"of the elements $x_{i,j}$ of matrix (X) and a sequence of logical proceedings.

According to this briefly stated analysis of the circuit and the equations 1, 2, we will develop the used algorithm in solving the variable topology circuit of figure 1.

$$(E_S) = (Z)(C_n)(I_n) \quad (3)$$

$$(E_n) = (R_n)(I_n) + (I_n)p(I_n) + (S_n)p^{-1}(I_n) \quad (4)$$

$$p(I_n) = (L_n)^{-1}(E_n)-(L_n)^{-1}(R_n)(I_n)-(L_n)^{-1}(S_n)p^{-1}(I_n) \quad (5)$$

where

$$p = \frac{d}{dt}, \quad S_{i,j} = \frac{1}{C_{i,j}}, \quad (E_n) = (C_n)^t(E_S)$$

$$(L_n)=(C_n)^t(L)(C_n), \quad (R_n)=(C_n)^t(R)(C_n) \text{ and } (S_n)=(C_n)^t(S)(S_n)$$

The solution of the integrodifferential equation (5) gives the values of the independent currents ($I_n$) of the instantaneous reduced graph. Then from equation (2) we find the independent currents (I) of the prime graph and then the branch currents (J) and branch voltages (U).

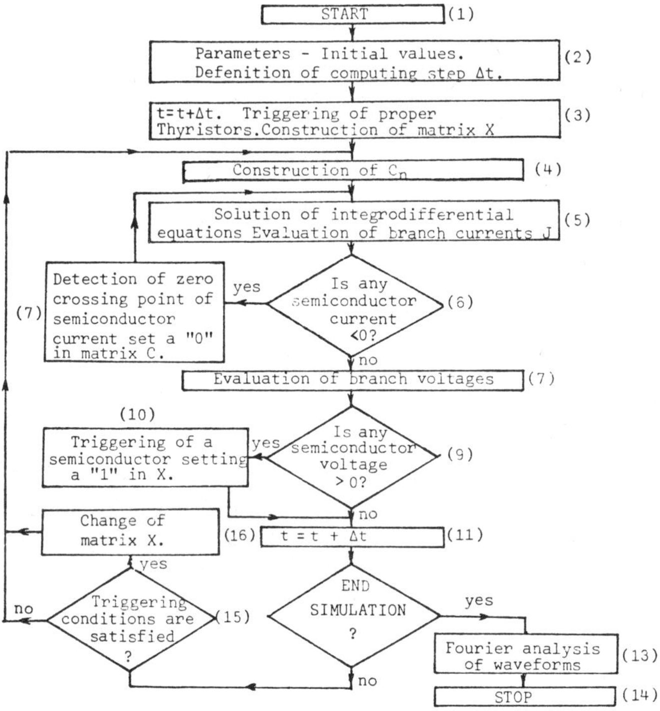


Figure 4: Flow chart

With the aid of the known values of current and voltage of each semiconductor and in according with the triggering method we construct the state matrix (X) and from that the "incidence" matrix

($C_n$) which corresponds to the new operation mode of the inverter.

With this process we have obtained to determine the response of the given circuit and consequently the output voltage and current waveformes are known. These waveforms are analyzed using fourier analysis.

All the computing process has been carried out using a digital computer. We used the Runge-Kutta improved by Gill method in solving the intergraldifferential equations and a method proposed by G. Goertzel in Fourier analysis.

The flow - chart of the used digital computer simulation program is given in figure 4.

## 4. ESTIMATION OF HARMONIC CONTENT OF OUTPUT VOLTAGE WAVEFORMES

We studied the phase voltage and current and the line voltage using the above mentioned digital computer simulation method. The waveformes of these functions are given in figures 5 and 6 for pure ohmic load and the conducting time of the main thyristor of a chopper as parameter.

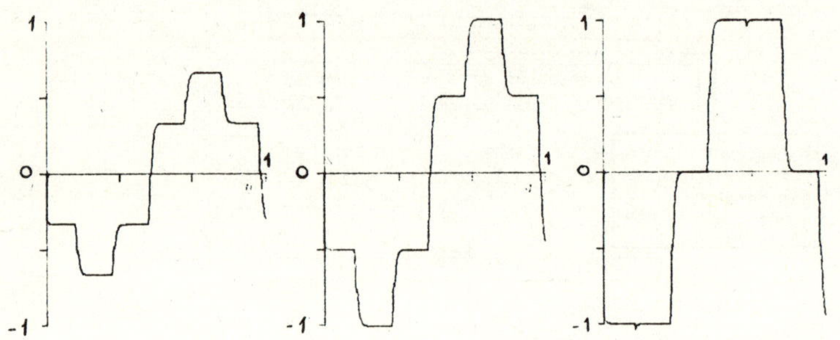


Figure 5: Line voltage, current and phase voltage for $t=173^{\circ}$

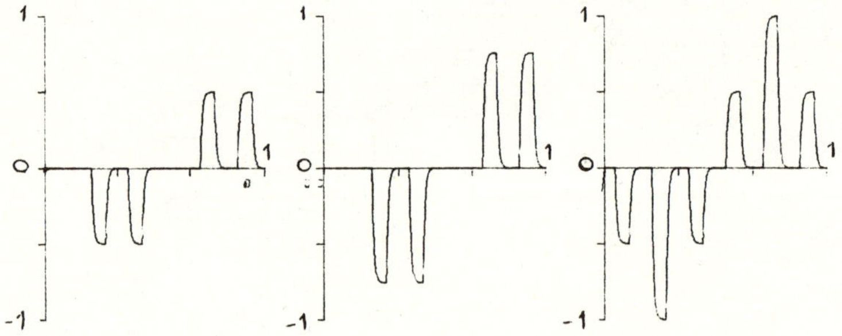


Figure 6: Line voltage, current and phase voltage for $t=83^{\circ}$

According to Fourier analysis we estimate the harmonic content and in figure 7 we show the variance of high order harmonics in accordance with the conducting time t. We can see that the high order harmonic content is strongly depended on the conducting time Additionally we must say that the chosen case of pure ohmic load is the worst one, in relation with the high harmonic content. This is explained from the fact that the waveformes of output voltages are strongly dependent on the conducting time t.

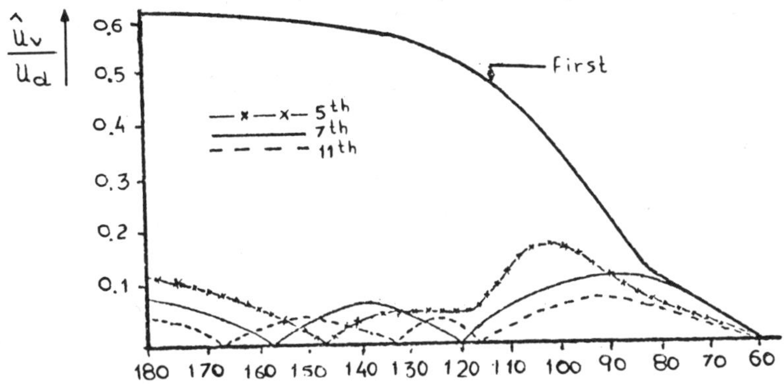


Figure 7: First and higher order harmonics of phase voltage in relation to conducting time t.

We can see that the even order and the multiples of order three are absent. Additionally the $5^{th}$ and $7^{th}$ harmonics are especially large.

At this point everyone can remark that we are obliged to search for other kind of triggering method in order to avoid the large variance of the percentage values of high order harmonics. All these methods are based on the fact that during the above mentioned conducting time t of a main thyristor $T_i$ we turn it periodically "on" and "of" many times.

The most suitable triggering method in minimizing the harmonics of low order, with very good results in various applications, is the "pulse width modulation" method. In figure 8 we can see the conducting period of each chopper and the output waveforms.

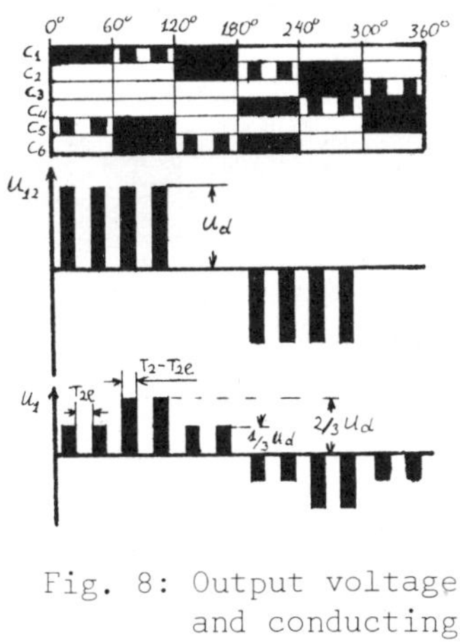


Fig. 8: Output voltage and conducting time.

During the half period $T_1/2$ six periodic pulses appears. The rms-value of the output waveform depends on the ratio $T_{2e}/T_2$.

The computer output waveforms for this triggering method are shown in figure 9, 10 and the dependence of high order harmonics from the conducting time t in figure 11.

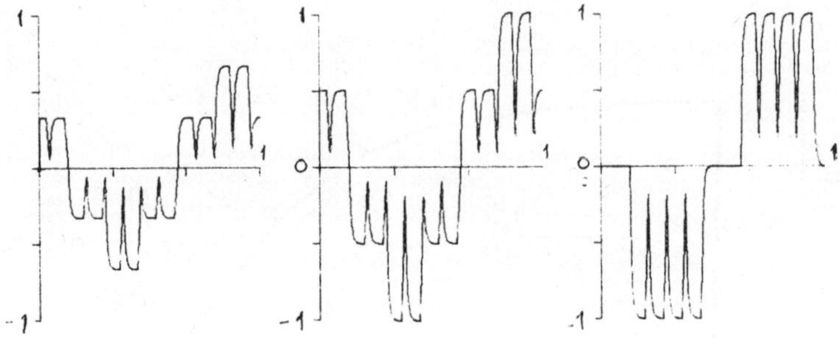

Figure 9: Line voltage, current and phase voltage for t=0.6

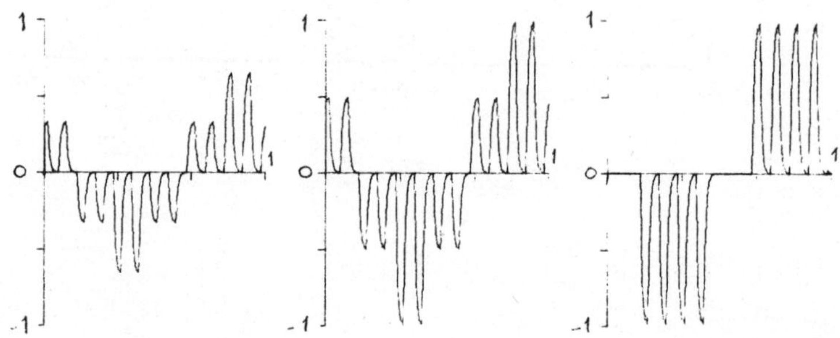

Figure 10: Line voltage, current and phase voltage for t=0.17

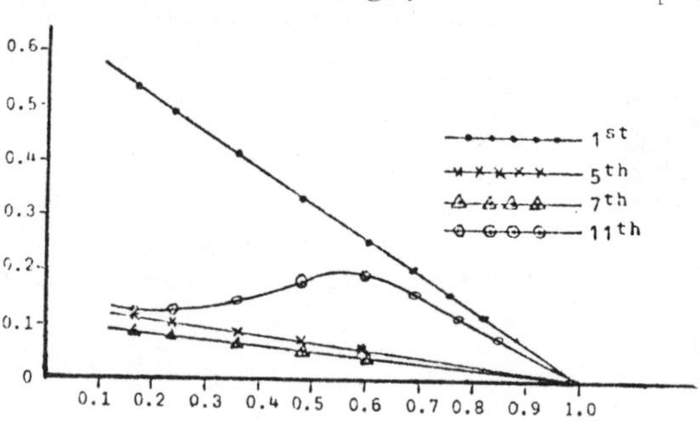


Figure 11: Variance of high order harmonics in dependence of conducting time t.

We can see that the percentage values of the $5^{th}$ and $7^{th}$ harmonics remain constant while the values at $11^{th}$ and $13^{th}$, $17^{th}$ and $19^{th}$ vary. However these high order harmonics don't affect the inductive - ohmic loads.

To minimize the percentage value at higher than $7^{th}$ order harmonics we must use a triggering method with frequency $f_2=18f_1$.

## 5. CONCLUSIONS

Now at the end of the above analysis we can say that if we need a prescribed rms - value output voltage from a three phase inverter we have two capabilities:

a: Continuous conduction of each chopper in a half period
b: Periodical interrupted conduction in a half period.

Additionally we can't generally say in which of these two cases the percentage values of the high order harmonics are greater.

The "pulse width modulation" method has the advantage that with a given value of frequency $f_2$, the values of some prescribed high order harmonics varies in prescribed limits independently of the rms - value of output voltage.

For example: with $f_2=12f_1$ the ratio $U_5/U_1$ and $U_7/U_1$ remain constant, where $U_1$ value of first, $U_5$ the value of fifth, $U_7$ the value of seventh harmonic.

With $f_2=18f_1$ the ratio $U_{v5}/U_{v1}$, $U_{v7}/U_{v1}$, $U_{v11}/U_{v1}$, $U_{v13}/U_{v1}$ remain constant.

Of course, it is easily understood that we can't make frequency $f_2$ too great. Large values of $f_2$ create problems to thyristors because of losses during transient phenomena.

When the frequency of output voltage is too large, we prefer the first method because frequency $f_2$ is limited by the various components of each chopper. Also the first triggering method is preferred in the case of strong inductive load because this kind of load smooths the currents.

## REFERENCES

(1) GIAVASSOGLOU Kl., A. SAFACAS: Digitale Simulation eines selbstgeführten Drehstromwechselrichters mit Einzellöschung. E. und M., under publication.

(2) MILIAS-ARGITIS J., G. GALANOS: Dynamic simulation of HVDC transmission systems. IEEE, WINPWR 76 Abstr. A76 118-0 T-PAS 76 May/June 759....760 (1A10).

(3) WILLIAMS S., I. R. SMITH: Fast digital computation of three phase thyristor bridge circuits. Proc. IEE, Vol, 120 (1973) Nr.7.

# BOUNDED APPROXIMATION ALGORITHMS. AN APPLICATION TO SOCIAL MEDICINE

A. Aiello^, R. De Carlo^, U. Giani+

^Laboratorio di Cibernetica, C.N.R. Arco Felice, Naples Italy
+Cattedra di Puericultura, II Facoltà di Medicina, Università di Napoli, Naples, Italy

## INTRODUCTION

The task underlying the research was the interpretation of the information contained in the answers given by a set of 369 children to 35 questions on social and sanitary topics.
The largest part of those answers were either numbers, e.g. the age, or easily numerizable as 0's or 1's being yes or not answers. Only two questions determined non numerical answers, but they were treated each as a complex of five mutually disjunctive and exhaustive questions requiring yes or not answers. So we were facing a set of data representable by means of 369 points (the individuals) in the fourty-three dimensional cartesian space $R^{43}$, each dimension representing one answer variable.

At that point, before any statistical manipulation, some reduction of the information contained inside the set of data was necessary in order to permit the formulation of any hypothesis about relationships (if any) among individuals, or among variables, or among individuals and variables. The strategy which characterizes the present approach to data analysis is the reduction of redundant information by means of Bounded Approximation Cluster Analysis Methods rather than by standard methods of dimensionality reduction.
Among the various ways of performing Cluster Analysis, those in which the procedure is formalized by the extremization (maximization or minimization) of a preassigned criterion function have certainly the most scientific favour in spite of the fact that there is not yet any general purpose clustering criterion which could be universally accepted, because really each criterion described in the literature depends on the specific application it



*D. G. Lainiotis and N. S. Tzannes (eds.), Advances in Control, 550–557.*

has been used for. In any case, the optimum classification involves unavoidably the inspection of all the possible partitions of the set under analysis. It is sufficient to recall that the number of partitions $S(n, k)$ of $n$ objects into $k$ nonemptly classes is given (1) by:

$$S(n,k) = \frac{1}{k!} \sum_{i=0}^{k} \binom{k}{i} (-1)^{k-i} i^n$$

and that (2), for example,

$$S(25,10) = 1.203.163.392.175.387.500$$

to justify the adoption of approximate clustering procedures. Moreover, since sophisticated clustering methods usually bring to approximate solutions whose relationship with the optimal solutions is hard to determine at all (3), in what follows we limit our interest to a very naif basic scheme of clustering procedure which in turn yields approximate solutions whose "goodness" can be formalized quite rigorously.

## THE BASIC CLUSTERING ALGORITHM

Up today, the only kind of optimum classification approximable within known bounds (4) is that which, as clustering criterion, requires the minimization of the within-cluster-sum of distances. According to the language used in the field of complexity of computer computation (5) that problem is called K-MINCLUSTER and is formalized as follows:

K-MINCLUSTER = divide a set N of points in k disjoint clusters $S_1, \ldots, S_k$ in such a way as to minimize the criterion function

$$(1) \qquad \sum_{i=1}^{k} \sum_{\{x,y\} \in S_i^2} d(x,y)$$

where a given distance function

$$d: (x,y) \in N^2 \rightarrow (0,\infty)$$

in some quantitive measure of dissimilarity within the point of N.

In our case, as a measure of dissimilarity between two individuals, we adopt the euclidean distance $d(\vec{x},\vec{y})$ between the two points of coordinates $(x_1,\ldots,x_{43})$ and $(y_1, \ldots,y_{43})$ representing the two individuals in the fourty-three dimensional cartesian space $R^{43}$:

(2) $$d(x,y)=\sqrt{\sum_{j=1}^{43} h_j(x_j-y_j)^2}$$

There the $h_j$'s are scale factors representing the possibility of changing, arbitrarily and indipendently from each other, measure unities of non-homogeneous variables.
K-MINCLUSTER belongs to the class of "NP-complete" combinatorial problems, all equivalently intractable from the computational point of view. Those problems concern the determination of properties of graphs, integers, finite family of finite sets, boolean expressions and are so strongly related one another that if there exists a polynomial-time-bounded algorithm for solving one of them, the same occurs for every other NP-complete problem.
Until now, the practical solvibility of NP-complete problem, is an open question. However, the following algorithm yields bounded approximate solutions in $O(n^2)$ computational time:

Basic Clustering (k, N, d)

COMMENT. k... number of disjoint sets, $S_i$, into which the points of the set N = 1, ...,n is partitioned; SET(i).... the cluster to which the point i has been assigned (SET(i)=0 .... the point i is not assigned); DT(i) ... the increase of the sum of distances if the point under examination is added to the cluster i; Q.A. ... the value of the within-clusters sum of distances.

Step 1 [Initialize]

<u>if</u> k ≥ n <u>then</u> any assignement of one point to one cluster is optimal;

QA ← 0; <u>stop</u>; <u>end</u>

<u>else</u> <u>do</u>

SET(i) ← i, 1 ≤ i ≤ k; SET(i) ← 0, k+1 ≤ i ≤ n;

QA ← 0; DT(i) ← 0, 1 ≤ i ≤ k; j=k+1

<u>end</u>

Step 2 [Consider the assignement of point j to the various clusters]

<u>for</u> each point m <u>do</u> <u>if</u> SET(m)≠0
<u>then</u> DT(SET(m))←DT(SET(m))+d(j,m)
<u>end</u>

Step 3 [Find the most promising assignement of j]

look at DT(a), 1 a k, and determine the cluster i such that DT(i) is minimal.

Step 4 [assigne the point j and update for next point]

```
SET(j) ← i;  QA ← QA + DT(i);

do DT(a) ← 0, 1 ≤ a ≤ k, end; j = j+1

if j ≤ n then go to step 2
   else terminate algorithm
end Basic Clustering.
```

In simpler words, Basic Clustering begins assigning the first k individuals one for each cluster, then the assignement of any other individual is done in such a way as to cause the least "instantaneous" increase of the within-cluster sum of distances. The strategy is inspired by the sensible philosophy according to which, when things are to be done step by step, it is worth to assure that each step is done in the best direction recognizable at the moment of the choise. However the true value of Basic Clustering can be better formalized as follows:

Let J, J* and $\hat{J}$ be respectively the values that the criterion function (1) assumes in correspondence of the approximate solution yielded by the Basic Clustering, in correspondence of the optimal solution (the minimal) and in correspondence of the worst solution (the maximal) just for the same set of data. The following inequality holds (4):

$$\frac{J - J^*}{\hat{J} - J^*} < \frac{1}{k}$$

That inequality really assures a quantitative rigorous measure of goodness of the Basic Clustering solutions. The first member equal to zero means that J = J*: the approximate solution was near to the worst possible solution: actually the given bound on the evaluating ratio guarantees that the Basic Clustering solutions does not exceed the optimum more than (1/k)-th of the whole range of the possible approximate solutions. The evaluating ratio above has such properties that we decided to adopt it systematically instead of the other more frequently encountered in the literature (7,8): it is enough to mention here that it is invariant under linear transformations of the optimizing function J.

## ADAPTING SCALE FACTORS

Any solution of K-MINCLUSTER depends strongly on the scale factors applied in computing the distances between individuals by the expression (2): considering all the $h_j$'s as equal one another is equivalent to accept to use unities proportional to those adopted by the people who collected the data and, therefore, to adhere the same unknown criteria which, though they could find many motivations, certainly are well far from being just those bringing to the most significant clusterizations. But, in order to avoid vagueness, let us begin to formalize the concepts by giving the following statement: clusterizations are significant if they are obtained taking "unbiassedly" into account of any variables involved in the characterization of individuals.

This is some kind of null hypothesis reflecting the situation in which the data analyst ignores any hierarchy of variables (if any). That requires adequate manipulation of the data in hand since the people who collect the data were not aware of giving really decisive bias to the clusterization choosing the measure unities. The relevance of this aspect can be realized thinking for example that usually ages are given in years. There should be no difference in giving them in days or centuries: nevertheless the weight of the variable age would become in correspondence three hundreds and sixty-five times higher or one hundred lower in determining the differences between individuals as computed by expression (2).

So we face the following questions: i) how to recognize and measure unbiassedness of variables? ii) how to get suitable scale factors in order to assure significant clusterizations? Usually arguments like those above are briefly cut off by standard normalization techniques though nobody knows really why and how normalizations could propitiate significant clusterizations. Instead, our way is based on the quantitative evaluation of the clusterizing power of each variable, on adaptive adjustments of the scale factors and on the iterative application of the Basic Clustering.

Let us suppose that the Basic Clustering was applied to our set of data and k clusters $S_1,\ldots,S_k$ were obtained (for the first application of Basic Clustering, we can put all the scale factors equal to one). For each cluster $S_j$ consider the central point

$$\vec{c}_j = (c_{j,1},\ldots,c_{j,43})$$

that is the point of $R^{43}$ whose coordinates are obtained by averaging the corresponding coordinates of all the points belonging to the cluster $S_j$: a profitable reduction of redoundant information is got if clusterization comes out to be significant. But just reversing the previous consideration, and taking it as a definition, we can add a further attribute to the concept of significant clusterization: a clusterization is significant if the center of the clusters are "good representative" of the individuals inside the respective clusters:

an evaluation of the representativeness of the centers can be obtained comparing the scattering of the individuals inside the relative clusters and the scattering of the centers themselves. As a measure of scattering we will adopt the standard deviation from the mean. So, indicating by $S_{j,1},\ldots,S_{j,43})$ the standard deviation vector of the individuals inside the cluster $S_j$ from the mean vector $C_j$, by $Z = (Z_1,\ldots,Z_{43})$ the standard deviation vector of the centers from their mean, a quantification of the representativeness of the center $c_j$ can be given by the vector $r_j = (r_{j,1},\ldots,r_{j,43})$ in which the i-th component $r_{j,i}$ is given by the ratio $S_{j,i}/Z_i$. Of course, lower are the components of $\vec{r}_j$ more representative is $c_j$.

If the i-th component of $\vec{r}_j$ is greater then or equal to one, it means that the scattering of the i-th variable for the individuals inside the cluster $S_j$ is greater then or equal to the scatter of the same variable of the centers. That is, the i-th component of the center $\vec{r}_j$ cannot be considered representative. Generally, it can happen to get either low or high values for the various components of the same vector $\vec{r}_j$, indicating that the correspondent variables had different influence in determining the cluster $S_j$. A measure $v_i$ of the overall clustering influence of the i-th variable can be given averaging the i-th components of the $\vec{r}_j$'s:

$$v_i = \frac{1}{k} \sum_{j=1}^{k} r_{j,i}$$

Like the $\vec{r}_j$ s, lower is the value $v_i$ greater is to be considered the overall influence of the variable i-th.
After that, the two requisites that clustering must satisfy in order to be significant are formalizable as follows:

1) the standard deviation D of the measure $v_i$'s of influence of variables from their mean must be as low as possible; D will be called the bias degree.

2) the mean M of the $v_i$'s (that is, the overall mean of the ratios $s_{j,i}/z_i$ performed with respect to both variable index and cluster index) must be as low as possible; M will be called the representativeness degree.

The first statement reflects unbiassedness of variable as requisite of significant clustering, the second formalizes representativeness of centers in order to get acceptable reduction of information, that is the main analyst's task. According to the foregoing rules we have implemented the following algorithm:

RAU(K,N,d)

Step 1. Put each scale factor equal to one and perform the first clusterization by applying Basic Clustering. Put an iterations counter $I = 1$,

Step 2. Compute the cluster centers $c_j$, $1 \le j \le k$ , the variables influence degrees $v_i$, $1 \le i \le 43$, the representativeness degree $M$ and the bias degree $D$. If $M$ and $D$ are lower than preassigned figures $M'$ and $D'$ or equal to those gotten at the previous iteration (if $I > 1$) or $I$ is greater than a preassigned limiting value then stop RAU;

Step 3. For each $i$, adapt the scale factor $h_i$ multiplying it for some preassigned non decreasing function $f$ of the i-th variable influence degree $v_i$.

Step 4. Apply the Basic Clustering modifying Step 1 as follows: the first point to be assigned to the various clesters are the centers computed at Step 2 of RAU. Go to Step 2. End RAU.

Algorithm RAU aims to reduce the degree D by suitable adaptation of the scale factors at Step 3. The reduction of the figure M is left to the iterative nature of the algorithm itself. Step 3 is without any doubt the most delicate aspect of algorithm RAU. Indeed effectiveness and convergence of the algorithm depend merely on the choise of the adapting function f. We have tested many variants of RAU obtained changing f and it seems the best results are gotten assuming

$$f(v_i) \sim \left(\frac{M+D}{M'}\right)^{\frac{v_i/M-1}{I}}$$

Let us remark that at Step 3, being $M'$ the preassigned figure still lower then M (in the opposite case the algorithm should have stopped before) $f(v_i)$ is greater than, equal to or less than unity according to being $v_i/M$ respectively greater than, equal to or less than unity. This means that at Step 3 scale factors are increased or decreased according to being the corresponding variables less or more responsible, with respect to the mean of the clusterization obtained at the previous step. In any case, as the exponent in f decreases with the number of iteration I, adjustments become more and more fine while the precedure goes on.

## CONCLUSIONS AND ACKNOWLEDGEMENTS

The problem of Cluster Analysis finds a complete rigorous formalization when it is posed in probabilistic terms [9] . Unfortunately, all the sosphisticated techniques of statistical inference become more and more useless while one moves from the ideal conditions (the probabilistic structure is almost completely known) to concrete cases like those we usually encounter (the probabilistic structure is almost completely unknown). However there are several heuristic clustering methods which, though lies on the weakest foundation, are acceptable just since no reasonable statistical hypothesis can be formulated.

Here we have introduced the basic concepts which led to the implementation of RAU, an adaptive-iterative clustering computer program which the authors can send to anybody interested with it. The peculiar character of that package is that, besides the often too hard to handle clusterization of the input, it provides both representatives of the clusters and measures of their representativeness: namely the centers $c_j$'s and the ratios individuals centers variables standard deviations $\vec{r}_j$'s . Overall evaluating measures like the bias degree D and the representativeness degree M are also yielded as well as the scale factors responsible of the eventual clusterization.

Special thanks are due to Mr. G. Della Monica who assisted the authors in implementing and testing RAU on UNIVAC 1106.

REFERENCES

1) W. Feller: An Introduction to Probability Theory and Its Applications, Vol. I , p. 58 (John Wiley, New York, Second Ed. 1959)
2) Y. Lechevallier: Techn. Report n. 200,IRIA Laboria, Rocquencourt, France, 1976
3) R.O. Duda and P.E. Hart: Pattern Classification and Scene Analysis, John Wiley, 1973
4) A. Aiello, E. Burattini, A. Massarotti and F. Ventriglia: Evaluating Approximation Algorithms for Optimization Problems, 5th Symposium "Algorithm 79", Vysoké Tatry (Czechoslovakia), April 1979
5) R.M. Karp: in Miller and Tatcher , pp. 85-104, 1972
6) A. Aiello, E. Burattini, A. Massarotti and F. Ventriglia: A New Evaluating Function for Approximate Algorithms, Informatica 77 - Bled, October 1977
7) G. Ausiello: On the Structure and Properties of NP-complete Problems and their Associated Optimization Problems (MFCS 1977)
8) S. Sahni and T. Gonzales: Journal of ACM, vol. 23, n. 3, July 1976, pp. 555-565
9) Luminita State: Rev. Roum. Math. Pures et Appl., Toma XXIII, n. 2, pp. 291-297, Bucarest 1978

# FAST APPLICATIONS OF SUFFIX TREES

Alberto Apostolico

Istituto di Scienze dell'Informazione, University of Salerno, Salerno (Italy)

ABSTRACT

Efficient algorithms, based on suffix trees, are presented to solve statistical problems on character strings and to implement nonsequential textfile compression schemes. The optimality of the compression achieved is not a primary concern here: rather, the complexity inherent to frequently used suboptimal strategies is analyzed under assumptions about the nature of the input well suited to the formalism and tools of algebraic coding theory.

## 1. INTRODUCTION

Suffix trees and related data structures (1,2,3) were introduced to serve as auxiliary indices in pattern matching problems that arise in text editing,data retrieval and symbol manipulation in general. Although linear time and space algorithms are presently available for obtaining such trees, their inherent dependency from alphabet size and the often exhorbitant space requirement prevent their use in a large number of instances. This work aims at showing that some extensions of the notion of suffix tree might be fruitfully adapted to solve problems naturally posed about a textfile.

Indeed, as pointed out, elsewhere (4),a suffix tree embodies most statistical information about the related character string. Some of these can be retrieved straigthforwardly in linear time,



*D. G. Lainiotis and N. S. Tzannes (eds.), Advances in Control, 558–567.*

while some other involve more tight assumptions about the nature of the input word.

As the natural support for substring statistics, suffix trees can be used to implement efficiently data compression schemata of the kind often devised to seek suboptimal solutions to an inherently exponential problem. The steadily lowering cost of main memories suggest that RAM based compression techniques can do much to alleviate the more restricting resources of on line secondary storage space. Some such applications are also discussed in the paper.

## 2. PRELIMINARY DEFINITIONS AND RESULTS

Let I be a finite alphabet and $I^*$ the free monoid generated by I. For any string (or word) $x \in I^*$, the notation $|x|$ shall be used to denote the length of x. The empty word is denoted by $\lambda$ and it is assumed that $|\lambda| = 0$. Further, the set $I^+$ is introduced in the usual way by letting $I^+ = I^* - \{\lambda\}$.

A string $x \in I^*$ is fully specified by writing $x = a_1 a_2 ... a_n$ where $a_i \in I$ $(i=1,2,...n)$. In the following, it shall be assumed that RAM resources are available to store any given string $x = a_1 a_2 ... a_n$ into an array structure x $(1,...,n)$. A substring w of x is specified by writing $w = a_1 a_2 ... a_d$ where $a_i \in I$ $(i = 1,2,...,d)$. A factor $\underline{w}$ of x is specified by giving a substring of x and its starting position in x $(1,...,n)$. The notation x $(i,j)$ shall be used to denote the factor of x that starts at position i and ends at position j, inclusive, in x $(1,...,n)$. A left (right) factor of x is a prefix (suffix) of x.

The set of all distinct nonempty substrings of x is called the vocabulary of x and denoted by $V_x$.

Example 1: x = a a b b a a b

$$V_x \equiv \{a,b,aa,bb,ab,ba,aab,abb,bba, aabb,abba,bbaa,baab,aabba, abbaa,bbaab,aabbaa,abbaab, aabbaab\}.$$

Let \$ be some extra symbol not included in the alphabet I. A data structure suitable for organizing the words in $V_x$ is the so called suffix tree (3) T for x \$. As is well known, such a tree is a multiway PATRICIA tree (5) collecting together all suffixes of x \$: if the string x \$ is stored into a common reference array x ( 1,...,n+1) , then each arc of T may be labelled simply by some ordered couple (i,j) that identifies the factor x (i,j) of x \$.

Example 2:

x \$ = a b b a a b b \$
1 2 3 4 5 6 7 8

T =

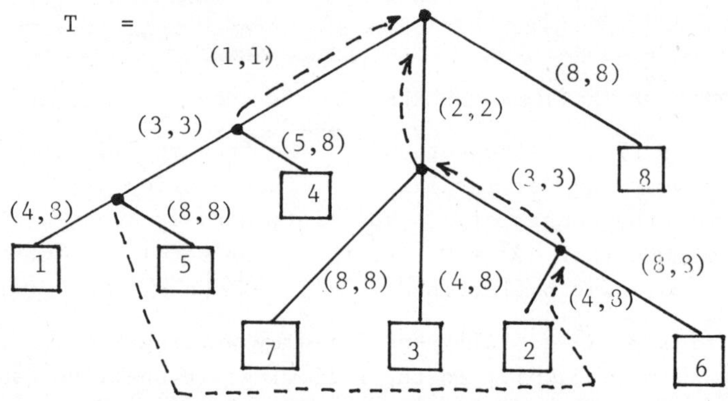


It is easily seen that $w \in V_x$ if and only if there is a downward path from the root of T through the arcs labelled $(i_1, j_1)$ $(i_2,j_2)$ $\ldots(i_d,j_d)$ such that

$$w = x\,(i_1,j_1)\; x\,(i_2,j_2)\ldots x\,(i_d,j) \text{ with } i_d \leq j \leq j_d$$

Finally, for any $x \in I^+$ a weighted vocabulary of x is any couple $(V_x, W)$ where $W : V_x \rightarrow N$ is a mapping that associates with each string in $V_x$ a natural number k.

The following definitions and results deal with the notions of periodicity and overlap. Two factors x (i,j) and x (k,d) of x are said chain if it is $k \leq j < d$ or $i \leq d < j$. If, moreover,it is x (i,j) = x (k,d) = w the two factors are said to overlap.

Further, p is a period of x if x (i) = x (i + p) ( i = 1,2,... x - p). On the other hand x is periodic if it has a period of size not larger than $|x|/2$. Let x be periodic and let p be its smallest period. The prefix (suffix) of x of length p is called the left (right) root of x with respect to p. The following well known lemma relates periodicity and overlaps in a string.

Lemma 1 (6,7): Let $x = a_1 a_2 ... a_n$. Then $x(1,d) = x(j,n)$ iff there are $k \geq 2$, $s \in I^+$, $t \in I^*$, such that $x = (st)^k s$ with $|st| = (j-1)$.

In general, a string x can have more than one period. The following lemma disciplines the way different periods can be hosted by the same string.

Lemma 2 (6,7): If x has periods of sizes p and q and $|x| \geq p+q$, then x has period g.c.d.(p,q).

## 3. SUFFIX TREES AND STRING STATISTICS

In this section, two weighting maps are introduced and the corresponding vocabulary construction is addressed. More precisely, let $w_1$ be such as to associate with each $w \in V_x$ the number of distinct factors of x that are equal to w. Let further $w_2$ be such as to associate with each $w \in V_x$ the maximum number of distinct factors $x(i_1,j_1)\, x(i_2,j_2) ... x(i_k,j_k)$ such that $x(i_d,j_d) = w$ $(d = 1,2,...,k)$ and there are words $w_1, w_2, ..., w_{k+1}$, $w_d \in I^*$ $(d = 1,2,...,k)$ such that $x = w_1\, w w_2\, w ... w_k\, w w_{k+1}$. The following problems are considered:

1) Input: $x \in I^+$
   Output: $(V_x, W_1)$
2) Input: $x \in I^+$
   Output: $(V_x, W_2)$

It can be shown very easily (4) that there is an algorithm to answer the first of the problems above in time linear in the length of x. Some more definitions need to be recalled. First, the node at the end of the path from the root that spells out some substring w of x\$ shall be called the locus of w. Moreover, a value $\delta$ is associated with each internal node $\nu$ in the tree, called the depth of $\nu$, that corresponds to the length of the factors of x of which $\nu$ is the locus. The level of $\nu$ is the number of edges traversed in order to get from the root to node $\nu$.

As a multyway PATRICIA tree, a suffix tree has O(n) nodes. Therefore, a brute force approach to its construction can require $O(n^2)$ steps in the worst case. Linear time algorithms have been set up (2,3) for the construction of this as well as closely related auxiliary indexes, all of them exploiting suffix links (dashed lines in Example 2) connecting each internal node $\nu$ to the locus of the longest suffix of the substring whose locus is $\nu$ . Finally, it should be mentioned that it might turn out to be convenient to store a suffix tree in form of the binary tree naturally (5) corresponding to it.

The linear time computation of $(V_x, W_1)$ stems from the following lemma whose proof is left to the reader.

Lemma 3: Let $w \in V_x$ correspond to some proper suffix in the form $x(i_1,j_1)\ x(i_2,j_2) \ldots x(i_d,j)$ of the factor of x on the path from the root through arcs $(i_1,j_1)$, $(i_2,j_2) \ldots (i_d,j_d)$, and let $\nu$ be the locus of $x(i_1,j_1)\ x(i_2,j_2) \ldots x(i_{d-1},j_{d-1})$. Then the number associated with w under $W_1$ equals the number of leaves in the subtree of T noted at $\nu$.
As a consequence $W_1$ can be computed in situ on T (used to store $V_x$) in a rather straightforward way: this involves the evaluation of the simple Bottom Up locally defined function (5) implicitely defined in the statement of lemma 3.

The computation of $(V_x, W_2)$ is not as easy. Even resorting to some tree to organize $V_x$ and to support the weights in $W_2$, it is not obvious in general that O(n) nodes shall suffice to carry all needed information. In the worst case (no two substrings along the same path have correspondingly equal weights in $W_2$) each arc should be split into a chain of unary nodes, thus leading the overall number of nodes to $O(n^2)$ for an highly unbalanced tree. More precisely, let an augmented suffix tree (AST) be introduced as a suffix tree such that, for any given word w in $V_x$, whenever w has no defined locus in T there exists a $w_1$ $w = w'_r$ such that w' has a defined locus and $k(w) = k(w')$, $k \in W_2$.

Example 3: Let x = st st st st sts, s,t $\in I^+$ and consider the portion of suffix tree and an eligible augmented suffix tree shown below. Notice that some arcs must be necessarily split in the augmented suffix tree as there is no way to store the weights in $W_2$ otherwise.

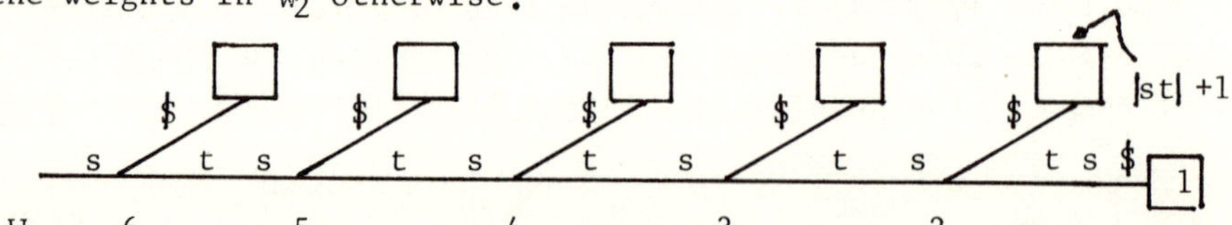

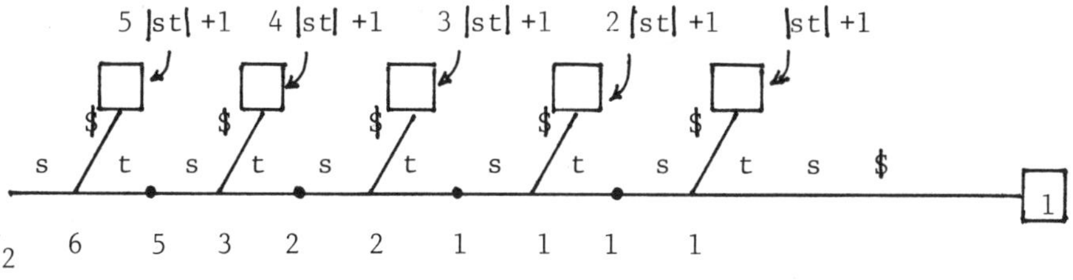


As the example suggests, intermediate nodes are needed in an AST due to the occurrence of overlapping factors that represent the same word in $V_x$. This circumstance seems also to prohibit the existence of a O(n) nodes AST for the most general string. However, it can be shown that under practically fairly reasonable assumptions such an AST can be derived from a suffix tree and then used to compute $W_2$ in linear time. The following definitions and results need to be recalled from ref.(4).

Theorem 1 (4): Let $x = (st)^k s$ with $|st| = p$ its smallest period and $k \geq 2$. Then there exists an AST for x with O(n) nodes.

Example 4: In the example developed above for the string $x=(st)^5 s$ it is easy to see that $|x|/(2|st|)$ intermediate nodes are needed at most along each of the first $|st|$ suffix paths in the tree. Thus $|x|/2$ nodes are to be added in the whole tree.

A word $w \in I^+$ is primitive iff $w = s^k$ for some $s \in I^+$ implies that $w = s$ and k=1. A word $w \in I^+$ is strongly primitive iff all the substrings in $V_w$ are primitive.

Example 5: w = aabaabc is primitive. It is not strongly primitive, though, because w (1,6) = a a b a a b is not primitive.

Finally, let w be some periodic word, p its smallest period. Then w is simple iff the left roots with respect to p of its first p-1 suffixes are strongly primitive. Alternatively, a periodic factor w of x is simple iff any equation in the form $w = s_1 (s' t')^k s' s_2$, $s' \in I^+$, $t' \in I^*$, $k \geq 2$ implies that $|s' t'| = rp$ for some positive integer r. This quickly derives the following lemma.

Lemma 4 (4): If x (i,j) and x (d,k) are distinct simple periodical factors of x then either $j < k$, $i < d$ or $i > d$ and $j > k$.

The interest in introducing simple periodic words lies in the fact that two such factors of a string x either do not share any position in x (1,...,n) or else they are chaining factors.

In fact lemma 2 enables to derive the following

Theorem 2 (4): Let $w_1 = x\ (i,j)$ and $w_2 = x\ (k,d)$ be two consecutive simple periodic factors of x with smallest periods p and q, respectively. Then it is $k \geq j-p$.

Finally, suppose that x be such that, whenever w is a periodic factor of x, then w is simple. The following theorem can be proved:.

Theorem 3(4): There is an algorithm to construct the couple $(V_x, W_2)$ in time linear in the length of x.

The reader is referred to ref.(4) for a detailed description of one such algorithm. Essentially, it is shown there that O(n) extra nodes need to be inserted in T to build a sufficient AST and that the corresponding node insertion and weighting process can be carried on in time proportional to the nodes eventually in the AST.

## 4. SUFFIX TREES BASED COMPRESSION SCHEMATA

In this section, the results illustrated before are used to develop off line data compression techniques for a textstring x which is supposed to satisfy the assumptions in theorem 3. As is intuitive, most text files would agree with such request, while more general signal streams such as grey tones in digitized pictures would probably not. The schemes proposed here are based on the principle of finding redundant strings or patterns and replacing them by a pointer to a common copy: when no two substrings that are factored out in this way correspond to chaining factors of the textstring x the process is reducible to an encoding of x. Finding an optimal solution to the compression of x has been recognized as an NP-complete problem (8). Suboptimal strategies have been suggested (9,10) that roughly rely on the following scheme.

Procedure 1

begin
    Let S = (1,2,...,n) be the set of positions in x(1, ...,n) filled with symbols from I.

while
        S is non empty do
    begin choose the word w in $V_x$ for which the ratio

$(|w| + k(w))/(|w| \cdot k(w))$ is a minimum.
Set $V_x = V_x - \{w\}$.
Replace each occurrence of w in x with a pointer to a common copy of w.
Delete from S all indices d such that $w = x(i,j)$ and $i \leq d \leq j$.

<u>end</u>

<u>end</u>

To improve the compression performance in this scheme it is desirable that the weights k be taken from the map $W_2$.

<u>Theorem 4</u>: It takes $O(n\log_2 n)$ time to run Procedure 1 on a RAM machine

<u>Proof</u> Assume that T, the AST associated with x by theorem 3 is used to store $V_x$ for x. By theorem 3 it takes O(n) steps to construct T. Since the depth and the level of each node are known at the time a node is inserted then each node could be weighted directly with its relative compression ratio at the insertion stage. Moreover, a heap (5) can be used to sort the nodes in T according to their associated compression. Since T has O(n) internal nodes then it takes O(nlogn) steps to construct the heap. Now the first instruction in the while loop takes constant time. The second is executed by throwing out of the heap the node currently extracted. In order to execute the third instruction, it is sufficient to visit the subtree rooted at the current node. Let d be the number of leaves found in this subtree. A subset of k(w) leaves such as to correspond to nonoverlapping instances of w can be easily found in O(d) time.

Indeed, if two factors corresponding to w overlap on x then there must be two leaves consecutively introduced in T whose labels (i.e., the starting positions of the corresponding suffixes of x $) differ by less than the current node level. It is easy to see that each leaf is considered only once, provided that no extension of the form wr of a string w in $V_x$ has got a better score than w in the compression ratio on T. For the actual replacement of each w by a pointer some more conventions are required. First, the structure of a pointer must be defined: assuming external targets, a pointer is defined by the couple ( (w),j) where (w) is the address of target w and j is the starting position in the original array. The length of a pointer is assumed equal to 1. Next, an auxiliary array y(1,...,n) is introduced to store, at each step in the while loop and for each position i, the position in x to the right of i that is not occupied by a pointer neither covered by its range. Initially, $y(i) := i$ $i = (1...n)$. Whenever a word w in $V_x$ is selected which occurs in positions $i_1, i_2, ..., i_d$ the corresponding location in $y(1...n)$ are tested one at a time: if position $y(i_j) = i_j$ then a pointer $((w), i_j)$ is placed on $x(i_k)$ and starting at $i_j$, a sequential scanning of y is initiated until either a position m is found such that $y(m) \neq m$ or else $|w| + 1$ moves are made on y. Let $i_j + t$ be the index of y at which this process is arrested and let $y(i_j + t) = r$ then it is set $y(i_j) = y(i_j + 1) ... = y(i_j + t - 1) = r$. On the other hand, if position $y(i_j) = r \neq i_j$ and $r - i_j \geq |w|$ then no pointer can be allocated on $x(1,...,n)$: otherwise, if $r - i_j < |w|$ then position r is reached and a scanning such as above is initiated. Finally, a pointer $((w), i_j)$ is introduced in $x(r)$. The array y can be used to ensure the termination condition: more precisely, let initially be $c := 0$; then whenever $y(i)$ is changed from i to some $r \neq i$ it is also set $c := c+1$. The condition that $c = n$ corresponds then to S empty. Each position on y is accessed directly exactly once and its content is modified also once. An analogous argument can be developed for x, thus the total effort is proportional to n. This completes the proof for theorem 4. It should be noted that the performance achieved is also a lower bound for Procedure 1.

A major drawback of this procedure, however, is in that the effort spent in constructing the minimal AST may be offset by the possible chaining on x of factors aimed to by pointers. One possible way of avoiding this is offered by Procedure 2. It is easy to describe roughly such procedure in terms of the first one: in fact it is sufficient to substitute there the words "pointer to a common copy" with "codeword". Since no two codewords could possibly overlap on x, one can get rid of the array y while implementing procedure 2. On the other hand, the involved complexity is higher. Indeed, it is not difficult to imagine that more work is needed to re-adjust note weights at each stage; since this process needs at least n steps, a rough bound of $O(n^2 \log n)$ could be reasonably traced.

## REFERENCES

1) Aho,A.V.,Hopcroft,J.E.,Ullman ,J.D.(1974): The Design and Analysis of Computer Algorithms, Addison-Wesley,Reading

2) Weiner,P. (1973): Linear Pattern Matching Algorithms, IEEE Symposium on Swtching and Automata Theory, 14,IEEE,New York, 1-11

3) Mc Greight, E.M.(1976): A Space Economical Suffix Tree Construction Algorithm, Journal of ACM,23,262-272

4) Apostolico,A.(1979): Some Linear Time Algorithms for String Statistics Problems, Quaderni dell'IAC,CNR, (to appear)

5) Knuth,D.E. (1973): The Art of Computer Programming,3: Sorting and Searching, Addison-Wesley,Reading

6) Lyndon,R.C.,Schutzenberger,M.P.(1962): The Equation $a^M=b^N c^P$ in a Free Group, Michigan Math.J.,9,289-298.

7) Lentin,A.,Schutzenberger,M.P.(1967): A Combinatorial Problem in the Theory of Free Monoids, Combinatorial Mathematics and its Applications (Bosa and Bowling Eds.), University of North Carolina Press,128-144

8) Storer,J.A.,Szymanski T.G. (1977): The Macro Model for Data Compression, Princeton University TR 235

9) Rubin,F. (1976): Experiments in Text File Compression,Communications of the ACM,19,617-623

10) Mc Carthy,J.P. (1974): Automatic File Compression, International Computing Symposium 1973, North Holland,Amsterdam, 511-516

---

Work supported in part by the National Group for Mathematical Information Sciences of the National Research Council.

# INFORMATION SCIENCES APPLIED TO COMPUTERIZED ELECTRON BEAM ANALYTICAL SYSTEMS

A. Christou
Naval Research Laboratory
Washington, D.C. 20375

## ABSTRACT

Metallurgical analysis of semiconductors, metals and insulators utilizes a combination of scattering techniques such as secondary electron microscopy (SEM), x-ray spectroscopy, Auger electron spectroscopy and material work function studies. The present paper summarizes the concurrent use of these techniques all interfaced with a minicomputer for automated operation, data collection, reduction, signal processing and image processing.

Information system analysis is applied to the data handling system in addition to appropriate signal processing. Certain characteristics of the SEM signal have been determined. First, power Fourier spectra from typical images reveal that some information is being carried at frequencies as high as the display cutoff frequency. This high frequency information is indistinquishable on an absolute basis from noise. However, signal processing is capable of discriminating against noise on a probability basis.

## INTRODUCTION

Image processing has seen increased utilization in recent years[1-3] in the area of microanalysis and electron microscopy. Signals generated by the interaction of an electron beam of energy up to 50 keV with a solid are of the order of $10^{-10}$ Amps. Processing the information included in such a signal becomes critical. This paper will describe electron beam systems, and in particular, scanning electron beams (SEMs). To be emphasized will be signal storage and enhancement techniques for the SEM, SEM image processing and results obtained from image processed-computerized images.



*D. G. Lainiotis and N. S. Tzannes (eds.), Advances in Control, 568–575.*

A scanning electron beam or ion beam is an information system when one includes the types of interactions which occur with a solid. As shown in Figure 1, information concerning basic characteristics of the solid are contained in the emitted secondary electrons, x-rays, backscattered electrons and Auger electrons. In a computerized system, the beam is created and focused in an electron optical column. In the example shown in Figure 1, six different types of information is processed and at which point signal storage, signal enhancement and image processing takes place. Normal computer graphics allows the investigator to choose and mix signals of interest.

## Signal Storage and Signal Enhancement

Specific image processing in the SEM involves three areas: correcting geometric distortions, enhancing signal to noise ratio (SNR) and enhancing resolution. Modern electron beam systems can generate information at a high rate: $10^6$ bits per second. However, since we have $10^6$ or more picture elements and 10 gray levels, on-line data reduction becomes difficult. In the last information science conference in Patras, we presented studies of the most probable gray level content of two adjacent picture elements.[3,4] The probable gray levels of two adjacent elements are shown in Figure 2 where $(X_1, X_2)$, $(Y_1, Y_2)$ are coordinates in two-dimensional space. The most likely subpictures are shown by the dots and are contained within the ellipse.[5] A picture element is enhanced by a two dimensional convolution of the original picture element or signal. The function h (u-j, v-k) has the desired properties for signal enhancement.

## Image Transformation in the SEM

The enhanced image g(u,v) is the result of a transformation of nxn picture elements by either a Fourier transformation or a Hadamard transformation, which in terms of matrix elements we may write as:

$$g(u,v) = \sum_{i=1}^{n} \sum_{j=1}^{n} A_{uvij} X_{ij} \qquad (1)$$

The nxn array of picture elements (pels) is arranged into an nxn matrix X. The coefficient A is obtained by the Fourier transformation:

$$A_{uvji} = \frac{1}{n} \exp\left[-2\pi \sqrt{-1}\,(ui + vj)/n\right]$$

or the Hadamard transformation:

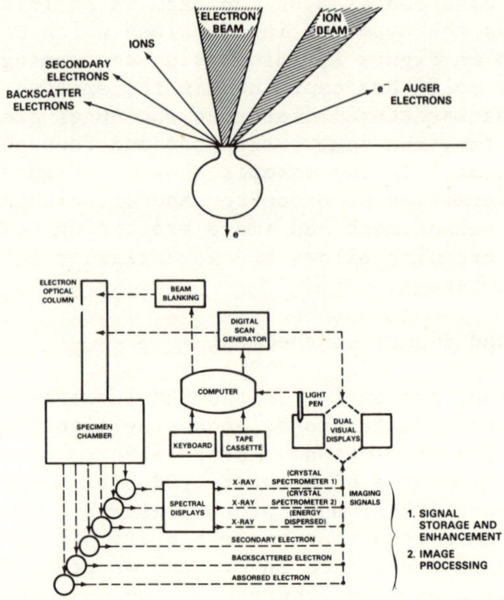


Figure 1. Electron beam system utilized for signal storage and enhancement.

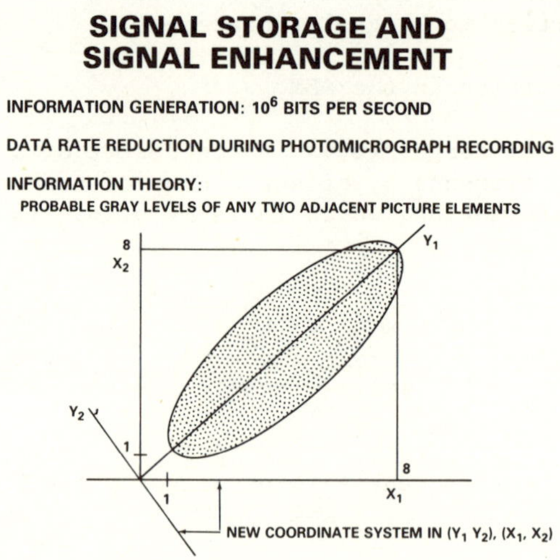


Figure 2. The most probable gray levels of two adjacent picture elements shown in two dimensional space.

$$A_{uvij} = \frac{1}{n} (-1)^b$$

$$b(uvij) = \sum_n (b_u(U)b_v(V) + b_n(i)b_n(j))$$

Typically the X matrix is 1000 X 1000. The set of basis pictures are chosen so that the coefficients are more independent than the pels. In addition, by choosing uncorrelated coefficients, one can minimize the mean square error. In the present investigation, we ave concentrated on the Fourier transformation. The final enhancement depends on the transformation selected (.2 bit/pel for n = 8), the quantization strategy (1 pel = 256 gray levels), the subpicture size and shape. The SEM direct image $X_{ij}$ is digitized and then processed off or on-line.

Experimental Set-up, Image Processing Functions

There are two different ways of accomplishing electron beam image enhancement. The first is an electrostatic memory[6,7] with a scan converter tube. Present systems can resolve 2000 raster lines. The read back of information can be performed under the control of the SEM scan generator. The second method uses a digital scan generator which is used with a magnetic tape recorder. The matrix density is capable of being adjusted up to 8,192 X 8,192. The scan generator digitizes any picture element into an 8 bit word and each into 256 gray levels.

In the analog mode, the image from the SEM is written onto an insulating storage mesh and stored in the form of a charge pattern. The image is read out by the scanning beam. A second read beam is also included. The modulated reading beam is collected at a second mesh in order to produce the converted video signal. The reading beam is scanned at normal TV rates. Typical application of electrostatic storage is in the correction of SNR by noise averaging. Figure 3 shows that random noise spikes in neighboring picture elements can be partially eliminated by spreading their effects across several adjacent picture points. This is accomplished by partially defocusing the read beam. Repeated noise processing further improves the image. In addition, poor SNR can be improved by integration of the stored image. Integration, however, uses a strictly digital technique with fast on-line ADC electronics.

The electrostatic storage system has been used to correct geometric distortions as shown in Figure 4. Foreshortening effects are present in images when a specimen is tilted with respect to the electron beam. In this application, the image is read with an inversely distorted raster. As shown in Figure 4, the image can be corrected for foreshortening and askew.

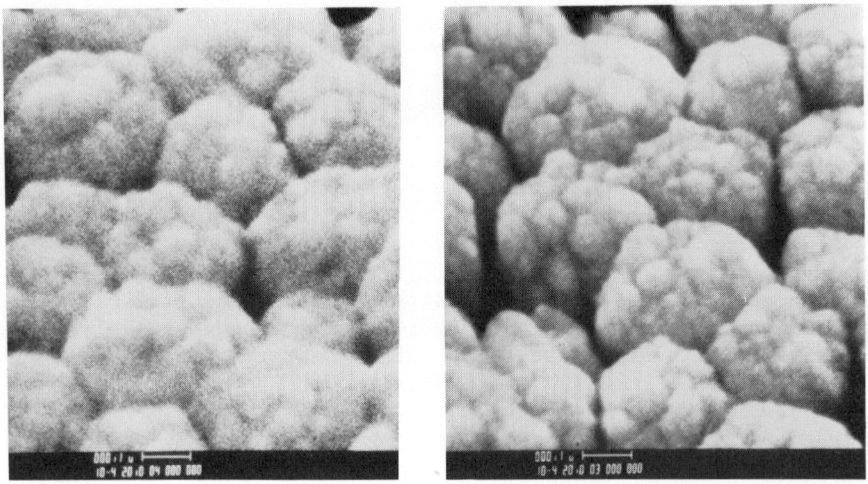

Secondary electron image of opal after integration, 20° tilt.

Figure 3. Secondary electron images showing noise averaging as a technique of improving image signal to noise ratio.

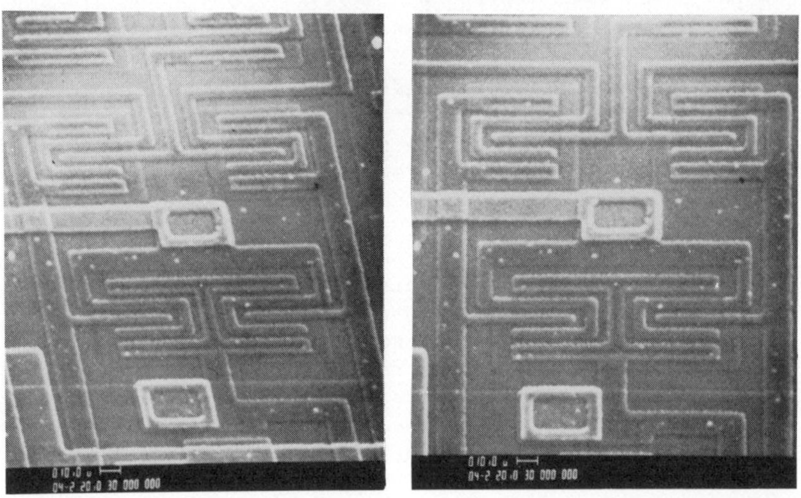

Portion of integrated circuit at 45° tilt. Fully corrected image of Figure 3a, 45° tilt.

Figure 4. Secondary electron images showing corrections for geometric distortions.

## Results From a Digital Matrix Scan Generator

As mentioned previously, the second experimental set-up consisted of digitizing the image utilizing a digital scan matrix. The image shown in Figure 5 has been digitized in a 1256 x 1256 matrix in order to show that a dense matrix photograph can be obtained. Again, the initial image was processed in order to eliminate tilt effects. A significant application, however, of digital linear fitering is in improving recorded image resolution. Figure 6 shows an unprocessed SEM image with resolution of the order of 200 - 250Å. Increased resolution and contrast has been obtained with linear digital image processing. The high resolution images, 100Å, show elimination or significant reduction of noise. Alone, noise reduction results in a resolution improvement. Details previously obscured by noise can now be observed in the image of a fracture surface. A beam current of $1X10^{-12}$ Amps was utilized with the original image digitized to a 2000 x 2000 matrix. A fast Fourier transform of the type shown in equation (1) provided the processing scheme.

## SUMMARY AND CONCLUSIONS

In conclusion, significant signal storage and enhancement techniques using both digital and analog electronics have seen great utilization in scanning electron microscopes. The digital techniques consist of linear digital processing with Fourier transformations. These techniques allow the greatest versatility since one may choose the coefficients in order to minimize the mean square error. The SEM direct image is digitized and then processed off or on-line.

The analog technique uses an electrostatic memory with a scan converter tube. The read back of information is performed under the control of the SEM scan generator. The greatest advances in image enhancement have been made with the digital scan generator and an interfaced computer.

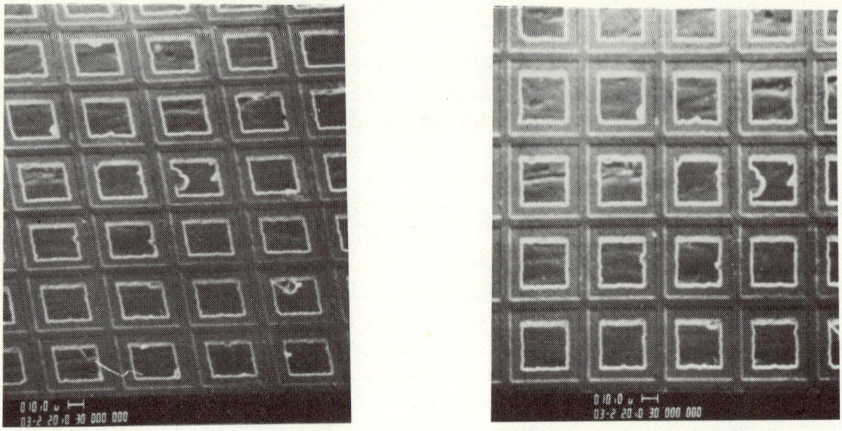

Figure 5. Application of digital scan generator as a method for eliminating tilt effects.

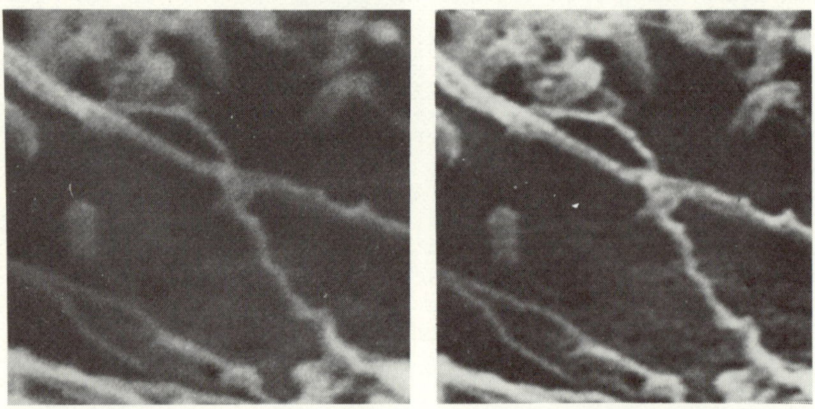

Unprocessed Scanning Electron Micrograph

Computer Processed Scanning Electron Micrograph

Figure 6. Scanning electron images of computer processed micrographs obtained by digital linear filtering.

## REFERENCES

1. Fiori, C. E., Yakowitz, H., and Newbury, D. E., in SEM/74, IITRI, Chicago, 162-174.
2. Heinrich, K. F., Fiori, C. E., and Yakowitz, H., Science, 167, 1970, 1129-1131.
3. Christou, A., First International Conference on Information Sciences, Patras, Greece, August 26, 1976.
4. Yew, N.C., and Pease, D. E., in SEM/75, IITRI, Chicago, 192-196.
5. O'Handley, D. A., and Greene, W. B., Proc. of the IEEE, 60 (7), 1972, 821-828.
6. Catto, J. D., and Smith, K. C., in SEM/72, IITRI, Chicago, 41-48.
7. Herzog, R. F. Lewis, B. L., and Everhart, T. E., in SEM/74, IITRI, Chicago, 175-182.

# COMPUTER SEARCHABLE REPRESENTATIONS FOR GEOGRAPHICAL CURVES AND REGIONS

F. Warren Burton

University of East Anglia

Abstract

Many queries to a geographical information system necessitate searching the representations of long curves or large regions. Such queries include finding the points of intersection of two curves (which in turn may be a step in finding the intersection or union of two regions represented by their boundaries), finding the minimum or maximum distance between objects, or testing a point for inclusion in a region. It is often possible to avoid exhaustive searches by selecting a suitable data structure. There are a number of ways to order points in two-space. Several searchable data structures, for representing curves and regions, which take advantage of various orderings are considered.

## 1. INTRODUCTION

It is often necessary to use a computer to search graphical information. For example, in a geographical information system one may wish to determine which points (houses) are contained in a region (administrative district) or which regions (properties) intersect another region (proposed road). As we will see later, problems such as finding the intersection or union of two regions or determining whether a point is inside a region may be reduced to search problems. Graphical search problems also arise in image processing, graphics and other areas.

When a list of numbers or character strings is to be searched, it is often sorted. Once sorted, a binary search may be used. At each stage in a binary search, the list is divided into two. Usually



*D. G. Lainiotis and N. S. Tzannes (eds.), Advances in Control, 576–582.*

one sublist may be discarded. If the other list contains more than one entry, it is searched further. With range queries it sometimes may be necessary either to continue searching both sublists or to continue searching one sublist while recognizing that every entry in the other sublist must be a hit. Many graphical searches may be viewed as generalized range queries.

There is no natural order for points in the plane (e.g. no one-to-one bicontinuous mapping onto a line). However, there are several useful ways to order points. We will consider three approaches to ordering points.

An *object* space ordering is determined by the values to be ordered. For example, points on a curve have a natural order which may be used to advantage. An *image* space ordering is an ordering of all points in the plane and is independent of the data under consideration. A mixed *image/object* space ordering involves an image space ordering in one direction and an object space ordering in the other. All three approaches have been used in practice.

## 2. PRELIMINARIES

Before exploring the advantages to be gained by using various orderings of points in the plane, let us consider some simple representations and algorithms for graphical data.

There are two basic approaches to representing curves. In the *continuous* space approach a curve may be piecewise approximated by a mathematical curve. The simplest and most common version of this is a polygonal approximation. In the *discrete* space approach, the plane is viewed as an array of discrete points. A rectangular grid may be used to produce a discrete approximation to a curve from a continuous one. Each time the curve crosses a grid line, the nearest grid point is noted. The discrete approximation consists of the noted grid points. Variations on this method are given by Montanari, 1970.

A continuous polygonal approximation may be represented as a list of vertices. Storage requirements tend to vary directly with the length of a curve.

A discrete approximation may be represented in a couple of different ways. The *raster* representation consists of an array of bits. The bits and grid points, called pixels, are in one-to-one correspondence (or sometimes the grid squares and bits correspond). A one bit indicates a grid point associated with the curve. Storage requirements are independent of the length of the curve but vary directly with the square of the precision required.

The *chain-encoded* representation is based on a polygon linking the grid points associated with a curve in the proper order. Since each side joins a grid point to one of its eight nearest neighbors, only three bits of storage per side are required. For an irregular curve, about four chain elements are required for each side of an unrestricted (continuous space) polygonal approximation (Merrill, 1973: Montanari, 1970) Details of the chain-encoded representation are given by Freeman, 1974.

Regions can be represented by their boundary curves. If the raster representation is used, then a region may be represented so that interior points are indicated by a bit value of one and exterior points by a value of zero.

We will now consider two search problems. The first is chosen because it involves searching a single curve or region. The second involves searching a pair of curves together.

To determine whether a point is inside a region bounded by a closed curve, the line crossing-crossing method (Shimrat, 1962; Hacker, 1962) makes use of a ray going from the point in some arbitrary direction. The point is inside the curve if and only if the ray crosses the curve an odd number of times. Hence, the problem reduces to a search for crossing points. With a polygonal representation, we can search exhaustively for crossing points by considering the sides one at a time.

With a raster representation of a region, we can determine whether a point is in the region by checking the value of the bit corresponding to the point. If we have a raster representation of the boundary, then we can consider a ray going from the point to the right. To count crossing points we search exhaustively one partial row of the representation. (For simplicity, we are ignoring a couple of special cases here.)

The problem of finding the points of intersection of two curves involves searching two curves at once. With a polygonal representation, each pair of sides can be considered in turn. For two $n$-sided polygons this involves $n^2$ steps. With a raster representation, each and every pixel must be considered in turn. The result of the intersection operation is the *logical and* of the two bit arrays.

With the above search algorithms, points on a curve are considered in some particular order. With the polygonal representation of a curve, we consider sides one at a time in the order they occur along the curve. (The infinite set of points on a single side is considered collectively since at most one of these points can be of interest.) With the raster representation we can consider the points by rows or by columns or in any of several other orders.

With polygons we have an object space order and with rasters an image space order. Let us see if we can use these to advantage.

## 3. FAST SEARCHING

### 3.1. Object Space Order

It is possible to perform a binary search on a polygonal representation of a curve by considering sections (sets of consecutive sides) collectively. Initially the entire curve is considered as a single section. If it is known that a section lies entirely within some bounding rectangle then it is often possible to eliminate the section from further consideration (just as we discard certain ranges of values in a binary search). If a section can not be eliminated and it contains more than one side then it must be split into two subsections which can be considered independently.

For example, to find the points of intersection of two curves we first see if their bounding rectangles overlay. If the bounding rectangles do not overlap then the curves can not intersect. If they do overlap, the curve with the larger bounding rectangle is split into two subsections. The points of intersection of the other curve and each subsection are found in a recursive manner. If two single-sided sections are found to have overlapping bounding rectangles then their point of intersection, if it exists, may be computed directly. Typically the points of intersection of two $n$-sided polygons can be found in $O(\log n)$ time.

In order to determine whether a point is inside a closed curve, we can consider a ray going to the right from the point. If we find that the ray passes through the bounding rectangle of a section, then we know that the ray intersects the section an odd number of times if, and only if, the endpoints of the section lie on opposite sides of the ray. Only those sections with bounding rectangles containing the point need to be split. Normally this search does not go full depth. The expected processing time is essentially independent of the number of sides in the polygon.

A data structure to support this type of searching is described by Burton, 1977. To store an $n$-gon requires $O(n)$ storage. Typically, storage requirements are about twice those of a polygon represented as a list of vertices. A discrete variation of this representation (Burton, 1978; Burton, 1979a) also exists. About 5 bits of storage per chain element are required.

### 3.2. Image Space Order

The raster representation of a curve is a locally dense

sparse array. One compact representation for such arrays has been attributed to Dijkstra (Dahl et al., 1972). We will consider a variation of this representation.

Let us suppose that the array of pixels has $2^i$ rows and $2^i$ columns for some integer $i$. (An array of pixels can always be extended to one of this form.) We can represent the pixel array as a 2 x 2 array of pointers where one pointer corresponds to the upper left of the pixel array, one to the upper right, one to the lower left and one to the lower right. That is, each pointer corresponds to a $2^{i-1} \times 2^{i-1}$ subarray of pixels. If a subarray contains no part of the curve (i.e. each pixel value is zero) then it is represented by a nil pointer. Otherwise, the pointer corresponding to a subarray must point to locally dense sparse array representation of the pixel subarray. This continues down to the point where we have a single pixel subarray. A special pointer value (a second nil pointer) is used to denote a pixel with the value one.

A region also can be represented in this way. The second nil pointer (denoting 1) is used to indicate any subarray of pixel which is entirely within the region.

This representation is a tree of degree 4 and height $i$. The storage required for an $n$ pixel curve (or region with an $n$ pixel boundary) is $O(n)$. Under favorable circumstances, the time required to compute the union or intersection of two $n$ pixel curves or regions having $n$ pixel boundaries is $O(\log n)$. Details are given by Burton, 1979b.

### 3.3. Mixed Space Order

Merrill, 1973, has devised a data structure, called a TCB (tightly closed boundary), which uses an image space order in one direction and an object space order in another.

A TCB is based on a discrete approximation to a curve. (Regions are represented by their boundaries.) To construct a TCB, a list of the pixels through which a curve runs is made. (Certain pixels are listed twice to take care of special cases which we will not consider here.) Each pixel is defined by an $(x,y)$ pair of co-ordinates. The list is sorted, with the $y$ co-ordinate value as the major key and the $x$ co-ordinate value as the minor key, and then indexed by the $y$ co-ordinate value. A set of points having a common $y$ co-ordinate value is called a $y$ partition.

In order to determine whether a point is in a region one first finds the appropriate $y$ partition (which is directly available through the index) and then counts the number of $x$ co-ordinate values to the right of the point.

Finding the points of intersection of two curves or regions is fairly straight-forward, but involves considering all $y$ values which are common to both objects.

Merrill's representation requires $O(n)$ storage for an $n$ pixel curve. The expected time for determining whether a point is in a region is independent of $n$. The worst case time, with a suitable algorithm is $O(\log n)$. The expected and worst case time for finding the points of intersection of two curves is $O(n)$ since no common $y$ value can be skipped.

## 4. CONCLUSION

We have considered several ways to represent curves and regions to facilitate efficient searching. Each representation is based on an object space, image space, or a mixed object space and image space ordering of points. The basic approach, in each case, can be extended to collections of graphical objects.

## References


Burton, W.: 1977, Representation of many-sided polygons and polygonal lines for rapid processing. Comm. ACM. 20, pp. 166-171.

Burton, W.: 1978, A fixed grid curve representation for efficient processing. *Proceedings of the Fifth Annual Conference on Computer Graphics and Interactive Techniques*. Atlanta, Georgia, U.S.A., August, 1978.

Burton, W.: 1979a, Implementation of the binary searchable grid chain representation for curves and regional boundaries. Geo-Processing 1, pp. 37-52.

Burton, W.: 1979b, A data structure for fast processing of spatial geographical information. School of Computing Studies, University of East Anglia, Norwich, England.

Dahl, O.J., Dijkstra, E.W., and Hoare, C.A.R.: 1972, *Structured Programming*. Academic Press, London, pp. 154.

Freeman, H.: 1974, Computer processing of line-drawing images. Comp. Surveys 6, pp. 57-97.

Hacker, R.: 1962, Certification of algorithm 112: Position of point relative to polygon. Comm. ACM. 5, pp. 606.

Merrill, R.D.: 1973, Representation of contours and regions for efficient computer search. Comm. ACM. 16, pp. 69-82.

Montanari, U.: 1970, A note on minimal length polygonal approximation to a digitized contour. Comm. ACM. 13, pp. 41-47.

Shimrat, M.: 1962, Algorithm 112: Position of point relative to polygon. Comm. ACM. 5, pp. 434.

## AUTHOR INDEX

## SUBJECT INDEX